EVALUATING MARKETING ACTIONS AND OUTCOMES

ADVANCES IN BUSINESS MARKETING & PURCHASING

Series Editor: Arch G. Woodside

Recent Volumes:

Volume 5:	Mapping How Industry Buys
Volume 6:	Handbook of Business-to-Business Marketing Management
Volume 7:	Case Studies for Industrial and Business Marketing
Volume 8:	Training Exercises for Improving Sensemaking Skills – With Solutions
Volume 9:	Getting Better at Sensemaking
Volume 10:	Designing Winning Products
Volume 11:	Essays by Distinguished Marketing Scholars of the Society for Marketing Advances

ADVANCES IN BUSINESS MARKETING & PURCHASING
VOLUME 12

EVALUATING MARKETING ACTIONS AND OUTCOMES

EDITED BY

ARCH G. WOODSIDE

Wallace E Carroll School of Management, Boston College, Massachusetts, USA

2003

ELSEVIER
JAI

Amsterdam – Boston – Heidelberg – London – New York – Oxford – Paris
San Diego – San Francisco – Singapore – Sydney – Tokyo

ELSEVIER SCIENCE Ltd
The Boulevard, Langford Lane
Kidlington, Oxford OX5 1GB, UK

First edition 2003

Library of Congress Cataloging in Publication Data
A catalogue record from the British Library has been applied for.

ISBN: 0-7623-1046-4
ISSN: 1069-0964 (Series)

♾ The paper used in this publication meets the requirements of ANSI/NISO Z39.48-1992 (Permanence of Paper).
Printed in The Netherlands.

CONTENTS

LIST OF CONTRIBUTORS

Per Andersson	Department of Marketing, Distribution and Industry Dynamics, Stockholm School of Economics, Sweden
Stanley G. Aungst	General Studies Building, Mont Alto Campus, PA, USA
Russell R. Barton	J. Business Admin Bl, University Park, PA, USA
Salvatore F. Divita	Department of Marketing, George Washington University, USA
Daniel J. Flint	University of Tennessee, Knoxville, USA
Hans Kjellberg	Department of Marketing, Distribution and Industry Dynamics, Stockholm School of Economics, Sweden
Oliver Koll	Institute of Marketing – Strategy Consulting, Grillparzestr., Austria
Michael W. Preis	Long Island University, Brookville, USA
Pauline Ratnasingam	Harmon School of Business Administration, Central Missouri State University, Warrensburg, USA
Marcia Y. Sakai	Tourism/Economics, School of Business Administration, University of Hawaii, USA
Maria Anne Skaates	Department of International Business, Aarhus School of Business, Denmark
Amy K. Smith	George Washington University, USA
David T. Wilson	College of Business Administration, Pennsylvania State University, PA, USA

Robert B. Woodruff — Department of Marketing, Logistics and Transportation, College of Business Administration, University of Tennessee, USA

Arch G. Woodside — Boston College, Carroll School of Management, MA, USA

INTRODUCTION: OVERCOMING OVERCONFIDENCE AND OTHER PROCESSES IN SHALLOW THINKING

Arch G. Woodside

What's really happening? For an organization this question contains at least four sub issues:

- What actions are being done now help to increase the organization's performance?
- What actions are wasted motions – what are we doing that does not contribute and wastes our time?
- What actions harm the organization's performance – what actions are counterproductive in helping the organization achieve what really needs to be accomplished?
- What actions are we not doing now but really should be doing to increase the organization's performance?

A fifth, related, sub issue is how to go about finding out what's really happening – what research methods should executives use, as well as avoid using, to go about finding out what's really happening? An implicit mental model (Senge, 1990) too often used by executives ("executives" includes you and me), is to believe that what first comes to mind is accurate. "It seems that the process of interpretation is so reflexive and immediate that we often overlook it. This, combined with the widespread assumption that there is but one objective reality, is what may lead people to overlook the possibility that others may be responding to a very different situation" (Gilovich, 1991, p. 117).

Evaluating Marketing Actions and Outcomes
Advances in Business Marketing and Purchasing, Volume 12, 1–12

ISSN: 1069-0964/doi:10.1016/S1069-0964(03)12012-1

META-THINKING TRAINING

A possibly surprising observation is that most executives do not acquire formal training in *meta-thinking*, that is, thinking about thinking and deciding including training in effective prescriptive tools to increase effective thinking and deciding. Two reasons may be responsible for this lack of training. First, the overconfidence bias is widespread – most executives tend to rely too often on their unconsciously driven automatic thoughts (see Bargh, Gollwitzer, Lee-Chai, Barndollar & Troetschel, 2001; Wegner, 2002). The tendency is natural to assume that our intuitive beliefs are accurate and that relying on external heuristics (i.e. written checklists, explicit protocols) are unnecessary. Even when presented with hard evidence that formal external searches of relevant information sources and the use of explicit decision rules result in more accurate decisions than intuitive judgment alone, the executive's first response often includes disbelief and resentment – resentment in the implicit loss of her/his authority to evaluate and decide.

Second, the scientific field of study of evaluation research is new; even early in the 21st century specific courses on how to think and decide are rarely found in university curriculums – this view may appear unfairly to discount formal training in accounting auditing practices. Unlike related fields of study (e.g. biology, sociology, psychology), formal research on meta-thinking and meta-evaluation (i.e. evaluating evaluation processes) became widely recognized as a field of study only recently (e.g. see Shadish, Cook & Leviton 1991).

AVAILABLE TOOLS FOR IMPROVING THE QUALITY OF DECISIONS

Gilovich (1991, pp. 41–42) offers the succinct observation that relates to the quality of executive decisions and actions, "A fundamental difficulty with effective policy evaluation is that we rarely get to observe what would have happened if the policy had not been put into effect. Policies are not implemented as controlled experiments, but as concerted actions. Not knowing what would have happened under a different policy makes it enormously difficult to distinguish positive or negative outcomes from good or bad strategies. If the base rate of success is high, even a dubious strategy can be seen as wise; if the base rate is low, even the wisest strategy can seem foolish."

However, several useful tools are available now for improving our sense-making capabilities in judging the impact of an executive policy/action (see Baron, 2000). These tools include estimating what would happen if the policy had not been put into effect (e.g. Campbell, 1969) as well as software programs to help us structure problems and test the impact of alternative problem structures (see Clemen & Reilly, 2001).

Two Training Exercises in Overcoming Ignoring of Base Rate Information

Not noticing and not using base rate (i.e. general background likelihood of success of an action) information in thinking and deciding concerns most of us. Here are two exercises and one tool (i.e. decision-tree analysis) to help you solve both.

The Taxicab Accident
A cab was involved in a hit-and-run accident at night. Two cab companies, the Green and the Blue, operate in the city. You are given the following data:

(a) 85% of the cabs in the city are Green and 15% are Blue.
(b) A witness identified the cab as Blue. The court tested the reliability of the witness under the same circumstances that existed on the night of the accident and concluded that the witness correctly identified each one of the two colors 80% of the time and failed 20% of the time.

What is the probability that the cab involved in the accident was Blue rather than Green?
Please write your answer here: _____ % before reading further.

Pricing a New Product by Kiwi Fencing Executives (adapted from Woodside, 1999)
Plaswire is a new fencing wire made from polyethylene terephthalate; it is designed to be used as a replacement for galvanized steel wire in permanent fencing construction. The President/CEO of Kiwi Fencing requests the Assistant Sales Manager to implement a pricing strategy for Plaswire that will help achieve national distribution. The President/CEO believes that pricing Plaswire substantially lower (30% less) than the competing steel wire price will help gain distributor's acceptance of the product because some farmers and livestock station managers may be sensitive to a low price level of the new product. The CEO feels certain that steel wire's management (the competitor) will not respond by lowering its prices in response to a low introductory price on Plaswire because the cost of galvanized steel raw material is 300% higher than plastic raw materials used for manufacturing Plaswire.

The assistant sales manager (ASM) for Plaswire knows from looking at many pricing lists of industrial distributors of agricultural products that, when manufacturers introduce new products at retail prices well below competing products, their competitors respond most of the time (three out of four times) with price reductions that match or more than match the new products' prices. The final result was failure or very low market share for new products in 92 of the 96 cases the ASM was able to find where a new product was introduced with a lower price than competing products and the competitors reacted by lowering their prices.

However, the ASM also knew that the President often predicts correctly what competitors are going to do. The President had been correct in his predictions in two of the three recent cases concerning competitors' responses to new product prices.

The ASM favors pricing Plaswire to match the current price of competing steel wire products. A quick payback period and substantial profits and the competitor would be somewhat less likely to react with such a parity pricing strategy.

The marketing manager (MM), recommends pricing Plaswire 10% above the current price of steel wire. He feels that few agricultural customers will buy the new permanent boundary wire because of a low price and the competitor will lower its prices to match or beat the price of Plaswire. The MM notes that in most cases of new product introductions with prices higher than competitors' prices on

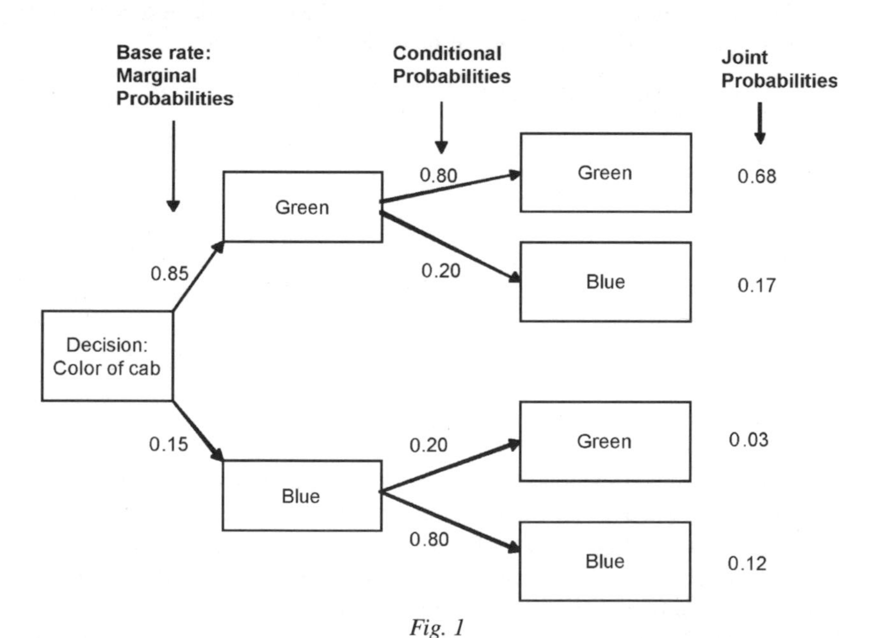

Fig. 1

Decision Tree for Taxicab Color Problem. *Note:* The subject will say Blue 29% of the time ($0.17 + 0.12 = 0.29$); the subject will be accurate 41% of the time that he/she says Blue ($0.12/0.29 = 0.41$). Thus, when the subject says, "Blue," the chances are still greater than 50-50 that the cab was Green.

existing products, only one in 10 competitors reacted by lowering the price on an existing competing product. The new products were still available for most of the other nine products introduced at prices higher than competitors' prices.

What decision do you recommend? What is the likelihood of success (i.e. the new product is still being marketed five years after market introduction and it is profitable) of your strategy? Before reading further, provide your answers here. Your recommendation, circle one:

(a) Price Plaswire 30% below steel wire
(b) Price Plaswire equal to steel wire's current price
(c) Price Plaswire 10% above steel wire.

The likelihood of success of your strategy is, circle one:

(a) 15%
(b) 25%
(c) 45%
(d) 75%
(e) 100%.

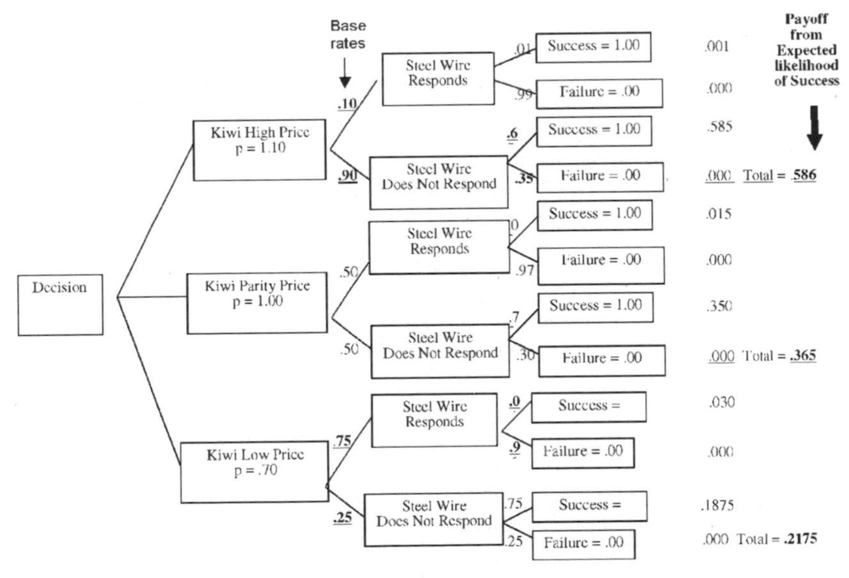

Fig. 2

Decision Tree of Kiwi Fencing Price Alternatives. *Note:* Probability estimates shown as **bold** are given in case with "most cases" estimated as **0.65** survival rate. For payoff, survival after a few years is judged success and is assigned a value of 1; death = failure = 0.

Answering the Taxicab Accident Problem

Most subjects (i.e. problem solvers, executives) say that the probability is over 0.50 that the cab involved in the accident was Blue, and many say that it is 0.80 (Tversky & Kahneman, 1982, p. 156). Such a decision focuses mainly or entirely on the conditional probability that the witness accurately predicts a cab's color given that the color is known, and ignores the base rate marginal probability of cabs being Blue vs. Green. The correct answer is 0.41. Structuring the problem in a decision tree (see Fig. 1) is helpful for making use of the base rate probabilities in solving it.

Answering the Plaswire Pricing Problem

Being mindful when solving the Plaswire problem requires giving some reasonable probabilities for some alternative's not given in the problem description. Using all reasonable estimates leads to the same recommendation: price higher than steel wire results in the highest likelihood of success. Figures 2 and 3 include the probabilities given in the problem description (in *bold*) as well as one set

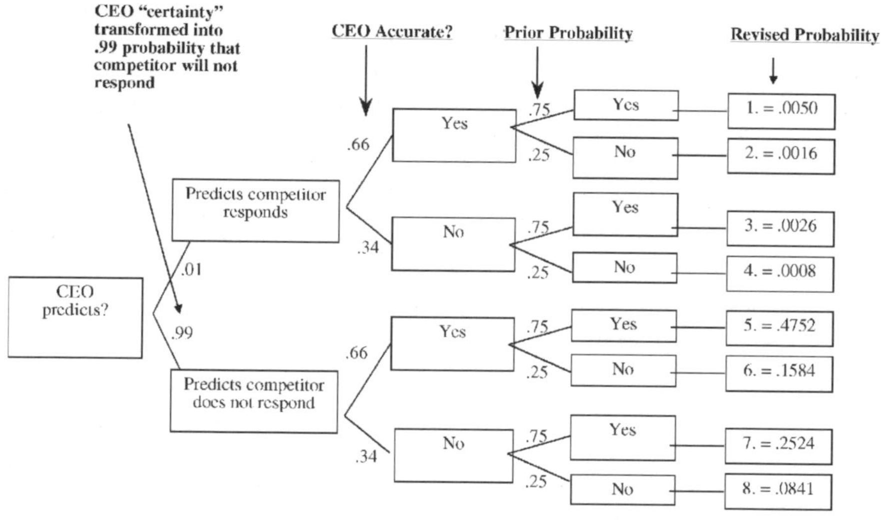

Fig. 3

CEO Predictions, Accuracy, and Outcomes. *Note:* Posterior probability competitor responds = 0.050 + 0.0026 + 0.4752 + 0.2524 = 0.74. Posterior probability competitor does not respond = 0.0016 + 0.0008 + 0.1584 + 0.0841 = 0.26. Conclusion: Because CEO is not highly accurate and prior probability is very high that competitor will respond (0.75), the posterior likelihood of the competitor responding is close to the same as the prior probability.

of reasonable probabilities for the other alternatives. Most executives advocate adopting the CEO's recommendation to price lower than the competing steel wire – they do so without considering the base rate probability that competitors usually (0.75) react when a new product is introduced at a lower price than their products' prices.

FUNDAMENTAL DIFFERENCES BETWEEN SCIENTIFIC AND EXECUTIVE THINKING

Adopting the following meta-thinking view is useful. Executive thinking differs fundamentally from scientific thinking in at least three ways (cf. Kozak, 1996). First, scientists (e.g. academic researchers) get to choose the problem, whereas in organizations, the problems (and symptoms of problems) are often thrust upon the executive (e.g. see Mintzberg, 1978). Second, scientists focus on a limited number of problems at a time; whereas executives are confronted with a vast number of potential problems and a myriad of possible presentation problem frames (see Wilson, McMurrian & Woodside, 2001). Third, scientists have the relative luxury of time to explore the problem at hand, whereas executives, particularly CEOs, do not.

Recognizing and accepting such distinct thinking paradigms aids in taking the next step in thinking about designing prescriptive tools for executives. "The mind that relies on its own processing capacities or intuition for clinical [executive] decision making cannot establish a sound basis for evaluating, in a combinatorial way, all the possible correlations and cross-correlations that exist between patient- [customer or manager] provided features of a given problem and all the possible causes of those features. That is, the mind, unaided by information tools, dealing with complex problems cannot elicit and couple all of the features of the problem to what is known from the relevant scientific literature about those features. Because of this inability to exhaust combinatorial reasoning, clinicians [executives] are forced to make judgments without having considered all the possible evidence that bears on the problem (i.e. they jump to conclusions). Such limitations of cortical processing in clinical [executive] decision making inevitably lead to prejudice and bias . . . (Kozak, 1996, p. 1335).

Unfortunately, the natural tendency in decision-making includes:

- drawing conclusions based on very limited amounts of information;
- being over-confident that one's own initial conclusions are accurate;
- not looking for disconfirming evidence;
- discounting disconfirming evidence if it pops-up;

- being hostile to the belief that using decision tools (e.g. computer software programs such as "Problem Knowledge Coupler (PKC)," see Gaither, 2002) results in more accurate problem framing than one's own judgment;
- not "thinking outside of the box" and considering all theoretically possible, even if seemingly implausible, combinations of events and their outcomes;
- and implementing a decision with limited consultation with knowledgeable others.

The conclusion is compelling (and findings supported the conclusion have been replicated in several fields of behavioral science) that designing and applying decision tools and simple heuristics lead to more accurate decisions than that made by intuitive thinking alone (see Gaither, 2002; Gigerenzer & Selten, 2002; Gigerenzer, Todd & the ABC Research Group, 1999). Equally compelling is the finding that the use of disruptive technologies (including using decision tools such as PKC) meets with great hostility by entrenched executives (see Christensen, 2000; Christensen & Bower, 1996). We need to train ourselves in the tools available for "actively open-minded thinking" (Baron, 2000) because our minds prefer to limit cognitive effort (see Payne, Bettman & Johnson, 1993). Consequently, being "mindful" (see Langer, 1989; Weick & Sutcliffe, 2002) requires training and, while highly valuable, it is an unnatural mental state. The words of Karl Weick are apt here: "Complicate yourself." Forcing yourself to learn new skills in applying decision tools and reading volumes now before you, are steps to achieving Weick's advice.

WHO SHOULD READ VOLUME 12?

The intention is for this volume to be read by executives wanting to learn how to reduce overconfidence, and to become more mindful, in making decisions and in learning how to scientifically evaluate the quality of outcomes that follow from implementing decisions. What's really happening? How can we accurately evaluate the quality of marketing decisions and outcomes? These issues are the central focus of all the contributions appearing in Volume 12.

Where is the Action?

After this introduction, the first paper by Andersson and Kjellberg asks how we can go about accurately reporting business action. The authors provide the basic tools of analysis in storytelling research and theory to enable you to find out what is really happening. Detailed examples are found in the paper that illustrates using these tools.

Narrative and Case Process Research

The second paper includes an excellent literature review on case research for understanding B2B marketing relationships. Andersson summarizes his more than 30 years of case study research and the major contributions of the European-based IMP Group. Andersson's work helps in particular in overcoming a one perspective bias in finding out what's really happening.

Integrating Marketing Models with Quality Function Deployment (QFD)

Quality function deployment and the "house of quality" are decision tools useful for designing products and services that customers prefer to buy. In the third paper Aungst, Barton, and Wilson describe how to collect estimates of customer preferences (i.e. utilities) for alternative design features in constructing "House One" via QFD. Along the way, they review relevant literature since Hauser and Clausing's 1988 *HBR* "House of Quality" article.

Stakeholder Value Creation and Firm Success

Does an empirical analysis of firm performance support the view that taking a stakeholder perspective outperforms setting objectives to considering a stockholder perspective only – should the CEO be inclusive in thinking about setting multiple objectives that serve suppliers, employees, customers, and the public, as well as profit objectives? In the fourth paper Koll provides evidence supporting the conclusion that CEO's should think actively about setting minimum levels of responsiveness to all stakeholder groups.

Building Effective Buyer-Seller Dyadic Relationships

How do salespersons' and customers' individual temperaments and personal value systems affect customer interpersonal satisfaction and repurchase intentions? In the fifth paper Preis, Divita, and Smith delve deeply into this issue. Because buyer-seller exchanges are often essential for marketing-purchasing, careful reading of this report is warranted. Liking one another does matter and nuances when interacting affect liking. The authors offer concrete prescriptions after presenting detailed findings on how to build positive personal buyer-seller relationships.

Trust and Business-to-Business E-Commerce Communications and Performance

Within E-commerce relationships, how is trust built and destroyed? How much does trust really matter in E-commerce relationships? In the sixth paper Ratnasingam reports evidence on how the actions of both the buyer-seller affect trust, and in turn how trust affects the ultimate outcomes of the relationship. She recommends that trust is activity based and thus, develops in stages and from promises delivered rather than promoted.

Internationalizing the Professional Services Firm

What are the nitty-gritty steps that a professional services firm takes to become successful internationally? In the seventh paper Skaates provides detailed answers as well as a review of the substantial literature on how to manage the professional services firm. Her synergy of the European and North American literature is a unique contribution to building knowledge into the theory and practice of managing services.

Business-to-Business Customer Value and Satisfaction

In the eighth paper Woodruff and Flint offer an expansive view on managing customer value. This contribution provides a succinct review on how to measure and manage changes in customer value. The authors' report complements and extends the findings that Aungst, Barton and Wilson describe in the third paper.

Meta-Evaluation

In the ninth and final paper, Woodside and Sakai review theories and methods of evaluation. The authors suggest the need to evaluate evaluations and apply their suggestion to evaluating performance audits of tourism marketing programs. They find widespread lack of knowledge of evaluation research among performance auditors as well as tourism marketing executives.

CONCLUSIONS

The literature on meta-thinking and deciding is robust and useful in providing tools helpful for reducing problem ambiguity and evaluating marketing actions and outcomes. Reading the literature helps overcome and keep in check one of life's biggest fallacies; that we usually know what we are doing without using structured search and thinking tools. Reading Baron (2000), Gilovich (1991), Gigerenzer and Selten (2002), and Weick (1995), as well as the contributions in this volume is humbling for learning how what we know that isn't so, and beneficial for learning to use prescriptive tools for improving decisions.

REFERENCES

Bargh, J. A., Gollwitzer, P. M., Lee-Chair, A., Barndollar, K., & Troetschel, R. (2001). The automated will: Non-conscious activation and pursuit of behavioral goals. *Journal of Personality and Social Psychology*, *71*(September), 230–244.

Baron, J. (2000). *Thinking and deciding* (3rd ed.). Cambridge, UK: Cambridge University Press.

Campbell, D. T. (1969). Reforms as experiments. *American Psychologist*, *24*, 409–429.

Christensen, C. M. (2000). *The innovator's dilemma*. Cambridge, MA: Harvard Business School Press.

Christensen, C. M., & Bower, J. L. (1996). Customer power, strategic investment, and the failure of leading firms. *Strategic Management Journal*, *17*, 197–218.

Clemen, R. T., & Reilly, T. (2001). *Making hard decisions with decision tools*. Pacific Grove, CA: Duxbury.

Gaither, C. (2002). What your doctor doesn't know could kill you. *The Boston Globe Magazine* (July 14), 12 ff. Boston: Globe Newspaper Company.

Gigerenzer, G., & Selten, R. (2002). *Bounded rationality: The adaptive toolbox*. Cambridge, MA: MIT Press.

Gigerenzer, G., Todd, P. M., & the ABC Research Group (1999). *Simple heuristics that make us smart*. Oxford, UK: Oxford University Press.

Gilovich, T. (1991). *How we know what isn't so*. New York: Free Press.

Kozak, A. (1996). Local clinicians need knowledge tools. *American Psychologist* (December), 1335–1336.

Langer, E. J. (1989). Minding matters: The consequences of mindlessness-mindfulness. In: L. Berkowitz (Ed.), *Advances in Experimental Social Psychology* (Vol. 22, pp. 137–173).

Mintzberg, H. (1978). Patterns in strategy formation. *Management Science*, *24*(4), 934–948.

Payne, J. W., Bettman, J. R., & Johnson, E. J. (1993). *The adaptive decision maker*. Cambridge, UK: Cambridge University Press.

Senge, P. (1990). *The fifth discipline*. New York: Doubleday.

Shadish, W. R., Jr., Cook, T. D., & Levitton, L. C. (1991). *Foundations of program evaluation*. Newbury Park: Sage.

Tversky, A., & Kahneman, D. (1982). Evidential impact of base rates. In: D. Kahneman, P. Slovic & A. Tversky (Eds), *Judgment Under Uncertainty: Heuristics and Biases* (pp. 153–160). Cambridge, UK: Cambridge University Press.

Wegner, D. M. (2002). *The illusion of conscious will*. Cambridge, MA: Bradford Books, MIT Press.

Weick, K. E. (1995). *Sensemaking in organizations*. Thousand Oaks, CA: Sage.

Weick, K. E., & Sutcliffe, K. M. (2002). *Managing the unexpected: Assuring high performance in an age of complexity*. Ann Arbor, MI: University of Michigan Press.

Wilson, E. J., McMurrian, R. C., & Woodside, A. G. (2001). How buyers frame problems: Revisited. *Psychology and Marketing*, *18*(6), 617–655.

Woodside, A. (1999). Evaluating alternative pricing strategies for market introduction of new industrial technologies. In: *Advances in Business Marketing and Purchasing* (Vol. 8, pp. 181–186). Amsterdam: JAI Press, Elsevier.

WHERE IS THE ACTION? THE RECONSTRUCTION OF ACTION IN BUSINESS NARRATIVES

Hans Kjellberg and Per Andersson

ABSTRACT

Taking a set of studies about business action as the empirical starting-point, this paper looks at the various ways in which action is represented. The overall research question can be stated as follows: how is business action reconstructed in our narratives? The texts analyzed are collected from research on exchange relationships in the field of marketing. To analyze how these texts depict business action, four narrative constructions are focused: space, time, actors, and plots. The categorization and analysis are summarized and followed by a set of concluding implications and suggestions for narrative practice aiming to reconstruct business action in the making.

INTRODUCTION

How is action reconstructed in research narratives about business life? The origin of this text was a reflection concerning texts claiming to depict or represent "business action," i.e. business operations, processes and activities. We were puzzled by the great variation in the extent to which different texts were able to reach out to us as representations of business action. Even more so since the texts contained few reflections concerning the translation process, i.e. the way in which researchers

Evaluating Marketing Actions and Outcomes
Advances in Business Marketing and Purchasing, Volume 12, 13–58

ISSN: 1069-0964/doi:10.1016/S1069-0964(03)12001-7

approach the object in focus, business actions, and how they subsequently translate their field studies into stories about business action.

It is important to stress that in analyzing how business action is narrated, we will not be concerned with the actual character of the events that the narratives claim to re-present (see Smith, 1978). Neither will we be concerned with the concept of business action as such. A brief overview of theoretical work on related concepts, e.g. human action, social action and organizational behavior, indicates that several issues are subject to controversy. For instance, the link between action and intentionality (Schick, 1991) and the possibility of collective actors (French, 1983; Garret, 1988).

Concerning the substantive content of business action, we have found no explicit typologies. It is however possible to impute such typologies for various traditions. Thus, decision making and the execution of decisions seem central to business action as depicted in the behavioral theory of the firm (see e.g. March, 1988), while analysis, planning, implementation and control have similar positions in managerial schools, e.g. marketing management (Kotler, 1999). As a heuristic, we will call business action *any undertakings that can be seen as serving to generate market exchange* (see Snehota, 1990).

METHOD

Our work has followed the logic of abduction (Alvesson & Sköldberg, 1994; Peirce, 1934); that is, we have combined the reading of inspiring theoretical texts with the reading of business action narratives (our empirical material). As part of this process, we have been selecting, ordering and labeling our examples. In so doing, we have gradually formed what one may call potential explanatory hypotheses/normative statements concerning the narration of business action.

As a theoretical guide for this exploration, we have used four dimensions borrowed from narrative theory (developed further in the next section): *time*, *space*, *actors* and *plots*. Our empirical material, that is, the business action narratives that we have studied, has been selected on the basis of two criteria: first, the studies are all published Ph.D. theses; second, they are part of the Scandinavian tradition of research on exchange relationships in industrial markets. In all, we make use of 18 such theses, including our own, in the "empirical" section of the paper. The selection undoubtedly contains biases. However, given the explorative character of our work, it is the degree to which this material can supply interesting ideas, rather than its potential for generalization that should be of interest. We argue that the chosen narratives offer interesting observations that may be further developed and used as input in a more encompassing study.

SOME TOOLS FOR ANALYSING NARRATIVES

Many studies of business action in the field of industrial marketing are conceptual, relying on empirical fragments to support their conceptual constructions. When these empirical fragments are turned into longer narrative texts, the "outside, retrospective hindsight" view seems to be common. That is, the researchers stand at the sidelines presenting their narratives. They tell their stories retrospectively and chronologically about events that happened during specific and delimited periods of time, more or less distant from the so-called present. In so doing, they employ a rather limited set of perspectives and devices for constructing their narratives.

But what are the alternatives available to us when writing (and reading) narratives about business action? To assist us in our analysis, we have turned to narrative theory for a general characterization of what a narrative is and what it is made up of.

Stories and Discourses – the Primacy of the Narrative

Following narrative theory, we make a basic distinction between contents and presentation when discussing narratives. According to O'Neill (1994), the problem when approaching texts is that in order to discover what "really happened," e.g. in an account of a real world event, it will be necessary to sift through the *account* of what happened.

> . . . we find much to support the contention that even on its apparently simplest and most uncomplicated level, that of what 'actually happened' in a given story – whether that story is fictional or non-fictional, literary or non-literary – narrative is always and in a very central way precisely a game structure, involving its readers in a hermeneutic contest in which, even in the case of the most ostensibly solid non-fictional accounts, they are essentially and unavoidably off balance from the very start (O'Neill, 1994, p. 34).

Basically, then, the world of the *story* – what "really happened" – can be reached only through the *discourse* that presents it. And since the same series of events, real or imagined, can be presented in many different ways, the narratives, or discursive presentations we are offered, will have considerable import on our reading of the world.

In fact, some authors claim, the process of narration is a deeply ideological one.

> . . . the text, however mimetic it may seek to be, cannot only deal with real-life problems as these are set up in ideological representations – which also, of course, propose their own range of solutions. However impelling the textual issues may appear to be, they are problems posed in a particular way – though the narrative may actually have the power to question the assumptions in such a setting up. This argument about ideology suggests that what the text does not say is as important as what it does say (Tambling, 1991, p. 91).

The final point made by Tambling concerning absent aspects in texts seems to us to suggest an important tool for exploring narratives. Especially since the conceptual schemes for analyzing business narratives are underdeveloped.

Elements of Narrative

What then are the resources available for a narrator? What are the dimensions of a narrative? From *narratology*, the branch of contemporary narrative theory focusing specifically on the analysis of narrative structure, we learn that:

> One of the most obvious tasks of narrative discourse is clearly to select and arrange the various events and participants constituting the story it sets out to tell. Initially this might well seem to be a relatively straightforward affair, since stories essentially amount to the doings of particular *actors* involved in various *events* at particular *times* and in particular *places*, and narrative discourse is thus merely a matter of saying who did what, and when, where, and why they did it. Different types of narrative may well privilege one or another of these elements, but most ordinary readers (or listeners or viewers) will feel themselves reasonably entitled to expect all four of them to play an, at least, implicit role in any narrative (O'Neill, 1994, p. 33).

Here, it is suggested that four elements are part of any narrative, namely time, space, actors and events. In some cases there are also important interactions between the elements, for instance, it has been suggested that the temporal and spatial dimensions of a narrative combine to form a *chronotope*, or time-space (McQuillan, 2000).

In a slightly different vein, the fourth element, *events*, is closely linked to the concept of *plot*, i.e. the arrangement of isolated instances into *stories* and narrative *discourses*. But a narrative discourse is not only that which arranges the *events* (including both *actions* and *happenings*) and the *existents* (including both *characters* and *setting*) into a plot and a final story; it also arranges the manner in which its reader will react. In the words of Hayden White (1978): Is the *mode of emplotment* of our story a case of romance, comedy, tragedy or maybe satire?

In the following, we discuss each of the basic narrative components and their interaction in the final narrative. We end each section with a few basic questions that are central to an analysis of business action narratives.

TIME

A narrative will always entail elements of time; particular actors will be involved in various events at particular times. To this must be added the complexities and the multiplicity of social times and individual proper times that intersect in each given situation (Nowotny, 1994).

In the process of narrating a story, a synthesis is created between the open and theoretically indefinite succession of incidents and another type of time – one characterized by integration, culmination and closure – which configures the story (see Ricoeur, 1984).

> In this sense, composing a story is, from the temporal point of view, drawing a configuration out of a succession. We can already guess the importance of this manner of characterizing the story from the temporal point of view inasmuch as, for us, time is both what passes and flows away and, on the other hand, what endures and remains (Ricoeur, 1984).

> However hard and fast (or otherwise) the ostensible facts of the world of story may be, they all exist in at least one real-world dimension, namely that of time. Narrative structure is both syntagmic (as regards the linear temporal sequence of the story) and paradigmatic (as regards the shape of the particular discourse chosen to relate the story). Nowhere has the relationship between the two been worked out more systematically than as regards the treatment of time. The distinction between story-time (erzählte Zeit), measured in temporal units (days, months, years) and discourse-time (Erzählzeit), measured in spatial units (words, lines, pages) has long been a staple concept of narrative theory (O'Neill, 1994, p. 42).

There are considerable similarities between the views of Ricoeur and O'Neill here. They both indicate that a central task for a narrator is to synthesize the sequentiality of events with the particular configuration of the final narrative.

When it comes to the way in which a narrative reflects story-time, it is possible to distinguish three temporal aspects (Genette, 1980): *order* (When?); *frequency* (How often?); and *duration* (How long?). Since each of these aspects can be handled in more than one way, there are several alternative modes of temporal re-presentation available to a narrator. Furthermore, there are important interrelations between the modes of re-presentation chosen for each aspect.

Temporal *order* can be used to contrast the "real" chronological order in which, for example, a set of business actions were taken, with the narrative order in which they are presented in a particular narrative discourse. Perhaps the most common mode of re-presentation in this respect is the "neutral" order, which mirrors an implied "real" chronological order of the story (see example 1 in Fig. 1). Narratives that break this "real" order make use of some form of anachrony. Two important examples (see examples 3 and 4 in Fig. 1) are the backwards looking flashbacks (*analepsis*) and the forward looking flash-forwards (*prolepsis*) (Genette, 1980).

The use of different narrative ordering techniques is complicated by the fact that the actors under study can be assumed to employ their own shifting ordering techniques as they engage in business action (Pieters & Verplanken, 1991).

> People are time travelers; they take different time perspectives to reflect on their past, present and future behavior. These perspectives are used in planning new behaviors, in anticipating new situations that one may encounter in the future. These reflections are necessary in evaluating past behaviors, and in planning future behaviors on the basis of the evaluations of the past. These reflections are necessary in executing behaviors. Our ideas of what we are doing depend in part

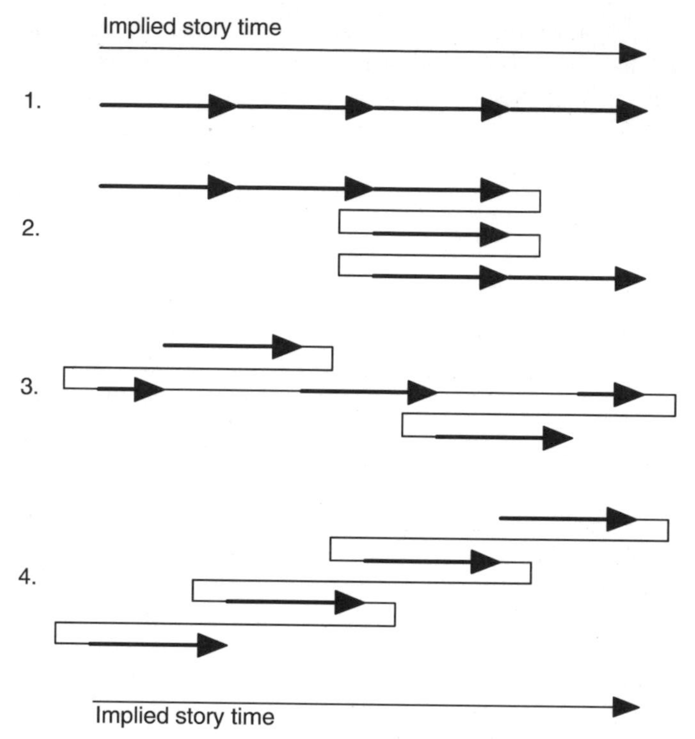

Fig. 1. Four Examples of Temporal Narrative Structures. (1) Singular, Natural Order. (2) Natural Order with Repetitions. (3) Singular Anachronistic Order (Flashback and Flashforward). (4) Singular Anachronistic Order with Multiple Consecutive Flashbacks.

> on the time perspective taken and on the time horizon. If we travel in the future and retrospect on our present behaviors, we might gain knowledge that is instrumental in determining what we are doing today, and what we did yesterday (Pieters & Verplanken, 1991, pp. 63–67).

Hence, the use of different time perspectives – *vantage points* (from the past, the present or the future) and *viewing directions* (towards the past, present or future) – is not only an option for the author creating a narrative, but can also be expected from the actors involved in various business actions.

The second temporal aspect, *frequency*, contrasts the number of times an event "really" happened in a story and the number of times it occurs in the narrative. Genette identifies four common modes of re-presentation (Genette, 1980):

- *singular* re-presentation (recounting once what happened once);
- *repetitive* re-presentation (recounting more than once what "really" happened only once, see example 2 in Fig. 1);

- *iterative* re-presentation (recounting once what really happened more than once);
- *irregular* re-presentation (recounting what happened a different number of times than it "really" happened).

Genette's (1980) third temporal aspect, *duration*, has been described as the amount of "real" time that elapsed in the story and the amount of discourse-time (i.e. space in the final text) involved in presenting it. Here, Genette suggests five major *tempos* that can be found in narratives:

- *ellipsis* (not reporting relevant events at all, i.e. maximum discourse speed in that discourse time is zero);
- *summary* ("real" story time is much greater discourse time);
- *scene* (story time largely equals discourse time, e.g. like in "real time" like dialogues);
- *stretch* (slow motion where story-time is less than discourse-time);
- *pause* (minimum discourse speed, e.g. as in long passages of reflective nature in the middle of a narrative with certain temporal characteristics).

It is important to underscore that despite these categorizations and the division between what we call the "real" (story) time and the narrative (discursive) time, there is no such "real" time against which we can check narrative time.

> . . . however you consider real-life time, you must think of it in some represented, narrative form – even a sense of time as linear is a representation of it, just as the word 'time' is an attempt to conceptualize something felt about the nature of reality (Tambling, 1991).

What then, are the issues of interest to us as regards the use of the temporal dimension in business action narratives? The general question we ask is simply: How is the temporal dimension of business action reconstructed in our narratives? More specifically, using the temporal aspects suggested by Genette, we ask: How do narrators make use of order, frequency and duration in their reconstruction of business action?

PLACE: SPACE AND SETTING

Compared to the multifaceted temporal dimension, narrative space is held to be more difficult to use and manipulate in fictional literature (O'Neill, 1994, p. 47). However, this does not mean that space lacks importance or that it cannot be used as a narrative device:

> It none the less offers a number of possibilities, since space can evidently be described in more or less detail, in a more or less orderly fashion, with more or less consistency, and with more or less emphasis on its allegorical or symbolic or ironic possibilities. Most obviously,

> perhaps, narrative space as setting can be used to establish a particular mood effectively and quickly . . . (O'Neill, 1994, *ibid.*).

Fictional narratives often use pre-fabricated settings, i.e. "typical" settings used in certain types of genres that effectively put the reader in a specific mood (e.g. Kafka's descriptions of cramped, ill-lit indoor spaces to establish a certain mood of pervasive oppressiveness). Anthropological and ethnomethodological narratives also seem to use such devices, affecting the moods of their readers by including more detail in spatial descriptions. Two examples are Clifford Geertz's account of the Balinese cockfights (Geertz, 1973) and John van Mannen's little story about a police unit in pursuit of a young car thief (Mannen, 1988).

In addition to the issues related to the positioning of actors and actions, the narration of space in accounts of business actions also has an interesting conceptual corollary. In the classical French drama, a major concern was to maintain the unity of space (as well as the unity of time); all actions were to be situated in clearly circumscribed spatial locations. Business action, on the other hand, is often associated with the spanning of space, most evidently so in the international business literature. During the past 15 years, work within the sociology of science and techniques has drawn attention to the link between action and space by discussing the problem of long-distance control (Law, 1986). Is it possible to act at a distance? For our purpose, then, it is relevant to study how the spanning of space is handled in business action narratives. To sum up, we will address two questions related to the spatial dimension of narratives in our analysis of business action: First, how do non-fictional business narratives make use of spatial narrative devices? Second, how are narrative devices used to connect different spatial locations?

CHARACTERS/ACTORS: THE ACTING SUBJECTS

In narratology, the acting subjects are labeled *characters*. A central issue concerns the way in which narratives deal with the process of *characterization*, i.e. the way in which the actors/characters acquire a *personality*. O'Neill argues that this involves three intersecting processes: a process of construction by the author, a process of reconstruction by the reader, and, pre-shaping both of these, a process of pre-construction by contextual constraints and expectations (O'Neill, 1994).

When constructing a narrative, there are two basic types of textual indicators of character, the direct *definition* and the indirect *presentation*. A direct definition simply tells the reader what the actor/character is like. This is common, for instance, in children's tales where good and evil is quickly sorted, e.g. "in the house lived an evil witch." In an indirect presentation, the reader is shown what the actor/character is like through its actions (events, time and space/setting).

Irrespective of the type of textual indicator chosen, however, the role and situation of the actors being put into the narrative world are largely at the hands of the narrator.

> What, finally of the actors who, as creatures of discourse, inhabit this world of story? Whether they realize it or not, they live in a world that is, in principle, the world of a laboratory rat. Their world is entirely provisional, it is fundamentally unstable, and it is wholly inescapable. The world of story is an experiment, a provisional reality under constant observation 'from above' on the part of those by whom it is discoursed. It is the world of a specimen in a display case, a prisoner in a bell jar, the world wished for by all authoritarian systems, a world whose inhabitants have no secrets . . . (O'Neill, 1994, p. 41).

If a narrator can control the actors in the narrative to the extent that O'Neill suggests, there is reason to expect a wide variety of ways of presenting actors in narratives. However, it is also possible that certain limitations are brought in simply by assuming the role of a narrator; i.e. that there exists a few more or less clearly explicated role-models for presenting actors.

O'Neill further elaborates on the role of the *narrator*:

> In any narrative, the most obviously indispensable narrative role is that of the *narrator*, for a story can become a story only by being told. When we talk of the narrator here, however, we might remember that we are really talking about the concretization of a necessary *narrative actant* . . . For in some narratives the narrating voice is external, belonging to a world outside of that occupied by the characters of the narrative text (O'Neill, 1994, p. 77).

This suggests that when we talk about "actors" in narratives there can be many involved in the production of the text; the author, an implied author, a narrator, a character, a narratee, an implied reader, and a reader.

To sum up, we will address three questions concerning the actors: Firstly, how are actors characterized in business action narratives? Secondly, to what extent do such narratives make use of direct definitions or indirect presentations of actors? Thirdly, to what extent and in what shapes do the narrators appear in the narratives?

SINGULAR EVENTS AND COMPOSITE PLOTS

The fourth analytical dimension put forward in narrative theory concerns the actual activities, the verbs of business action, and their interrelations. We start off by briefly discussing the individual actions, or *events*, and continue with a discussion of how these events are merged into a composite whole, or *plot*.

The Events: The Actions and Happenings

A central problem for a narrator is that irrespective of how he or she defines an *event*, no matter how big or small, the amount of information that he or she decides to include in, or exclude from the narrative must always be to some extent arbitrary. This applies to . . .

> . . . all event-labels, from the broadest in scope, such as 'Napoleon marched on Moscow' to the most specific such as 'Jim walked to the door'. The latter, for example, involves all of 'Jim decided to walk to the door', 'Jim shifted his weight to his left foot', Jim advanced his right foot', 'Jim planted his right foot on the floor, 'Jim shifted his weight to his right foot' and so on, not to mention an indefinitely large number of even more minutely differentiated activities as well. Each of these, moreover, is itself at least potentially, an entirely full-fledged event that could be absolutely vital given the appropriate narrative context (O'Neill, 1994, p. 39).

In principle, then, all events can be deconstructed into an infinite series of constituent events. The practical question for the narrator is when to call it quits? Both as readers and writers, we consider some events to be more necessary for our understanding of the story than others. Such considerations, it seems to us, are formed on the basis of our explicit or implicit definition of the situation that is being narrated (see Goffman, 1986 (1974)).

Lack of agreement regarding how to define the situation being narrated is likely to cause problems for any narrator trying to reach out to some imagined reader. Moreover, this problem will present itself both macroscopically and microscopically. First, at what point will a continued progression towards the "Napoleon marched . . ." type of narrative, make readers lose sight of the action? Second, at what point will a continued regression into the "minutely differentiated activities" prevent the reader from seeing the forest for all the trees? That is, when do additional details no longer contribute to a narrative?

Plot – The Unfolding Sequence of Events

In any final narrative, the individual events will be linked into a composite structure or *plot*. For our purpose, we will use the term plot to refer to a coherent and meaningful *order*, a pattern that arises out of the combination of purposive individual actions (O'Neill, 1994). The arrangement of events, or "what happens," is what turns action into a plot. In narrative theory, a distinction is generally made between the story and the plot (or rather, the "plot-structure").

> But surely the historian does not bring with him a notion of the 'story' that lies embedded within the 'facts' given by the record. For in fact there are an infinite number of such stories contained therein, all different in their details, each unlike every other. What the historian must bring to

> his consideration of the record are general notions of the *kinds of stories* that might be found there, just as he must bring to consideration of the problem of narrative representation some notion of the 'pre-generic plot-structure' by which the story *he* tells is endowed with formal coherency (White, 1978, p. 60).

By a specific arrangement of the events reported and without offence to the truth-value of the facts selected, a given sequence of events can be *emplotted* in a wide variety of ways. As Barthes says: The narratives of the world are numberless (Barthes, 1977, p. 79).

Still, there exist a number of fairly well-developed types of plots, such as the four modes of emplotment – romance, comedy, tragedy and satire – suggested by White above. These carry with them *genre* constraints that will establish certain expectations on behalf of the readers. This can be clearly seen from the discourse setting, the textual setting in which the discourse more or less self-referently places itself. For example, in accounts emanating from companies about business events that they themselves are involved in, we expect, more or less overtly, to be drawn into a *romantic* story ("success story"). Other narrators with other perspectives, we would expect as readers to use other discourses and discourse settings, e.g. an investigating journalist might give the story a *mystery* form.

Given our focus on academic work, an influential ideal-type is likely to be the *factual account*, i.e. an account that implicitly or explicitly claims to convey what actually happened. It should perhaps even be regarded as a super-structure, inside of which other plots may be placed. A problem of particular importance for a narrator working on a factual account is how to *authorize* his or her version as *that* version which can be treated by others as *what has happened* (Smith, 1978, p. 33).

The choice of plot is important also from the perspective of the kind of explanation that the resulting story will provide. By distinguishing between "story" and "plot" in a narrative it is possible to "specify what is involved in a 'narrative explanation' " (White, 1978).

> One can argue, in fact, that just as there can be no explanation in history without a story, so too there can be no story without a plot by which to make of it a story of a particular kind...
>
> Thus far I have suggested that historians interpret their materials in two ways: by the choice of a plot structure, which gives to their narratives a recognizable form, and by the choice of a paradigm of explanation, which gives to their arguments a specific shape, thrust, and mode of articulation. It is sometimes suggested that both of these choices are products of a third, more basic interpretative decision: a moral or ideological decision (White, 1978, pp. 62 and 67).

If differences between narratives are linked to the explanatory models preferred by the narrators, as suggested by White, then the question of where narrators turn for inspiration when plotting becomes relevant. A common distinction in this respect is between using the logics offered by actors involved in the story under narration, often referred to as an inductive or grounded approach, and using the logics offered by actors in the business of doing research.

Possibly, this distinction is discernible only in principle. First, any logic offered will be both "*ideo-logical*" that is, infused with ideas, and *practical*, that is, associated with the performance of specific activities in specific contexts. Second, the process of business action narration necessitates movements across the imagined border between the world of business action and the world of business research. Furthermore, it must also handle the variety of perspectives found among business actors and business researchers themselves. Hence, rather than attributing the differences between narratives to a fairly coarse dichotomy, we find it more interesting to study how the process of narration weaves together ideas from different worlds.

Our brief overview suggests that there are a number of issues of interest concerning the treatment of individual events and their interrelation in business action narratives: First, how are individual events chosen and depicted? Second, how are these events combined into some meaningful order, or plot? Third, how do the narrators attempt to authorize their accounts? Fourth, and finally, how is the multitude of perspectives that may be applied to both the individual events and the overarching plot handled in the narratives?

A SUMMARY OF NARRATIVE CONCEPTS

Before moving on to our exploration of business action narratives, let us sum up the narrative concepts we will use. First of all, narrative theory suggests that any narrative can be discussed in terms of four dimensions: time, space, actors and plots.

When exploring the *temporal dimension*, we will distinguish between three temporal aspects: order, duration, and frequency. In terms of the spatial dimension, we will look at if and how the narrators construct particular *spatial* settings, potentially affecting the mood of the reader. Moreover, we will explore if and how the narratives handle actions that transgress localities.

Actors emerge in narratives as results of a process of characterization, combining the narrator's efforts to construct acting entities, the reader's efforts to re-construct the result of the narrator's efforts, and the pre-shaping of both these efforts by contextual constraints and expectations. In the final narrative, actors are re-presented either by the use of a direct definition or through an indirect presentation (through what they do).

In addition to the description of time, space and actors – the existents of a narrative – all narratives necessarily involve a selection of individual events (which are arbitrary and subject to infinite regress and progress). These events must finally be brought together into a whole through the use of some *plot*.

Besides their "internal" complexities, the narrative dimensions may also be combined in various ways, e.g. producing a specific time-space combination. Building on the work of Mikhail Bakhtin, McQuillan (2000, p. 53) has used the term *chronotope* as a label for "the intrinsic connectedness of temporal and spatial relationships that are artistically expressed in literature":

> Time, as it were, thickens, takes on flesh, and becomes artistically visible; likewise, space becomes charged and responsive to the movements of time, plot and history. This intersection of axes and fusion of indicators characterizes the artistic chronotope. The chronotope in literature has an intrinsic generic significance. It can even be said that it is precisely the chronotope that defines genre and generic distinctions, for in literature the primary category in the chronotope is time. The chronotope as a formally constitutive category determines to a significant degree the image of man in literature as well. The image of man is always intrinsically chronotopic (McQuillan, 2000).

Hence, in narratives, we can expect a strong connection between time and space, and the actors and their actions. As stated by O'Neill (1994, p. 53), "times, places and characters interact in a complex fashion in the narrative transaction."

Next, we will make use of these generic, narrative variables when exploring a set of business narratives. First of all, we will look at how these narratives reconstruct the four dimensions of a narrative suggested in narrative theory, i.e. time, space, actors and plot. Specifically, for each dimension we will look for something we have chosen to call *modes of re-presentation*. In some cases, we are able to make use of concepts suggested to us from narrative theory. In others cases, where we find ourselves wanting of such concepts, we put our own labels on the observations we have made.

SITUATING BUSINESS ACTION IN TIME

How is the temporal dimension of business action reconstructed in our narratives? In what ways do we account for the location and unfolding of business actions in time? As we started to look for ways in which time was represented, we found that temporality was nearly always part of the accounts, albeit in very different ways. In the following we present three modes of representing time: the anonymous, the chronological and the relative. For each we also discuss various narrative techniques in terms of the concepts suggested by Genette (above).

Anonymous Temporality

A first mode of re-presentation of temporality is that which situates business action in an anonymous temporal dimension. In these narratives, events are unfolding "before," "during" or "after" other events. Historical influences are inferred through the use of expressions such as "originally" or "from the start," etc. Here is one example:

> [The company] was originally established as a subcontractor for a large local manufacturing company. Nowadays the proportion of annual sales to this customer has decreased significantly from a record 90% to approximately five percent at the moment. During this long history as a supplier of the large firm, the company has collected experience of inter-firm cooperation . . . (Nummela, 2000, p. 130).

Accounts such as this one create a relative temporal space within them. By so doing, they implicitly ascribe a general importance to temporality. They tell us that it matters that the events occurred in a certain temporal *order*. And while the order in which events are presented may differ from the chronological order in which they supposedly occurred, chronology is never far away.

These accounts do not, however, provide the reader with any tools for evaluating the temporal relations. Consequently, we cannot assess whether a certain change was quick or slow, or whether it was continuous or discrete. Further, since the events are not related to chronological time, the reader can neither translate them to his or her own temporal perspective, nor relate them to other events outside the account. The reader becomes, so to speak, temporally captive.

In terms of their *tempo*, these narratives tend to be summaries, bordering on ellipses. That is, they proceed at high speed in relation to the unfolding story, skipping over large intervals of time. Further, they often re-present actions iteratively, i.e. recounting once what supposedly happened several times. This practice asks the reader to accept first, that several individual actions can be seen as instances of the same type, and second, that each individual action is relatively unimportant as compared to their aggregation.

Business Action and Chronological Time

The most common way of situating business action in chronological time seems to be by simple reference to a year.

> In 1972, knowledge regarding the possibilities and prospects of image processing together with the introduction of mini-computers and microprocessors incited the establishment of several research ventures in Sweden (Lundgren, 1991, p. 122).

Here, the narrator makes use of a singular mode of re-presentation. The formation of knowledge about image processing and the introduction of computers, are presented as events that led to the establishment of research ventures. Moreover, these events are reported to have taken place in 1972. It is our contention that this relatively coarse use of chronological time endangers the credibility of an account.

Asking the reader to accept "the formation of knowledge about image processing" and "the introduction of computers" as singular events taking place in 1972, is asking a lot. What happened in 1972? How were those events related to preceding ones?

In many cases, the re-presentation of individual events is expanded into a *chronology*, thus situating business actions in a more detailed temporal setting. By displaying a series of discrete events along a timescale or in a list, the reader is given an overview of their temporal interrelation (see Fig. 2).

In many ways, this mode of re-presentation resembles that of anonymous temporality. The predominant narrative *order* is natural, i.e. events are presented chronologically; the narrative tempo is relatively quick, offering summaries of events; and the *frequency* is often singular. Still, these narratives may handle some of the shortcomings of the anonymous temporality discussed above. Primarily,

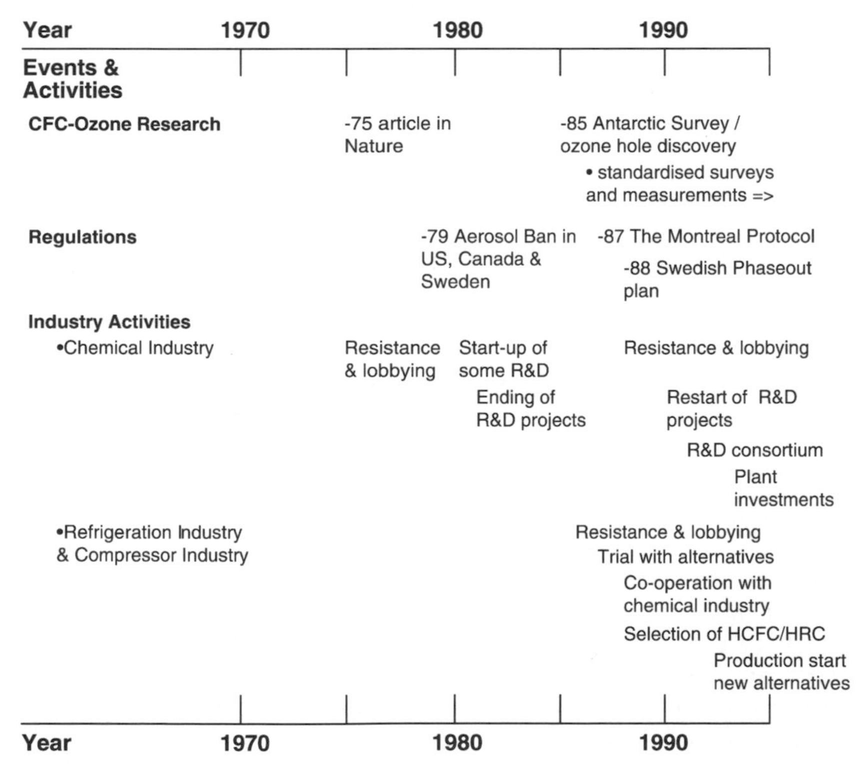

Fig. 2. A Chronology: Critical Events and Industry Activities to Replace CFCs in Refrigeration. *Source:* Sweet (2000, p. 139).

a chronology allows the reader to locate and relate events to their own temporal perspective.

Still, a chronology does in fact place a single temporal perspective on the events. This is so because it does not allow differences in how the involved actors viewed the development. "Were they early or late in introducing . . .? – They did it in 1978." "Was the development rapid or slow? – It took two years."

Implicitly, these narratives also promote a form of blind temporal causality. They direct attention only to the chronological sequence in which the actions unfold. Without additional information, then, chronological time becomes the only link between events. There is simply no plot in a chronology but the chronology itself. Hence, one can argue that most chronologies lack the integrating and closing temporality of a plot.

Relative Temporal Perspectives on Business Action

Reliance on chronology and sequentiality renders a narrative a rather mechanistic character. We have found that this character is reduced in narratives that, besides accounting for the chronological succession of events, also inform of different time perspectives that involved actors have as they engage in business action. What are the future-oriented perspectives of the actors? How do actors make use of historical developments when engaging in business action?

One way of checking the risk of only appealing to blind temporal causality is by expanding an underlying chronology. Consider the following excerpt taken from the same text as Fig. 2:

> In May 1987, the Federation of Swedish Industries and the Federation of Swedish Wholesalers submitted a letter to the Governmental Department of Environment and Energy giving a proposal for a phase-out of the CFC use. The industry did not object to the CFC phase-out stipulated by the Montreal protocol, but objected to the proposed plans of an accelerated phase-out in Sweden. Later during the fall . . . (Sweet, 2000, p. 150).

Here, temporality as well as the different perspectives that actors may have on it, emerge as important aspects of business action. The suggested pace at which certain actions were to be undertaken is said to have triggered other actions. By re-presenting the temporal links between individual actions in a more direct and concrete fashion, the narrative provides possible explanations while simultaneously creating expectations on behalf of the reader.

Another way of creating a more complex temporal re-presentation is through the use of remnants. This allows the narrator to "fold" time within the narrative.

> Towards the end of January an information and technical procurement meeting was held at the Atlas Copco Export Centre in Nacka. . . . Both the meeting and the project plan were important

> activities at the prescriptive stage. The discussions concerned different solutions/methods for different stages of the project implementation. . . . In the revision there were also suggestions concerning the order in which the access-tunnel should be excavated.
>
> "Further we suggest that one starts with the access-tunnel towards the flow-shaft and that one proceeds with part of the access-tunnel from this end while simultaneously working on the path in the flow-shaft. (Also here, the small height and width of the Hägghauler is very advantageous.)"
>
> Revised project plan, Atlas Copco MCT, xup28/80.
>
> The project-plan contained a number of revised blueprints for the tunnels that should be excavated and a zero-base specification of the proposed Atlas Copco-equipment (Liljegren, 1988, pp. 240–241 (Trans.)).

Here, the event is not only located in time, but the reader is transported in time through the use of a remnant. The excerpt from the revised project plan discloses a scenario; it offers the reader a temporally situated view of how one actor, Atlas Copco MCT, attempts to direct the unfolding of events in the future. By attending to such details, these narratives acquire a slower *tempo* than chronologies usually do. Although many are still *summaries*, they occasionally slow down and describe particular *scenes*.

The slower tempo resulting from a more detailed narrative seems to encourage modes of temporal re-presentation that make use of the different temporal perspectives of the actors.

> This time, the issue was brought up by Harald Mörck in connection to a discussion of "Our economic direction" and in particular the decreased turnover reported for 1949 and the ambition to reach MSEK 175 in turnover during 1950:
>
> "*The Chairman* pointed out that the results for the past year had not been completely favorable... Without departing from the principle of "the new deal", which is correct, we must perhaps discuss some telephone-sales of goods that are hard to sell, and which burden our stocks, e.g. rice. Such a sales method should only be used to balance the stocks" (Minutes of the local directors' conference, March 13, 1950, p. 1).
>
> Obviously things were not altogether good. This time, the need for sales measures was not due to any provincial circumstance perceived by the local directors. This time, it was in the light of a decreasing turnover for the entire company – something that had not occurred since the 1920s (Kjellberg, 2001, p. 259).

In this excerpt, the chosen quote provides a contemporary account of an event. Moreover, the actor providing the account discloses his temporal perspectives on the issue, reaching both forward and backward in time. The narrative tempo takes on a scenic character which is further underscored by the narrator's use of a brief flashback to provide a historical perspective on the event.

Some Concluding Remarks about Temporality

It is often taken for granted that humans can assume different temporal *vantage points* (from the past, the present, or the future) and *viewing directions* (towards the past, the present, or the future) to describe, understand and predict their own behavior (see Pieters & Verplanken, 1991). However, our overview suggests that this variety becomes more limited in narratives about business actions and behaviors.

There seems to be a great potential in exploring different ways of representing time in business action narratives. For instance, we have found few narratives that systematically seek to convey how actors make use of different temporal perspectives, both oriented towards the past, the present and the future. Moreover, the temporal vantage points from which certain views on events are derived are often obscured in the final narratives. That is, the reader remains ignorant as to whether a certain characterization of an event is to be understood as a retrospective view or a characterization made at the time of the event. Needless to say, this drastically reduces the reader's ability to interpret the actions accounted for.

The choice of temporal form for the narrative may also affect our impression of action. By using past tense, the narrative acquires a reminiscent character. The events accounted for are forever gone. We can remember and reflect over them with hindsight. The question is how our perceptions of such past actions can change if the author manages to create an illusion of contemporariness? Will this allow the reader to "take" a more future-oriented perspective on the event?

Although we have detected some, more or less deliberate, variations in *tempo*, most business narratives we have studied employ a summary mode. Occasionally, they approach the fast-moving ellipsis, or the more prosaic scene. A few narrators allow their accounts to be interrupted by reflective pauses, but so far, we have not found a single account that *stretches* the reported events into slow motion.

In terms of *frequency*, the dominant mode of re-presentation in business narratives seems to be the *singular* (recounting once what happened once). We have also found examples of *repetitive* narratives, where the narrator describes more than once what "really" happened once. For instance, some accounts describe the same business episode from the perspectives of two or more interacting companies (like the repetition of the shopping-mall scene in "Jackie Brown" from the perspectives of the involved actors).

Finally, the *iterative* mode of re-presentation, where the narrator recounts once what really happened more than once, seems to be relatively common. The practice of presenting several individual actions as if were they a single one has obvious advantages, not least avoiding to kill the reader with boredom. However, it also de-emphasizes any import that could be ascribed to the repetitive nature of these business interactions and processes. In addition, even if they were highly routinized, such business actions need not be completely identical. If nothing else, they will have to take place concurrently in many different places, bringing us to the second narrative dimension: space.

SITUATING ACTION IN SPACE

How is the spatial dimension reconstructed in business action narratives? What use is made of narrative devices such as settings? Are actions presented as local? If so, do the actions determine these localities? How are instances of acting at a distance represented? These were questions raised as we started to think about spatiality and action. Largely, we found space to be a relatively neglected dimension in business action narratives.

Business Action as Spatially Independent

Most narratives we have studied roughly situate the involved *actors* in space, e.g. in terms of their nationality. This mode of re-presentation provides a spatially static backdrop for business action. An actor's existence appears to be tied to a particular location. This image is further underscored by the fact that the specific instances of business action accounted for, often lack explicit location in space. Instead, the accounts make use of a general mode of re-presentation that renders them an air of spacelessness.

> In order to further specify the image transmitter according to the needs of the users, Hasselblad continued to discuss with AFP and other representatives of the global press.... A "dummy" of the new transmitter was presented to AFP and Hasselblad declared themselves willing to perform the development within a certain time if AFP signed a letter of intent to acquire a specified number of transmitters at a fixed price. AFP agreed and Hasselblad had taken on a new challenge (Lundgren, 1991, p. 161).

Implicitly, by sustaining the action in a peculiar non-space, the account conveys an image of business action as spatially independent. That is, despite whatever spatial distance there is between the actors, their inter-actions flow smoothly.

A problem with these accounts is the difficulty of assessing whether their de-emphasis of spatiality is empirically or theoretically motivated. This seems crucial for the image of business action they provide us with. If the spatial independence is put forward as an empirical observation, the account tries to tell us something interesting about business action. If, on the other hand, the spatial independence can be read as an effect of the theoretical perspective used – something we have found to be common – the narrative not only loses much of its suggestive power, but one can even raise doubts concerning the account.

Business Actions that Link Localities

Our interpretation of the variation in importance attributed to spatiality as a theoretical effect, is supported by the observation that situated accounts of business action are more common in narratives about international business action. In such narratives, the spatial dimension is ascribed theoretical importance through concepts such as *physical* and *psychic distance* (Johanson & Wiedersheim-Paul, 1974).

> In 1976, ASG realized that the situation in Belgium had to change. The traffic that AMA and ASG had together could not possibly continue to develop under the conditions that existed. Personnel in AMA engaged in the Swedish traffic contacted the ASG European Representation Office in Brussels and asked for help. ASG tried to find a solution, looking at the different alternatives such as buying a local company, setting up its own office or finding another agent . . .
>
> In 1977, ASG set up a company of its own and at the same time took over some of the personnel in AMA responsible for the Swedish traffic (Hertz, 1993, p. 115).

While the spatial dimension *is* accounted for in narrative texts such as this one, it is most prominent in passages describing the *effects* of business action. More seldom do these narratives specifically *locate* business action in space.

Acting at a Distance – Intermediaries

The modes of re-presentation discussed above are based on an implicit assumption that it is possible to act at a distance. Distance is no bar to action. Some narratives, however, try to account for how this is possible by explicating the various intermediaries involved in business action.

> It was decided that the vice president of Cantel should visit Ericsson in Sweden, in the spring of 1984. During this visit, he met the president of the Ericsson Group and the president of ERA, and found Ericsson's products and competence very interesting. However, after he had returned to Canada, ERA did not hear from Cantel for two months (Blankenburg Holm, 1996, p. 390).

In this excerpt, the spanning of space becomes important. The interaction between the two companies depends on an intermediary that links their localities. In this case, the linking is ascribed to "the vice president of Cantel" and his actions. Compared to the account above, where the inter-action between the parties was presented as completely unproblematic, we feel that we learn something more about business action. Still, we remain ignorant as to what it is that the intermediary in question transports across space. There is something missing in the account of that which is moving. Perhaps, this something would have shed light on the concluding remark made in the excerpt?

Some accounts suggest that large numbers of intermediaries may be needed to link localities.

> It was Atlas Copco's "ambassador", Göran Orwell, who tipped Larsgösta Almgren... The information from Orwell came by telex to Bill Sundberg who gave it to Almgren. Orwell was able to get this initiated information through his personal contact with Odd Hansen who was working for Höyer-Ellefsen on the South-American continent (Liljegren, 1988, pp. 206–207 (Trans.)).

Here, at least four individuals and two telex-machines were involved in spanning the distance between Höyer-Ellefsen in South-America and Atlas Copco in Sweden. By making explicit the number of intermediaries involved, this account provides a flavor of the amount of work that may be necessary to act at a distance. This suggests that if such acting is to be possible on a routine basis, a reliable set of intermediaries must be put into place. This directs attention to the way in which intermediaries affect business action. How is information transformed during transport? How do actors attempt to increase the fidelity of the intermediaries they use?

Ways in Which to Represent Localities

The narratives that do account for the spatial dimension use various devices to represent the localities. The most common way is by reference to geographical space – countries, regions, cities, etc. As Fig. 3 suggests, some authors also make use of maps to visualize the spatial dimension in their accounts.

In the narrative, the map in Fig. 3 is presented as part of a contract between two parties and put forward as illustrating their agreement to make a geographical division of a market. This suggests that business action may be closely linked to spatial considerations, or rather, that space may be subject to business action.

Concerning the mode of re-presentation as such, we view this as an attempt to transport the reader out of the text and into some other locality. A locality which is put forward as relevant to the account. The use of the map, we argue, plays on an implicit assumption of correspondence: the reported event refers to something external to the text and the map provides the reader with a route through which he or she can access this something and then return to the text. Others make similar use of blueprints and layouts of factories, stores, warehouses, etc.

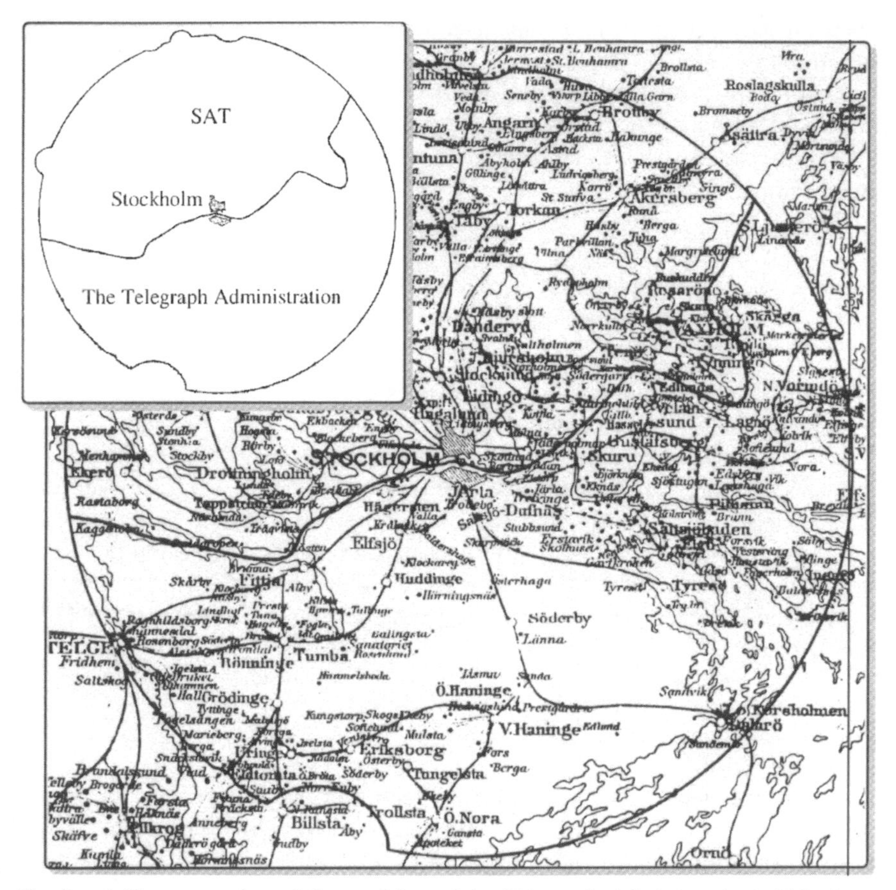

Fig. 3. A Representation of Space: Map of the Telegraph Administration's Telephone Network from the 1904 Contract with the Division Agreed upon Indicated. *Source:* Helgesson (1999, p. 68).

Concluding Remarks on the Narration of Space

Our general impression is that business action narratives make poor use of the spatial dimension. Few of them even get close to providing a spatial setting for the events they recount. Given this state of affairs, the spatial dimension should offer many opportunities, e.g. the use of spatial settings to enrol the reader in the narrative.

Many social observers have argued that the notion of inter-action is central to action at large (e.g. Giddens, 1984; Weick, 1979). At first, inter-action would seem to imply co-presence in both time and space. But this is not necessarily the case. To understand this, it is necessary to take into account the various mediators that are used in inter-action. For inter-action is generally supported by a host of intermediaries, e.g. texts, tokens, and technical devices. Such intermediaries seem to allow action at a distance. This links the spatial dimension to the question of acting entities, for how should such intermediaries be characterized in our business narratives? Are they simply neutral carriers of the will of man? Although few would subscribe to such a view in principle, many business narratives display a strong deterministic streak in their treatment of these intermediaries.

THE ACTING SUBJECT(S)

The third narrative dimension concerns the acting entities. Within industrial marketing there is wide theoretical agreement that business action is undertaken by actors, and that these actors come in different shapes and in different sizes (see, e.g. Håkansson & Johanson, 1992; Lundgren, 1991). Still, narratives differ markedly in how and to what extent they reconstruct these acting entities. So: How are actors re-presented in business action narratives? How are they characterized? These questions bring up another, related issue, namely: What are the consequences of the narrator's view of who the "actor" is for the study of business action?

As with time and space, accounts of business action are not always explicit concerning who the focal actors are (neither subjects nor objects of business action). In the following we will look at seven modes of re-presentation, starting with those that pay relatively little attention to the acting subjects and working our way up to those that reconstruct actors as highly problematic entities.

The Absent Actor

A first mode of re-presentation is that which routinely suppresses the acting entities. Not that the acting entities are absent as such; these stories are often full of potential actors. But their characteristic trait is the reluctance to associate specific actions with specific actors. The passive form is their hallmark.

> The merger of the international pharmaceutical sales operations is started in June–July 1990.... Organizational adaptations caused by the merger take place during 1990 and 1991.... The planning, the actual plans produced, the communication and the fusion process come to be heavily dominated by the striving to take a radical step towards the formation of a new, technologically diversified, Sweden-based, global pharmaceutical giant ... The practical

> integration and change process, based on the initial corporate analysis, is started during the second half of 1990 and continues with varying intensity throughout 1991 (Andersson, 1996, pp. 51–53).

To us, this mode of re-presentation de-emphasizes agency. Things appear to just happen; without effort, without controversy. In this sense, the narratives point in another direction than business action. They point towards forces over which specific actions and individual actors have no say, towards structures and macroscopic processes.

So, what is so appealing with this form? One obvious possibility is that the author is interested in constructing an explanation that goes beyond individual actions. If so, de-emphasizing agency is a pre-requisite rather than an unfortunate consequence. This would also place the account offside in relation to our present concern.

A second possibility is simply lack of information regarding the event. This could be due to the fact that the author is not really interested in the event, and that he or she consequently chose not to pursue it further. Assuming that the author is interested, it could be due to the event being controversial and that he or she was unable to establish anything more than the fact that something had occurred. It seems to us, that there are a number of alternative ways of accounting for the event that would not impoverish it as does the use of the passive form. One way would be a clear indication of the sources used and the way in which these differ.

A third possibility is that the author perceives the passive form to free him or her of the responsibility of distributing responsibility (blame, credit, etc.). By omitting the actors, then, the account might become less contestable. What we would like to point out, is that the account also might become less credible as an account of business action.

Clumping

A common mode of re-presentation that similarly dodges the question of responsibility is the use of some form of functional reduction to characterize the actors. That is, to use a direct definition of the actor based on its function in relation to a specific other (often the focal firm). One example of this is the narrative in which actors are routinely "clumped together" (Woolgar, 2002) into functional aggregates such as "the customers" or "the suppliers." Here, although the form may be active, the specific actions and actors remain impossible to identify.

> The key decision makers in the large business segment were used to personal sales representatives and account managers. This was what they knew from the mainframe years of IBM.

> Many of the CIOs felt pressure to reduce IT related costs, which were increasing year by year. In many organizations, the CIOs did not have as much control of PCs as they did of other computers and systems. Individual staff in our units bought whatever computer they liked and central directives on standards was often neglected (Kaplan, 2002, p. 230).

It is worth noting that this mode of re-presentation is necessarily linked to the use of an iterative temporal re-presentation. When individual actors are being grouped together, so also must individual actions.

A possible justification for *clumping* with respect to business action would be that it is used to reflect a way in which some actor speaks of a particular situation. In such cases, the aggregation of actors is part of the story and potentially important for our understanding of the recounted actions. However, since these narratives often lack explicit justification as to why a certain group or category should be regarded as a single entity, we suspect that this is not the only reason for utilizing this mode of re-presentation. More often, we suspect, the theory used to interpret the story told promotes *clumping* by positing that certain aggregates share characteristics that make them amenable to analysis as a group.

Functional Reduction and Entities Capable of Cognition

Related to *clumping* is the anonymous functional reduction of individual actors:

> In 1990, there was a specific event which caused the business relationship between the group sales subsidiary and the customer, and also the relationship between the group sales subsidiary and the Swedish supplier, to slacken. At that time the supplier was also selling the same product through another channel in Germany: its own sales subsidiary. The customer made, as it usually does, several inquiries concerning the purchase. One inquiry was made to the group sales subsidiary and another to the supplier's own sales subsidiary in Germany. The Swedish supplier gave the same price to both subsidiaries. The group sales subsidiary then put 10% on that price and made an offer to the customer. The supplier's own sales subsidiary, on the other hand, made an offer to the customer without any margin (Havila, 1996, p. 115).

The advantage of this mode of re-presentation over *clumping* is its capacity to account for specific actions. By refraining from aggregation, then, business action is given a more prominent position. The relative poverty of these accounts has to do with the perspective from which they are told. The actor is identified in terms of its function vis-à-vis some other actor(s), usually the focal one(s). Without additional information, the actor becomes very circumscribed and stereotypical.

In some cases, the actors in these accounts are also identified by name:

> Gyssens & Co, in the main a small airfreight company, had office space available in the right location (Hertz, 1993, p. 115).

Whether equipped with a name or not, the actors in this mode of re-presentation are most often restricted to "the company level." That is, business action is presented as a something that exclusively involves formal organizations. The narrator thus assumes, and/or asks the reader to take for granted the monolithic quality of the acting entities.

The scope of the above-mentioned assumption becomes more clear when "cognitive capacities" are attributed to these entities.

> Scania would like to see the development time of new coolers being reduced, implying less room for trial & error as is custom today. In bringing the development of oil coolers to a higher level, it is not Scania's intention to decrease its expertise in this field . . . (Wynstra, 1997, p. 96).

Here, the reader is asked to make a similar assumption about the monolithic quality of the acting entity. The difference is that the character of the assumption is made more clear through the explicit attribution of cognitive capacities. Of course, the identification of action with intentional behavior is an important heritage from western philosophy (Davidson, 1980) and is, as such, not surprising. More so, however, is the self-evidence of the attribution, given the prolonged debates about the extent to which collectivities, such as business firms, can be assumed to possess such capacities (French, 1983; Garret, 1988; Mahmoodian, 1997).

Hence, in relation to these types of re-presentations, many readers may pose critical questions such as: who did the narrator speak to in order to be able to state that "it is not Scania's intention to . . .?" It appears, then, that these modes of re-presentation also involve a form of "clumping." At least, they do not provide any reasons for treating this or that company as an acting entity in the specific situation.

Multiple Constitutions

The "company level" narratives discussed so far are sometimes developed into a mode of re-presentation that allow for "multiple constitutions" of actors. This multiplicity is most often made use of when highly specific (and important) events are represented. Further, it almost invariably results in the appearance of human individuals. When the plot thickens, the humans arrive . . .

> Suddenly problems of volume losses arose for Torsmaskiner. Since the margins were poor, only minor changes in volume were needed for the operations to generate a loss. The operations in Belgium, Blekinge and Torsås all reported losses of 4 MSEK each, which meant that 25 years of collected profits for Torsmaskiner were wiped away. The development was perceived as dramatic by the management of Torsmaskiner and led to reflection.

> "All the good years vanished in a stroke. Then, Ove [Eklund] and PeO [Wetterllind] went in and cleaned it up and structured the operations" (Olle Segerdahl, Purchasing manager, Torsmaskiner).
>
> Lilliecreutz (1996, p. 91 (Trans.))

Compared to the strict "company level" narratives, we find that the introduction of human individuals into the narratives generally promotes the impression of action. It seems that by identifying the (most important) individuals involved, it is possible to create an impression of being "close to" the business actions in focus.

Sometimes, these "actor levels" (individual, company) are mixed (including also collective/individual levels), appearing as counterparts in the focal events (at times also including collective feelings and thoughts).

> . . . Saima's officers were sincerely interested in developing the alliance project. However, they were discouraged from exhibiting more proactive behavior by the relational problems with Nedlloyd . . .
>
> . . . Nedlloyd . . . impaired the Saima officers' trust in the sincerity of Nedlloyd's intentions . . . Nedlloyd managers began to think of themselves as . . . (Ludvigsen, 2000, pp. 206–208).

Here, the asymmetry between how the focal actor(s?), i.e. "Saima's officers," and its(their?) counterpart are described, i.e. "Nedlloyd," raises questions. Does this reflect how the subjects of business action define their counterparts? This would be a plausible interpretation given that also Saima was characterized as a more monolithic entity from the perspective of Nedlloyd's managers in the final sentence.

Material Heterogeneity

Besides humans and companies, some narratives ascribe actor-like qualities to technologies, products and artefacts of different types.

> As so many others, those responsible for the tests at Billerud must acknowledge that refining untreated wooden chips lead to "a devastatingly bad result with pulps full of splinters throughout."
>
> "The chip-pulp is bad from all points of view. The refinement must be taken to a higher degree of mincing for the pulp to become at all competitive with the ground pulps" (Research report Billerud AB, Göran Annergren, 62.10.16).
>
> When the chips are given chemical pre-treatment the result is much more positive. Despite that the capacity of the refiners is not enough to produce a pulp resulting in sheets of paper that are smooth and even enough, there is still a completely new character in the chemically pre-treated chip-pulps. The kinship with mechanical pulp is great, with characteristics such as high yield (approx. 80–85% of the wood input), good light-distribution and opacity (Waluszewski, 1989, p. 77 (Trans.)).

Here, a number of non-humans appear as part of the narrative, e.g. chips, splinters, pulps, chip-pulps, refiners, sheets of paper, a research report, etc. Although they are mainly attributed passive roles, the use of words such as "kinship" suggests that they share certain actor-like qualities.

In our concluding discussion concerning space, we argued that intermediaries of various sorts may be involved in business actions. Texts, tokens and technical devices, can obviously act as mediators bridging both time and space. The question, then, is how such non-human entities should appear in the final narrative. Depending on the role attributed to them, it is quite possible to let non-humans appear as actors on equal terms with humans and human collectivities in business actions.

Emergence

So far, the actors have been characterized in a fairly static way. Some narratives do, however, allow the actors to change their configuration and/or characteristics during the course of the narrative. In narrative theory, this is sometimes described in terms of how "the character," i.e. the characteristics of the actor, is allowed to emerge via descriptions of, e.g. its actions, via descriptions of the setting, or via direct descriptions of its characteristics.

Examples of the reverse development also exist: What initially is described as an actor in a certain situation, can fall apart and dissolve into a number of actors in the next situation, even during the course of events belonging to the same situation. Alternatively, several actors can converge into one. These variations in the composition of actors are not only part of a process over time. At the same time, an entity which is characterized as an actor vis-à-vis a certain counterpart, can by another counterpart, be treated as part of a collective that is ascribed agency. Thus, agency is put forward as something that emerges during the course of the story. The idea that actors have *variable geometries* (Latour, 1996), also suggests that important aspects of business action are lost in many narratives, due to the use of predetermined, taken-for-granted, categorizations of actors.

Who Attributes Agency?

In most of the excerpts above, the narrator has been clearly responsible for attributing agency to the various entities. This attribution, in turn, seems to have been the result of a blending of theoretical and empirical matter. For instance, the rationale for limiting the account to a pre-specified level, e.g. companies, may be put forward as theoretical, whereas the identity of the specific actors is put forward as empirically derived, e.g. in a "snowballing" fashion.

In some texts, however, the actors themselves are left to define *who or what* is an actor in the business actions they are involved in.

> In early June Axel Hultman and Herman Olson were in Antwerp, staying at the Weber Grand Hotel. On 7 June 1913 a telegram was sent to the Administration's general director, Herman Rydin, asking for permission to immediately order the test-exchange. According to the telegram, the manufacturer agreed to have the exchange erected within 9 months, but requested on the other hand a guarantee that the Administration did not later manufacture switches of that system without the specific agreement of Bell Telephone Manufacturing Co. In a letter to the general director, Axel Hultman the same day reported that they had obtained what seemed to him to be good terms, and that "... Olson had got all his demands on the system fulfilled" (Helgesson, 1999, pp. 163–164).

By sending the telegram to Rydin, Hultman and Olson were ascribing agency. In fact, Rydin was seen as a spokesperson for "the Administration." Further, the telegram also ascribed agency to Bell Telephone in relation to the Administration. It seems to us that this practice of allowing actors to describe and define each other may be one way of achieving a credible correspondence between actors and actions. We do not, however, mean that this allows the narrator to reduce his or her involvement in the text. Rather, we are discussing a narrator's possibilities of making credible that some entity should be awarded agency in a particular situation.

Regarding the extent to which the author lets him/herself become part of the narrative, there are also many differences between texts. Some authors are clearly present, appearing as one of several voices involved in the narrative. Still, the impression of presence in the actual business action described might be low.

The Narration of Business Actors, a Summary

We have identified seven more or less distinct ways in which the actors are represented in business action narratives. First, it is common to suppress the acting entities in the narrative by using passive form. Second, actors are often clumped together into functional aggregates in the narratives, e.g. "the customers." Third, individual actors are often described only in terms of their function vis-à-vis a focal actor, something we call functional reduction. Fourth, many narratives are relatively flat, recognizing actors only at a specific level, e.g. either individuals or companies. Fifth, some narratives break this flat description by allowing for variation in the constitution of the actors. Sixth, in a few narratives, the constitution of the actors varies not only in terms of "level," i.e. individuals or collectivities, but also in terms of the materials which they consist of, i.e. humans and non-humans. Finally, although we have seen few examples of this, it should be possible to allow the acting entities to change over the course of the narrative, that is, to present them as emerging entities.

This final mode of presentation is linked to the use of indirect presentations, i.e. characterizing an actor through its actions. Based on our readings, this seems to be very unusual. Possibly, the use of functional reduction rests on this idea: "a supplier" should then be assumed to be an actor that "supplies." However, this approach suffers both from the lack of specific details and the singular perspective from which the label is attributed; a supplier may also be a customer, a partner, a competitor, etc.

EVENTS AND PLOTS

So far, we have focused on three basic dimensions of business action. In this fourth section, we will take a look at how the various ways of re-presenting these dimensions are combined into full-blown narrative accounts.

The Narration of Events

We start by looking at the way in which specific events are accounted for in the narratives. How is the event described? How does the author handle the problem of infinite regression and progression. What is the underlying definition of the situation?

A first mode of re-presentation we call the *summary report*. It accounts for the focal event without additional information concerning the circumstances under which it took place or the way in which it unfolded.

> AFP had not yet commenced the development of image transmitters and they communicated the need for an instrument capable of transmitting digital images to Hasselblad. In 1982 Hasselblad initiated a development of a digital image transmitter (Lundgren, 1991, p. 161).

Here, the two events are grouped together in a sequence: first A, then B. This is a common trait of a summary report; each individual event is presented as a link in a chain. In some cases the chain is put together to lead the reader to an event of greater importance than the preceding ones. In other cases, there are no dramatic changes in the import attributed to individual events, and the chain itself emerges as the more important narrative construction. An important challenge for the narrator is to make the succession of events convincing. If it is, a reader may infer from it *why* an individual event occurred, particularly if the chain leads to a clear finale.

A reader of a summary report cannot, however, grasp *how* an individual event occurred and only rarely will an actor's motives be presented. In fact, the account denies that these are at all issues. The choice of events making up the chain is made out as self-evident and seems to be based on a relatively strict definition of business action. The issue of importance is (implicitly) *that* the events have taken place and not *how* they took place.

A second mode of re-presentation that can be identified is the *focused report*. Here, the narrative is concerned with the circumstances under which a specific act is undertaken by some actor.

> To reduce the irritation among channel members, Compaq Sweden tried to re-organize the channel by tying different channel members to different customer segments. The channel members of Compaq Sweden were becoming increasingly diverse as Compaq sold computers to large, medium, and small businesses, the public sector including education and health, and private individuals. Compaq Sweden identified five principal categories of channel members and decided that they should be treated differently: (1) Resellers; (2) Retailers; (3) Distributors; (4) Solution Partners; and (5) System Specialists (Kaplan, 2002).

Similar to the summary reports, these accounts rest on relatively clear definitions of business action. In the example above, it is the effort "to re-organize the channel" that is put forward as the business action of interest. Here, however, auxiliary sentences are used to inform us not only that A did X, but also *how* and *why*. We learn about the actor's motives, "to reduce the irritation among channel members," and also something about form and content, the differentiation of channel members according to category. The differences between a focused report and a summary, we argue, mainly serve to underscore the importance of the focal act.

This changes when we encounter a *situational report*, the third mode of re-presentation we have noted. As implied by the label, these accounts are concerned with more than a single act. A number of individual actions and the details surrounding them are presented as interlocking into a whole, always involving more than one actor.

> In 1987, IBM and ABB signed a contract for the initiation of a six-month-long joint study aimed at testing the feasibility of the project and of developing a prototype of the expert system. The contract was quite detailed and defined the parties' obligations concerning the resources to be devoted to the project, the costs and the criteria for the transfer of critical information. IBM provided for the "electronic environment" (the so-called shell) and a team of five people with competence in software, artificial intelligence, software design and systems engineering. ABB devoted three people, among them, the key actor was the so-called "domain expert", i.e. a former director with responsibility for testing activities on electric engines who was supposed to transfer knowledge about the specific field in which he had worked (Tunisini, 1997, p. 81).

Here, the situation is put forward as important and problematic. Due to spatial limitations we cannot reproduce the entire passage, but the text from which the excerpt is taken goes on to discuss details about the actual study referred to, as well as problems encountered by the parties. The reader is asked to digest and resolve the situation – attending to the interlocking of non-trivial individual actions making it up – before moving on.

As a summary, we have identified three major ways of accounting for individual events in the narratives that we have studied: the summary report, the focused report and the situational report. These methods result in accounts that differ as to the answers they provide the reader with (what happened, how, why). They also draw on different implicit definitions of the events or situations accounted for.

The Use of Narrative Perspectives

A second issue concerning events has to do with the perspectives from which they are accounted for. As suggested in our discussion about the temporal dimension of narratives, a narrator can assume at least three different temporal vantage points (from the past, present or future) and apply at least three different viewing directions (towards the past, present or future). Given a possible variation also in the location of the narrator vis-à-vis the events being narrated (external or internal), we can identify at least 18 perspectives from which a story can be told. Despite this, the narratives we have overviewed make use of a rather limited set of perspectives.

Almost all narratives are told from the same position: the external present. The most common viewing direction applied from this position is the retrospection (as illustrated by the excerpts taken from, e.g. Hertz, 1993; Lundgren, 1991; Nummela, 2000; Sweet, 2000). Occasionally, some narratives also offer views towards the present and the future from this position (see, e.g. Wynstra, 1997 above).

It is much more unusual for the narrator to assume a position inside the story:

> Isn't this conversation rather silly? The wordings are formal and sometimes unfamiliar. Why would anyone spend time and effort on such a futile attempt to seek formal restitution? The only explanation I can think of is . . . (Kjellberg, 2001, p. 247).

The effect of this kind of active intrusion into the train of events being accounted for can be considerable. Suddenly, there is someone in the text who is addressing the reader directly with a reflection as to what is going on. The reader is thus asked to increase her or his involvement in these events by reflecting over them as were they happening right before her/his eyes. This practice, however, does not come without costs. Depending on the agenda of the narrator, these intrusions may just as well detach the reader from the story. Both the frequency at which they occur, the length of the intrusions, and their contents must be carefully considered by the narrator.

Despite its difficulties, we argue that the most important issue with respect to the narrative perspective is not whether the narrator has chosen the "right" one. We are convinced that it is possible to tell good stories of business action from many perspectives. Rather, we believe that the challenge lies in mixing different vantage points and viewing directions in the narrative. Since a static perspective so rarely can be successfully applied to situations outside a text, a strict application of any single perspective might do more harm than good to a narrative. After a while, then, a persistent use of a singular perspective will "throw the reader out of the story." Or, to speak with Goffman (1986 (1974), p. 346): a singular narrative perspective reduces a reader's ability to become *engrossed* in the story.

We have found that this effect is alleviated in narratives that permit the perspectives of the involved actors to interrupt the otherwise static narrative perspective. Some narratives thus ascribe to actors both internal retrospection and internal prospection from the past, e.g. a remnant is called upon to speak on behalf of an actor, disclosing either a past view concerning some previous event or a past vision about things to come (Helgesson, 1999; Liljegren, 1988). In other cases external retrospection or prospection from the past is used, e.g. an actor is reported to have had certain experiences or to have made some commitment for the future (Kaplan, 2002; Lundgren, 1991).

To sum up, we have found that few narratives make use of more than a very limited number of the narrative perspectives available. Moreover, we argue that such poverty of perspectives can dissuade a reader from accepting an account. But the narrative perspective is hardly the only factor affecting the credibility of an account, an issue which we will discuss in the next section.

The Narration of Factual Accounts

In our discussion of narrative plots, above, we suggested that many of the narratives chosen for our study were likely to be modeled, implicitly or explicitly and to varying degrees, as factual accounts. In this section we discuss two areas which we have found to have particular import on the credibility of the chosen narratives: the treatment of individual "facts" and the use of an explanatory plot.

The first area is one that we have already touched upon, namely how individual "facts" are established. Our first observation is inspired by Dorothy Smith's (1978, p. 35) argument that "it is the use of proper procedure for categorizing events which transforms them into facts. A fact is something which is already categorized, which is already worked up so that it conforms to the model of what that fact should be like." An example of this is the way in which many of the narratives that we have studied establish, without further ado, the existence of *exchange relationships* between companies. We suggest that for readers familiar with the theoretical concepts used in industrial marketing, such a statement simply conforms to the model of how a relation between two companies should be characterized.

In most narratives, however, there will be statements that cannot draw on such established categorizations. Here, the most obvious tool is the reference. That is, by indicating the source from which of a given statement emanates, its facticity is underscored. The reference creates a link between the text and an external entity. The author is no longer alone before the reader, but has been joined by others to whom the reader is advised to turn if in doubt.

A further development of the simple reference, which in some cases seems to be particularly effective, is the reproduction of some part of the entity to which the author refers, e.g. the use of an excerpt from a text, a graphical representation, a photograph. Supposedly, in this way, the reader no longer has to leave the text to assess the veracity or facticity of a statement. The external world has, so to speak, been brought to the reader by folding it into the text. The reader can now, if only for a brief moment, in the comfort of his or her armchair, travel out of the text and into some other time and location. Underlying such a trip, of course, would be an assumption that the particular item not only is accurately re-presented, but that the source from which it is taken, in turn, is an accurate re-presentation of . . . Hence, what the narrator must establish is a *circulating reference*, a chain made up of a series of transformations that transports truth (Latour, 1999).

In our view, there are many alternative styles of referring that will produce the same kind of effect, i.e. a sense of increased facticity. It seems to us that the "facticity-effect" depends more on the consistency with which a certain style is used than on the choice of style itself. That is, if certain statements in the text are singled out as needing external support to establish their facticity, then a reader may rightly question the facticity of similar statements for which such support is lacking.

The second area concerns the mode of emplotment used for the narrative. Here, we have found traces of "typical plots" that can be associated with the research tradition. For instance, a narrative about technological development may confirm to the following, highly stylised, minimal story:

> The focal company had an established exchange relationship with a customer (or supplier). The customer started to require adaptations in the products supplied by the focal company. Various ideas were tested, and some dead ends were encountered along the way. In the end, these efforts produced a quite unexpected innovation.

By conforming to an explanatory scheme that is already accepted within the research tradition, the facticity of the account is underscored. Thus, constructing the narrative in accordance with an established story can be expected to affect the perceived facticity of the account positively. The effect is even more apparent in narratives where the expected explanatory scheme is incomplete, e.g. "What about the relations to customers and suppliers, why are they left out?"

Plotting Business Action

Finally, we have reached the level of the story at large and the issue of how a coherent and meaningful order arises from the combination of individual actions. As we have tried to show above, the choice of plot is closely connected to the question of facticity. This turns a conscious use of plot into a resource in narrating business action.

Very few of the narratives that we have studied are emplotted in a way that serves to highlight business action. Nor do the authors make such claims. The narratives simply have other objectives, e.g. telling stories about technological development, the ordering of markets, the organizing of certain functions, etc. These objectives have made it necessary for the authors to account for business action, but they do not require them to structure their narratives around it. It is thus difficult for us to identify modes of emplotting business action.

A common trait that we have observed in the narratives, however, is that the plots tend to be heavily influenced by the authors' theoretical concerns. Some of these plots are mainly derived from previous theoretical work, and the resulting narratives follow the logic of the pre-specified theoretical framework. One example is Barbara Henders' use of the concept of network position to structure her account of the marketing of newsprint in the U.K. (Henders, 1992).

Other conceptually influenced plots are explicitly derived from the author's own attempt to induce a conceptual model from the events. One example of this is the plot used by Anders Lundgren, which consists of three distinct phases, derived from the study of the emergence of a digital image-processing network in Sweden (Lundgren, 1991). Largely, this amounts to inventing a new type of plot which supposedly can be used to tell other stories of the same kind.

Besides this aspect of plot-structure, we have observed a variation in terms of the number of perspectives represented in the plots. A first group of plots are told from supposedly neutral ground, allowing an "objective" account of the events.

A second group of narratives tell the story from the perspective of a focal actor. Many of these accounts mix the subjective perspective with an objective account of counterparts, etc. One example is Martin Johanson's (2001) study of how a Russian printing house handles relationships with, e.g. authorities during the transition from planned economy to market economy. Such a combination requires that the author makes credible why the reader should accept an "objective" view of some actors when a subjective view is offered for another.

A third group of plots combines a multitude of subjective views on a series of events. A prosaic example can be found in Virpi Havila's study of triads in international business, which at times accounts for the perspectives of all three actors involved (Havila, 1996). The resulting narratives vary widely in quality, but our general impression is that it can be more difficult to achieve a coherent plot in these stories. This difficulty becomes more pronounced when the plot combines not only several of the involved actors' perspectives, but also other voices, such as the author's own, those of other authors who have written on the subject, those of various theoretical scholars, etc. Since this multitude may result in a bewildering cacophony rather than a coherent plot, it will require a very skilled narrator. These observations have led us to believe that there might be a tradeoff between achieving a strong explanatory plot and offering multiple perspectives.

DISCUSSION

Does it Work? Is it Beautiful? Who is it for?

Our reason for writing this text was the varying degrees to which different texts were able to reach out to us as representations of business action. When analysed in terms of the basic dimensions discussed in narrative theory, this variation between narratives does not seem so strange. The use of various narrative devices to account for time, space, actors, events and plots, may very well explain why some narratives seem more "convincing" or more "real" than others.

Since we are dealing with representations, however, it can be argued that we must evaluate performance by other means than the degree of correspondence to "reality":

> In scientific ethos, the answer seems relatively simple: good scientific writing is true writing . . . [But if we assume] that no linguistic items correspond to any non-linguistic items, then these criteria are neither appropriate nor well grounded. We are left with the 'performance criteria' of beauty and good use. The question is no longer 'Does it correspond to outside reality? Does it observe the rules of formal logic?' but 'Does it work? Is it beautiful? Who is it for?' (Czarniawska, 1995, p. 26).

Our discussion above has indicated that some narrative devices might promote favorable answers to the question "Does it work?" Such a view is also supported by Stern (1973) who elaborates on ways to achieve a realistic effect without getting trapped in naive realism (an "it-is-true-because-I-was-there" realism). He argues that there are at least three ways of understanding realism in literature:

> A way of depicting and describing the situation in a faithful, accurate, "life-like" manner; or richly, abundantly, colorfully; or again mechanically, photographically, imitatively (Stern, 1973, p. 40).

Other researchers have pointed to ways in which the use of words can make narratives "work," that is, using language to achieve a realist impression. Karl Weick (Weick, 1979), for example, argues that we must "think -ing" (and "write -ing," we assume), when trying to visualize the continuous flux associated with the reality of organizations:

> . . . Notice first, that there are both spatial and temporal aspects . . . . The spatial aspects are preserved by nouns . . . and the temporal aspects by verbs and verb forms ("moving", "extending", "pulling"). The verbs and verb forms in the descriptive statement capture the process features . . . . Motion, change, and the flow of time would not be apparent without the verb and the verb forms . . . . Verbs anticipate sequences of events and bind together the various changed appearances that occur when the object becomes transformed. The process of moving requires time, and it involves change. Without verbs, people would not see motion, change, and flow; people would see only static displays and spines. The point that verbs and verb forms are crucial in process descriptions can be broadened to include the argument that process descriptions rely heavily on relational words of all kinds . . . . Whenever people talk about organizations they are tempted to use lots of nouns, but these seem to impose a spurious stability on the settings being described (Weick, 1979, pp. 43–44).

Our impression from studying the narratives is that passive and inactive verb forms are very common (e.g. "were established . . .," "the establishment of"), despite the frequently stated ambition to capture aspects of business processes. Over time, however, it is possible to identify an increasing use of active forms (e.g. "they were changing their ways of doing . . ."). Still, the choice of form seems to be closely linked to some implicit definition of business action on behalf of the author. That is, only verbs that are regarded as reflecting business action are given an active form. Before making such choices, we feel that the question of which those verbs are must be made an issue. To this end, i.e. in order to come closer to business processes and actions, the suggestion made by John Law (1994) is instructive: we have to try to do without nouns.

Ways of Making it Work Some Suggestions on the Use of Narrative Structure

Returning to our discussion above, from the perspective of our four narrative dimensions, we asked how action is re-constructed in different types of narratives from the business field. There seemed to be ways in which a realist effect could be achieved, an experience of being close to business action in the making, with the help of narrative structure. With the possibility of actions to span across time and space, with a view of temporality as multidimensional, and of actors as variable, a number of opportunities to tell our stories and write our narratives about business actions are opened up. Given the relativity of the experiences of closeness in narrative accounts and the rich source of tools for achieving it, it is impossible to provide an exhausting list of narrative practices. We suggest a few that seem to us to be conducive to achieving an impression of presence and hence produce convincing stories about business action:

- giving voice to actors' different temporal perspectives (past, present, future), for instance through the use of remnants (Time).
- experimenting with the use of the three basic temporal categories in narratives: order, duration and frequency (Time).
- describing the spatial setting of business actions to convey moods (Space).
- attending to how actors transgress localities through mediation by representatives and representations (Space).
- reflecting over how the re-presentation of the chronotope (the integration of time and space) affects the character of both actors and actions (Time & Space).
- clearly accounting for who or what is acting and unto who, or what the action is directed (Actors).
- making credible that entities have been/should be awarded actor status in the situations you account for (Actors).
- testing various narrative means to achieve polyphony, involving several subjects in the production of the text, e.g. an author (or several!), an implied author, a narrator, a character, a narratee, an implied reader, and a reader . . . (Actors).
- reflecting over the definition of business action underlying your choice of events (Events).
- constructing a succession of events not by simple mechanical analogy or chronology but by providing credible links between events (Events).
- trying alternative means for authorizing your narrative, e.g. established procedures, consistent references, typical plots, etc.
- making the narrative polyphonic, allowing the integration of concordant and discordant processes and the intersection of different plots in one event (Plot).

Does it work? We do not know. Still, we suggest that these practices might be well worth trying, when attempting to make narratives "work" and be "beautiful."

From the list above, it is possible to get the impression that we are unconditionally promoting more detail in our narratives. We do not. We believe that it is important to be symmetrical when discussing the potential effects of a more reflexive narrative approach. There is undoubtedly a tradeoff that each narrator

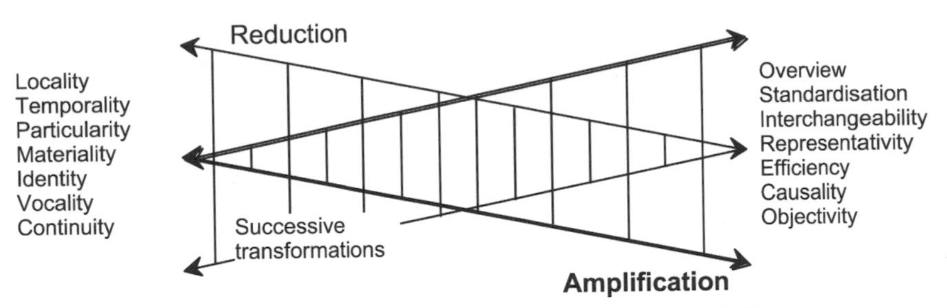

Fig. 4. A Narrative Tradeoff (inspired by Latour, 1999).

must consider. This tradeoff can be illustrated (see Fig. 4) in terms of two processes working in opposite directions: *reduction* and *amplification* (inspired by Latour, 1999).

A narrator can be thought of as choosing to position his or her narrative on a scale from the particular and local to the general and global. This choice will always include a tradeoff; a narrative cannot be both these things. Starting from an "empirical account" (left side in Fig. 4), the narrator will, through a series of transformations, be able to gain (amplification) the ability to, e.g. speak on behalf of, to generalize, and to establish causality. At the same time, the narrator will, in each transformation, lose (reduction) the ability to, e.g. use multiple voices (vocality), temporal perspectives (temporality) and achieve a continuity in the flow of events. Of course, starting from a "theoretical account" (right side in Fig. 4) and moving to the left, will have the opposite consequences.

We argue that there are many ways of writing good narratives about business action. What we would like to point out, is simply that you cannot do everything at once (confer the tradeoffs discussed by Weick, 1979). Given a decision concerning what you want to do, reflections over the narrative techniques applied will be necessary. So also will reflections over who the narrative is for.

Is it Beautiful and Who is it for? On Interpretation and the Relativity of the Real

The relativity of what we have termed "a realist effect" becomes obvious when attempting to set up criteria for achieving beauty in narratives. What is experienced as "real" and "realism" by one reader need not be seen as such by others. Whether a narrative works and is beautiful, must ultimately be determined by the reader.

> It is impossible, however, to establish *intentio operis* of a given text once and for all. Intentions are being read into the text each time a reader interprets it. Again, this does not mean there is

> an unlimited variety of idiosyncratic interpretations. In a given time and place, there will be *dominant* and *marginal readings* of the same text . . ., which makes the notion of *interpretative communities* very useful (Czarniawska, 1997, p. 201).

Czarniawska argues that the interpretation of a narrative is affected both by the relation between the individual reader and the text, and by the relation between different readers. This suggests that what we, i.e. the authors of this text, have interpreted as a "distance" between us and some accounts of business action, might for a reader belonging to some other interpretative community bring on a strong sense of "closeness" to those actions. Consequently, the author of any narrative must handle the question of who the narrative is for.

In this respect, we suggest that the chosen *mode of emplotment* is particularly important. The reason for this being our observation that the mode of emplotment is often closely connected to a research tradition and/or theory. Such traditions might, in turn, be taken as indicators of interpretative communities within which certain methods for achieving a sense of reality in narratives will be more effective than others. The theory embedded in the narrative will then contribute to establishing its authority, since the way in which to read the text is already known to the reader (see Smith, 1978). It is thus likely that theory-driven plots, i.e. narratives that are strongly connected to some underlying theoretical paradigm, will create a feeling of being close to the action for some readers – the members of the corresponding interpretative community – while distancing others.

At the same time, if research traditions in the field of business can be found to rely only on a few dominant methods for achieving a sense of reality in their narratives, then other areas of research should offer authors of business narratives a rich source of narrative ideas. For example, anthropological research (Geertz, 1988) and ethnomethodology (Boden, 1994) may offer interesting variations to explore. More specifically, there are many relatively distinct forms that may be employed in an attempt to achieve a sense of presence, for example *polyphonic realism* (Latour, 1996), *microrealism* (Silverman & Jones, 1976) and *ironic realism* (Kunda, 1992).

Returning to the issue of who it is for, Czarniawska (1997, p. 200) argues that all readings are "uses," different uses for different readers. While the *methodological reading*, for example, of a business narrative, is controlled by the reader and the purpose at hand, the *inspired reading* is a reading that changes the reader and the purpose as much as it changes the text.

This prompts a few reflexive comments also on this text, i.e. this particular "methodological narrative," and the potential readings and "uses" of it. Just as other narratives will be subject to both methodological and inspired readings, so will this one. As a methodological narrative, we admit that this text is written with the expectation of a *methodological reading*, much in line with how we have perceived many of the empirical business narratives quoted above.

So, reflecting on this particular narrative, what do we see? Or, in our own words, *where is the action*? On the whole, the text can be "accused of" following the same narrative pattern as so many other texts in its genre. That is, in many readers it will most likely create the sense of "distance" which signals the adoption of a "methodological reading." Following an obvious, theoretically-based, plot structure (built around the four concepts time, space, actors and plots), it has many similarities with some of the business narratives quoted: Time is anonymous or dominated by retrospectives (looking back and accounting for already existing texts). Space is non-existent both concerning the text extracts and the factual text (except for the sparse address information linked to the authors' home university). The actors, i.e. the narrators ("we") are fairly anonymous, as are the text extracts accounted for and the authors behind them. The events are mainly singular and the plot structure is overly theory driven.

What we as authors can hope for, though, is a few inspired readings by readers that – despite the strong connections between the text and a more or less predetermined interpretative community – through their readings will change themselves, the purpose, and the text. In the best of cases, these readings will open up for creating many other, new and complementary, forms of narratives in the genre of methodological texts.

CONCLUSIONS AND SUGGESTIONS FOR BUSINESS ACTION RESEARCH

We have identified a number of narrative practices that we feel are likely to improve business action narratives in general. However, rather than a list of normative statements concerning the narration of business action, our major conclusion is that authors of such narratives need to reflect over the consequences of their choices. The use of different narrative devices *will* affect the way in which a text is perceived, and therefore deserves at least some basic consideration.

If our discussion would contribute in some way to such an increased reflexivity, or even a desire to experiment with alternative modes of re-presentation, we would be pleased. These issues, we feel, are simply too important to be governed by routine and tradition alone.

Given this, we would like to suggest a slightly different agenda for future research, rather than proceeding towards increased formalization and quantified results, we would favor trying! There are, in our opinion, a number of interesting avenues to explore with a more reflexive approach to business action narration.

Some Areas for Narrative Experiments

Since most of the empirical selections above have been collected from the so-called IMP tradition of research on exchange relationships in industrial markets, our suggestions will be related to this tradition. One source of ideas on central topics in this tradition is Ford's (2002) comprehensive coverage of the work within the IMP Group, including the Interaction Approach to understanding and managing business networks. The first book by the IMP Group (Håkansson, 1982) has provided additional ideas.

The dominating levels of analysis in this tradition are business exchange episodes, business interactions, long-term business relationships, and business networks. Our first general suggestion is that narrative devices might be useful for achieving convincing descriptions of the connections *between* these analytical levels. For example, narrative tools might help us to better re-present how business actions are associated with the central assumption of *connectedness*, and how actions within individual relationships affect the overall network (see Anderson, Håkansson & Johanson, 1994).

Here, the use of *multivocal plots* could be explored. It has been suggested that an actor's network context is that part of the network which lies within its horizon and that the actor considers relevant (Håkansson & Snehota, 1989). Applying polyphony to capture how different, but connected actors construct their respective "horizons," could be one path to explore when attempting to capture the embedded nature of a single relationship and the business actions connected to it. Accounts of the *links between events* through, e.g. situational reports where the individual business actions and the details surrounding them are presented as interlocking into a whole, could also promote our understanding of embeddedness and connectedness. As for time perspectives, *repetitive accounts* of the same business actions might be useful to make convincing the concurrence in networks, and to match, e.g. shifts in the spatial dimension when alternating between perspectives.

Many issues within the IMP tradition can be said to concern the micro dynamics of business exchange episodes and recurrent interactions. For example, the control and coordination of personal contacts between suppliers and customers, e.g. marketing and purchasing controlled and coordinated contact patterns (Cunningham & Homse, 1986). Of central importance in such research is the fact that continuous and *mutual adaptations* between firms involved in business exchange successively lead to, and are a central part of, the emerging relationships between firms (Håkansson & Gadde, 1994).

Increased narrative reflexivity might be conducive to a better understanding of the dynamics of negotiations, coordination, organization and general mutual adaptation between marketing and purchasing representatives in a buyer-seller relationship. Here, the experimentation with narratives that make *convincing the actual links between episodes* in the plot should be an important tool for any author. This should include the use of matching *temporal and spatial accounts that bridge the successive line of adaptations in time and place*. This could also include conveying a sense of mutual orientation between the actors by creating *polyphonic* accounts with multi-vocal views of *the acting subjects*. Employing detailed *place* accounts could be important if the goal is to create a sense of presence in both the technological and organizational adaptations on both sides in a relationship. *Time* descriptions that use, e.g. *stretch* (i.e. where story-time is less than the so-called discourse time) could bring the reader closer to the business actions associated with the details of adaptation processes.

Within the IMP tradition, it has also been argued that each business relationship has a certain *atmosphere*, i.e. it is a setting encompassing certain degrees of power/dependence, cooperation, closeness and expectations (Håkansson, 1982). The actions of companies in a relationship signal a certain degree of mutuality, mutual orientation.

Here, detailed re-presentations of spatial *settings*, combined with plots that focus on the actors' vantage points and viewing directions in *their* accounts of the business actions connected to the relationship, are worth employing. How far into the future do the acting subjects on both sides look upon the relationship? Further, the exploitation of narrative space may allow an author to shed light on the mood surrounding certain business actions, in this case the relationship atmosphere.

Lastly, an important research area within the IMP tradition since its conception, has been the process of internationalization, where either the dyadic or the network perspective is in focus (see, e.g. Håkansson, 1982; Johanson & Mattsson, 1988).

Here we could make use of spatial modes of re-presentation that make apparent how actors *act at a distance*, i.e. spanning a certain space. The narratives on internationalization could better re-present these particular business actions by accounting in detail for how this is possible, *explicating the various intermediaries involved* in business action. If a large number of intermediaries is needed to link localities (in the shape of humans or non-humans), narratives might need to be multi-vocal in order to make explicit the roles of these entities. This directs attention to narrative devices for both actors and space that allow us to explicate such aspects.

Although they are merely a few rough sketches, we believe that our suggestions indicate the rich opportunities offered by a more reflexive narrative approach for researchers with an interest in business (inter-)actions and processes. Opportunities that hopefully will allow us to say more frequently: There is the action!

REFERENCES

Alvesson, M., & Sköldberg, K. (1994). *Tolkning och reflektion*. Lund: Studentlitteratur.

Andersson, P. (1996). The emergence and change of Pharmacia Biotech 1959–1995: The power of the slow flow and the drama of great events. Ph.D., The Economic Research Institute, Stockholm School of Economics, Stockholm.

Anderson, J. C., Håkansson, H., & Johanson, J. (1994). Dyadic business relationships within a business network context. *Journal of Marketing*, *58*(October), 1–15.

Barthes, R. (1977). *Image-music-text*. S. Heath (Trans.). New York: Hill and Wang.

Blankenburg Holm, D. (1996). Business network connections and international business relationships. Ph.D., Dept. of Business Studies, Uppsala University, Uppsala.

Boden, D. (1994). *The business of talk: Organizations in action*. Cambridge: Polity Press.

Cunningham, M. T., & Homse, E. (1986). Controlling the marketing-purchasing interface: Resource development and organizational implications. *Industrial Marketing and Purchasing*, *2*(1), 3–27.

Czarniawska, B. (1995). Narration or science? Collapsing the division in organization studies. *Organization*, *2*(1), 11–33.

Czarniawska, B. (1997). *Narrating the organization. Dramas of institutional identity*. Chicago: University of Chicago Press.

Davidson, D. (1980). *Essays on actions and events*. Oxford: Clarendon Press.

Ford, D. (Ed.) (2002). *Understanding business marketing and purchasing* (3rd ed.). Padstow: Thomson Learning.

French, P. A. (1983). Kinds and persons. *Philosophy and Phenomenological Research*, *XLIV*(2), 241–254.

Garret, J. E. (1988). Persons, kinds and corporations: An Aristotelian view. *Philosophy and Phenomenological Research*, *49*(2), 261–281.

Geertz, C. (1973). *The interpretation of cultures*. New York, NY: Basic Books.

Geertz, C. (1988). *The anthropologist as author*. Stanford: Stanford University Press.

Genette, G. (1980). *Narrative discourse: An essay in method*. J. E. Lewin (Trans.). New York: Cornell University Press. Original edition, Discours du récit, in figure III, 1972.

Giddens, A. (1984). *The constitution of society, outline of a theory of structuration*. Cambridge: Polity Press.

Goffman, E. (1986) (1974). *Frame analysis*. Boston, MA: Northeastern University Press.

Havila, V. (1996). International business-relationship triads: A study of the changing role of the intermediating actor. Ph.D., Dept. of Business Studies, Uppsala University, Uppsala.

Helgesson, C.-F. (1999). Making a natural monopoly: The configuration of a techno-economic order in Swedish telecommunications. Ph.D., The Economic Research Institute, Stockholm School of Economics, Stockholm.

Henders, B. (1992). Positions in industrial networks: Marketing newsprint in the U.K. Ph.D., Dept. of Business Studies, Uppsala University, Uppsala.

Hertz, S. (1993). The internationalization process of freight transport companies. Ph.D., The Economic Research Institute, Stockholm School of Economics, Stockholm.

Håkansson, H. (Ed.) (1982). *International marketing and purchasing of industrial goods: An interaction approach*. Chichester: Wiley.

Håkansson, H., & Gadde, L.-E. (1994). *Professional purchasing*. London: Routledge.

Håkansson, H., & Johanson, J. (1992). A model of industrial networks. In: B. Axelsson & G. Easton (Eds), *Industrial Networks: A New View of Reality*. London: Routledge.

Håkansson, H., & Snehota, I. (1989). No business is an island: The network concept of business strategy. *Scandinavian Journal of Management, 5*(3), 187–200.

Johanson, M. (2001). Searching the known, discovering the unknown: The Russian transition from plan to market as network change processes. Ph.D., Dept. of Business Studies, Uppsala University, Uppsala.

Johanson, J., & Mattsson, L.-G. (1988). Internationalization in industrial systems: A network approach. In: J.-E. Vahlne (Ed.), *Strategies in Global Competition*. New York: Croom Helm.

Johanson, J., & Wiedersheim-Paul, F. (1974). The internationalization of the firm – Four Swedish case studies. *Journal of Management Studies* (3), 302–322.

Kaplan, M. (2002). Acquisition of electronic commerce capability: The cases of Compaq and Dell in Sweden. Ph.D., The Economic Research Institute, Stockholm School of Economics, Stockholm.

Kjellberg, H. (2001). Organizing distribution: Hakonbolaget and the efforts to rationalise food distribution, 1940–1960. Ph.D., The Economic Research Institute, Stockholm School of Economics, Stockholm.

Kotler, P. (1999). *Marketing management: Analysis, planning, implementation, and control* (Millennium ed.). Prentice-Hall series in marketing. Upper Saddle River, NJ: Prentice Hall.

Kunda, G. (1992). *Engineering culture: Control and commitment in a high-tech organization*. Philadelphia: Temple University Press.

Latour, B. (1996). *Aramis, or the love of technology*. Translated by C. Porter. Cambridge, MA: Harvard University Press.

Latour, B. (1999). *Pandora's hope, essays on the reality of science studies*. Cambridge, MA: Harvard University Press.

Law, J. (1986). On the methods of long-distance control: Vessels, navigation and the Portugese route to India. In: J. Law (Ed.), *Power, Action and Belief: A New Sociology of Knowledge*. London: Routledge & Kegan Paul.

Law, J. (1994). *Organizing modernity*. Oxford: Blackwell Publishers.

Liljegren, G. (1988). Interdependens och Dynamik i Långsiktiga Kundrelationer: Industriell försäljning i ett Nätverksperspektiv. Ph.D., Department of Marketing, Stockholm School of Economics, Stockholm.

Lilliecreutz, J. (1996). En leverantörs strategi – från lego-till systemleverantör. En studie av köpare- och leverantörsrelationer inom svensk personbilsindustri. Ph.D., Department of Management and Economics, Linköping University, Linköping.

Ludvigsen, J. (2000). The international networking between European logistical operators. Ph.D., Economic Research Institute Stockholm School of Economics, Stockholm and Institute of Transport Economics, Oslo.

Lundgren, A. (1991). Technological innovation and industrial evolution, the emergence of industrial networks. Ph.D., The Economic Research Institute, Stockholm School of Economics, Stockholm.

Mahmoodian, M. (1997). Social action: Variations, dimensions and dilemmas. Sociologiska institutionen Univ. distributör, Uppsala.

Mannen, J. van (1988). *Tales of the field*. Chicago: University of Chicago Press.

March, J. G. (Ed.) (1988). *Decisions and organizations*. Oxford: Basil Blackwell.

McQuillan, M. (2000). *The narrative reader*. London: Routledge.

Nowotny, H. (1994). *Time. The modern and postmodern experience*. In: N. Plaice (Ed.). Cambridge: Polity Press. Original edition, Eigenzeit Entstehung und Strukturtierung eines Zeitgefühls, Frankfurt am Main: Suhrkamp Verlag, 1989.

Nummela, N. (2000). *SME commitment to export co-operation, Turun kauppakorkeakoulun julkaisuja. Sarja A, 2000:6*. Turku: Turku School of Economics and Business Administration.

O'Neill, P. (1994). *Fictions of discourse: Reading narrative theory*. Toronto: Toronto University Press.
Peirce, C. S. (1934). *Collected papers of Charles Sanders Peirce, Vol. 5: Pragmatism and pragmaticism*. Cambridge, MA: Belknap Press of Harvard University Press.
Pieters, R., & Verplanken, B. (1991). Changing our mind about behavior. In: G. Antonides, W. Arts & W. F. van Raaij (Eds), *The Consumption of Time and the Timing of Consumption – Toward a New Behavioral and Socio-Economics*. Amsterdam: North-Holland.
Ricoeur, P. (1984). *Time and narrative* (3 vols, Vol. 1). Chicago: University of Chicago Press.
Schick, F. (1991). *Understanding action, an essay on reasons*. Cambridge: Cambridge University Press.
Silverman, D., & Jones, J. (1976). *Organizational work*. London: Collier Macmillan.
Smith, D. E. (1978). 'K is mentally ill' The anatomy of a factual account. *Sociology*, *12*, 23–53.
Snehota, I. (1990). Notes on a theory of business enterprise. Ph.D., Dept. of Business Studies, Uppsala University, Uppsala.
Stern, J. P. (1973). *On realism*. London: Routledge & Kegan Paul.
Sweet, S. (2000). Industrial change towards environmental sustainability. Ph.D., The Economic Research Institute, Stockholm School of Economics, Stockholm.
Tambling, J. (1991). *Narrative and ideology*. Milton Keynes: Open University Press.
Tunisini, A. (1997). The dissolution of channels and hierarchies: An inquiry into the changing customer relationships and organization of the computer corporations. Ph.D., Dept. of Business Studies, Uppsala University, Uppsala.
Waluszewski, A. (1989). Framväxten av en ny mekanisk massateknik – en utvecklingshistoria (The emergence of a new mechanical pulp technology). Ph.D., Dept. of Business Studies, Uppsala University, Uppsala.
Weick, K. E. (1979). *The social psychology of organizing* (2nd ed.). New York: Random House.
White, H. (1978). *Topics of discourse. Essays in cultural criticism*. Baltimore: Johns Hopkins University Press.
Woolgar, S. (2002). Introduction: Five rules of virtuality. In: S. Wollgar (Ed.), *Virtual Society? Get Real! The Social Science of Electronic Technologies*. Oxford: Oxford University Press.
Wynstra, F. (1997). Purchasing and the role of suppliers in product development. Licentiate thesis, Dept. of Business Studies, Uppsala University, Uppsala.

NARRATIVES AND CASE PROCESS RESEARCH

Per Andersson

ABSTRACT

The paper discusses some of the central features of IMP and industrial network research. Different types of empirical phenomena that are in focus of this research are presented. The paper also comments on epistemology, acknowledging some of the underpinnings of industrial network research and how they affect the use of case studies. Examples of case or narrative methodology are provided, taking a starting point in a set of chosen doctoral theses. In addition, a condensed version of the author's own experiences from a case research and case-writing process covering a period of more than five years is provided (Andersson, 1996a, b). Literature support is brought in for the fact that case writing and the creation of narratives is often a long and ambiguous process of finding a final plot which merges the theoretical with the empirical. The conclusions and comments summarize some of the main implications and ideas emerging from the text, and points also to some emerging discussions in social science on the importance and status of narrative knowledge.

INTRODUCTION

The traditional methodology in the field of industrial networks research, being inspired by the first so-called IMP (Industrial Marketing and Purchasing) projects, has been a mixture of relationship or focal organization surveys and case research. When the IMP Group was formed in 1976 by researchers from five European

Evaluating Marketing Actions and Outcomes
Advances in Business Marketing and Purchasing, Volume 12, 59–88

ISSN: 1069-0964/doi:10.1016/S1069-0964(03)12002-9

countries, one of the first major studies was a comparative survey of a large number of dyadic buyer-seller relationships across Europe. The richness of stable, long-term buyer-seller relationships was captured in a large number of in-depth case studies following the first IMP study. In parallel with the emerging notion and conceptualizations of industrial networks and the view that single buyer-seller relationships can be seen to be embedded in and connected to other relationships, case study narratives came to form an important methododological foundation. Although the first IMP study was to be followed by more large sample surveys – some of which were to be closely connected to the methods of the U.S. marketing research tradition (e.g. Hallén, Johanson & Seyed-Mohamed, 1992) – single, or several comparative, case narratives became the dominant methodological tools by which industrial network phenomena were studied.

Although the match between case study methods and the epistemological assumptions underlying much of Industrial Networks and IMP research (from here abbreviated NW) has been discussed and defended, apart from the methodological discussions in doctoral theses method discussions have not been of prime concern in NW texts (Easton, 1995). Easton is one of the few, suggesting a strong link between the case method as used in NW research and what he calls a *realist* epistemological orientation. He also presents a set of possible explanations for the fact that non-NW readers might find the methodological discussions very sparse. One might be that researchers in all paradigms are so ensconed in their own particular paradigm that they simply take for granted that the methodologies they employ are correct because they are doing what everyone else is doing. (An explanation that could apply also to the marketing research presented in the prestigious U.S. journals, heavily dominated as they have been by questionnaire surveys, operationalization of variables and randomness of large scale samples.) Another explanation forwarded by Easton is that to the researchers themselves, the epistemological justification of methodology might be seen as self-evident. (The choice and domination of one methodological approach might hide a lack of methodological training.)

Hence, broadly conceived, "case research" methodology has a central position in the NW-oriented part of the European marketing research tradition, and has in relation to much of U.S. marketing research in the same areas, assumed a partly different epistemological and methodological tradition. This we know, and we also know some of the explanations for this development, related to epistemology, to the choice of empirical phenomena under study, to views on the role of theory testing and theory building, etc. This paper discusses some of these explanations, but also takes a closer look at the case methodology-in-use in the NW tradition. "Case research" is for several reasons not a very useful concept, as it tends to conceal a very rich variety of narratives presented within the NW tradition.

Furthermore, such generic characterizations seldom reveal much about the actual research processes.

The aim of this paper is to describe some of the "how" aspects of this sometimes obscure "thing" that we often choose to collect within the boundaries of one single term – "case study." Taking NW research as a starting point, we elaborate on the casing process. How do NW researchers look upon the process of generating a case narrative? What are the major methodological problems and challenges and what are the major experiences made of such casing processes? The paper is triggered by two general problems that the author has had when approaching case study research: Firstly, there is the confusion that terms like "case study" and "case research" can convey. It will be suggested towards the end of the paper that case research in marketing could dissolve some of this confusion by linking to methodological discussions on the status of "narratives" in other areas of social science, e.g. organization theory and history (see e.g. Czarniawska, 1995, 1998). Secondly, there is the confusion that descriptions of case research in methodological text books can convey to its readers. Some of texts have become central sources in wide areas of marketing research. The stepwise accounts of how cases are created and emerge from the research process, often with strong links to the epistemology of a positivist research tradition, make a bad match with many researchers' actual experiences from casing processes, as well as with the actual processes under study. Both of these problems will be touched upon later in the article.

STRUCTURE AND SOURCES

The paper begins with a description of some of the central features of NW research (i.e. IMP and industrial network research grouped together). We begin by taking a look at the types of empirical phenomena that are in focus of this research. Assuming a strong link between the process under study and the actual study process, we need to take a deeper look at some typical phenomena under study. An underlying assumption is that the characteristics of the phenomena studied will leave marks on the actual research processes and on the structure, contents and characteristics of the final case narrative.

The paper continues with a comment on epistemology. In a note on the case method as applied in NW research, Easton (1995) has discussed the strong links between what he labels a *realist* epistemology and some of the case method applied in NW research. Linking to the first part, we acknowledge some of the underpinnings of NW research and how they affect the use of case studies.

Next, we bring in examples of case or narrative methodology in use, taking a starting point in a set of chosen NW-related doctoral theses. (Emphasis is put on describing doctoral theses from the Scandinavian part of the NW research tradition). The large bulk of NW-related articles in academic journals, in conference proceedings and in working paper series most often is dominated by conceptual discussions. Sometimes they are complemented with condensed versions of longer empirical cases. In doctoral theses, we are more likely to find more detailed accounts of cases and of the methods applied. The examples chosen here are used to *illustrate* some specific ideas concerning methodology and should not be viewed as typical, or as representatives of the many different ways of presenting texts within the large bulk of research here labeled NW research.

This fact becomes even more obvious in the following section, which is a condensed version of the author's own experiences from a case research and case-writing process covering a period of more than five years (Andersson, 1996a, b). One of the ideas permeating the original version of this text is that "the mode of emplotment," that is, the creation of a plot for the final narrative, is a process which is very much colored by the author's context, including the changes in this context. The condensed version brings in some literature support for the fact that case writing and the creation of narratives is often a long and ambiguous process of finding a final plot which merges the theoretical with the empirical.

Finally, the conclusions and comments summarize some of the main implications and ideas emerging from the text. They also point to some emerging discussions in social science on the importance and status of narrative knowledge.

The paper builds mainly on four types of sources from the rich material of NW research: literature reviews and internal analyses of articles within the paradigm, NW papers and conference proceedings (for the interested reader there are proceedings from 14 annual conferences with a constantly growing number of contributions), a selection of (mainly Scandinavian) doctoral theses within or with strong links to the paradigm, and finally, literature on the way narratives can be applied in social science research.

BUSINESS PHENOMENA, RESEARCH ISSUES, AND RESEARCH METHODS

There are several stereotypes of the NW approach to marketing. Gemuenden (1997) lists some of them, for example: that it is empirically-based, is internationally oriented, has a holistic approach, focuses on relationships and networks, and also that it uses case studies as a primary tool. Based on an analysis of IMP conference proceedings from 1984 to 1996, he shows that some of these are stereotypes. He shows that NW research (judging from the conference proceedings only) has a large number of purely theoretical papers, has less and less international themes, is in fact a forum for holistic frameworks, and is actually focusing on relationships and networks. Methodologically, Gemuenden's analysis also shows that case studies have become the most popular method. However, he also adds that "if we sum up quantitative descriptions and quantitative analyses, however, quantitative contributions are more frequent than purely qualitative contribution."(p. 9). He also shows that network-oriented researchers, in contrast to (dyadic) relationship researchers; prefer case studies to quantitative descriptions or quantitative analyses, and "therefore, the phenomenon studied by the researchers partly explains the method of data analysis used" (p. 9). More interestingly, over time another emerging trend is observed:

> The steady growth of methodological contributions and of positive theories with testable hypotheses indicates a cumulative tendency in IMP research. However, we also find a growing trend for frameworks and a decreasing trend for themes. This raises the question to which extent IMP has a cumulative research tradition (p. 8).

Despite this observation, research based on "case studies" has come to be very important in the NW tradition. Then, what are the types of business phenomena that have been studied with these "qualitative" methods? What research issues and focuses motivate the use of the methods? Before turning to ideas about the epistemological grounds of NW research in the next section, we will sum up some of the phenomena, issues and viewpoints that partly explain this importance of case studies. In the last part of the paper, we will elaborate on "cases studies" and "qualitative studies" in NW research. A tentative proposition is that these aggregate terms encompass a large variety of "narrative approaches" to the phenomena under study. In addition, although there are signs (especially in NW-related theses) that researchers are developing a much freer and liberal approach to case methodology, the text book way, "narratives" and the way we tell our marketing stories, deserve more attention. This goes for NW research, as well as other marketing and social science areas.

The frequent use of case studies has often been linked to the fact that much NW research has been occupied with the study of change phenomena,

> Change is a central feature of much of what is written about industrial networks. Networks are concerned with relationships and these cannot be conceived of in anything but dynamic terms. It is hardly surprising that the processes by which networks function have been a major preoccupation of workers in this field (pp. 21–22).

Processes of change in industrial networks, changing network structures, and steady state comparisons of network structures at different points in time, have all been described and analyzed in NW-based case studies. The cases, whose length and richness in depth and details, and whose contents (mixtures of quantitative and qualitative descriptions) vary considerably between studies, often have one thing in common: they serve to develop knowledge about ongoing change processes in industrial networks and other marketing contexts. Thus, understanding change and developing knowledge about change, including the interdependence between stability and change, has become one of the major driving forces of NW research, especially for the Scandinavian researchers within the paradigm.

Longitudinal case studies have for many NW researchers become the most important method to study some frequently recurring issues and general themes. For example, *innovation processes, technological development, and the management of product development processes*, form such a theme, and is covered in a large number of longitudinal cases. Complementing the issues on marketing and technological change, *marketing investments and market strategy*, *industrial purchasing*, *and distribution system and institutional change* belong to the main strands of issues and phenomena in NW research. In another main strand of research, the changing nature of *internationalization processes* has generated a large number of narrative descriptions. This strand of research encompasses issues on foreign market entry and foreign direct investment processes, the internationalization processes over time of companies of different sizes and of different industries. The processes whereby both single companies and the networks in which they are embedded, internationalize over time, is a recurring perspective in the internationalization cases emerging in NW research during the 1990s.

The main strands of business issues of NW research have been fairly stable over time. Johansson and Mattsson's (1994) description of the Swedish section of NW research confirms this:

> On a very general level, the current topics of network research concern conditions for firm competitiveness and for industrial change in a globally interdependent world characterized by technological and institutional change. Future topics will most probably belong under the same general heading. Given our belief in history dependence both in business and in research, we are not prepared to make a clear separation between current and future topics of industrial network research (p. 332).

One assumption in NW research is that stability, in a general sense, is a central, important outcome of interactions and emerging relationships between actors in networks. In line with this, there are two broad network themes that have drawn the attention of researchers applying case descriptions: *the formation of networks*, and *the change and restructuring of networks*.

During the 1990s, a number of case studies on network formation and network change themes began to apply *the methods of network structure analysts*, used in sociology (e.g. Wasserman & Faust, 1994). Network density and connectivity, the way clusters, blocks and cliques form in networks, are examples of characterizations of networks that emerged from the application of these methods. Some sociological, structural analysts tried also to capture relationship and network changes in their analyses (Iacobucci & Wasserman, 1988; Iacobucci, 1989). Some of these researchers constituted a bridge into marketing research, suggesting ways in which these methods could be applied in the study of marketing phenomena and industrial networks (Hopkins et al., 1995; Iacobucci & Hopkins, 1992; Reingen & Kernan, 1986).

These methods have been used to do positional analysis (cf. Johanson & Mattsson, 1992 for a discussion on network positions). Each actor in a network may be described in terms of what relations it has to other actors, or by the fact that it does not have any relations to certain types of actors. One important aspect of a position concerns *centrality*. The handling over time of a company's position defines opportunities and restrictions for the future strategic development of the company. The position changes can be captured in structural network analyses of how the centrality of individual actors changes and/or remains stable over time. Following Freeman (1979), some NW research has adopted three types of *point* (i.e. actor) centrality to study companies' positions and position changes. (The measures of these different types of centrality are termed *degree, rush*, and *distance*.)

During the 1990s, some NW researchers have used the methods to characterize changes over time in industrial networks (see e.g. Andersson & Nyberg, 1998; Lundgren, 1991). However, the use of these methods, as complements to other types of texts and descriptions in cases or as the dominant method, is still not widespread in NW research. The contents and dynamics of buyer-seller relationships and industrial network features and characteristics are not easily translated into the structural network measures and concepts.

NW researchers have also looked to other areas of scientific research for methods to capture the nature of network formation and network change processes. Hence, researchers have tried to capture the logic of relationship connections in more abstract ways, being inspired by recent developments in the *modeling of complex systems* (see Wilkinson & Easton, 1997; Wilkinson et al., 1998). Drawing on recent developments in the science of complexity (e.g. Kauffman, 1992), Wilkinson et al. try to show how different so-called *attractors* emerge. They describe how the ecology of relations and connections between relations produce different types of attractors, resulting in one of three broad types of regimes: fixed, edge of chaos, and chaotic.

Axelsson (1992) suggested that to demonstrate change and reaction patterns in networks, NW researchers could make use of work and experiences from using *experimental methods*, as used in sociological network analyses (e.g. Tichy & Fombrunn, 1979). Being based on idealized and simplified models of the world, these methods would give different results and demand different interpretations, as compared to the results of rich ethnographic studies, according to Axelsson. One of the challenges, "would be to make the content of interactions between actors more realistic and, indeed, to make it one of the experimental variables" (p. 247). So far, few NW researchers seem to have adopted these experimental network methods.

To sum up, NW researchers' view on marketing and business phenomena as complex, multifaceted, and embedded, coupled with the ambition to capture the contents of interactions and relationships in more "realistic" descriptions, has often made the adoption of thick ethnological cases the "natural" first choice. The need to develop in-depth knowledge on the multidimensional contents of inter-firm interactions and relationships has in some cases resulted in the use of a combination of cases with other methods, e.g. structural network analysis (e.g. Lundgren, 1991). Such combinations of methods have also been used to develop knowledge on the processes whereby interorganizational networks emerge and change.

Next, with the help of NW discussions on epistemology, we will argue that the explicit ambition among many NW researchers has been to develop and in-depth understanding of the nature of marketing and business phenomena. And this has been one of the reasons why case narratives have been accepted as the most important method.

EPISTEMOLOGY AND THE CASE METHOD IN INDUSTRIAL NETWORK RESEARCH

Easton (1995) discusses the strong links between a realist epistemology and the case methododology applied in NW research. In this section, we acknowledge some of the underpinnings of NW research and how they affect the use of case research. Comparing four basic epistemologies – realism, positivism, constructivism and conventionalism – with an emphasis on the comparison of the two first, Easton suggests that the first orientation, realism, is the one that matches the particular characteristics of case research and thus a large part of NW research. According to this orientation, there is a reality "out there," independent of us and waiting to be discovered, but this does not mean that "this reality is obvious or self-evident or easy to discover" (p. 373). Quoting Sayer (1984), Easton argues that our knowledge of that (real) world is fallible and theory laden. In his comparison with a positivist orientation, Easton concludes about the use of case research versus surveys:

> Realists should not however reject positivist methods but rather use the possibility of reversing the normal role of case and, for example, survey work. Surveys, given their superficial nature, might be used as the first stage of research to provide a broad overview of the research domain and to guide more in-depth explanatory studies: in a word, to provide something to explain. They offer the chance to identify the contingent variables. An example from industrial networks

> would be a survey of the links between a net of firms and subjecting them to a clique detection programmed. Case research would then seek to explain how, for example, one of the cliques came into being, how it has changed over time, and, of course, why (p. 383).

The connectedness between actors in industrial networks, the ambiguous, unclear and dynamic boundaries between nets in large networks, and the fact that networks can be perceived as dynamic forms are parts of the explanation why NW research has been driven more towards case research. According to Easton, the epistemological orientations of realism provide another justification for the drive towards rich and in-depth cases. Case studies, it is argued, are a powerful research method and one particularly suited to the study of industrial networks.

However, the frequent use of case research in industrial network studies has very rarely been justified on the grounds of particular epistemological orientations. Such justifications are mainly found in NW doctoral theses. In this respect, the NW tradition does not differ much from other marketing research traditions, which are founded on other epistemological orientations. Likewise, the researchers' own ontology is seldom presented to their readers, except in longer texts like doctoral theses. To exemplify from an industrial network thesis, Lundgren (1991) bases his doctoral thesis on an historical case study and commences his notes on methodology by stating:

> Even so, our ontological starting point is that there is a reality, which exists independent of our own consciousness, of which it is possible to get objective knowledge. The question is whether the study of this reality, the reconstruction of the past, can be conducted independent of our own consciousness, our own preconceptions and implicit or theoretical propositions. Can a story be told as it really was . . ..

This type of introduction to the researcher's ontology (which in this case continues and ties it to methodology) is not likely to be found in texts other than doctoral theses. In marketing traditions where research is driven by publication in key marketing journals, there is little room for elaborate discussions on epistemology or ontology. The institutionalized tradition for presenting texts in the established marketing journals locks out thick empirical case descriptions and theoretical presentations that are not purely conceptual and/or based on condensed, statistically testable survey data.

Like Easton above, who compares the positivist and the realist epistemology and views on method, many other NW researchers have since long defended and argued for the use of case studies. Such discussions are not unique for marketing. Similar debates and methodological defenses can be found in sociology and other social sciences. Sociologists, like NW researchers, often take the methodological discussion from the positivists' standpoint. To exemplify, Abbott's (1992) sociological defense of case methodology is in line with many other disciplines of social research, including marketing. The way that the positivists' position is accepted as a starting-point is part of this discussion:

> What then do cases do in standard positivist analysis? For the most part, they do little. Narrative sentences usually have variables as subjects. For anything unexpected and for the authorial hypothesis, the level of narrativity rises in various ways. When cases do do something, it is generally conceived as a simple rational calculation. All particularity lies in the parameters of calculation. Since only the parameters change, there are no complex narratives; narratives are always one-step decisions. There are no real contingencies or forkings in the road. There is simply the high road of variables and the rest – which is error.

This and similar methodological analyses and critique from NW researchers use the positivists' own method tradition as a stepping-stone. Alternatively, as above, views of how the positivist traditions look upon other, case-based, traditions are taken as a starting-point. Although methodological justifications and comparisons with positivist dominated traditions of marketing research used to be "natural" when positioning NW research and methods, over time such justifications and comparisons have become less important.

Comparisons of the empirical phenomena studied, of the underlying views on them, and comparisons of how knowledge about them is created have lost some of their former importance, as NW research has matured and established its own position in marketing and business research. Instead of defending and justifying case methodology as something different from an established research tradition, successively NW research has begun to leave this behind. For example, by bringing in methodological ideas about the status, structure and logics of "narratives," rather than "cases," some researchers have approached other areas of social science, like history and certain areas of organization theory. One reason is the problems that the term "case" and the mixture of ideas and terms linked to it bring about. Despite the fact that it is widely used, the term is not well defined in marketing and other areas of social science.

Marketing cases, both teaching cases and research cases, encompass a variety of narrative forms. One of the arguments here is that some of the most interesting methodological changes emerging in NW research are found in this area. As we continue, we successively leave out the term "case," using instead the term "narrative" to introduce some emerging ways in NW research to convey its often rich and multifaceted marketing and business stories.

METHODOLOGICAL FOUNDATION: A VARIETY OF RICH, LONGITUDINAL NETWORK NARRATIVES

Turning for a moment to the American marketing research tradition, for most marketing academics the need to publish research in prestigious marketing journals is frequently assumed to be highly correlated with the use of methods. More specifically, in order to get published, it is often required that articles are based on quantifiable survey data with randomness of samples, and operationalized variables linked to a priori theories.

In this line of reasoning and from the point of view of the American research tradition, Wilson (1994) comments on Johanson and Mattsson's (1994) description of parts of NW research. He describes it as having been heavily-weighted toward qualitative work and description of markets and networks. It has used behavioral theory to support research propositions induced from descriptive data, and has seldom let research be driven by a priori theory. He continues:

> Although North Americans show some interest in networks, they do not generally see them as an interesting or important research topic . . . .Unfortunately, the softness of the research and the (IMP) model caused this excellent body of insightful work to have a much lower impact than it merited. Networks have value as descriptors of markets, but theory acceptance for purposes beyond description is hampered by the lack of quantitative research supporting them and lack of basic theory. In North America, we are more concerned with the focal relationship between firms in the network. Our reward system still heavily reflects the need to publish in quality journals, and this makes studying networks, which are complex and open ended, also not very attractive. The lack of quantitative measures also inhibits network research (p. 346).

When NW research becomes theory driven and the "data-gathering problem" is solved, Wilson expects the network approach to become a "hot topic" in marketing research. Hence, methodological differences are linked to differences in institutionalised reward systems, differences in views on, for example, data, data-gathering and on the status and use of established theory vs. theory development in research. (It can be noted how the term "softness" indicates some of the differences.)

In comparison to this view, Easton (1995) and other NW researchers reverse the underlying epistemology and implicit evaluations and uses of surveys and case studies (see introductory quote above). Thus, NW views on the use, status and role of cases vs. surveys differ. American researchers admit that the choice of surveys is partly due to strong bonds to a positivist, epistemological tradition, partly due to the fact that surveys are better suited for a tradition where the publication in a prestigious journal is important. The pragmatism of the NW tradition is somewhat different. While epistemology is successively being discussed and made part of the explanations for the use of case studies, the reason given is that it is simply very difficult to find any survey methods that can capture the dynamics, the embeddedness, the complexity, the long-term perspective etc. of industrial networks.

Johanson and Mattsson (1994) argue that the NW tradition could not have evolved (e.g. in Scandinavia) the way it has if academic evaluators and academic leaders had not been so open to time-consuming, qualitative research, which is very difficult to publish in leading academic journals. Consequently, the many significant characteristics of the methodological traditions of NW research can be found in its doctoral theses. In a growing number of theses, many of the recurring marketing phenomena under study have been captured in longitudinal, historical case studies; technological development in networks (e.g. Lundgren, 1991; Waluszewski, 1989), companies' market strategy and network positioning processes (e.g. Andersson, 1994), companies' and networks' internationalization processes (e.g. Hertz, 1993), companies' marketing reorganization processes (e.g. Andersson, 1996a, b), longitudinal distribution channel changes (e.g. Tunisini, 1997) etc. Here, we also find some of the methodological ideas emerging from NW research. For example, an emerging theme in the methodological discussions in several doctoral theses emphasizes the fact that case studies are time-consuming. Casing processes often extend over a fairly long period of time, and methodological challenges and experiences become coloured by this and by the fact that researchers often work in a close "relationship" with their case(s).

For example, Lundgren (1991) in a study of a technological innovation process in an industrial network, described the method of analysis as a complex activity of analysis, genesis, and insight:

> It is a process of uniting empirical observation with theoretical insights: combining inductive and deductive reasoning: integrating unique observations, classification of hard data or articulated knowledge with theoretical insights, higher levels of understanding or expressions of tacit knowledge. The process of theorizing around dynamic processes, characterized as they are by multiple change and multiple causation, calls for true pluralism. In the successive abstraction of unique observation, in the pursuit of insight through analysis and genesis, anything goes. In the sequential reiteration of analysis, genesis and insight, pluralism is essential: poems, prayers or promises, theoretical reasoning, anecdotal evidence or quantitative methods of analysis, anything goes in the pursuit of insight (p. 76).

Arguing not for an extreme form of *pluralism*, but for pluralims within the realms of the existing line of reasoning and existing observations, Lundgren disarms criticism and meets the misconception that NW research regularly avoids quantitative analysis.

Hence, methodological reasoning is coloured by the fact that research often deals with longitudinal processes and with the fact that the research process in itself is longitudinal. (The methodological experiences from one such research process are presented in the next part of this paper.) Inevitably, the study of longitudinal network processes brings in history into research. This is made explicit in the methodological descriptions by Waluszewski (1989), in her historical network study of the emergence of a new production process technology.

Firstly, she has to link epistemology and ontological assumptions to her *historical methods* and the historical accounts of the longitudinal network process. For example, building on Croce (1941), she states that accounts of historical processes will always be seen from the perspective of the present. Therefore, any historical network process should be viewed as "contemporary history" (p. 154). The longer the time period under study, the more continuity will appear, and the more obvious will be the focus on the interdependencies between the continuity and the changes taking place. (This has also been suggested by Pettigrew (1985) in strategy research. In comparison, Lundgren (1991), building on David (1988) brings in the general ideas of *path-dependence* and path-dependent processes, some leading to lock-in by small historical events and some to path-dependent transitions.

Secondly, Waluszewski (1989) brings in from historical research, the more *practical, historical methods* for evaluating and scrutinizing various historical sources. The use of historical sources and remnants is not unproblematic, and NW case research often requires that some of the historian's methods and experiences are applied in the research process.

NW methodology is both pluralistic and dynamic. Its dynamic, i.e both stable and changing nature, is easy to observe in the somewhat longer texts of the doctoral theses. While "case studies" (in a very wide sense) constitute the methodological stability, the way the methodology and case narratives have been used and presented has changed over time. The focal unit of analysis in NW texts puts its mark on the structure and contents of the narratives. For example, longitudinal, in-depth descriptions of dyadic relationships (e.g. Liljegren, 1988), of single company's position changes, e.g. (Andersson, 1994), of networks of companies in new technologcal systems (e.g. Lundgren, 1991), and of single companies in overlapping networks (Hertz, 1993) have generated different narrative logics.

Methodological descriptions in NW research used to be based on a logic of justification that often appeared to be defensive. The case studies, the various types of narratives, had to be defended as something standing in sharp contrast to the ruling, positivist influences in marketing research. Leaving the defensive stance behind, focus is slowly turning to the narrative as such, to the text and to the circumstances of each particular study. While the defensive logic of methodological justification used to be generalized (defending case studies in general), the emerging logic of justification is successively becoming more "particularized," treating the narrative as a stable foundation. This also means an opportunity for NW research to open up for new methodological ideas (and critique) from science areas where narratives are important (in contrast to texts dominated by figures, lists, tables, etc.). Researchers connected to, and familiar with, NW research, and dealing with largely the same topics and phenomena, have begun to apply new methods from the sociology of science and technology. For example, building on the so-called "actor-network" ideas presented by Latour (1987) and others, NW-related researchers have begun to apply new views on and ways of presenting longitudinal narratives (e.g. Helgesson, 1999; Kjellberg, 1999). This includes, for example, the use of multiple voices to describe the enrolment of actors in networks, actors which can be both humans and non-humans, i.e artifacts.

One element of these new methodological foundations is the use, and the role of self-reflection. This is in line with another element of much NW research, the fact that studies and the writing processes often extend over a considerable period of time. It draws attention to the many methodological challenges of handling the integrated study and writing processes when they extend over time. From an organizational theory perspective, Czarniawska (1998) states that in social sciences "texts on methods have traditionally focused on the process of conducting the study, assuming that once discovered the text will write itself" and furthermore, that "the emphasis has shifted to writing and reading research reports" (p. 51). The process of reading and writing needs reflection in methodological justifications. Next, a condensed version of such a research and writing process is used to illustrate this.

REFLECTIONS ON A CASE STUDY RESEARCH PROCESS 1991–1996

The final section is a condensed version of a methodological text collected from a study by Andersson (1996a, b). The author's own experiences from a case research and case-writing process is decribed. A reorganization process in the marketing organization of a biotechnology company was followed during the first half of the 1990s. One of the ideas permeating the original version of the condensed text here is that "the mode of emplotment," i.e creating a plot for the final narrative in an industrial marketing and network study, is a process which is very much coloured by the author's context, including the changes in this context.

The Emergence of Case Narratives: A Long-Term Story-Building Process

A case narrative will be subject to innumerable choices of research paths, directing attention to processes rather than end products and results. In the often blurred boundaries between the object and the subject of research, a structured case study text successively emerges. The stability and change of methodological choices, the process of inquiry, the contents of the final text, emerge in a stable and changing context. Inevitably, there will be strong links between the study process, including the way it will be described, and the process under study. Research is a matter of coping with the ambiguities, tensions and contradictions of learning and of finally creating a story and a text.

What is presented in a narrative is a representation by a particular narrator from a particular point of view. The final text has often emerged from a study process which has sometimes diverged far from the (often rationalistic) instructions given in textbooks on case-study research. Researchers often describe an uneven and unpredictable process of research which has generated the final text, with several stops, reversals, back-tracks, and both deliberate and undeliberate choices. In method descriptions lies also attempts to make the narrator visible, to be self-reflective, in order, e.g. for readers to be warned that what they are reading is always one very particular version of a story. Like Burke (1991) states about historical writing and research:

> Historical narrators need to find a way of making themselves visible in their narrative, not out of self-indulgence but as a warning to the reader that they are not omniscient or impartial and that other interpretations besides theirs are possible (p. 239).

What is chosen to be included as central in methodological descriptions of emerging case narratives, often reflects ideas, thought patterns, theories, empirical insights, etc. found elsewhere in the text. Methodological "stories," like all stories, will be colored by our preconceptions and ambitions. Popper (1986) states:

> We cannot free ourselves of past experience and we enter into a new study with specific ambitions. The way out of this difficulty is to introduce a 'preconceived selective point of view into one's history; that is, to write the history which interests us' (p. 150).

Building on the assumption that "timing is everything," methodological stories will inevitably be colored by the methodological, theoretical, and empirical insights gained during the research process and when the methodological description is written. In addition methodological stories will be colored by ambitions to present a coherent text and to unravel links between the object and the subject of study. This is often the starting point when researchers reflect on the many methodological endeavors to present a final text.

Connecting the description of the study process with the concepts and viewpoints supporting the processes under study, these were some of the underlying methodological ideas and underpinnings shaping the structure and contents of the story presented in Andersson (1996a, b).

A historical case narrative will be colored by the context in which it emerges, including the way this context changes during the course of the study process. A historical case research project will be difficult to separate from the context of other ongoing projects. There will be both stability and change in the study process, concerning both the factual interactions of the researcher (in the research community and in the organizational study context) and as concerns the realization steps in the process. The process is born in a context and transforms in a context, and the text and the research problem will bear marks of its emergence in a historical context. During the course of the process there will be both stable and changing elements in the emerging contents of the case narrative. Over time, important contradictions will emerge in the writing process (leading to ambiguities for the researcher), between the emerging structure – encompassing and reflecting theoretical ideas – and the successively accumulating narrative contents – the empirical insights. Burke (1991) states about the dispute between "narrators" and "analysts" in historical writing that:

> One might begin by criticizing both sides for a false assumption which they have in common, the assumption that distinguishing events from structures is a simple matter (p. 237).

These contradictions lead to change actions and transitions; new paths are successively chosen for presenting the text. Contradictions and the choice of new paths also emerge as a consequence of breaking with parts of the researcher's history. Breaking free from and leaving behind some of the old theoretical ideas and old empirical material becomes a central, difficult part of the process. (In addition, methodological descriptions – "stories" – like all methodological presentations, is a narrative where the reproduction of "what really happened" in the process will often be a difficult goal to achieve.)

A number of concerns always emerge in a case research process, often connected to the fact that plots intersect. What is the plot, i.e. what is the structure of the story that puts the events in a causal relationship to each other? How does the case narrative handle the fact that different interrelated events take place at different organizational levels? What is the most appropriate way to delimit the case? Both concurrent and historical events put their obvious mark on the changed actions in the focal events. Concerns about the multidimensional character of temporality can also enter the research process. Time moves differently for different actors in different contexts, and the researcher needs to know how to conceptualize this temporality. Parts of an earlier chronology might have to be discarded, leaving behind preconceptions about "appropriate" structures of single case narratives heavily indebted to the westernized view of temporal linearity and consecutive chronologies. History matters, in different ways for different marketing actors. What theory, perspective, idea, structure, etc. concerning temporality should guide the case story? Concepts, modes of presentation and methodology of a contextualist approach might have to be adopted.

With the entry of history and the frequent use of secondary sources (corporate records and statements from archives, internal corporate letters, newspaper articles, travel reports, other research reports, etc.), enter also other methodological problems. Historical data has to be compared and scrutinized, comparing various written and oral descriptions with often diverging versions.

Methods for critical analyses of the sources have to be adopted (Thurén, 1986; Waluszewski, 1989).

In the process of matching ideas and evidence, of finding a story and plot, of presenting a convincing imagery and plausible narrative, and of finding theoretical concepts and ideas that are both dynamic and general, the case is delimited (in time and place) and an idea of what it is a case of emerges. The researcher's understanding of what the case is a case of can often change during the process.

When it is a matter of understanding and generating knowledge about marketing change processes from an interactional, network-oriented perspective, it often requires theories embracing moving contexts at varying organizational levels, history and various aspects of time and temporality. The acknowledged methodological problems with case narratives – intersecting plots, periodization and events at different contextual levels – need to be integrated and made central parts of the case story and the theoretical reasoning (Fig. 1). In the search for a story, the researcher needs to decide to settle with a final version despite the fact that lack of understanding can remain, concerning, for example, the interplay and shifts between different temporal contexts.

Despite the decision to finally choose a narrative structure, a number of ambiguities and uncertainties will often remain. The narrative contents and style, and the structure of the case narratives are central parts of the rhetorics. Theoretical pluralism and empirical diversity will often leave the author with a number of unanswered questions: Is the case narrative only a study of the focal research problem or can it be defined and classified in other ways? Does the story only open up for more ambiguities concerning what can be defined as the focal problem area? Will different readers – irrespective of science area – regard the writer as "intruder," "defector" and "dilettante?" The problem of what the case can be perceived to be a case of partly remains, for good and for bad (Ragin, 1992):

> Strong preconceptions (of what the case is a case of) are likely to hamper conceptual development. Researchers probably will not know what their cases are until the research, including the task of writing up the results, is virtually completed. What *it* is a *case of* will coalesce gradually, sometimes catalytically, and the final realization of the case's nature may be the most important part of the interaction between ideas and evidence...The less sure that researchers are

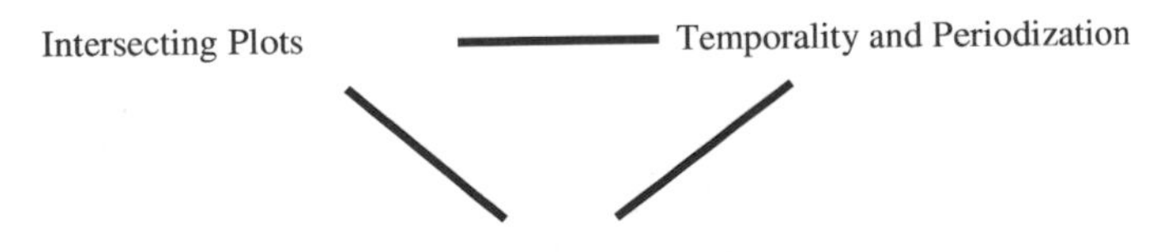

Fig. 1. Case Narratives: Three Major Methodological Problems.

> of their answers, the better their research may be. From this perspective, no definitive answer to the question 'What is a case?' can or should be given, especially not at the outset, because *it depends*. The question should be asked again and again, and researchers should treat any answer to the question as tentative and specific to the evidence and issues at hand (p. 6).

There can be uncertainty of the complexity of the narrative and also of the future reception, and cases may be multiple in a given piece of research. What the case is may change both in the hands of the researcher (during the course of the research and when the results are presented) and in the hands of the researcher's audience. There is also the question how the constructed story will be perceived by people in the focal organization(s) of the narrative. The author also has to handle the problem of creating a coherent text in the ongoing process of matching theory and empirical evidence, relying on pluralism, concerning theory, method as well as data generation. Lundgren (1991) describes this necessary pluralism:

> The process of theorizing around dynamic processes, characterized as they are by multiple change and multiple causation, calls for pluralism. In the successive abstraction of unique observation, in the pursuit of insight through analysis and genesis, anything goes. In the sequential reiteration of analysis, genesis and insight, pluralism is essential: poems, prayers or promises, theoretical reasoning, anectdotal evidence or quantitative methods of analysis, anything goes in the pursuit of insight. The only action that would be totally wrong would be to refrain from trying additional methods or alternative modes of reasoning, that is, to refrain from pluralism. Note here that what I am advocating is not an extreme form of pluralism, but a pluralism within the realms of existing observations and within the actual line of reasoning. The basic scientific criterion employed is that of internal consistency: the consistency of the theoretical reasoning and the consistency between the empirical findings and the theoretical reasoning (p. 76).

Sometimes, concerns about the external validity can remain, e.g. related to unique, single case narratives, and to the connections between external validity and unique observations (Merriam, 1988). A language and construct generating case narrative – formed within the realms of a language-building ideal (Glaser & Strauss, 1968) – will partly leave the question of applicability of the theoretical findings unanswered. Case studies and thick descriptions will be a natural choice in the process of generating useful language.

For the author, the learning process can generate a better understanding of, for example, the many ambiguities surrounding marketing and purchasing actions and processes in moving industrial network contexts. A difficult step in this process is when transferring the learning to the text, making it persuasive, trying to create universal knowledge from the specific. The process to provide rich and plausible, "thick descriptions" of the processes (Geertz, 1973) can be long. But, this detailed knowledge about the very specific can make it possible to appreciate similarities when transferred to new and unknown contexts.

In this long process of creating a narrative, there can remain concerns also regarding the language, that is, regarding the narrative style of the story and of the complete text. The narrative style becomes wrapped in arguments, being part of the internal consistency and, thus, of the "results" of the study. The narrative style will be intimately connected to the rhetorics (e.g. as the classical rhetorics encompassing invention, arrangement and style). An emerging research text is a conversation encompassing meaning (Czarniawska-Joerges, 1992). What *type of* historical plot does the narrative style finally reveal? White (1973) simplifies the discussion by arguing that there are only four kinds of historical plots in social research: tragedy, comedy, romance and irony. Is the style in accordance with the rest of the arguments? If not, is there anything that can be done about it? There will remain concerns about whether the style of the narrative text in the story cohere with or contradict the rest of the reasoning. Research is writing and there is a need for the writing to cohere.

Reflections and Comments on the Casing Process

The transformation process will continue after the case narrative is concluded (Giddens, 1993):

> Human beings transform nature socially, and by 'humanizing' it they transform themselves; but they do not, of course, produce the natural world, which is constituted as an object-world independently of their existence. If, in transforming that world they create history, and hence live *in* history, they do so because the production and reproduction of society is not 'biologically programmed' (p. 168).

The case emerges as a multiple way station, an intermediate product where an important part of the continuing process remains with the readers, while for the author/researcher the question what the case is a case of can keep coming back. As stated by Walton (1992):

> If the analysis accomplishes anything, it does so by pursuing the deceptively simple question "A case of what?" Any case, of course, may offer a variety of answers to the question "A case of what?" Rather than arguing that there is a single or ideal answer, I am saying that any answer presumes a theory based on causal analogies. What constitutes the best answer at a given time will be decided by those communities of social scientists who confirm today's theoretical fashion and will surely change as new questions are put to old cases (p. 135).

Despite the emerging closure on the disorder of events in the final story, there can still be uncertainty about remaining ambitions and intentions. Some of these can be coupled to difficult, but necessary choices and reductions made early in the research process. The actor focus chosen in the narrative can leave open questions concerning how certain processes were experienced by other important actors in the story. The fact that the intentions and ambitions to include other versions were there from the beginning does not help much; the curiosity can remain with the feeling that a much more interesting story might be there, waiting to be told.

One of the main conclusions from our focal case research process 1989–1996 from which our experiences are drawn is that cases – here, a single case narrative – are fundamentally dynamic. The methodological problems and issues in the process of linking the theoretical with the empirical are determined by the unequivocal fact that the single case narrative is inseparably linked with process and successive transformation (Abbott, 1992):

> . . . within the case/narrative approach transformation in attributes can be so extreme that a case which began as an instance of one category may complete a study as an instance of another . . . (pp. 63–64).

What the study, the case, is a "case of" will shift during the process (Walton, 1992):

> My argument is that cases are 'made' by invoking theories, whether implicitly or explicitly, for justification or illumination, in advance of the research process or as its result. This interpretation supports a renewed appreciation for the role of case studies in social research and offers a fruitful strategy for developing theory. The argument derives from a research project in which my own understanding of what the case study was a 'case of' shifted dramatically in the process of pursuing the study and explaining its results. The research, begun as a study of one thing, later proved to be a study of something quite different. I believe that there is a general lesson here . . . (p. 121).

And the delimitation of the subject of the case will also change (Abbott, 1992):

> The first step of the single-case narrative is delimiting the case itself, what historiographers have called "the central subject problem" (Hull, 1975). There are many varieties of central subjects in historical case studies, for subjects need not be social actors or groups. They can be events, social groups, or even states of affairs. The crucial difficulty (a subject of much historiographical conflict) lies in drawing boundaries around the central subject given the continuous character of the social manifold. Note how this difficulty is avoided by the population/analytic approach, which tends to work with populations where cases are unambiguously distinct . . . (pp. 62–63).

Identifying the events is only one step in creating a single case narrative, the very first step as described by Abbott (1992):

> Describing what the case does or endures is what philosophers of history call the colligation problem. It has several subparts. The first is identifying the events involved. Events, like concepts in more familiar methods are hypotheticals. Every historian considers dozens of indicating occurrences when deciding whether a given event has taken place . . .. Moreover, these hypothetical events have varying duration and visibility (p. 64).

Finding ways to deal with problems of handling the concurrent participant observation and "immersion" into the organizational context becomes part of the process (Giddens, 1993):

> Immersion in a form of life is the necessary and only means whereby an observer is able to generate such characterizations. 'Immersion' here – say, in relation to an alien culture – does not, however, mean 'becoming a full member' of the community, and cannot mean this. To 'get to know' an alien form of life is to know how to find one's way about in it, to *be able* to participate in it as an ensemble of practices. But for the sociological observer this is a mode of generating descriptions which have to be mediated, that is, transformed into categories of social-scientific discourse (pp. 169–170).

The methodological lessons to be learned will be inseparably connected to the fact that the contents, the processes and the context of the single, longitudinal case study will be both stable and changing. When stable contacts have been established, and the researchers are immersed into the micro dynamics of the objects of the study, organizational change processes can cause opposite problems. The involvement reduces the understanding of the moving macro context (Vaughan, 1992):

> We develop very personal research styles that become comfortable, and while we may get better at finding the information we seek and interpreting what we find, we may be blind to other sorts of information because our ability to see it is undeveloped. We readily "see" either micro or macro, but not both (p. 183).

In a dynamic, moving industrial network context, with sometimes blurred boundaries between the object and the subject of research, a structured case study text successively emerges (Czarniawska-Joerges, 1992):

> To begin with, there are many constructions of reality (thus many realities) which will always lead to divergent inquiry results. Therefore, the outcomes of a scientific inquiry can contribute to an increased understanding, not to prediction and control – just like *belles letters*. Also, there is no obvious border between a "subject" and an "object" of research, the "objects" being human and therefore subjects themselves. Inquiry is an interaction – just as a text is, between the author and the reader. Knowledge is idiographic in character – attempts at nomothetic descriptions of the human world are in vain. Social phenomena are over determined (a statement by Freud, applied to organization theory by Weick, 1979) and fruitless efforts at establishing simple causation should be abandoned in favor of detecting patterns of actions, events, processes.... Lincoln... points out that the "case study reporting mode," that is, the one closest to a traditional novelistic narrative, renders itself best to the service of the emergent paradigm (Lincoln, 1985, pp. 10–11).

The emergence of a case narrative is an indivisible, continuous process where the story and plot has to be found successively when linking theory with empirical insights. The case writing process is also one with basically arbitrary boundaries (Lundgren, 1991):

> The scheme through which we generate knowledge from historical processes is a complex activity of analysis, genesis and insight, actively constructed by the mind of the investigator in order to understand, predict or control the complex social processes of reality. The process is never finished. New insights set the scene for reclassification of observations: calling for

> re-analysis of the changing parts: producing opportunities for more insight. And there is always plenty of room for new insight (p. 75).

This way a case story will always be an intermediate product (Ragin, 1992):

> Casing is an essential part of the process of producing theoretically structured descriptions of social life and of using empirical evidence to articulate theories. By limiting the empirical world in different ways, it is possible to connect it to theoretical ideas that are general, imprecise, by dynamic verbal statements. In this perspective a case is most often an intermediate product in the effort to link ideas and evidence. A case is not inherently one thing or another, but a way station in the process of producing empirical social science. Cases are multiple in most research efforts because ideas and evidence may be linked in many different ways (p. 225).

The "process of casing" in a case narrative does not involve a search for causes but a search for a story; a story which comes wrapped in arguments, claims, regularities and theories. The embedded rhetorics can also emerge from the intended audience (Platt, 1992):

> The ways cases are chosen, analyzed, amalgamated, generalized, and presented are all part of their use in argument. It is assumed that an argument is designed to reach a conclusion which the reader (and the writer) will find convincing. It is thus always relevant to consider the intended audience, and the use of cases may be treated as part of a work's rhetoric (p. 21).

The process of producing a plausible story from social processes in terms of a complex case with its own plot(s) draws attention to the methodology of successive case transformation. Like in this particular case, three general problems will recur, over and over again: the intersection of plots, periodization, and the fact that events take place at different contextual "levels" (Abbott, 1992):

> The idea that we ought to think about social processes in terms of complex cases going through plots has its own problems, however. There are three principal ones. The first is that plots intersect. A given event has many immediate antecedents, each of which has many immediate antecedents, and conversely a given event has many consequents, each of which has many consequents . . . A related problem is the implicit assumption that subsections of the social process have beginnings, middles, and ends, rather than simple endless middles . . . the issue of plot as having beginning, middle, and end – the issue of periodization – is a major problem . . .. How in fact do narratives explain? In a curious way the answer to this question doesn't matter much, for narrative is where positivists turn when reasoning in variables fails, and of course a particular narrative is what is rejected if an entailed set of variable relationships is implausible or incorrect . . . And the roving focus of the case/narrative approach has another distinct advantage over the population/analytic approach. It need make no assumption that all causes lie on the same analytical level . . . (pp. 65–68).

A plot can emerge (Abbott, 1992):

> . . . events must then be arranged in a plot that sets them in the loose causal order that we generally regard as explanatory (p. 64).

There emerges a story based on historical understanding, successively fusing different time horizons of the immediate past, the present and the immediate future (Gadamer, 1960):

> The horizon of the present is continually formed, in that we have continually to test all our prejudices. An important part of this testing is the encounter with the past and the understanding of the tradition from which we come. The horizon of the present cannot be formed without the past. There is no more an isolated horizon of the present than there are historical horizons. Understanding, rather, is always the fusion of those horizons we imagine to exist themselves (p. 258).

The events are put in a temporal sequence, a chronicle, a "prenarrative discourse" with a number of events as the "raw material" of history. Callinicos (1995) comments on White's discussion (*The Content of The Form*,1987): "White thus oscillates between treating events as the chaotic raw material of history, nothing without the form-giving intervention of narrative, and positing a prenarrative discourse, chronicle, which offers unmediated access to these events" (pp. 74–75).

Thus, a perspective on temporality will be needed, a theory, to guide the narrative (Callinicos, 1995):

> As we have seen, modern historical discourse depends on an awareness of temporality, and of the radical differences it introduces between societies and epochs. Marc Bloch calls history ' "the science . . . of men in time." The historian does not think of the human in the abstract. His thoughts breathe freely the air of the climate of time.' But if this is right, then historical enquiry requires some conception of how human beings relate to their variable social contexts, and of the nature of and the differences between these contexts. In other words, it requires a social theory (p. 91).

Using the words of Ricoeur, the process of constructing the final story, the operation of emplotment, will be "a synthesis of heterogeneous elements" (Ricoeur, 1991, p. 21). It will be a synthesis between the multiple events and incidents and the unified and complete story. The plot that emerges serves to transform the many events into one story (Ricoeur, 1991):

> In this respect, an event is more than something that just happens; it is what contributes to the progress of the narrative as well as to its beginning and to its end (p. 21).

And with Ricoer's view, the emplotment is a synthesis in a more profound sense, because the final narrative composition will entail a *temporal* totality, synthesizing two types of time (Fig. 2).

As stated by Ricoeur (*ibid.*):

> We could say that there are two sorts of time in every story told: on the one hand, a discrete succession that is open and theoretically indefinite, a series of incidents (for we can always pose the question: and then? and then?); on the other hand, the story told presents another temporal aspect characterized by integration, culmination and closure owing to which the story

An Open and Discrete Succession of Events

("A Series of Incidents", theoretically indefinite)

A Configuration Drawn Out of A Succession

("Integration, Culmination and Closure", theoretically determined)

Fig. 2. Case Narratives: Two Types of Temporal Totality in Narratives.

> receives a particular configuration. In this sense, composing a story is, from the temporal point of view, drawing a configuration out of a succession. We can already guess the importance of this manner of characterizing the story from the temporal point of view inasmuch as, for us, time is both what passes and flows away and, on the other hand, what endures and remains (p. 22).

In the final narrative can be collected the various complexities and the multiplicity of social times (Adam, 1995):

> Emphasis on the complexity of social times brings together the personal and the global, the technological and the literary, the bodily and the scientific, totalizing tendencies and local particularities, coevalness and difference. It binds into a unified but conceptually unconventional whole what constitute antinomies, contradictions and incompatible categories in the traditions of Enlightenment thought (p. 150).

However, having stated this, there remains the question whether the complexity and problematic elements of time can ever be resolved by narratives. Having examined the relations between time and narrative, Ricoeur reflects upon the inscrutability of time itself (Ricoeur, 1988, pp. 241–274). It is commented on by Wood (1991):

> What perhaps we should remember here is that time is not just an enveloping beyond to all our little bubbles of narrative order; it is more like the weather, capable of gentle breezes and violent storms. For all our ability to breed domestic forms of time, it also holds in reverse the apocalyptic possibility of dissolving any and all of the horizons of significance we have created for ourselves (p. 10).

Nevertheless, we can assume that in practice it should be possible to grasp the properties of, for example, social time in organizations that are most relevant to the actors constructing them, constructing concepts and narratives in the interaction between the actors and the researcher (Czarniawska-Joerges, 1994). And being aware of the fragile character of concepts and narratives, and "that they are temporary and contingent on place and circumstance" (*ibid.*, p. 10).

Suffice to say, and following the reasoning of Ricoeur, in the hands of the reader the final narrative can be a path to increase our self-knowledge and our understanding of humans and their experiences of time:

> Do not human lives become more readily intelligible when they are interpreted in the light of the stories that people tell about them? And do not these "life stories" themselves become more intelligible when what one applies to them are the narrative models – plots – borrowed from history or fiction? . . . It is thus plausible to endorse the following chain of assertions: self-knowledge is an interpretation; self-interpretation, in its turn, finds in narrative, among other signs and symbols, a privileged mediation; this mediation draws on history as much as it does on fiction, turning the story of a life into a fictional story or a historical fiction . . .

The plot is there to be discovered and invented in the perpetual dialogue between evidence and theory (Hoeg, 1994):

> Jason had discovered that in reality there was no story/plot. Reality consisted of those endless circles which he was now on his way to begin; life emerged from an unclear beginning, into infinity. Endings, which are necessary for all stories, are inventions . . . The major part of the literature, he said, has a limited number of different beginnings, a predictable ending, and in-between a story. Reality has none of this" (pp. 235, 238, Trans.).

Herein lies some of the methodological and explanatory powers and problems of the single case narrative. It is a matter of entering a process of incalculable transformation (*ibid.*):

> The history of science in Europe has not been able to settle the discussion between the followers of Aristotle and Galenos. The question whether he who sees passively receives an optical impression of reality or is himself shaping what he sees. In front of the woman on the floor I realized that this dialogue had always been meaningless, because the question was wrongly posed. It assumes that there is a stable reality to observe. There is not. In the moment that we start observing the world, it starts to change. And so do we. To observe reality is not a matter of understanding a static structure. It is a matter of undergoing and beginning an incalculable transformation" (p. 267, Trans.).

The single case narrative has a "transformative power" (Vaughan, 1992):

> The transformative powers of this approach lie not only in having lots of facts, but in the radically different kinds of facts that varying cases can produce, which result in three major benefits for theory. First, because shifting units of analysis can produce qualitatively different information, case comparison can generate startling contrasts that allow and, in fact, demand us to discover, to reinterpret, and ultimately to transform our theoretical constructs. Secondly, selecting cases to vary the organizational form sometimes permits varying the level of analysis. Because of the different sorts of data available from micro level and macro level analysis, choosing cases that vary both the unit of analysis and the level of analysis, when possible, can lead to the elaboration of theory that more fully merges micro understanding and macro understanding. Third, this method can be particularly advantageous for elaborating theories, models, and concepts focusing on large, complex systems that are difficult to study (pp. 176–177).

Finally, it is also a matter of accepting the fact that no matter how many versions of the case narrative, it will only be an intermediate "way station" in the process of creating knowledge. In this process of transformation there is a need for both stability and change. The methodology of the single case narrative – the methodological problems and dilemmas, strengths and weaknesses, opportunities and constraints, etc. – cannot be separated from the dynamics inherent in the move to a way of seeing realities as fuzzy, and with autonomously defined complex properties (Abbott, 1992, p. 65). The single case narrative can leave the researcher in unknown terrains, in which the uncertainty only successively is replaced by more stable facets, structures and ideas:

> There are no predetermined rules or procedures that can be followed during the course of the study, neither concerning planning nor data collection and analysis . . . You have to realize that 'the right way' is not always that obvious. It is precisely this lack of structure which is so attractive for some researchers, as it makes it possible for the researcher to adapt to unpredictable events, and thereby change path in the search for meaning . . . To be confident beforehand that the method of analysis chosen shall prove to be acceptable is impossible in this kind of research. The case study thus leaves the researcher in practically unknown terrains (p. 51).

Some of the initial, necessary stability can be provided by the researcher's ontology, philosophical standpoints, history, and social network context in the research community and elsewhere. But the overarching focus on process and transformation in longitudinal case narratives can also leave these factors open for change.

CONCLUDING COMMENTS

In the process of writing a case narrative, for example on an industrial marketing and network phenomenon, one of the methodological lessons that can be learned concerns the handling of different perspectives. Writing a case is a process of finding a form of narrative structure which is appropriate for the need to move between the details of cross-cutting processes of the parts and the movements of the larger whole, between different actor perspectives, and also between different time perspectives. Finding a model for the narrative, which in thick descriptions of micro narratives can illuminate the movements and social constructions of the emerging structures, is one of the major challenges faced in many NW studies. One of the ways for enriching theory generation in industrial marketing and network research in general concerns the search for influences and ideas for developing the narratives.

Experimenting with form for the purpose of synthesizing events and structure, narratives and theory, may be one of the biggest methodological challenges facing industrial marketing and network research.

From where could new ideas be collected? Some of the potential sources have already been indicated above. Not only NW's methodological tradition but marketing research in general can be enriched by, for example, the methodological ideas and the ways of presenting narrative texts developed in the sociology of science and sociology of technology (Latour, 1987). Telling stories of the many "micro processes" of marketing and purchasing that form the interactions, relationships and networks of business, based on these ideas, is one such road for further research, maybe bringing in technology and artifacts as actors and giving them a voice of their own.

These and other influences from various science areas should probably not be seen as anything but that, as sources for inspiration and for new ideas. If we instead look upon the results of our marketing studies as narratives, we might end up with the same conclusion as Czarniawska (1998) when introducing a narrative approach to organization studies:

> The narrative device does not predetermine in any sense how the material is to be constructed or collected. In more traditional parlance, there is no obvious connection between the narrative approach and any specific method of study. It is doubtful whether there is any method in social science studies, at least in the sense of a prescribed procedure that brings about foreseeable results. There is a bunch of institutionalized practices on the one hand and individual experimentation accompanied by self-reflection on the other . . . (p. 19).

Also NW research contains institutionalized practices. Instead of using methods and method references that are considered "appropriate," in the future NW research will continue to find new influences. NW research will also find new ways of collecting and constructing the narratives and the marketing stories. This also means that the rationalistic, prescriptions on case study research designs and the action plans that are presented in books on case study research (e.g. Yin, 1994), including underlying epistemologies, can be difficult to combine with some of these new ideas and influences. Like in all maturing research areas, marketing researchers in the IMP and industrial network tradition will continue to look for new ways of presenting their marketing texts. The influences and the narrative devices are there, waiting to be explored.

ACKNOWLEDGMENTS

This text is a rewritten and extended version of the second part of Chapter 10 in a doctoral thesis on industrial marketing change processes. The original text has undergone revisions. Parts of the same text have earlier been part of the theoretical discussion of a thesis of two volumes: (1) Andersson, P., *Concurrence, Transition and Evolution. Perspectives of Industrial Marketing Change Processes*,

Stockholm: The Economic Research Institute, Stockholm School of Economics, 1996; (2) Andersson, P., *The Emergence and Change of Pharmacia Biotech 1959–1995, The Power of the Slow Flow and the Drama of Great Events*, Stockholm: The Economic Research Institute, Stockholm School of Economics, 1996. This is a compilation of the original text, focusing on the theoretical contents of the thesis.

REFERENCES

Abbott, A. (1992). What do cases do? Some notes on activity in sociological analysis. Chap. 2 in: C. C. Ragin & H. S. Becker (Eds), *What is a Case? Exploring the Foundations of Social Enquiry*. Cambridge: Cambridge University Press.

Adam, B. (1995). *Time watch – The social analysis of time*. Cambridge: Polity Press.

Andersson, H. (1994). Ett industriföretags omvandling (The transformation of an industrial company) (Trans.) doctoral thesis, The Economic Research Institute, Stockholm School of Economics, Stockholm.

Andersson, P. (1996a). *Concurrence, transition and evolution. Perspectives of industrial marketing change processes*. Stockholm: The Economic Research Institute, Stockholm School of Economics.

Andersson, P. (1996b). *The emergence and change of Pharmacia Biotech 1959–1995, the power of the slow flow and the drama of great events*. Stockholm: The Economic Research Institute, Stockholm School of Economics.

Andersson, P., & Nyberg, A. (1998). Marketing cooperation in automotive strategic alliances. *Journal of Business-to-Business Marketing*, *4*(3), 43–74.

Axelsson, B. (1992). Network research – future issues. In: B. Axelsson & G. Easton (Eds), *Industrial Networks – A New View of Reality* (pp. 237–251). London: Routledge.

Burke, P. (Ed.) (1991). *New perspectives on historical writing*. Cambridge: Polity Press.

Callinicos, A. (1995). *Theories and narratives*. Cambridge: Polity Press.

Croce, B. (1941). *History as the story of liberty*. London: George Allen and Unwin Limited.

Czarniawska-Joerges, B. (1992). *Realism revisited: Historical and contemporary connections between the novel, social sciences and organization theory* (Vol. 15). Lund: Lund University.

Czarniawska-Joerges, B. (1994). *The three-dimensional organization – a constructionist view*. Lund: Studentlitteratur.

Czarniawska-Joerges, B. (1995). Narration or science? Collapsing the division in organization science. *Organization Articles*, *2*(1), 11–33. London: Sage.

Czarniawska-Joerges, B. (1998). *A narrative approach to organization studies. Qualitative Research Methods*, *43*. Thousand Oaks: Sage.

David, P. A. (1988). Path-dependence: Putting the past into the future. Stanford University, Institute for Mathematical Studies in Social Sciences, Economic Series, Technical Report No. 533 (Nov).

Easton, G. (1995). Case research as a methodology for industrial networks: A realist apologia. Paper presented at The 11th International IMP Conference. Manchester, (Sept 7–9), 370–391.

Freeman, L. C. (1979). Centrality in social networks: I. Conceptual Clarification. *Social Networks*, *1*, 215–239.

Gadamer, H.-G. (1960). *Truth and method*. (*Wahrheit und methode*). Tubingen

Geertz, C. (1973). *The interpretation of cultures*. New York: Basic Books.

Giddens, A. (1993). *New rules of sociological method* (2nd ed.). Stanford, CA: Stanford University Press.

Glaser, B. G., & Strauss, A. L. (1968). *The discovery of grounded theory*. Chicago: Aldine.

Hallén, L., Johanson, J., & Seyed-Mohamed, N. (1992). Interfirm adaptations in business relationships. *Journal of Marketing*, *55*(April), 29–37.

Helgesson, C.-F. (1999). The making of a natural monopoly. Dissertation manuscript, Stockholm: The Economic Research Institute, Marketing Department, Stockholm School of Economics.

Hertz, S. (1993). The internationalization of freight transport companies. Doctoral, thesis, Stockholm: The Economic Research Institute, Stockholm School of Economics.

Hoeg, P. (1994). *Berättelser om natten* (*Tales of the Night* (Trans.)). Stockholm: Norstedts Förlag.

Hopkins, N., Henderson, G., & Iacobucci, D. (1995). Actor equivalence in networks: The business ties that bind. *Journal of Business-to-Business Marketing*, *2*(1), 3–31.

Iacobucci, D. (1989). Modeling multivariate sequential dyadic interactions. *Social Networks*, *11*, 315–362.

Iacobucci, D., & Hopkins, N. (1992). Modelling dyadic interactions and networks in marketing. *Journal of Marketing Research*, *29*, 5–17.

Iacobucci, D., & Wasserman, S. (1988). A general framework for the statistical analysis of sequential dyadic interaction data. *Psychological Bulletin*, *103*, 379–390.

Johanson, J., & Mattsson, L.-G. (1992). Network positions and strategic action – An analytical framework. In: B. Axelsson & G. Easton (Eds), *Industrial Networks – A New View of Reality* (pp. 205–217). London: Routledge.

Johanson, J., & Mattsson, L.-G. (1994). The markets-as-networks tradition in Sweden. Chap. 10 in: G. Laurent, G. L. Lilien & B. Pras (Eds), *Research Traditions in Marketing* (pp. 321–342). Kluwer Academic Publishers.

Kauffman, S. (1992). *Origins of order: Self organisation and selection in evolution*. New York: Oxford University Press.

Kjellberg, H. (1999). Organizing distribution – The making of a modern model of distribution. Dissertation manuscript, Stockholm: The Economic Research Institute, Marketing Department, Stockholm School of Economics.

Latour, B. (1987). *Science in action*. Cambridge: Harvard University Press.

Liljegren, G. (1988). Interdependens och dynamik i långsiktiga kundrelationer. Industriell försäljning i nätverksperspektiv.(Interdependence and dynamics in long-term customer relationships. Industrial sales from a network perspective (Trans.). Doctoral, thesis, Stockholm: The Economic Research Institute, Stockholm School of Economics.

Lincoln, Y. S. (Ed.) (1985). *Organizational theory and inquiry: The paradigm revolution*. Beverly Hills, CA: Sage.

Lundgren, A. (1991). *Technological innovation and industrial evolution – The emergence of industrial networks*. Doctoral thesis, Stockholm: The Economic Research Institute, Stockholm School of Economics.

Merriam, S. B. (1988). *Case study research in education*. San Francisco: Jossey-Bass Inc.

Pettigrew, A. (1985). *The awakening giant – Continuity and change in ICI*. Oxford: Basil Blackwell.

Platt, J. (1992). Cases of cases . . . of cases. In: C. C. Ragin & H. S. Becker (Eds), *What is a Case? Exploring the Foundations of Social Enquiry*. Cambridge: Cambridge University Press.

Popper, K. (1986). *The poverty of historism*. London: Ark Paperbacks.

Ragin, C. C. (1992). Casing and the process of social inquiry. Chap. 10 in: C. C. Ragin & H. S. Becker (Eds), *What is a Case? Exploring the Foundations of Social Enquiry*. Cambridge: Cambridge University Press.

Reingen, P. H., & Kernan, J. B. (1986). Analysis of referral networks in marketing: Methods and illustration. *Journal of Marketing Research, 23*, 370–378.

Ricoeur, P. (1988). *Time and narrative* (Vol. 3, Trans.). Chicago: University of Chicago Press.

Ricoeur, P. (1991). Life in quest of narrative. Chap. 2 in: D. Wood (Ed.), *On Paul Ricoeur – Narrative and Interpretation*. London: Routledge.

Sayer, A. (1984). *Method in social science*. London: Routledge.

Thurén, T. (1986). *Orientering i källkritik* (*Introduction to source critical methods* (Trans.)). Stockholm: Almqvist & Wiksell.

Tichy, N., & Fombrunn, C. (1979). Network analysis in organizational settings. *Human Relations, 32*(11), 923–965.

Tunisini, A. (1997). The dissolution of channels and their hierarchies. An inquiry into the changing customer relationships and organization of the computor corporation. Doctoral thesis, Uppsala: Department of Business Studies, Uppsala University.

Vaughan, D. (1992). Theory elaboration: the heuristics of case analysis. In: C. C. Ragin & H. S. Becker (Eds), *What is a Case? Exploring the Foundations of Social Enquiry*. Cambridge: Cambridge University Press.

Walton, J. (1992). Making the theoretical case. In: C. C. Ragin & H. S. Becker (Eds), *What is a Case – Exploring the Foundations of Social Enquiry*. Cambridge: Cambridge University Press.

Waluszewski, A. (1989). *Framväxten av en ny massateknik – En utvecklingshistoria*. Uppsala: Department of Business Studies, Uppsala University.

Wasserman, S., & Faust, K. (1994). *Social network analysis*. Cambridge, NY: Cambridge University Press.

White, H. (1973). *Metahistory*. Baltimore: Johns Hopkins University Press.

Wilkinson, I. F., & Easton, G. (1997). Edge of chaos II: Industrial network interpretation of Boolean functions in NK models. In: F. Mazet & J.-P. Valla (Eds), *Interaction, Relationships and Networks in Business Markets* (13th IMP Conference, Vol. 2). Groupe ESC Lyon.

Wilkinson, I. F., Araujo, L., Easton, G., & Georgieva, C. (1998). On the edge of chaos: Towards evolutionary models of industrial networks. In: H. G. Gemuenden & T. Ritter (Eds), *Interaction, Relations and Networks*. Thousand Oaks: Sage Publications.

Wilson, D. T., Commentary on: Johanson, J., & Mattsson, L.-G. (1994). The markets-as-networks tradition in Sweden. Chap. 10 in: G. Laurent, G. L. Lilien & B. Pras (Eds), *Research Traditions in Marketing* (pp. 343–346). Kluwer Academic Publishers.

Wood, D. (1991). Introduction: Interpreting narrative. In: D. Wood (Ed.), *On Paul Ricoeur – Narrative and Interpretation*. London: Routledge.

Yin, R. K. (1994). *Case study research. Design and methods* (2nd ed.). Thousand Oaks: Sage Publications.

INTEGRATING MARKETING MODELS WITH QUALITY FUNCTION DEPLOYMENT

Stan Aungst, Russell R. Barton and David T. Wilson

ABSTRACT

Quality Function Deployment (QFD) proposes to take into account the "voice of the customer," through a list of customer needs, which are (qualitatively) mapped to technical requirements in House One. But customers do not perceive products in this space, nor do they not make purchase decisions in this space. Marketing specialists use statistical models to map between a simpler space of customer perceptions and the long and detailed list of needs. For automobiles, for example, the main axes in perceptual space might be categories such as luxury, performance, sport, and utility. A product's position on these few axes determines the detailed customer requirements consistent with the automobiles' position such as interior volume, gauges and accessories, seating type, fuel economy, door height, horsepower, interior noise level, seating capacity, paint colors, trim, and so forth. Statistical models such as factor analysis and principal components analysis are used to describe the mapping between these spaces, which we call House Zero.

This paper focus on House One. Two important steps of the product development process using House One are: (1) setting technical targets; (2) identifying the inherent tradeoffs in a design including a position of merit. Utility functions are used to determine feature preferences for a product.

Evaluating Marketing Actions and Outcomes
Advances in Business Marketing and Purchasing, Volume 12, 89–140

ISSN: 1069-0964/doi:10.1016/S1069-0964(03)12003-0

Conjoint analysis is used to capture the product preference and potential market share. Linear interpolation and the slope point formula are used to determine other points of customer needs. This research draws from the formal mapping concepts developed by Nam Suh and the qualitative maps of quality function deployment, to present unified information and mapping paradigm for concurrent product/process design. This approach is the virtual integrated design method that is tested upon data from a business design problem.

1. THE MARKETING PROBLEM

Global competition is changing the face of business. Organizations are flatter as restructuring has removed layers of management. The goal is to emulate Jack Welch's vision for General Electric, "I wanted GE to run more like the informal plastics business I came from – a company filled with self-confident entrepreneurs who would face reality everyday" (Welch & Byrne, 2001). The flatter more agile model of business operation places a premium on communication between all elements of the enterprise and particularly marketing and engineering are essential for success in the market place (Griffin & Hauser, 1994; Product Manager, 1999). Linking market knowledge about customer needs to technical knowledge of engineering is a fundamental step to achieve winning product designs. Unfortunately, highly focused specialists in marketing and engineering do not always communicate. Engineering and marketing have grown apart becoming unaware of what the other area has to contribute to product creation. Decreased integration and communication between these two critical business areas can lead to the customer requirements being poorly communicated. The ability to combine skills to develop and produce successful products decreases (Griffin & Hauser, 1994; Product Manager, 1999).

When engineering and marketing fail to cooperate we have product launch disasters. Engineering designs and builds the product and "throws it over the wall" to marketing who must take it to market. Cooperation between marketing and engineering throughout the entire development process is required for successful product development. Marketing and Engineering responsibilities in new-product development are neither independent nor static; they cannot be analyzed separately but must be viewed as a whole (Griffin & Hauser, 1994; Product Manager, 1999). The availability of new technological solutions, customer needs' changes, new competitive products and changing governmental and environmental constraints impact marketing and engineering responsibilities in the new product development process which in turn challenges engineering and marketing to cooperate.

The need for managing information flows across marketing and engineering boundaries first became critical in the mid-1980s and it has continued to be a critical variable in the formula for success perhaps even more as organizations seek to become flatter and more responsive to the market. Vigorous competition manifested itself in intense pressure to reduce new-product-development cycle times and manufacturing lead-times. Leaner (flatter) management organizations are called to increase success rate for new-product introduction combined with less wasted (Griffin & Hauser, 1994). Firms have responded by experimenting with flatter management structures, cross-functional teams (concurrent engineering), cellular manufacturing, physical relocation of R&D and engineering, and cross-discipline management processes.

Research on product development found that U.S. firms with successful product development programs have achieved more integration between marketing and engineering than firms with less successful programs (Cooper, 1995; Griffin & Hauser, 1994). In a study of both marketing and engineering managers, it was found that the more successful firms (profitable) achieve product success and more integration when they worked on the following tasks together:

- analyzing customer needs;
- generating and screening new ideas;
- developing new products according to market needs;
- analyzing customer requirements; and
- reviewing test market results.

Cooperation and communication face validity have been linked to new product success (Griffin & Hauser, 1994). However, the study did not identify solutions to a specific integration problem. Knowledge about integration and communication in the new product development process will likely contribute to making it more competitive and profitable.

As a general rule marketing and engineering do not always communicate resulting in the loss of the "voice of the customer" in product design (Day, 1993). Without a means to transfer knowledge and the ability to combine expertise across all the functional areas constrains the firm's ability to develop and produce successful products. The failure of engineering and marketing to communicate increases the chances of a product launch disaster. The "voice of the customer" links marketing and engineering. Ideally, marketing and engineering should be viewed as a whole. Product development programs with more integration between marketing and engineering have a greater success rate than product development programs with less integration. The more successful firms as measured by profits, (profitable firms) integrated marketing and engineering in the design tasks.

Quality function deployment provides a framework for integrating marketing and engineering (Day, 1993). Decisions made in different domains impact other domains in terms of costs, performance and ultimately the overall profitability of the firm. QFD in the past has provided only qualitative links of the domains of customer needs, technical requirements, design parameters, manufacturing processes and process monitoring and management. A formal House of Quality (HOQ) will lead to integration of decisions made in these domains. Formalizing the HOQ replaces the qualitative data input to the houses with quantitative models developed from field research that is shared with marketing, engineering, and management science.

Aungst, Barton and Wilson (2002) have developed a methodology for mapping from the customer perceptual space to the customer requirements space based on House Zero. This paper focuses on quantitative mapping techniques for House One. In the next section we review the overall five-house QFD mapping strategy. Section 3 describes the purpose for House One and the formal maps needed at this stage. Section 4 presents the procedures used to obtain the technical requirements for House One. A numerical example is presented in Section 5. The final section identifies the value and benefits of the approach, and future work.

2. A FORMAL MAPPING PARADIGM FOR PRODUCT DESIGN

2.1. QFD – Beyond the House of Quality/4 House Approach

The "House of Quality" (HOQ) is one step in the overall process of quality function deployment. The HOQ links the customer needs to the technical requirements. Such a link is essential to relating product characteristics to customer needs in a way that assists engineers in designing the product.

However, we must address more than the customer needs/technical requirements integration, if we are to have a design that is ready for production. For example, suppose the design team for a new copier decides that the *number of pages per minute of output* is a critical design need and that the relevant technical requirement is *speed*. Setting a target value for speed gives the design team a goal but it does not give the team an efficient copier. To design a copier, the design team must define the technical specifications, the product design that will meet these specifications (motor, gears, belts, cabinet, metal or plastic, etc.), the processes to manufacture these parts, the processes to assemble the copier, and the production plan to have it built. QFD links each of these activities using a series of four maps or houses, as shown in Fig. 1.

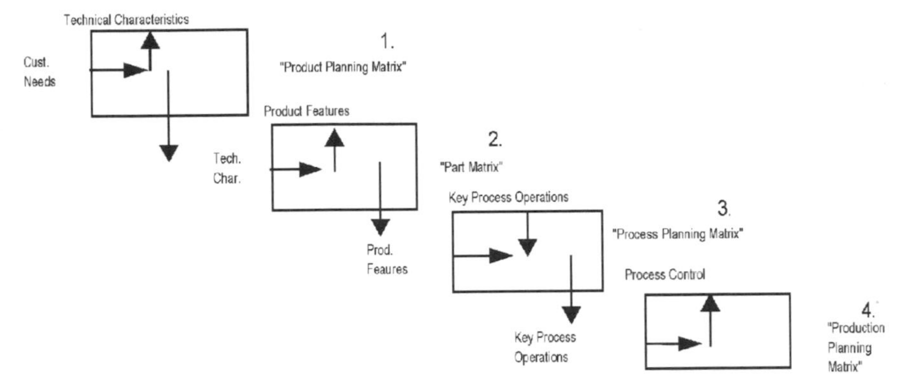

Fig. 1. Four House Approach to QFD.

The First House, known as the House of Quality, links the customer needs to the technical requirements but in a crude fashion: a graphical key indicates qualitatively the relationships as positive or negative.

The next task is to take the target values for the technical requirements, which is called the technical characteristics and links them to physical attributes: part characteristics or product features. QFD's approach to this task is to use a house diagram again, but in a revised manner. Technical characteristics such as *speed* become the rows of this Second House (Part Matrix) while part characteristics such as motor type, rotor diameter, etc. become the columns. Again, the relationship is described qualitatively using symbols similar to those in the first house.

Next, product features such as motor type become the rows of the third house while process design parameters such as the r.p.m. of the copper armature machine become the columns. The relationships are qualitative and subjective. This house is sometimes called the Process Planning Matrix.

Finally, in House Four, the process design parameters become the rows and production requirements such as process control, operating training, and maintenance become the columns. These relations, again, are subjective and qualitative.

The use of a QFD approach links customer needs to technical requirements and manufacturing decisions so that products can be designed effectively and manufactured efficiently. By the links provided in these Houses, the design team assures that the customer needs are deployed through to manufacturing. For example, a process control setting of 7.5 gives the copper winding machine a speed of 500 RPM, to give a reproducible coil diameter for the motor, which assures a certain motor speed. This, in turn satisfies the customer need to out-put copies.

2.2. Weaknesses of the QFD Four House Approach

There are at least four benefits to QFD that will be provided by the incorporation of a more formalized methodology. First, better organizational integration is enabled by the link that perceptual maps provide between the marketing and design functions. Second, the number of activities in the design process requiring subjective judgment is reduced. Third, the perceptual mapping formalism facilitates the documentation of design decisions. Fourth, the use of auxiliary models of cost and revenue assist the design team in assessing profitable design decisions early in the product development life cycle.

There are at least six specific problems with the QFD process (Hazelrigg, 1996; Shin & Kim, 1997):

(1) Targets that are set based on the information contained in the HOQ alone are unrealistic. Using only customer and competitor information to set targets can result in targets that can never be achieved in practice, leading to time-consuming iterations until an achievable specification is reached.
(2) The manner of describing the *coupling* between the design variables in the HOQ does not adequately reflect the tradeoffs that must be made in a real design problem and can lead to highly inappropriate and undesirable designs. The roof of the HOQ cannot represent the complex coupling between the technical requirements because it requires the classification of the coupling into one of four categories (strong or weak). In reality, the variables may be coupled differently with respect to different levels of other variables and, hence, using a qualitative classification is usually an oversimplification.
(3) The subjective ranking of customer requirements, technical requirements (1–9), and measures of performance are misleading and can cause an unending circle of inappropriate, infeasible product designs. Using subjective ratings for customer and technical requirements may not represent the reality of the design problem and can result in targets that can never be achieved in reality.
(4) In each House of QFD, the row and column relations are established only qualitatively without any formal methodology. The interrelationship between customer requirements and technical requirements (body of the matrix) is totally subjective and most likely does not reflect the true tradeoffs and interaction in a real design problem.
(5) QFD's large matrices especially in the number of technical requirements make the methodology more cumbersome and complex.
(6) QFD does not consider other product risks such as market share, contribution margin, or potential profit calculations.

2.3. Formalizing Relationships in the QFD Four House Approach

In each House of QFD, the row and column relations are established only qualitatively, without any formal methodology. This paper develops a formal mathematical model that can be applied to each house. Further, since the output vector of one mathematical model is the input vector to another model, the models can be combined to form an integrated design environment. The general approach is motivated by existing work by Bose (1973), Barton, Barton, Blecker, Lyons, Pitts, Stein and Woolson (1982), and Suh (1990).

This research recognized the need to develop empirically-based mathematical models relating the measured properties across different spaces including statistical models relating consumer preferences to technical measures of product performance. Each house in QFD provides qualitative models connecting two design spaces, for example house one provides a connection between the customer requirement space and the technical requirement space. The mathematical maps between abstract spaces developed by Bose (1973), Barton et al. (1982), and Suh (1990), parallel the tabular maps provided by QFD House One and House Two, but did not extend to the other houses. In this paper, we will develop this relation and extend it across all houses in the QFD framework.

2.3.1. Formal Mapping Techniques for QFD (Design Spaces)

Figure 2 illustrates a sequence of design spaces by perceptual mapping techniques, and other formal mathematical models. The new model is a sequence of spaces linked by formal mathematical models. This formalized method could be regarded as a new quantitative design architecture as a tool to facilitate new product design.

This paper proposes a new mathematical, formalized, integrated method for product design research. New methods for customer-perceived value are developed and linked to engineering (design and manufacturing) through the use of abstract vector spaces. The strategy includes the components shown in Fig. 2. House Zero is a new house which contains the customer perceptual dimensions which are linked to the customer requirements. House One matches the customer requirements to the high level technical requirements. House Two compares the technical requirements to the physical product features. House Three compares the physical product features to the key manufacturing processes and finally the key manufacturing processes are compared to process controls in House 4.

Use of a customer perceptual space where customer perceptual mapping, preference methods for strategic product positioning, market segmentation and design concepts evaluation, and value elements are determined.

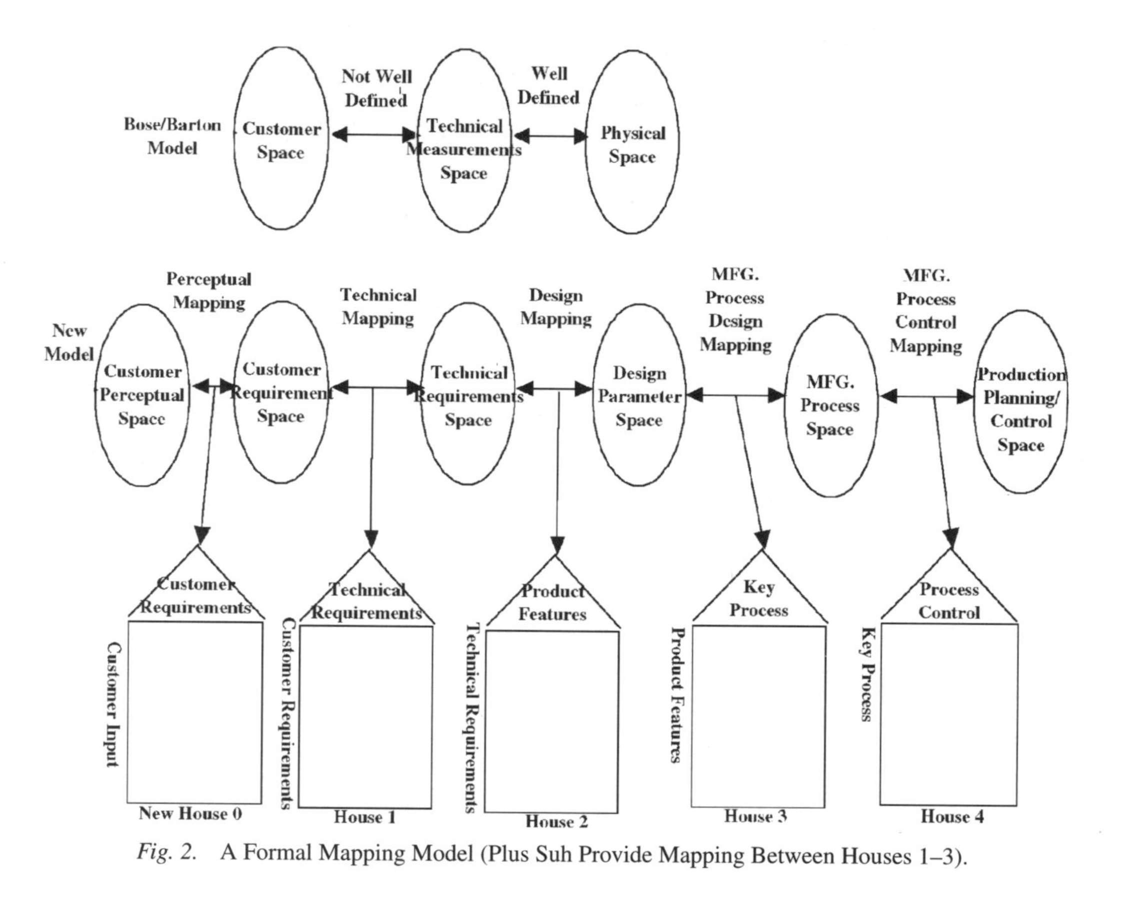

Fig. 2. A Formal Mapping Model (Plus Suh Provide Mapping Between Houses 1–3).

Mathematical linking enables the design team to take a vector in customer perceptual space and map it to an input vector in customer requirement space for House One.

(1) The mathematical linking continues with the output technical specification vector required to meet the customer requirement vector as input to the technical requirements space where engineering measurements of product performance are derived and quantified.
(2) The engineering modeling space (a product description space), CAD and visual representation of the product is presented. Product design in this space will balance cost from the manufacturing and production planning space (next two spaces) and performance/value from the previous three design spaces. The output vector(s) of this space is the P.D.S./C.D.S. (product design space/component design space). Mathematical mapping to the technical requirement space is via CAE Mapping to the next space, the manufacturing process space is through DFM/DFA.
(3) Finally, after the key processes are identified process control charts are generated along with an automated process plan for the production planning/control space.

2.4. Brief Description of the Five House Formal Mapping Paradigm (Five Houses)

The five houses of augmented QFD methodology provide a framework for product design. Each house provides a map linking two different views of the product, each view in a different space. For example, House One links the customer needs space (attribute space) with the technical requirements space.

Product design must address multiple objectives. They include market share, profit and product reliability, for example. These objectives arise naturally in different spaces: market share can be determined from models using information from perceptual space, profit requires knowledge of information from perceptual space (sales price) and manufacturing space (production and distribution cost), product reliability can be determined from information from design space (materials and form), technical requirement space (product performance requirements) and perceptual space (expected product use and environment). The computation of these objectives requires *auxiliary functions* or maps that may have inputs from one or more spaces.

After the design team produces a perceptual map and a new product position opportunity is identified, preference models along with value analysis and cost data from the other houses will be used to calculate a figure of merit based on profit, revenue, or some combination. Design decisions in the perceptual space result in obtaining one vector in perceptual space, which maps into a customer requirement vector in the customer requirement space.

In the customer requirement space, the vector gives product attribute scores on each customer requirement. From this input vector of customer requirements, the design team determines the technical requirements necessary to meet each customer requirement.

The customer needs (dimensions) are linked to engineering characteristics, the technical characteristics of the product provide insight to the design team in setting performance targets without inhibiting creativity. These links have been identified only qualitatively in traditional QFD. Instead, this research proposes the use of formal mathematical models. Factor analysis, design and analysis of experiments, principal component analysis, and conjoint analysis and lab experiments may be used to map *high level* technical requirements in technical requirements space (where performance is tested). These statistical models will also allow the design team to determine which technical requirements affect key primary perceptual dimensions, and to what degree.

The technical requirements vector in the technical requirements space provides performance targets (metrics) to assist in technical concept testing and candidate product evaluation. This vector must be mapped to the design parameter space. This mapping is difficult. The reverse map, from design parameters space can be performed in many cases using a CAD/CAE model (if possible) or engineering equation. Recently, however inverse design techniques have been developed to map from requirements to design parameters (Dulikravich, 1991). The design parameter vector must be mapped to the manufacturing process space where the product and component design specifications are related to manufacturing process cost via DFM and DFA tools (Boothroyd & Dewhurst, 1989). These key manufacturing processes positioning the product in the manufacturing process space must be used to map the design to the final space, the production planning and control space where the production and process control plans are determined.

An integrated design methodology can be developed if it is possible to develop mathematical representations for each of these maps. Existing mathematical representation tools for each house are shown in Fig. 3. In each of the following sections, we present formal mathematical maps between each of the spaces, to replace the qualitative maps provided by the "Houses" of QFD.

2.5. *Perceptual Space versus Customer Requirement Space*

Marketing efforts to define a product "position" often manipulate a small set of product characteristics from which detailed customer requirements can be inferred.

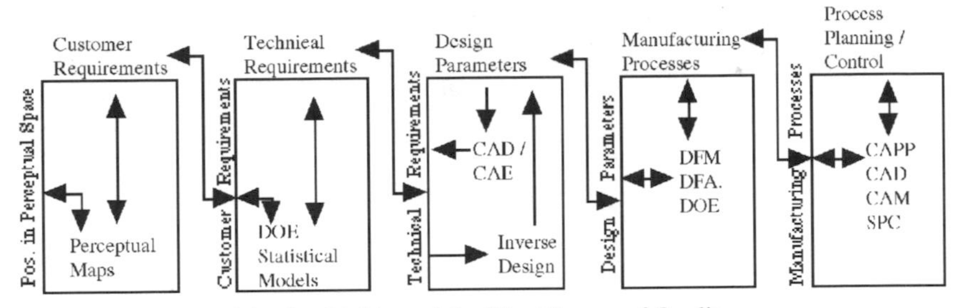

Fig. 3. Linking of the Five Houses of Quality.

The authors call this space *perceptual space*. For automobile design, for example moving along the *luxury vs. economy* axis in perceptual space has implications for seating quality, interior noise, trunk space, ride quality, and other customer requirements. Generally, perceptual space provides a lower-dimensional space in which strategic decisions about the product features can be made, and in which consumer preferences studies can be designed, to construct auxiliary models of expected market share, risk, and sales volume (Hsu & Wilcox, 2000; Lilien & Rangaswamy, 1998).

After the design team identifies the appropriate perceptual space and maps to/from customer requirements, they can identify new product position opportunities using preference models that operate in perceptual space, along with value analysis and cost data from the other houses. These costs and revenues data can be combined to calculate a figure of merit (profit, revenue, market share, etc.).

Perceptual Space is essentially a linear composite of the original customer requirements space. Existing technology for mapping from perceptual space to customer requirements space uses linear maps.

2.5.1. Mapping from Perceptual Space to Customer Requirement Space

Design decisions on product position in perceptual space can be mapped into a customer requirements vector in the customer requirement space. In order to map from perceptual space to customer requirement space, we use the relation

$$x^T = z^T V^T \tag{1}$$

where x is the customer requirement vector, z is the product position vector, and V is the matrix of eigenvectors resulting from factor analysis or principal components analysis applied to different quantitative levels of customer requirements (Aungst, 2000). For example, a copier must have enough speed to maximize out-put (customer requirement) and still have good resolution (customer requirement). Thus, this equation will give the design team that level of quality (perceptual

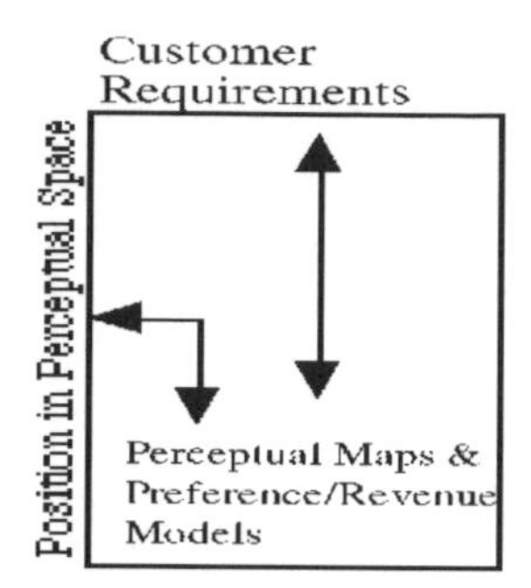

Fig. 4. Customer Perceptual and Requirement Space and Revenue Models of Fig. 3.

dimension) and still meet the customer requirements to maximize out-put and have good resolution.

Preference measures, both direct ranking (expectancy) and revealed (preference regression), indicate which perceptual dimensions are most important to customers and assist the design team to establish a *target* position for a new product concept. Another auxiliary model, a *feature-based* preference measure (conjoint analysis) enables the design and manufacturing engineers to focus on the technical requirements and the most efficient means to attain this target position. Marketing identifies whether or not to focus on a niche of the market or the overall market. The strategic role of product positioning is focused in this new customer perceptual space. The final output from the new house zero is the resulting customer requirement vector in customer requirement space. See Fig. 4.

2.5.2. Mapping from Customer Requirement Space to Perceptual Space

We can map back from the target levels in the customer requirement space to the target levels in the customer perceptual space by using the equation $z^T = x^T V$. Thus, we can add or delete customer requirements, change customer requirements levels, perform value analysis, calculate revenue in the customer requirements space and then map back to the customer perceptual space, focusing on the customer's needs (Fig. 3).

2.6. Customer Requirements versus Technical Requirements

The customer requirements are linked to technical requirements, these technical specifications of the product provide insight to the design team in setting performance targets (metrics) without inhibiting creativity in our technical performance space Statistical tools can be used to establish the relationship between the *high*

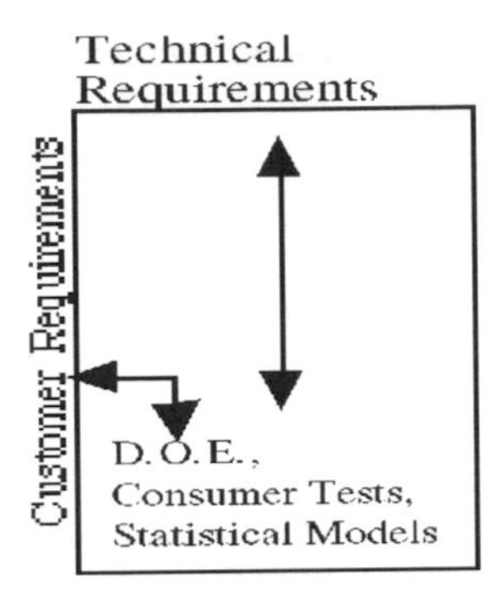

Fig. 5. Customer Requirements Maps Technical Requirements Space of Fig. 3.

level technical requirements for our technical requirement space and corresponding attribute scores in customer requirement space. These statistical models assist in setting realistic performance targets, and by House Zero allow the design team to determine which technical requirements affect key primary perceptual dimensions (see Fig. 5).

2.6.1. Mapping from Customer Requirements Space to Technical Requirements

The maps from customer requirements to technical requirements based on statistical models using design and analysis of experiment (D.O.E.) and consumer tests similar to Bose (1973) and Barton et al. (1982).

2.6.2. Mapping from Technical Requirements Space to Customer Requirements Space

A common technique to determine technical requirements is conjoint analysis. Conjoint analysis uses data on customers' overall preferences for a selected number of product attributes and decomposes these overall preferences into utility values that the customer associates to each level of each attribute. Simple regression analysis is used to obtain these utilities. An overall preference for the different levels of attributes is also calculated to determine a preference for different combinations and levels of attributes for a particular product.

After determining the technical requirements using conjoint analysis, we then calculate the customer's utility for each level of the customer requirements. We then calculate the slope and then use the point slope formula to calculate the utility for a certain level of a customer's requirement between two points. This will be covered in detail in the numerical example in Section 4.

2.7. Technical Requirements Space versus Product Parameter Space (Design Parameter Space)

Parallel to customer needs are the technical requirements to meet the level of customer needs. For example, for a portable mixer, *speed* is an important technical requirement in order to perform an efficient mixing. One measurement of this technical requirement is revolutions per minute (r.p.m) a performance metric. The Cusinart mixer has a certain motor type with a mixing speed of 4000 r.p.m. (a 220 watt motor, with a technical requirement capability of 4000 r.p.m.). The customer needs to be able to mix cookie dough, mix cake batter, mix mash potatoes, and to mix whip cream; the engineer fulfills those customer needs, via the technical requirements, at the technical requirement performance target level of r.p.m. The technical requirements are identified by the design team by defining all dimensions of a product that represent each customer need that are required to specify the product technically.

The technical requirements space provides performance targets to assist the design team in evaluating candidate products. Mapping these technical requirements to the design parameter space is difficult.

2.7.1. Mapping from Technical Requirements Space to Design Parameter Space (Inverse Design Equations)

The technical requirements vector in the technical requirements space provides performance targets (metrics) to assist in technical concept testing and candidate product evaluation. This vector must be mapped to the design parameter space, but this mapping is difficult. Recently, however, inverse design techniques have been developed to map from requirements to design parameters (Dulikravich, 1991) and work has been started to develop forward and inverse approximation maps simultaneously (Barton, Meckesheimer & Simpson, 2000). In inverse design, technical requirements are given (e.g. lift and drag for air craft wing design) and used to derive design parameter values by applying engineering equations and optimization methods (e.g. values for lift and drag will generate the geometry parameters for the wing). The inverse design methodology gives the design team the ability to map from the technical requirements space back to the design parameter space, a critical capability for customer-driven design that originates in perceptual space.

2.7.2. Mapping from Design Parameter Space to Technical Requirements Space (CAD/CAE)

The reverse map from product design space to technical requirements space is the core of traditional engineering design. An extensive set of specialized

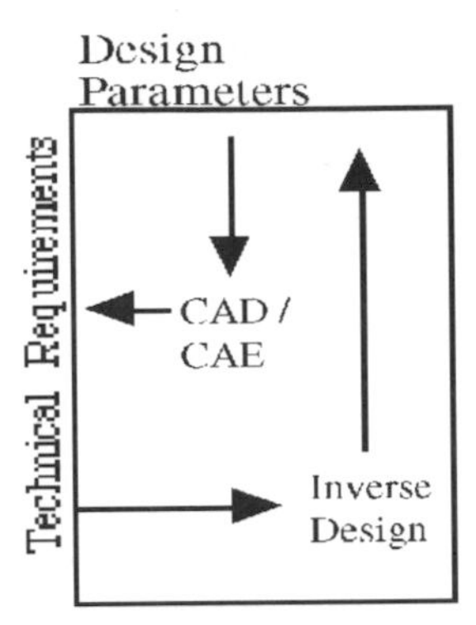

Fig. 6. Mapping Between Technical Requirements and Design Parameters.

mathematical models has been developed in this area. Often, this mapping can be accomplished using a CAD/CAE model (if possible) or engineering equations. Computer assisted engineering (CAE) software typically provides numerical solutions to a set of equations that relate the design variables (physical characteristics) to the performance metrics of the technical requirements used to quantify performance of a product. These models are used to decide whether the designs are feasible, to explore the performance envelope of a design without actually building a physical prototype, and to study the tradeoffs involved in the design. Figure 6 shows these maps as part of Fig. 3.

2.8. Incorporating Cost and Estimating Costs in Product Parameter Space

The decision to proceed with a new product development project must incorporate costs, benefits, and expenses. It may appear at first that it is difficult to quantify the costs, benefits, and expenses of the various candidate products, but approximation of these dollar metrics in the design phase is important. There are four primary areas to be considered early in new product development decisions concerning cost and benefits:

- *Product features* determine product performance and set the cost of performance, which is an important determinant of market success. The design team should analyze cost/benefits to help them decide if a new product feature(s) should be added.
- *Product cost* is the total cost of the product over its life cycle. This includes manufacturing, distribution, warranty, and logistical costs. Some product parameters

(such as material type and volume) affect cost directly; others must be mapped to manufacturing/distribution space to determine cost impacts.

- *Speed of product development* determines the time to market-from the time the need for the product is realized to the point it is delivered or sold to the customer. This can be critical to the success of the product. It involves all of the company's departments, not just design engineering.
- *Development expenses* are the one-time costs associated with development of the product. Although seldom an overriding concern, these expenses must be justified to upper management early in the design phase.

The goal of identifying and calculating cost cannot be realized by the design engineers alone. Other departments' expertise also needs to be utilized to identify and minimize cost – at the very minimum, purchasing, process planning and, cost engineering (value analysis/value engineering) are also required. It is important to understand that different components of cost are calculated in different spaces. Material costs are calculated using an auxiliary function whose domain is design parameter space. Manufacturing costs are calculated using an auxiliary function in manufacturing process space. Revenue is calculated using pricing and market share auxiliary functions in customer perceptual space.

2.8.1. Designing for Cost

The needs for *Designing for Cost* (D.F.C.) are:

- *Providing cost information.* Current cost of parts and assemblies are needed, along with information on cost-influencing parameters.
- *Reaching a cost decision.* The most cost-favorable solution must be chosen, on the basis of material used, the production process, the desired time to market and other deadlines, quality and other relevant factors.
- *Product design for cost.* Design for cost (D.F.C.) needs to be viewed in its totality. Not only the costs but also the deadlines and technical risks should be considered. Incorporating these costs into design is made possible by the use of auxiliary cost functions.

  All of this data can be obtained from mapping forward and backward from the design spaces which could then be entered in the houses for a design for cost activity.

2.8.2. The Task of Purchasing in Providing Cost Information

Cost information is a prerequisite for D.F.C. It is necessary not only to provide information about part prices, manufacturing cost, etc. but also to establish the cost structures and the interaction of the costs of the different products under consideration. Manufacturing costs can vary widely. Purchasing can help designers by:

- Providing specialized knowledge and information about typical purchasing items such as semi finished parts and housing.
- Providing the designers with suppliers' cost structures.
- Helping to build a cost database that can be used by the design team.

A software product by Parametric Technologies called Pro/MODEL is a high performance, low-cost viewer designed to increase access to Pro/ENGINEER product models across the manufacturing enterprise. A purchasing agent can allow outside vendors to bid on a long-lead item without releasing the complete design externally. Suppliers can access the item through their Netscape browser and Pro/MODEL. Suppliers can then check measurements, such as length, volumes, edge-to-edge distances, etc. and send electronic model mark-up to the purchasing agent with their bids (Parametric Technologies, 1999, http://www.com/srch/pronews/vol3–4/pg11.html, 1998).

2.9. Design Parameter Space versus Manufacturing Process Space

The design parameters must be related to the selection of manufacturing processes in order to complete the design of the product and process, and estimate manufacturing costs. So many factors come into the choice of how to manufacture a product that only the most general of overall guidelines can be specified early in the design of a product. What shape and size is a component, how strong, how accurate in dimension and what surface finish is required are just a few of the general guidelines (McMahon & Browne, 1998). In some cases the techniques are so well established that the choice is clear. For example, automotive bodies are almost invariably made by stamping and spot welding sheet steel. Nevertheless some general guidelines that do apply are:

- Select a process commensurate with the materials, and the required accuracy and surface finish.
- Select component dimension and surface finish parameters that allow the widest possible tolerance range and surface finish variation.
- Make a detailed comparative assessment of available manufacturing systems at the design phase; in particular, carry out an analysis of the sensitivity of part and assembly cost to production volume for different processes.

Process determination and validation drives the design team to investigate the potential of alternative processes as well as their capability under various candidate product scenarios and product volume mixes. From these analyses, processes can be selected that optimize the quality and cost objectives of the candidate products.

The use of computer-based technologies, such as solid modeling can enable free-form fabrication for analysis and process validation. This can cut weeks and even months from the overall design process (McMahon & Browne, 1998). Process validation is critical to verifying that production processes can best meet the costs and quality goals of a given design.

2.9.1. Mapping Design Parameters Space to Manufacturing Process Space

Traditionally, good manufacturing practice has been recorded in textbooks and in training programs (McMahon & Browne, 1998). A recent trend has been to incorporate these guidelines into expert systems for advice on the manufacturability of a design. Another recent development again, amenable to computer implementation has involved methods for systematically rating a product in terms of manufacturability and then suggest procedures for improving the rating (McMahon & Browne, 1998). This has been applied in particular to design for assembly (DFA) (Boothroyd, 1992; Boothroyd & Dewhurst, 1989). DFA has assumed increasing importance because assembly is so labor intensive: as process costs have reduced owing to improved machines and processes, so assembly costs as a proportion of the total have increased (McMahon & Browne, 1998). Design for manufacturing DFM: guidelines have been developed for practically every aspect of manufacture, but they may be broadly divided into four groups relating to the *general approach* to DFM, to *selection of manufacturing processes*, to *design for particular processes* and to *assembly*.

The most recent advancement in software for DFA and DFM is Pro/PARTNERS (Parametric Technology) with Bothrooyd and Dewhhurst. This DFA integrated software (DFA) provides a tightly integrated layer written in Pro/DEVELOP for generating DFA analysis files. The generated analysis files contain the complete assembly structure, including envelope dimensions for all assembly components parts. Also provided are functions to allow the display of the parts and assemblies as they are being analyzed, along with the ability to update the analysis directly from Pro/ENGINEER as changes are made to the Pro/ENGINEER assembly. A complete user-definable library capability allows the engineer to describe commonly used components in terms of DFA characteristics. The Pro/ENGINEER link will then use the library to construct future DFA analyses, thereby reducing further the steps required to complete the full analysis (Parametric Technologies, 1999, http://www.ptc/srch/partners/csp/partner_html/9.htn, 1998).

2.9.2. Mapping Manufacturing Process Space to Design Parameter Space

This can be done by reverse mapping from the manufacturing processes to the design features by using Pro/ENGINEER integrated software package. For example, a company may be in the plastics business and by using the reverse mapping procedure can determine what design geometry is required to manufacture a part using plastic injection molding.

2.9.3. Determining Manufacturing Process Costs

In manufacturing we understand the following operations:

- Production of individual parts; and
- Assembly of parts.

Thus, the process cost of production can also be expressed as the sum of two costs:

$$C_{\text{pr}} = C_{\text{pt pr}} + C_{\text{pt as}} \tag{2}$$

where $C_{\text{pt pr}}$ = cost of part production and $C_{\text{pt as}}$ = cost of part assembly.

Of all the DFX (Design For X = anything) practices, design for manufacturing has the greatest impact on product cost. Design for manufacturing directly includes the design of parts for production and assembly. Indirectly, it also implies designing under the constraints of time, cost and quality.

While manufacturing considerations are addressed present to a certain extent in the design parameters space (e.g. material costs) it is certainly in the manufacturing process space that concrete steps must be taken so that the product can be manufactured in a short time, at a low cost with high quality. Manufacturing cost estimates, which depend on the product's position in manufacturing space, allow alternative designs to be evaluated and assist in achieving the most cost-effective design.

Reducing the manufacturing process cost implies according to Eq. (2),

- Reducing the part production cost; and/or
- Reducing the assembly costs.

The most important considerations regarding cost of parts are:

- Materials and their costs;
- Manufacturing processes;
- The number of different fabrication steps;
- The quantities produced; and
- The use of standard parts.

Assembly costs arise primarily from:

- Time required for assembly;
- Equipment and tools required; and
- Workers' dexterity, in the case of manual assembly.

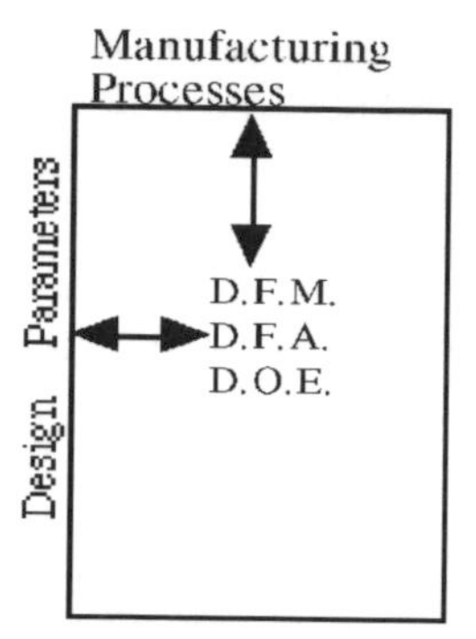

Fig. 7. Mapping from Product Parameters Space to Manufacturing Processes Space from Fig. 3.

Boothroyd and Dewhurst (2000) describe the various measures for design for assembly, and their software provides cost estimates.

In summary, Pro/ENGINEER and Boothroyd DFA software can be used by the design team in manufacturing process space. DFA is a quantifiable method for identifying unnecessary individual parts and for determining assembly times and costs. DFA results in the development of more elegant designs with fewer parts that are both functionally efficient and easy to assemble. The larger benefits of DFA are improved quality and reliability, reduced direct and overhead costs and shorter product development cycles. (Parametric Technologies, 1999) (http://www.ptc.com/srch/partners/csp/partner_html/9.htn, 1998). These maps are summarized in Fig. 7, as part of Fig. 3.

2.10. Manufacturing Process Space versus Manufacturing Management Space

Very little software exists to map between these two spaces. Enterprise resource planning software (E.R.P., SAP and Oracle Applications) are attempts to manage and map between these two spaces. In reality ERP software is a sophisticated product tracking system.

2.10.1. Mapping from Manufacturing Process Space to Manufacturing Management Space: Determining the Production Plan

After determining the most efficient and cost effective manufacturing processes and generating a process plan, this result becomes the input to our last space: the production planning and control space. Mapping between these spaces has little existing software analysis other than SPC software.

2.10.2. C.A.P.P.

In traditional process planning systems the process plan is prepared manually. The task involves reasoning about and interpreting engineering drawings, asking decisions on how many cuts should be made or how parts should be assembled, determining in which order operations should be executed, specifying what tools, machines and fixtures are necessary for the production of the finished product. The resulting process plan is therefore very much dependent on the skill and judgment of the planner. The type of plans which a planner produces depends on the individual's technical ability, the nature of his experience. The use of computer based decision support systems (CAPP) offers potential benefits in terms of reducing work loads of manufacturing engineers and also provides the opportunity to generate rational, consistent and optimal plans. Additionally, an integrated CAD/CAM system develops only when a system exists that can utilize design data from a CAD system and information from manufacturing databases to manufacture the part. CAPP seeks to provide this interface between CAD and CAM (McMahon & Browne, 1998). Figure 8 illustrates these maps, as part of Fig. 3.

2.11. Selecting Candidate Designs

Ultimately, commercial product design is based on projected profit. Profit is a calculation that all firms want to know. To answer this question, we will have to link the process planning and control space to the other four spaces. We will have to forward and reverse engineer using all of the formal maps, represented as QFD

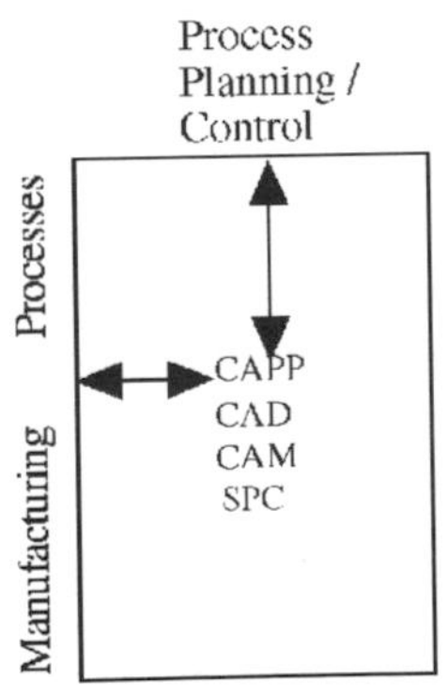

Fig. 8. Mapping Between Manufacturing Processes and Process Planning and Control as Part of Fig. 3.

houses, in order to analyze the contribution margins for each candidate product design for a particular customer requirement at a specific level of performance.

2.11.1. Value Engineering

Value Engineering is a design selection method that uses a proxy for profit value.

Value engineering (also called value analysis, value control, value management and VA/VE) aim to maximize the value of a product. The value of a product is defined by equation):

$$\text{Value} = \frac{\text{Product Function and Performance}}{\text{Product Cost}} \tag{3}$$

Value engineering aims to obtain maximum performance for minimum cost. VA/VE procedures employ cross-functional teams to evaluate each step in the product realization process – design, procurement, and manufacturing – to achieve its aims (Pugh, 1993).

2.11.2. Pugh's Conceptual Funnel

Experienced gained over many projects both in industry and associated with industry has led to the conclusion that matrices in general are probably the best way of structuring or representing an evaluation procedure, in that they give structure and control to a process (Pugh, 1993). A common facet of all matrices is the evaluation of alternative solutions against criteria that have been identified as being significant with the specific purpose of identifying those alternatives that best satisfy the criteria.

The purpose of any method of evaluation is to allow the design principles to emerge visibly in a context and to be articulated. The method of controlled convergence (Pugh, 1993) has been devised in practice to select actual concepts in practice with greater certainty of success.

According to Pugh, there are four stages to QFD:

(1) Customer requirement planning translates customer expectations "Voice of the Customer" in the form of market research, competitor analysis and technological forecasts into the desired and specific product characteristics.
(2) Product specifications convert the customer requirements plan for the finished product into its components and the characteristics demanded.
(3) Process and quality control plans identify design and process parameters critical to the achievements of the product requirements.
(4) Process sheets, derived from the process and quality control plans are the instructions to the operator.

Pugh's utilizes QFD as an auxiliary method to his design methodology. The four steps above are Pugh's interpretation of QFD's four-house approach (Pugh, 1993).

2.11.3. Profit and Revenue Maximization

The design team may perform sensitivity analysis by changing design parameters, in design parameter space, and then map forward and backward obtaining new data points still focus on customer needs. For example, the design team will recalculate all costs implied by these new design parameters, setting the price (customer input), and then calculating the amount that each candidate product contributes to profit (contribution margin). The next step in the profit analysis is to calculate market share. Our model for market share prediction uses the equation (Lilien & Rangaswamy, 1998):

$$m_n = \frac{\sum_{i=1}^{n} w_i p_{in}}{\sum_{n=1}^{N} \sum_{i=1}^{I} w_i p_{in}} \tag{4}$$

where $I =$ number of customers in the market study, $N =$ number of product alternatives for the customer to choose from, including the new product concept, $m_n =$ market share of product n, $w_i =$ the relative volume of purchases made by customer I with the average volume across all customers indexed to the value 1, and $p_{in} =$ proportion of purchases that customer i makes of product n (or equivalently, the probability that customer i will choose product n on a single occasion).

Confidence interval estimates for market share are discussed by Hsu and Wilcox (Hsu & Wilcox, 2000). Finally, this model is based on the notion that each product receives a share of the customer preferences:

$$p_{in} = \frac{q_{in}}{\sum_{n=1}^{N} q_{in}} \tag{5}$$

where q_{in} is the estimated utility of product n to customer i (Lilien & Rangaswamy, 1998). An example calculation is in Section 4.

Product design rarely depends solely on profit, at least not on a quantifiable characterization of profit. Other factors such as the development of corporate capabilities, establishment of market presence, etc., are considered (Wheelwright & Clark, 1993).

2.12. Summary of Mapping Paradigm

Previous approaches, including QFD linked the design and manufacturing processes qualitatively. This section outlined a strategy to replace these qualitative links with formal mathematical models. These statistical models and engineering equations assist the design team to understand the global implications of decisions made in any of the six domains. The methodology provides a quantifiable impact of alternative decisions across domains, which is a tool for communication between members of the design/decision making team.

3. FOCUSING ON HOUSE 1: EXISTING METHODS

In this section, we focus on formal maps between customer requirements space and technical requirements space. This map is captured by "House 1" in quality function deployment. This QFD map is sometimes referred to as the House of Quality (HOQ).

3.1. Customer Needs

The HOQ is a useful format for arranging data collected about competitors and potential customers. It assists the design team to systematically relate the customers' needs (customer attributes in the HOQ) to technical performance specifications (technical requirements in the HOQ). Customer needs collected through surveys and interviews are usually phrased in day-to-day language and are not suitable for direct use in a engineering project. The important objectives of the HOQ technique is to translate the customer needs into formal engineering targets and to store the information necessary for this translation in a readable understandable format (Hauser & Clausing, 1988). The authors will illustrate the techniques and benefits using the design of a network printer/copier as an example. We will give a detailed discussion of this example in the fourth section.

For reasons of confidentiality, the details of the case study were changed but the actual product features and market segment are similar. The company that was selected is a Fortune 100 firm. The company designs high tech electronic products for business applications. It is interested in developing computer hardware for the internet. The company has expertise in developing computer-related hardware. It faces tough competition from a dominant competitor in this market segment. Potential revenues are large and it is an important segment of the market for the company. The company believes it must position the product carefully because the difference in the selling price between a high end product and a low end product is $125,000.00. The company spent $500,000.00 to determine if there are any opportunities in this particular market segment for a similar product. The management was willing to help test the new methodology and architecture. The company provided confidential data, in the form of a pilot study rather than a full

Table 1. Customer Needs for Copier.

Customer Needs
1. Good pages of output
2. Resolution
3. Scanning quality
4. Scanner resolution
5. Stapler reliability
6. Number of scheduled maintenance calls
7. Is light in weight
8. Has a comfortable handle design
9. Easy to read control dials
10. Paper capacity
11. Fast scanning
12. Prints reliability
13. Has a low noise level
14. Is safe to operate

study. The company gave timely feedback to the end results of the field research and application. Table 1 contains the customer needs for a printer/copier for the internet.

3.2. Technical Requirements

Linked to customer needs are the technical requirements to meet the level of customer needs. For example, *print speed* is an important technical requirement in order to perform efficient copying. One measurement of this technical requirement is pages per minute (p.p.m.), a performance metric. The X copier has a certain print speed with a range of p.p.m. from 110 to 180 p.p.m. The customer needs to be able to *print efficiently and reliably* are linked indirectly to the design parameters decisions, via the technical requirements specifications. Customer needs are linked to a performance target level of p.p.m. that in turn will require the ability of the electric motor to handle industrial copying. The technical requirements are identified by the design team by defining all dimensions of a product that represent each customer need that are required to specify the product technically. Other examples are shown in Table 2. Technical requirements differ from the customer needs as they are directly measurable at a fine grained level that matches a softer measure of customer needs. These technical requirements are used by engineers throughout the development of the product and become the basis for the component (part features) and process specifications. In the detailed example, in Section 4 our approach uses linear interpolation from conjoint data; calculating the slope and

Table 2. Technical Requirements for Copiers.

1. Print speed (p.p.m.)
2. Image quality (d.p.i.)
3. Scanning resolution (d.p.i)
4. Stapling (number of sheets per set)
5. Scanner image quality (d.p.i.)
6. Scanner input hopper capacity (number of pages)
7. Scanning speed (p.p.m.)
8. Reliability (number of calls per month)
9. Paper hopper capacity (number of sheets)

using the point slope formula calculate the level of performance of the technical requirements.

Note that technical requirements are not product features (portable vs. non portable copiers vs. industrial copiers) nor are they part characteristics (steel vs. aluminum vs. plastic vs. composite materials). The product features and part characteristics are important to new product design and need ultimately to be linked to technical requirements. Thus, we clearly need a way to communicate customer needs to engineers so they can use their knowledge of the technical requirements at their disposal to design products that meet or exceed customers' needs. To enhance communication we need to describe the customer needs in the language of the engineer (technical requirements). Thus, we must identify the descriptions of the physical product that are measurable and quantifiable and that relate to customer needs.

The HOQ is most useful after determining the customer needs, and the customer's perception of competitive products with respect to those needs. The HOQ allows designers to focus on technical issues that relate to the most important customer needs. R&D provides creative solutions to improve the technical requirements that relate to those needs, and engineering produces the final product design that is consistent with the technical requirements and the customer needs. Most important, the HOQ encourages all of the functional areas to work as a team to produce a coordinated, physical product and marketing plan.

3.3. QFD Representation for Customer Needs and Technical Requirements

The first technical requirement to meet the customer requirement of good pages of output is print speed according to Table 2.

The question is: "What target should be set?" "How many pages of output per minute?" QFD focuses the design team on what customer needs are important

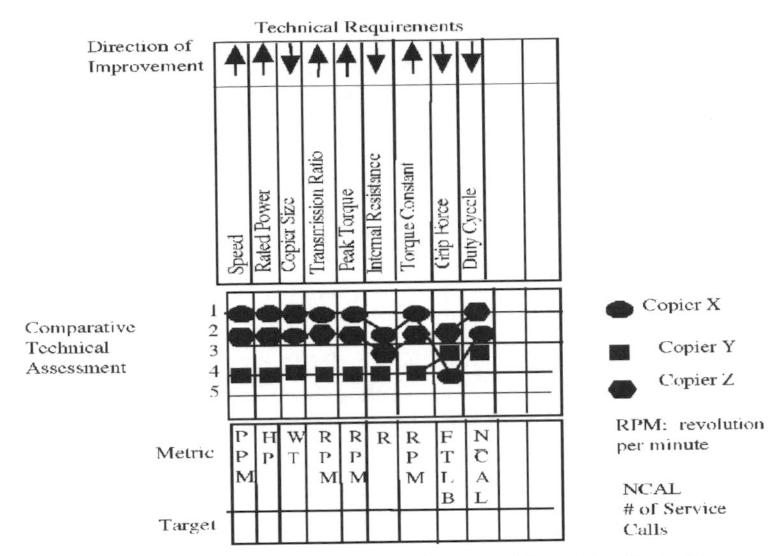

Fig. 9. Technical Requirements and Performance Metric for Copier/Scanner.

compared to its competitors, also on what technical requirements are important and to what performance level are we meeting the customers' needs.

Figure 9 shows the technical requirements, direction of improvement and technical competitive assessment for company X, company Y and company Z for industrial copiers.

The HOQ representations are qualitative and this limits the value of the HOQ. In the following sections, we examine existing quantitative mapping strategies from customer needs to technical requirements.

3.4. Nam's Suh's One Way Prescriptive Methodology

Suh divides product/process design into four domains as shown in Fig. 10 (Suh, 1998). The first is the customer domain where the customer needs are expressed as customer attributes (CA). The second is the functional domain, which has transformed the customer attributes into a set of functional requirements (FR). Once the FRs are chosen they are mapped in the physical domain as a set of design parameters (DPs) which together become the product design. Next, a manufacturing process for the product is designed in the process domain that is characterized by process variables (PV). Our focus here is on the first map, but the assumptions and restrictions of Suh's methodology apply to all maps.

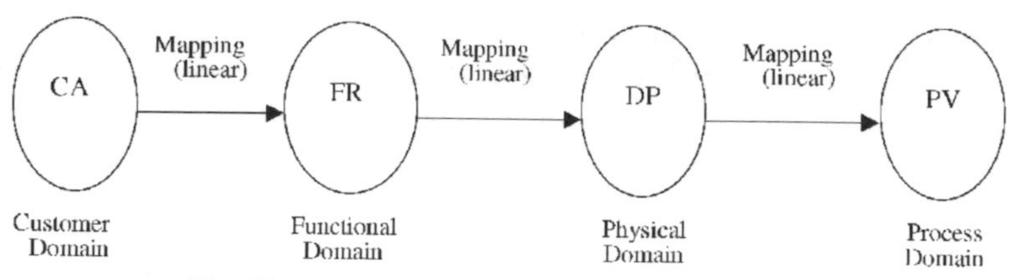

Fig. 10. The Design Spaces of Axiomatic Design.

3.5. Design Axioms

There are two design axioms: The Independence Axiom and the Information Axiom. They may be stated as follows (Suh, 1990, 1998):

The Independence Axiom (Axiom One). Maintain the independence of the FRs. Axiom One states that during the design process, as we go from the FRs in the functional domain to the DPs in the physical domain, the mapping must show that the design with the minimum information content is the best design.

The Information Axiom (Axiom Two). Minimize the information content. Axiom two states that among all the designs that satisfy the Independence Axiom (Axiom One), the design with minimum information content is the best design. The term "best" is used in a relative sense, since there are potentially an infinite number of designs that can satisfy a given set of FRs. In an acceptable design, the DPs and the FRs are related in such a way that a specific DP can be adjusted to satisfy its corresponding FR without affecting other functional requirements. Thus, the best design is a functionally uncoupled design that has minimum information content. This relationship may be represented mathematically (Eqs (1) and (2)). Also, the minimum information content (minimum design complexity) of the design is achieved by selecting the design parameters and manufacturing processes that meet the independence axiom most efficiently.

3.6. Mathematical Representation

The mapping process between the domains can be expressed mathematically in terms of the characteristic vectors that define the design goals and design solutions. The set of functional requirements (FRs) that define the specific design goals constitute a vector in the functional domain. Similarly, the set of design parameters (DPs) in the physical domain that are the "hows" for the FRs also constitute a vector DP. The relationship between these two vectors can be written as:

$$\text{FRs} = [A](\text{DPs}) \tag{6}$$

where $[A]$ is the design matrix that characterizes the product design. For the design of processes involving mapping from the physical domain to the process domain, the design equation may be written as:

$$\text{DP} = [B](\text{PV}) \quad (7)$$

where $[B]$ is the design matrix that defines the characteristics of the process design and is similar in form to $[A]$ (Suh, 1990, 1998). The methodology is linear, and the matrices $[A]$ and $[B]$ are not assumed to be invertible. As a consequence, the design method specifies a sequential process that moves from customer attributes to functional requirements to design parameters to process variables. We say that the method is prescriptive rather than descriptive because it enforces a procedure that differs from current practice – most design processes iterate between these domains rather than progressing unidirectional through them.

For our copier example the customer requirements of powerful motor, dependability, paper capacity etc. are derived by usual marketing research techniques such as interviewing, focus groups etc. and are called the customer needs or customer attributes (Suh, 1998). Then, the functional requirements must be determined in *solution neutral environment* – defining functional requirements without thinking about the solution. These functional requirements are the technical requirements of print speed, duty cycle, stapling, etc. It is important to Suh that the functional requirements (technical requirements) are the minimum set (number) of independent requirements that the design must satisfy. To satisfy this independence axiom, the design matrix must be either diagonal or triangular. When the design matrix $[A]$ is diagonal, each of the functional requirements can be satisfied by means of one DP (design parameter). Such a design is called an uncoupled design (Suh, 1998). For example, consider a square matrix $[A]$ where $m = n$ ($m = 3, n = 3$), $[A]$ may be written as

$$[A] = \begin{bmatrix} a_{11} & a_{12} & a_{13} \\ a_{21} & a_{22} & a_{23} \\ a_{31} & a_{32} & a_{33} \end{bmatrix}.$$

According to Eq. (6), Suh's design equation (Suh, 1990, 1998), the left-handed side of Eq. (6) represents "what the design team wants in terms of design goals," and the right hand-side of equation one represents the "how the design team hope to satisfy" the design requirements. The simplest case of design occurs when all the nondiagonal elements of $[A]$ are zero; that is $a_{12} = a_{13} = a_{21} = a_{23} = a_{31} = a_{32} = 0$.

Then Eq. (6) may be written as for the case where $[A]$ is $m = n = 3$ as

$$FR_1 = a_{11}\,DP_1$$
$$FR_2 = a_{22}\,DP_2$$
$$FR_3 = a_{33}\,DP_3$$

This design satisfies axiom 1, since the independence of the FR's is assured when each DP (design parameter) is changed. That is, FR_1 can be satisfied by simply changing DP_1, and similarly FR_2 and FR_3 can be changed independently without affecting any other FR's by varying DP_2 and DP_3, respectively. Therefore, the design can be represented by a diagonal matrix whose elements are only nonzero elements). This satisfies the independence axiom and is defined as an uncoupled design (Suh, 1990, 1998). The converse of an uncoupled design is a coupled design whose design matrix consists of mostly non-zero elements. A change in FR_1 cannot be accomplished by simply changing DP_1, since this will affect FR_2 and FR_3. Such a design violates axiom 1.

When the matrix is triangular, the independence of the functional requirements, can be guaranteed if an only if the DP's are changed in the proper sequence. Such a design is called decoupled design (Suh, 1998). Consider a special case of matrix $[A]$ which is triangular (e.g. $a_{12}, a_{13}, a_{23} = 0$). This can be represented as

$$\begin{Bmatrix} FR_1 \\ FR_2 \\ FR_3 \end{Bmatrix} = \begin{bmatrix} a_{11} & 0 & 0 \\ a_{21} & a_{22} & 0 \\ a_{31} & a_{32} & a_{33} \end{bmatrix} \begin{Bmatrix} DP_1 \\ DP_2 \\ DP_3 \end{Bmatrix}.$$

In this case the independence of the FR's can be assured if we adjust the DP's in a particular order; thus axiom 1 is satisfied. The sequence changing the DP's is important. If the design team vary DP_1 first, then the value of FR_1 can be set. Although it also affects FR_2 and FR_3, we can change DP_2 to set the value of FR_3 without affecting FR_1. Finally, DP_3 can be changed to control FR_3, without affecting FR_1 and FR_2. If the design team had reversed the order of change, the value of FR_3 would have would have changed while changing DP_2 (Suh, 1990).

All other designs violate the independence axiom and they are called coupled designs (Suh, 1998). For our copier example, according to Suh we want an uncouple product design characteristic (ideally one for each functional requirement) that is uncoupled (not compounded) with the other design characteristics and minimizes the amount of information. In other words, a simple design is best. Each design characteristic for the copier should meet each functional requirement (print speed, duty cycle, etc.) and be a simple design. If the design team wishes to maintain a simple design, (DFM, DFA, Design for Cost, etc.) it must adhere to these axioms and follow the methodology prescriptively. Suh also states the designers must change the process variables not the functional requirements or the design parameters for the design to be easily manufactured and assembled (Suh, 1990, 1998).

4. LINKING CUSTOMER NEEDS TO TECHNICAL REQUIREMENTS

The core of the House of Quality links technical requirement variables to the detailed customer needs. The diagram for our copier/scanner example is shown in Fig. 11. Because of the distinctive shape it is called the House of Quality. While the team is waiting for data to fill in the technical requirements performance targets, the design team – marketing, engineering, R&D, manufacturing etc. "build" the relationship matrix between the technical requirements and the customer needs (see Fig. 11). The team's decisions are recorded in the matrix using symbols to indicate the relationship. The + indicates a positive relationship between the technical requirement and a customer need. A – indicates a negative relationship. When determining the relationship strengths, it is important to work in columns. The design team looks at each technical requirement and moves down the column, asking "Would we work on this technical requirement to satisfy this customer need?"

When the relationship matrix is complete then the co-relationship, a triangular *roof* section is filled in to identify conflicting or cooperating technical requirements for the House of Quality (Fig. 11). Let us return to our copier example.

In the center of the house, a *relationship matrix* indicates how each technical requirement affects each customer need. For example, increasing the horsepower of a copier has a strong positive relationship on the customer's perception that the copier will be able to have an efficient copier output (speed in p.p.m., see Fig. 11). Also, increasing the horsepower of the copier may cause the copier to be harder to stack in the output tray and a negative relationship between the customer need of *output* (p.p.m.).

The values in the relationship matrix are generally qualitative, and obtained by consensus of the new product design team. In many cases this judgment is influenced by market research studies measuring the impact of different product designs on customer ranking or by customer evaluation of products with alternative features or technical performance.

Once the relationship matrix is complete, the design team compares the customer needs to objective measures. For example, if the copier with the most horse power (electrical motor) is perceived by the customer as having the most difficulty to maximize output, then the measures are faulty (metrics) or the copier has a image problem that might be addressed by a new marketing strategy.

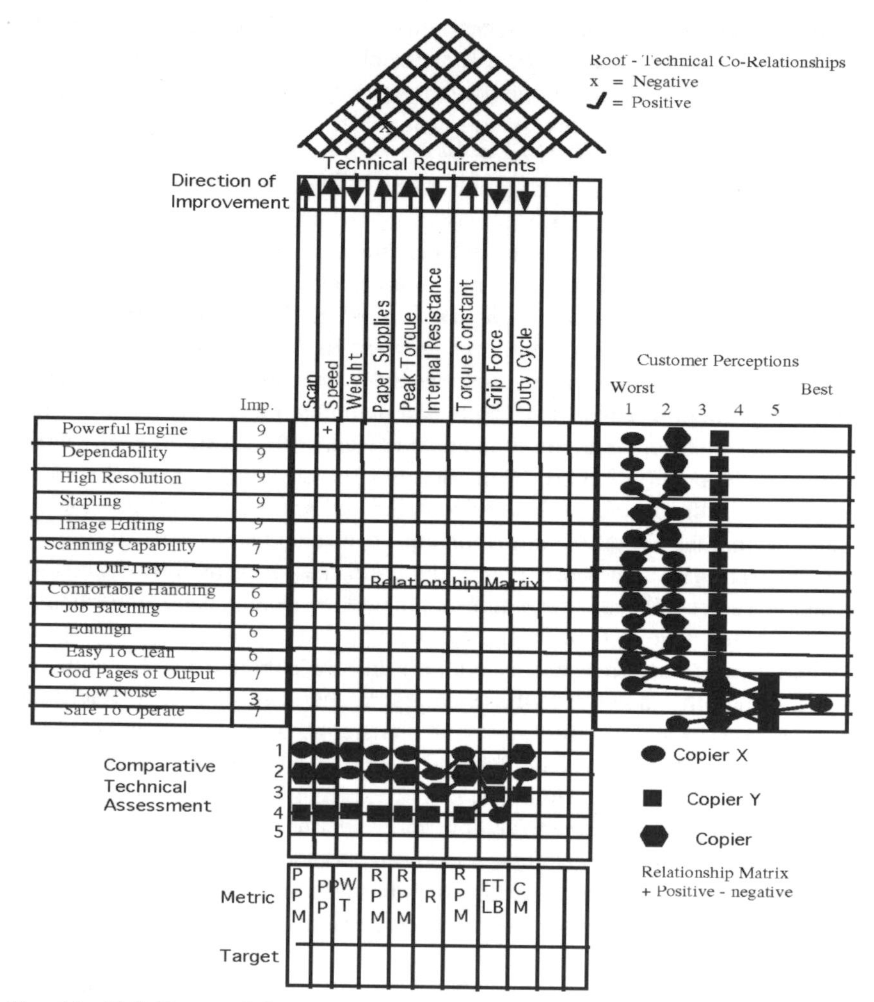

Fig. 11. Full House of Quality for Copier/Scanner, N. B. ?, ×, + and − are Only Illustrative Entries.

A change in the gear ratio for the paper driver may make the copier internal components smaller but may also reduce its peak output and reliability. If the design team decides to enlarge or increase the power of the motor the copier might be heavier, causing it to affect the overall weight (lbs.) of the commercial portable copier. The House of Quality represents these technical tradeoffs with the *roof* matrix (see Fig. 11) that links each technical requirement with every other technical requirement.

The roof matrix helps the design team balance technical changes to ensure that important customer needs are not inadvertently adversely affected. Together the relationship matrix and the roof matrix indicate the complex interrelationships among various product changes (see Fig. 11).

Consider for example, two technical requirements *peak torque* and *copier weight* in Fig. 11. These two technical requirements share many technical variables, including *motor choice*. In situations like this, where there are multiple design variables, there is a potential for two technical requirements to be coupled differently with respect to each common design variable. For example, if the design team increases the size of and power of the motor, the values of portable copier weight and peak torque go up which are undesirable and desirable changes respectively. Since they tend to conflict in terms of goodness an "x" is placed in the roof at the intersection of peak torque and copier weight.

Design of experiments and consumer tests can be used to create statistical maps between customers' needs space and technical requirements space. These maps may involve a large number of variables independent and dependent. Like Suh's method, they provide formal mathematical models linking the spaces. Unlike Suh's method, they can be invertible, and need not be linear.

4.1. Factor Analysis – Simplifying the Links Between Customer Needs and Technical Requirements

Factor analysis is a mathematical technique for simplifying the interrelationship among a large number of observed variables by certain underlying, unobserved common dimensions (factors) unrelated to each other. The purpose of factor analysis is data reduction and summarization; that is, summarizing the characteristics of a set of original variables into a smaller number of new common factors.

The general model of factor analysis (Jobson, 1992) is

$$(x - u) = Af + e, \tag{8}$$

where x represents p observed variables with mean vector $u(p \times 1)$ and covariance matrix $\sum(p \times p)$, A is a $p \times m$ factor pattern matrix consisting of unknown factor loadings, f is a $m \times 1$ vector of uncorrelated common factors and e is a $p \times 1$ vector of errors.

The factors $f = (f_1, f_2, \ldots, f_m)$ are common to all x variables, whereas the error e_i is unique to x_i. The common factors are assumed to have mean 0 and variance 1 and are mutually uncorrelated. The errors are assumed to have mean 0 and variance

$\text{Var}(e_i)$, $i = 1, 2, \ldots, p$. In addition it is assumed that all of the common factors are uncorrelated with the errors. Given these assumptions, the variance of each variable x_i is divided into two parts and can be written as

$$\text{Var}(x) = \sum_{j=1}^{m} \sigma_{ij}^2 + \text{Var}(e_i) \tag{9}$$

where the first part $\sum_{j=1}^{m} \sigma_{ij}^2$, called the communality of x_i, denotes the portion of $\text{Var}(x_i)$ related to the common factors, and the second term $\text{Var}(e_i)$ is the unique variance or specific variance related to the common factors.

Sometimes it is desirable to transform factors into new ones for easier interpretation, which is called factor rotation. There are two types of factor rotation – orthogonal rotation and oblique rotation. The choice between the two depends on the objective of the problem. When summarizing a large number of variables into a smaller number of uncorrelated variables for subsequent, orthogonal rotation is preferred, whereas oblique rotation is more suitable when the objective is to extract some meaningful grouping while permitting the factors to be correlated (Green, 1978).

In this paper, an orthogonal rotation using the varimax algorithm (Green, 1978) is employed because the objective is on simplifying the structure of the HOQ and reducing the number of customer requirements. The same methodology was used for reducing the number of technical requirements (Shin & Kim, 1997).

4.1.1. Determining the Number of Factors

Several criteria apply for determining the number of factors to be extracted:

- eigenvalue criteria;
- a priori criteria;
- percentage of variance criteria; and
- scree test criteria.

Eigenvalue criterion restricts selection of significant factors only to those having eigenvalues greater than 1 (Jobson, 1992). The a priori criteria fixes the number of factors before employing factor analysis. The percentage of variance criterion involves setting a specific percentage of variance level which in turn, determines the selection of successful factors. For example, if the level is set at 95%, successive factors will be selected until the cumulative percentage of variance is greater than or equal to 95%. Finally, the scree test criterion is a graphical procedure which is based on plotting eigenvalues. In summary, the number of factors to be extracted should be determined considering the trade-off between the degree of data reduction and the amount of unexplained variance.

4.1.2. Identifying the Factor Dimensions (Naming the Factors)
One objective in factor analysis is to replace the set of p attribute variables with a small set of z principal components. To name the principal factors, the coefficients of the principal factors in relationship to the original attributes must be examined. The coefficients of greatest magnitude can be considered as a group to give the design team insight in naming the factors. The copier example will be discussed in Section 5 since the common practice is to use principal component analysis to extract the factors and then factor rotation is used to identify and group the attributes.

4.1.3. Kim's Methodology Using CFA
Factor analysis is employed in constructing a new house one by restructuring a given HOQ. Suppose there are p original technical requirements (TC) and n customer requirements. For each customer *requirement* (CA) (for each row in the relationship matrix) there are p cell values representing the relationship between the CA and the p TA. The cell values will be used as input to the factor analysis (Shin & Kim, 1997). As a result of restructuring, m ($<p$) new "grouped TCs" are formed and new m new "grouped cell values" for each CA are obtained (Shin & Kim, 1997).

Kim's method first builds the HOQ by filling in the usual relevant information of the customer requirements, which is received from marketing. He also includes the technical requirements, the relationship matrix and the correlation matrix in the roof of the matrix. The correlation in the roof part of the house of the HOQ is estimated by calculating the statistical correlations with the real technical benchmarking data on the TCs. Alternatively, the design team can make subjective judgment when a reliable data set is not available (Shin & Kim, 1997).

The next step is to compute the factor pattern matrix and derive grouped TCs. The factor pattern matrix ($p \times m$) is computed using the model shown in Eq. (8), where m is the number of new grouped TCs and p is the number of original TCs. (A variable (x) in Eq. (8) corresponds to an TC in this discussion.) Most statistical packages (e.g. SAS, SPSS, and Minitab) have the capability of extracting the factor pattern matrix. A factor rotation can also be performed. After employing factor analysis, m grouped TCs are obtained. A grouped TC is essentially a weighted combination of the original TCs. The weights of the original TCs in the combination are determined by factor analysis in such a way that the resulting TCs are uncorrelated with each other (Shin & Kim, 1997).

The new grouped cell values are calculated as follows:

$$Y = XA \tag{10}$$

where Y is an ($n \times m$) matrix consisting of new group cell values, X is an ($n \times p$) matrix consisting of the original cell values (e.g. relationship matrix) and A is a ($p \times m$) factor pattern matrix.

Next, the new house is completed by calculating the absolute and relative importance using the new relationship matrix. The new house incorporates the information on the correlation among the grouped TCs into the relationship matrix and thus eliminates the need for a roof (Shin & Kim, 1997).

The purpose of Kim's work is to develop a systematic way to obviate two difficulties in QFD: the complicated correlations among the technical requirements and the large size of an HOQ, especially in terms of the number of technical requirements. Kim employs factor analysis to restructure a given HOQ to create a new HOQ. The new HOQ has the following characteristics: has a smaller number of technical requirements which reduces the size of the HOQ, the new HOQ also contains information about the correlation among the technical requirements, thus, it essentially removes the need for the roof part of HOQ. For the copier example this approach might reduce and group the 18 technical requirements into three or four grouped TCs via factor analysis.

4.2. Principal Components Analysis

Like factor analysis, principal component analysis is a statistical technique that linearly transforms an original set of product attributes into a smaller set of uncorrelated perceptual dimensions (principal components) that explain a large proportion of the total sample variance of the original product attributes. Unlike factor analysis, PCA does not include a unique or specific variance component in the model. The transformed product attributes (principal components) are orthogonal and uncorrelated.

For example, geometrically the X matrix of customer needs defines n points in a p dimensional space defined by p axes $x_1, x_2, \ldots, x_p$. In a full rank case the orthogonal transformation (rotation) defined by the matrix say V replaces the p x axes by p z axes which can be viewed as a orthogonal rotation of the x axes in p-dimensional space (Jobson, 1992). The first principal component z_1 represents the single axis, which most approximates the n data points. The estimates x_1 based on z_1 are the projections of n observations onto z_1 axis. The remaining principal components can similarly be defined as a sequence of mutually orthogonal axes each designed to most closely approximate the observation residuals from the previous components (Jobson, 1992).

4.2.1. Determining the Number of Dimensions

The eigenvector matrix Z and eigenvalue matrix V will be used to determine the new variables called principal components. Since we are using principal component analysis as a dimension reduction technique, we will retain those eigenvectors k that are dominant say that have eigenvalues > 1. We retain only the first k components $(k < p)$.

4.2.2. Identifying the Component Dimensions

Recall, one of the main objectives in principal component analysis is to replace the set of p attribute variables with a small set of k principal components. To name the principal components, the coefficients of the principal components in relationship to the original attributes must be examined. The coefficients of greatest magnitude can be considered as a group to give the design team insight in naming the components. When applied this methodology to a case study of the design of a printer/copier scanner for the internet, allowing us to experiment, observe, and learn about the problems encountered in a applied product development project. It provided information and insight concerning the practical implementation of the methodology and its cost effectiveness (Aungst, Barton & Wilson; Quality Engineering, forthcoming).

4.3. Conjoint Analysis

Conjoint analysis studies comprise three stages. In the first stage the product evaluation experiment must be designed. By experiment design, we mean determining the collection of attribute combinations to be presented to the customers for ranking. The second stage collects data from a sample of customers from the target market segment, by asking them to evaluate hypothetical products whose attributes vary according to the experiment design. The third stage explores the impact of alternative product decisions in terms of predicted utility scores and selecting a computational method for calculating utility functions (regression, ANOVA). Each of these stages is described in more detail in the following paragraphs.

4.4. Stage 1: Designing the Conjoint Study

The methodology for selecting attributes is the same for all attribute based analyses. The next step in designing the conjoint study is to select the levels of each attribute to be used. We can start by asking the design team what specific design options it is considering. In selecting the levels of attributes, the design team must keep in mind several conflicting considerations.

First, to maximize the relevance of the conjoint study, the design team should choose attribute levels that cover a range similar to that actually observed in existing products. The design team should include the highest prevalent attribute level (e.g. horsepower among competing chainsaws) and the lowest prevalent level (lowest horsepower). Second, the design team should include as few attributes and attribute levels as possible to simplify the customers' evaluation task. Typically conjoint studies use five to fifteen attributes and two to five levels for each attribute.

Third, to avoid biasing the estimated importance of any attribute, the design team should include the same number of levels for each attribute. Otherwise, some attributes may turn out to be more important simply because customers have more levels (options) to evaluate for those attributes (Lilien & Rangaswamy, 1998). The design team can equalize the number of levels in attributes by redefining attributes, combining two or more attributes or breaking up an attribute into two or more attributes.

The final step of stage one is to develop the candidate products to be evaluated. Here we define a product as a specified set of attribute levels. As in our copier example, it is unreasonable to expect a customer to evaluate every possible attribute and attribute level. The design team must choose a subset of all possible products that might be presented to the customer. A fractional factorial design is typically used to reduce the number of attributes and attribute levels to be evaluated by the customer. A common approach is to select orthogonal combinations of attribute levels to reduce the number of candidate products that customers must evaluate, and at the same time to permit the design team to measure the utility function contribution of each attribute.

To summarize stage one, the design team should use a range of attribute levels that are consistent with the observed marketplace and should try to have approximately the same number of levels for each attribute, both to simplify the evaluation task for the customers and to avoid misleading results on the importance of attributes. A fractional factorial experiment is set up to reduce the number of candidate products.

4.5. Stage 2: Collecting the Data

The first step of stage two is to obtain data from a particular market segment. Once the experimental design is set, the next stage is to obtain evaluations of the selected products from a representative sample of customers in the target segment(s). The design team can represent the products verbally, pictorially, or physically using prototypes. Using pictures has some advantages. Pictorial representations make the task more interesting and they are superior to verbal descriptions. Physical prototypes, while desirable, are expensive, and they are not often used in conjoint studies. Once the design team decides on the presentation mode, there are a number of ways to obtain data from the customer.

The most popular method is for the customer to rank order the candidate products. In this method the customer ranks the products presented, with the most preferred having rank 1 and the least preferred having the rank equal to the number of products presented. Ordinary regression analysis can be used to calculate the utilities, and is discussed in the Stage 3, Part 2, Computational Aspects, that follows.

Another method is to have the product profiles evaluated on a rating scale (e.g. on a scale of 0–100), with larger numbers indicating greater preference. Alternatively, the customers can allocate a constant sum (e.g. 100 points) across the products presented to them. The assumption is that the customers are able to indicate how much more they prefer one product profile to others. The advantage of this type of measurement is that the design team can again use ordinary least squares regression with indicator variables for attribute levels to compute utility functions. This approach is most convenient for management, and will be discussed in more detail in the following sections.

The final method to obtain data from customers is through the use of pair wise comparisons. The customer is presented two products at a time and asked to provide preference rankings for the candidate products. These ratings are obtained using a rating scale (1 = not preferred, 9 = greatly preferred). Then the next two products are evaluated in a similar manner. This process continues until all possible products are compared. This process has the potential of being very boring and cumbersome to the customer. For a small set of products and attributes this can be a very efficient way of obtaining the customer data.

4.6. Stage 3: Part 1, Obtaining the Utility Scores

In stage three, a computational method is employed to obtain utility values. The basic conjoint model may be represented by the following formula:

$$q_{in} = \sum_{p=1}^{P} \sum_{l=1}^{L_p} \alpha_{ipl} x_{npl} \tag{11}$$

where q_{in} = overall utility for product n by individual or group i; L_p = the number of levels of attribute p; P = the number of attributes; $x_{npl} = 1$ if product n has attribute p at level l, 0 otherwise; α_{ipl} = customer i's utility for the pth attribute at level l.

The importance of each attribute is defined as the difference between the highest and lowest utilities for that attribute: $I_{ip} = \{\text{Max}(\alpha_{ipl}) - \text{Min}(\alpha_{ipl})\}$ for each p. The importance metric informs the design team, the importance of that attribute to the other attributes in the study.

Each attribute's importance is normalized to ascertain its importance relative to the other attributes, w_{ip}:

$$w_{ip} = \frac{I_p}{\sum_{p=1}^{P} I_{ip}}$$

so that

$$\sum_{p=1}^{P} w_{ip} = 1.$$

4.7. Stage 3: Part 2, Computational Method for Obtaining Utility Scores

The utility coefficients, α_{ipl} can be estimated with MONANOVA that is monotonic analysis of variance or regression. The simplest and most popular method is multivariate regression with indicator variables. In this case, the independent variables consist of indicator variables for the attribute levels. If an attribute has L levels it is coded in terms of $L - 1$ indicator variables. If metric data are obtained, the rankings form the dependent variable for preferred products. If the data is nonmetric, the rankings can be converted to 0 or 1 by making paired comparisons between products. The common practice is to have metric ranking of preference for products and 1 or 0 to indicate whether the attribute level is present (1) or not (0).

The design team must also decide whether or not the data will be analyzed at the individual customer or focus group level. At the individual customer level, the responses of each customer are analyzed separately. If a focus group analysis is to be conducted, some procedure for grouping the responses must be devised. One common approach is to estimate individual utility functions first. Customers are then clustered on the basis of the similarity of their utility functions. Aggregate analysis is then conducted for each cluster then an appropriate model for estimating the parameters should be specified.

5. NUMERICAL EXAMPLE

Let us examine a portable copier/printer example and discuss the use of conjoint analysis to establish utilities for design attribute levels. Table 3 lists the product attributes and levels chosen for the experiment design. For this experiment, $L_p = 3$ for each attribute p.

Table 3. Potential Product Attributes in Conjoint Design.

	Speed ($p = 1$)	Resolution ($p = 2$)	Weight ($p = 3$)
$l = 1$	10 hp	26 dpi	150 lbs
$l = 2$	8 hp	22 dpi	100 lbs
$l = 3$	6 hp	18 dpi	50 lbs

Note, that there are $3 \times 3 \times 3 = 27$ candidate products, each of which can be described by a particular combination of product attribute levels. Let us number these products $n = 1, 2, 3, \ldots, 27$. One way to elicit preferences is to give each customer twenty-seven 3×5 cards, each of which describes a possible copier product, and ask each customer to rank these cards in order of preference or probability of purchase. Let r_{in} be the rank of individual i gives to product n (r_{in} = 1–27). To reduce the customers' evaluation task, a fractional factorial design was employed and a set of nine candidate products was constructed and ranked by a customer (see Table 4). Table 4 is a reduced set or product profile that explores different possible chainsaw products (combinations of hp., length of saw and weight). Some of these products may be actually products in the market others are hypothetical new product concepts. The customer is asked to rank these different combinations of products. Note, it should be mentioned that some of these products may be infeasible to offer because of lack of present technology, manufacturing constraints, pricing or lack of profitability. It also should be noted that the attribute levels correspond to those in Table 3.

Table 4. Portable Copier/Printer and Their Rankings.

Product Number	Attribute Levels			Preference Ratings
	Speed	Resolution	Weight	
1	1	1	1	5
2	1	2	2	8
3	1	3	3	6
4	2	1	2	8
5	2	2	3	9
6	2	3	1	7
7	3	1	3	5
8	3	2	1	4
9	3	3	2	5

In conjoint analysis, the dependent variable is usually preference or intention to purchase. In other words, customers provide rankings in terms of their preference or intention to buy. This will be related to the α values by a regression model. The conjoint methodology is flexible and can accommodate a range of other dependent variables, including actual purchase or choice.

In evaluating chainsaw products, customers were required to provide preference rankings for chainsaws described by nine candidate products in the evaluation set. These rankings are obtained using a nine point scale ($1 =$ not preferred, $9 =$ greatly preferred). Rankings obtained from the customer are shown in Table 4 as well.

Next, the data reported in Table 4 is analyzed using ordinary least squares (OLS) regression with indicator variables. The dependent variables are the preference rankings. The independent variables are six indicator variables, two for each attribute. The transformed data are shown in Table 5.

The model to be estimated may be represented as:

$$r_{in} = \beta_0 + \beta_1 d_1 + \beta_2 d_2 + \beta_3 d_3 + \beta_4 d_4 + \beta_5 d_5 + \beta_6 d_6 + \varepsilon_{in}$$

for

$$r_{in} = b_0 + b_1 d_1 + b_2 d_2 + b_3 d_3 + b_4 d_4 + b_5 d_5 + b_6 d_6 + \varepsilon_{in}$$

where d_1, $d_2 =$ dummy indicator variables representing speed; d_3, $d_4 =$ dummy indicator variables representing resolution; d_5, $d_6 =$ dummy indicator variables representing weight; $\beta_i =$ unknown coefficient for x_i; $b_i =$ least squares estimate of unknown coefficient.

Table 5. Portable Copier Data Coded for Indicator Variable Regression.

Preference Ranking (r_{in})	Attributes					
	Speed		Resolution		Weight	
	d_1	d_2	d_3	d_4	d_5	d_6
5	1	0	1	0	1	0
8	1	0	0	1	0	1
6	1	0	0	0	0	0
8	0	1	1	0	0	1
9	0	1	0	1	0	0
7	0	1	0	0	1	0
5	0	0	1	0	1	0
4	0	0	0	1	1	0
5	0	0	0	0	0	1

For speed, the attribute levels were coded as follows:

	d_1	d_2
level 1	1	0
level 2	0	1
level 3	0	0

The levels of the other attributes are coded similarly. The parameters are estimated using regression as follows:

$$
\begin{aligned}
b_0 &= 4.22\\
b_1 &= 1.000\\
b_2 &= 1.000\\
b_3 &= 1.000\\
b_4 &= 0.667\\
b_5 &= 2.333\\
b_6 &= 1.333
\end{aligned}
$$

Given the indicator variable coding in which level 3 is the base level, the coefficients are related to the utilities. Each indicator variable coefficient represents the difference in utilities for that level minus the utility for the base level. Thus, the indicator coefficients of the linear regression model gives the design team the slopes and breakpoints of the linear utility functions. For speed, we have the following:

$$
\begin{aligned}
\alpha_{111} - \alpha_{113} &= b_1\\
\alpha_{112} - \alpha_{113} &= b_2
\end{aligned}
$$

To solve for the utility, an additional constraint is necessary. The utilities are estimated on a interval scale, so the origin is arbitrary. Therefore, the additional constraint is of the form

$$\alpha_{111} + \alpha_{112} + \alpha_{113} = 0$$

These equations for the first attribute, speed, are

$$
\begin{aligned}
\alpha_{111} - \alpha_{113} &= 1.000\\
\alpha_{112} - \alpha_{113} &= 1.000\\
\alpha_{111} + \alpha_{112} + \alpha_{113} &= 0
\end{aligned}
$$

Solving these equations, we have

$$\begin{aligned}
\alpha_{111} &= 1.00 + \alpha_{113} \\
\alpha_{112} &= 1.00 + \alpha_{113} \\
(1.00 + \alpha_{113}) &+ (1.00 + \alpha_{113}) + \alpha_{113} = 0 \\
3\alpha_{113} &= -2.00 \\
\alpha_{113} &= -0.666 \\
\alpha_{111} &= 1.00 + (-0.666) \\
\alpha_{111} &= 0.334 \\
\alpha_{112} &= 1.00 + (-0.666) \\
\alpha_{112} &= 0.334
\end{aligned}$$

The utilities for the other attributes can be estimated similarly. For resolution, we have

$$\begin{aligned}
\alpha_{121} - \alpha_{123} &= b_3 \\
\alpha_{121} - \alpha_{123} &= b_4 \\
\alpha_{121} + \alpha_{122} + \alpha_{123} &= 0
\end{aligned}$$

For the third attribute, weight, we have

$$\begin{aligned}
\alpha_{131} - \alpha_{133} &= b_5 \\
\alpha_{132} - \alpha_{133} &= b_6 \\
\alpha_{131} + \alpha_{132} + \alpha_{133} &= 0
\end{aligned}$$

The relative importance weights are calculated based on ranges of utilities as follows:

$$\begin{aligned}
\text{Sum of ranges of utilities} &= [0.334 - (-0.666)] + [0.445 - (-0.445)] \\
&\quad + [1.111 - (-1.222)] = 4.22
\end{aligned}$$

$$\text{Relative importance of speed} = \frac{1.334}{4.22} = 0.24$$

$$\text{Relative importance of resolution} = \frac{0.89}{4.22} = 0.21$$

$$\text{Relative importance of weight} = \frac{2.333}{4.22} = 0.55$$

The results are summarized in Table 6.

Table 6. Utilities and Relative Importance for Portable Copier, $i = 1$.

Attributes	Product Number	Description	Utility	Relative Importance
Speed	3	6 hp.	−0.666	0.24
	2	8 hp.	0.333	
	1	10 hp.	0.334	
Resolution	3	18 dpi.	0.445	0.21
	2	22 dpi.	0.445	
	1	26 dpi.	0.445	
Weight	3	50 lbs.	0.442	0.55
	2	100 lbs.	1.111	
	1	150 lbs.	−1.222	

The estimation of the utilities and the relative importance weights provides the basis for interpreting the results and positioning a new product by conjoint analysis.

5.1. *Positioning a New Product by Conjoint Analysis*

For interpreting the results it is helpful to plot the utility functions. The utility function values are given in Table 6 and graphed in Fig. 12.

As can be seen from Table 6 and Fig. 12, this customer has the greatest preference for the 8.0 hp. or the 10.0 hp. when evaluating speed for a portable copier. The least preferred is the 6.0 hp. portable copier. An 18 dpi or 22 dpi for resolution is preferred to a 26 dpi resolution. Finally, a 50 lb. weight of portable copier is preferred over a 100 lb. or 150 lb. portable copier. In terms of relative importance of the attributes, we see that weight is number one (0.552). Second most important is speed (0.24) and least important is resolution (0.21).

Conjoint analysis is another method of selecting product attributes. The utility functions indicate how sensitive customers' preferences are to changes in attribute scores. By examining graphs like Fig. 12, the design team gains insight into which attribute levels to select for a new product. In general the best attributes are those that give the greatest gains in preference at the lowest costs.

Previously the authors applied our model to a case study of the design of a printer/copier scanner for the internet, allowing us to experiment, observe, and learn about the problems encountered in a applied product development project. It provided information and insight concerning the practical implementation of the methodology and its cost effectiveness (Aungst, Barton & Wilson, 2002).

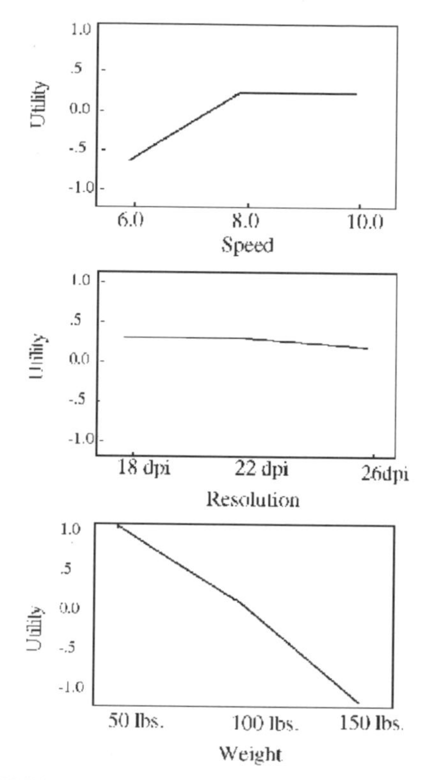

Fig. 12. Utility Functions for Portable Printer/Copier Example.

Constraints on time, resources, and technology forced us to limit the study to customer perceptual and customer requirements spaces at that time. We now proceed to the technical requirements of house one and mapping back to the customer requirements. To protect company confidentiality, the firm will be referred to as Company X, with two different product sets.

Figure 13 depicts Company X's qualitative map between customer requirements and technical requirements space. House One has 25 customer requirements with 18 technical requirements. Each technical requirement has a performance level (target).

The customer attributes were acquired through customer surveys and factor analysis was employed as the statistical technique. The technical requirements were given to us along with results of the conjoint analysis. The product technical requirements and utilities are:

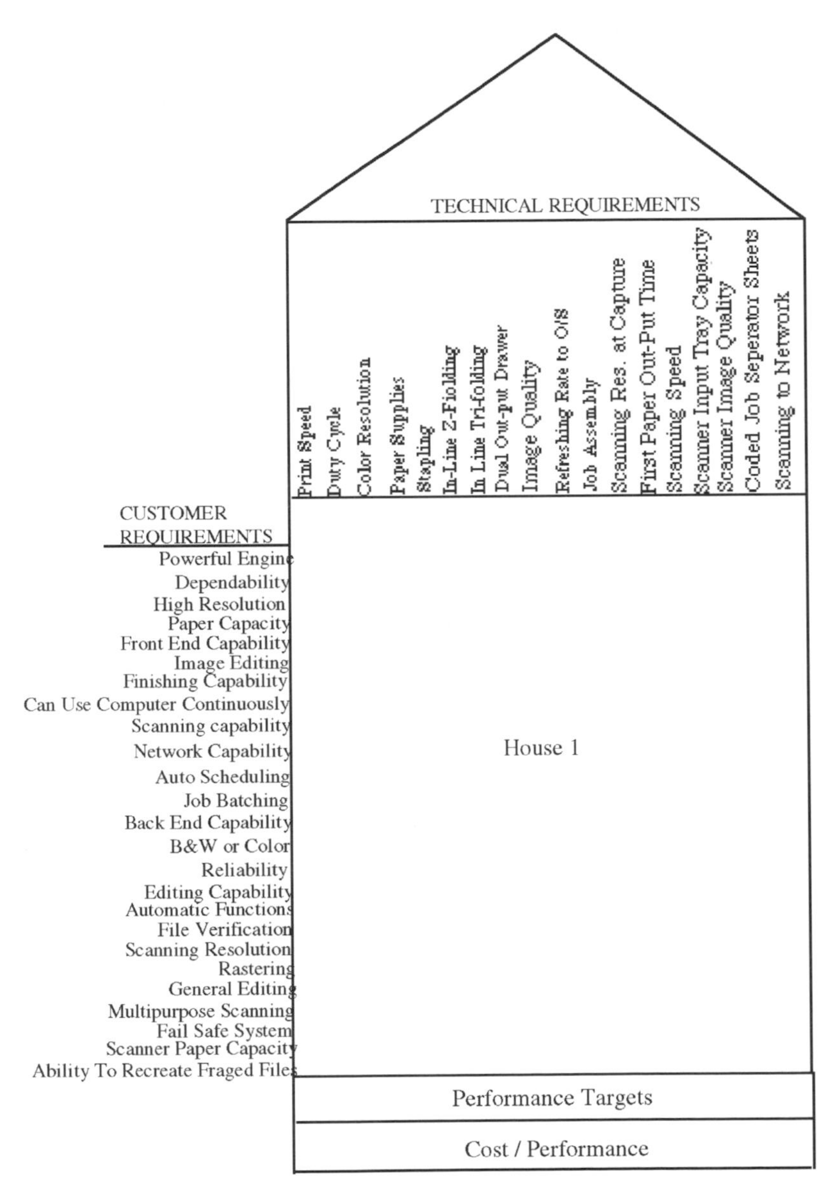

Fig. 13. House One for the Case Study.

Product Features	Level	Utility
1. Print Speed	110 ppm	0.00
	120 ppm	0.11
	135 ppm	0.26
	150 ppm	0.41
2. Duty Cycle	4 Unsched Calls	0.00
	3 Unsched Calls	0.17
	2 Unsched calls	0.31
3. Color Resolution	Blk Only	0.00
	Blk & Std. Color	0.37
	Blk & Delux Color	0.66
4. Paper Supplies	8.5 × 11–11.17	0.00
	5.5 × 8.5–11 × 17	0.18
	8.5 × 12–12 × 18	0.23
	5.5 × 8.5–12 × 18	0.40
5. Stapling	70 sheets, 2 staples	0.00
	100 sheets, 2 staples	0.15
	100 sheets, 3 staples	0.29
6. Z-Folding (In Line)	No	0.00
	Yes	0.20
7. Tri-Folding (In Line)	No	0.00
	Yes	0.28
8. Dual Output Drawer	No	0.00
	Yes	0.37
9. Refresh Rate to OS		0.00
	AT CP	0.25
	AT CP or Remote	0.56
10. Scanner Input Tray	No	0.00
	Simple	0.23
	Simple & Advanced	0.23
11. Scan to Network	No	0.00
	Scan & Store Internal	0.31
	Scan External	0.37
	Scan Int. & Ext.	0.51
12. Job Release (Sep).	No	0.00
	Yes	0.20
	With Opt.	0.28
	With Many Opt.	0.49

Product Features	Level	Utility
13. Job Assembly	No	0.00
	Simple	0.31
	Simple & Adv.	0.31
14. Scanning Speed	20 ppm	0.00
	60 ppm	0.21
	80 ppm	0.37
15. First Copy O.T.	30 sec.	0.00
	12 sec.	0.17
	6 sec.	0.33
16. Scan Image Quality	400 dpi	0.00
	600 dpi	0.29
	1200 dpi	0.44
17. Scanner Hopper	50 pages	0.00
	100 pages	0.15
	200 pages	0.32
18. Scan Sep. Sheets	Manual	0.00
	Auto. Set.	0.27

We assume 100% market awareness and availability, instant market penetration, equal marketing effectiveness and company X's two different products are on the market for the entire year. Company X decided to vary only two technical requirements *print speed* and *duty cycle*. The first product set has a print speed of 120 p.p.m. and a duty cycle of 4 unscheduled calls a year. The second product has print speed of 140 p.p.m. and a duty cycle of two unscheduled. Calculating the first product set customer utilities is straight forward using Eq. (11). The total customer utility for product set one is 3.7. The total customer utility for product set two is 4.02. We now calculate market share using Eqs (4) and (5). The market share for product set one varying only print speed and duty cycle is 48%.

The second product set has a print speed of 140 p.p.m. which is not part of the conjoint study and a duty cycle of two unscheduled calls. The first step is to use our methodology to calculate the utility for 140 p.p.m. We do know the print speed levels and utilities in technical requirements space for product set one.

Print Speed	Utility
110 p.p.m.	0
120 p.p.m.	0.11
135 p.p.m.	0.26
150 p.p.m.	0.41

We want to calculate the utility for 140 p.p.m. Solution is to calculate the slope between 135 and 150 and then calculate the *Y* intercept and finally use the slope intercept to find the utility for a print speed of 140 p.p.m.

$$\text{Slope} = \frac{y_2 - y_1}{x_2 - x_1} = \frac{0.41 - 0.26}{150 - 135} = 0.01$$

$$y \text{ intercept} \quad b = y - mx$$

$$b = 0.26 - 0.01 \times 135$$

$$b = -1.09$$

Finally, calculate the utility for a print speed of 140 (*X*) using the slope intercept form.

$$u(x) = y = mx + b$$

$$0.31 = y = 0.01 \times 140 - 1.09$$

The customer utility for a print speed of 140 p.p.m. is 0.31. Any position between 135 p.p.m. and 150 p.p.m. can be calculated using these equations. We return to calculating the total utility for product set two by Eq. (11). Recall, the total customer utility for product set two is 4.02. We now calculate market share using Eqs (4) and (5). The market share for product set two varying only print speed and duty cycle is 52%.

This is a high tech, very expensive market where company X can increase its market share by simply focusing in on two technical product parameters. It is certainly worth the effort to change these technical parameters.

This example is limited to the construction of House One. The example illustrates the proof-of-concept pertaining to the VDM. It is only one case but the products are real products in the market. Company X has designed and manufactured high tech computer-related products for several years. Previous approaches, including QFD linked the design and manufacturing processes qualitatively. The authors advocate the use of formal mathematical models in product design. Factor analysis, DOE, conjoint analysis and lab experiments may be used to map technical requirements in technical requirements space where performance can be further tested via engineering equations. These statistical models and engineering equations assist the design team to make technical performance decisions which affect primary perception dimensions in House Zero. The mathematical formalism (mapping) continues to other decision making domains such as manufacturing processing House Four and control House Five. Thus, the methodology provides a quantitative impact of all decisions across domains, which is a tool for communications between members of the design/decision making team.

6. CONCLUSION, VALUE AND FUTURE WORK

QFD is a subjective, primarily a qualitative structured methodology to document customer needs. The QFD matrix as a product design tool include complicated correlation among the technical requirements that need to be considered and this contributes to a large potential size of the HOQ. Our quantitative approach via factor analysis and conjoint analysis address the problem of subjective, qualitative analysis and decrease the size of the QFD matrix. The methodology allows the designer to be more quantitative in their analysis, and restructure the HOQ and obtain a new HOQ which is smaller in size. VDM is a quantitative method that brings the "Voice of the Customer" into the HOQ. VDM allows a virtual design before building the product. The design team can find an optimal design depending on trade-offs of the design parameters with cost/benefit analysis. VDM is an integrated modified QFD architecture for communications which identifies all tasks and improves the quality and quantity of communications within and among the functional areas of the organization leading to better design earlier, and is augmented by auxiliary maps that provide product and system characteristics such as performance, cost, reliability and market share. The next step is linking engineering models to House One.

REFERENCES

Aungst, S. (2000). *Applications of multidimensional mapping techniques for quality function deployment*. Ph.D. Dissertation; The Harold and Inge Marcus Department of Industrial and Manufacturing Engineering, The Pennsylvania State University.

Aungst, S., Barton, R., & Wilson, D. (2002). The vital integrated design methodology. *Quality Engineering*, forthcoming.

Barton, D. P., Barton, R. R., Blecker, M., Lyons, P. W., Pitts, K. A., Stein, P. G., & Woolson, J. R. (1982). VideoDisc testing at RCA laboratories. *RCA Review*, *43*, 228–257.

Barton, R. R., Meckesheimer, M., & Simpson, T. (January, 2000). *Experimental design for simultaneous fitting of forward and inverse metamodels*. Proceedings of the ASME Design Engineering Technical Conferences; DETC/DAC-1482, 1–11.

Boothroyd, G. (1992). *Assembly automation and product design*. New York: Marcell Dekker.

Boothroyd, G., & Dewhurst, P. (1989). *Product design for assembly*. Wakefield, RI: Boothroyd and Dewhurst.

Boothroyd, G., & Dewhurst, P. (2000). Pioneering prototypes. *Computer Graphics World*, 39–52.

Bose, A. (1973). Sound recording and reproduction, Part One: Devices, measurements and perception. *Technology Review*, 19–25.

Cooper, R. (1995). Developing new products on time, in time. *Research and Technology Management*, *38*(5).

Day, R. (1993). *Quality function deployment: Linking a company with its customers*. Milwaukee, WI: ASQC Quality Press.

Dulikravich, G. (1991). *Inverse design concepts and optimization in engineering sciences*. ICIDES-III, Washington, DC, October 23–25.

Green, P. (1978). *Analyzing multivariate data*. Hillsdale: Dryden Press.

Griffin, A., & Hauser, J. (1994). The marketing and R&D interface. *ISBM Working Series*, *14*, 4.

Hauser, J., & Clausing, D. (1988). The house of quality. *Harvard Business Review*, *66*(3), 63–73.

Hazelrigg, G. (1996). The implication of arrow's impossibility theorem on approaches to optimal engineering design. *Journal of Mechanical Design*, *118*, 161–164.

Hsu, A., & Wilcox, R. (2000). Stochastic prediction in multinominal logit models. *Management Science*, *46*, 1137–1444.

Jobson, J. (1992). *Applied multivariate data analysis: Volume II: Categorical and multivariate methods*. New York: Springer-Verlag.

Lilien, G., & Rangaswamy, A. (1998). *Marketing engineering*. Reading, MA: Addison-Wesley.

McMahon, C., & Browne, J. (1998). *CAD/CAM, Principles practice and manufacturing management* (2nd ed.). New York: Addison-Wesley.

Product Manager, Interview at Ameridrive, Erie, PA. (1999).

Pugh, S. (1993). *Total design: Integrated methods for successful product engineering*. New York: Addison-Wesley.

Shin, J. S., & Kim, K. J. (1997). Restructuring a house of quality using factor analysis. *Quality Engineering*, *9*, 739–746.

Suh, N. (1990). *Principles of design*. New York: Oxford Press.

Suh, N. (1998). Axiomatic design theory for systems. *Research in Engineering Design*, *10*, 189–209.

Welch, J., & Byrne, J. (2001). *Straight from the gut*. New York: Warner Books.

Wheelwright, S., & Clark, K. (1993). *Managing new product and process development*. New York: Macmillian.

W.W.W. (1999). Promodel, Parametric Technologies, http://www.ptc.com/products/proe/ind-ex.htm

W.W.W. (1999). Partners, Parametric Technologies, http://www.ptc.com/partners/esp/ind-ex.htm

STAKEHOLDER VALUE CREATION AND FIRM SUCCESS

Oliver Koll

ABSTRACT

Scanning both the academic and popular business literature of the last 40 years puzzles the alert reader. The variety of prescriptions of how to be successful (effective, performing, etc.)[1] *in business is hardly comprehensible: "Being close to the customer," Total Quality Management, corporate social responsibility, shareholder value maximization, efficient consumer response, management reward systems or employee involvement programs are but a few of the slogans introduced as means to increase organizational effectiveness. Management scholars have made little effort to integrate the various performance-enhancing strategies or to assess them in an orderly manner.*

This study classifies organizational strategies by the importance each strategy attaches to different constituencies in the firm's environment. A number of researchers divide an organization's environment into various constituency groups and argue that these groups constitute – as providers and recipients of resources – the basis for organizational survival and well-being. Some theoretical schools argue for the foremost importance of responsiveness to certain constituencies while stakeholder theory calls for a – situation-contingent – balance in these responsiveness levels. Given that maximum responsiveness levels to different groups may be limited by an organization's resource endowment or even counterbalanced, the need exists for a concurrent assessment of these competing claims by jointly evaluating

Evaluating Marketing Actions and Outcomes
Advances in Business Marketing and Purchasing, Volume 12, 141–262

ISSN: 1069-0964/doi:10.1016/S1069-0964(03)12004-2

the effect of the respective behaviors towards constituencies on performance. Thus, this study investigates the competing merits of implementing alternative business philosophies (e.g. balanced versus focused responsiveness to constituencies). Such a concurrent assessment provides a "critical test" of multiple, opposing theories rather than testing the merits of one theory (Carlsmith, Ellsworth & Aronson, 1976).

In the high tolerance level applied for this study (be among the top 80% of the industry) only a handful of organizations managed to sustain such a balanced strategy over the whole observation period. Continuously monitoring stakeholder demands and crafting suitable responsiveness strategies must therefore be a focus of successful business strategies. While such behavior may not be a sufficient explanation for organizational success, it certainly is a necessary one.

1. PURPOSE AND EFFECTIVENESS OF ORGANIZATIONS

1.1. Introduction

Studies dealing with the effectiveness of organizations have come to different, sometimes opposing, conclusions as to what is necessary to be effective. A substantial portion of these conflicting results arises because of the definition of organizations chosen by the respective researchers. These definitions often assume a certain organizational purpose which makes the assessment of effectiveness an obvious task but hinders the comparability of studies: By assuming different purposes for organizations, universally valid guidelines for how to be effective are difficult to discern. In addition, effectiveness itself has been conceptualized in a number of different forms which renders comparisons between research results cumbersome.

Therefore any study dealing with the effect of any explanatory variable on organizational effectiveness should clearly outline which organizational definition (theory of the firm) it employs and which approach to organizational effectiveness it takes. While theories of the firm state what organizations do and how they should do it, approaches to organizational effectiveness aim to determine how successful organizations are. Some theories of the firm – at least in their original conceptualization – prescribe the effectiveness approach to be taken (i.e. the economic theory of the firm calls for the pursuit of one goal: profit) while others do not. The "original" behavioral theory argues that organizations have no goals (thus ruling out the goal approach to organizational effectiveness), whereas some of its extensions try to integrate a variety of goals to be pursued by organizations.

Table 1. Important Features of the Three Most Prominent Firm Theories.

	Economic Theory of the Firm	Market Value Theory of the Firm	Behavioral Theory of the Firm
Origin	Adam Smith (1776)	Fisher (1930) Fama and Miller (1972)	Cyert and March (1963) Simon (1957)
Goal of organizations	Profit	Market value	No obvious goal
Approach to effectiveness	Goal-oriented	Goal-oriented	Varies
Strategy to be pursued	Maximize profit	Maximize Market value	Not predetermined
Reasoning	Deductive-normative	Deductive-normative	Inductive-realistic
Research community	Economics	Finance	Management
Preferred method	Econometrics	Econometrics	Econometrics observation
Purpose of the model	Instrumental	Decision support	Descriptive

1.2. Theories of the Firm

Theories of the firm try to establish the link between organizational goals and corresponding behavior and have comprehensively emerged in the disciplines of economics, finance and management (see Table 1 for a broad overview of the main distinctions between these theories). The first two theories of the firm are normative since they prescribe what purpose and goal organizations have and what they are to do in order to meet these goals. These goals are not inferred from actual behavior of organizations but deducted from the respective understanding of organizations. In contrast, the behavioral theory takes a descriptive perspective and denies the existence of an obvious organizational purpose.

The main reason why these three theories differ stems from the fact that they perceive the role of organizations differently and accordingly call for different strategies in fulfilling this very role. While the economic purpose of organizations is the core of both the economic and financial theory of the firm, the organization as a social entity is the focus of consideration in the model developed by management theorists. The next section takes a closer look at these theories and their respective implication on the measurement of effectiveness.

1.2.1. The Economic Theory of the Firm

This model of enterprises is the one proposed in most economic textbooks and views the firm as a single product firm operating in a competitive environment pursuing the maximization of single period profits where profit is the difference between receipts and the sum of fixed and variable costs. Therefore the owners of the firm will set output at a quantity where marginal costs equal marginal returns. While various extensions to the model have been added over time (like monopoly situations or multiple product firms), the prescribed purpose of organizations has remained the same: to maximize the profits of the current period. Although the theory itself is inconsistent with most other fields of business research and their research traditions its main goal has been adopted among others by researchers in marketing, production theory or logistics. Most marketing textbooks – although many different organizational purposes like satisfying customers, gaining market share or building customer loyalty are cited – treat the maximization of profit as the ultimate organizational purpose (Dibb, Simkin, Pride & Ferrell, 1997; Kotler, 1991). Kohli and Jaworski (1990) mention three pillars of the marketing concept, namely customer focus, coordinated marketing and profitability and Narver and Slater (1990) claim that market orientation consists of two decision criteria, long-term focus and profitability.

The criticism launched against the economic theory of the firm is threefold:

(1) The model is unable to serve as a decision-making aid.
(2) The assumption of rationality is unrealistic.
(3) The characteristics of the firm portrayed in the model contradict those of companies found in reality.

The first issue points at the failure of the theory to provide decision makers with a feasible criterion to make investment decisions (Solomon, 1969). The assumed corporate goal – single period profit maximization – does not allow for consideration of risk differences. If different projects vary in their level of risk, the economic model – by only considering the profitability of the options – will lead to suboptimal decisions (Copeland & Weston, 1992; Fama & Miller, 1972). Additionally, the theory shows a tendency to create behavioral problems when return on investment is used as a performance assessment device. Managers may be induced to cut back necessary long-term investments to increase return on investment, thereby boosting their own performance criterion but endangering the long-run health of the organization.

The rationality assumption in the theory of the firm implies that firms seek to maximize profits and that they operate with perfect knowledge. The first point has been questioned on two counts, namely whether profit is the only objective of business firms, and, if this was true, whether organizations really aim for maximization of their profits. Profit is not the only personal motive of entrepreneurs. Different others – like food, sex, and leisure – will intervene with the profit target and hence alter the firm's behavior (Katona, 1951). Even if profit were the one organizational goal, its maximization is considered an unlikely objective by organizational researchers (Gordon, 1948; Margolis, 1958; Simon, 1955). They argue that organizations set an aspiration level for profit rather than pursuing its maximization. This aspiration level may change over time, but defines a threshold for companies that divides profit rates into satisfactory or unsatisfactory ones. The second criticism concerning the rationality assumption – perfect knowledge – questions whether decision-makers know the shape of the supply and demand curves in their product market. Otherwise setting output where marginal costs equal marginal return is impossible (Cyert & March, 1963; Simon, 1955). Whether companies know the exact effect of price changes on both consumption and production level must seriously be questioned.

The final criticism points out that the firm described in the economic theory has little in common with the business firm as found in reality. The "economic" firm has no budget, no control problems, no aspiring employees and no problems of control. Real-world decision-making procedures differ substantially from the ones suggested by theory. For example, it will be hard to find companies that equate marginal costs and marginal revenues in deciding on either output or price. Rather, they will use some rule of thumb in determining a mark-up to cover all costs and earn some profit.

Adopting the organizational view outlined by the classic economic theory makes effectiveness measurement an obvious task: Given the normative nature of this theory, the definition of a business firm also includes its goals: Single-period profit is the one indicator of organizational success. Obviously, when comparing organizations of different size one would realistically relate it to another measure, like sales or assets or investment. The effectiveness of organizations without an economic purpose (not-for-profit organizations) by definition cannot be measured if this organizational model is employed. Return on investment, the most prominent measure of profitability, was developed by du Pont during the 1920s and has several attractive features (McCrory & Gerstberger, 1992). A single number is able to both reflect operating and asset performance, allows for comparisons among business units within a corporation and with other corporations, and can easily be generated from traditional accounting systems. However, its use has been criticized by a number of researchers because the ratio of net income over total assets or investments does not properly relate the stream of profits to the investment that produced it (Fisher & McGowan, 1983; Solomon, 1969). The numerator is a consequence of investments made in the past, while the denominator is likely to not only impact past and current earnings, but also future earnings. In addition, since return on investment is a number emerging from the accounting system, its relation with actual firm effectiveness may be spurious depending more on the accountant's skills or national accounting standards than actual performance.

1.2.2. The Market Value Model

The market-value model – developed by financial theorists – assumes competitive capital markets and insatiable human wants. It calls for the maximization of a firm's present market value which – in the case of a publicly held corporation – is the same as maximizing the price of a firm's stock. The assumption of a perfectly competitive capital market allows the company to pursue one single goal – maximization of the stock price – irrespective of the preferences of investors for future or current income. If they prefer more current income than the firm pays in dividends they can sell part of their stock. If they prefer less current income they can use their dividends to buy additional stock of the company. Thus for each investor utility is maximized by a strategy that maximizes the company's value. Fama and Miller (1972) showed that this objective is reached if all projects providing a positive net present value (NPV) are undertaken where

$$\mathrm{NPV}_j = \sum_{i=1}^{n} \frac{\mathrm{CF}_i}{(1 + k_j)^i}$$

with CF_i equaling the net after tax cash-flow in period i, n the expected life of the project and k_j the after-tax, risk-adjusted rate of return required by the market to compensate for the risk of project j. If the firm succeeds in establishing the accurate k_j of each project then the employment of this decision rule guarantees a maximization of the stock price. The discount rate k_j is usually established by using the Capital Asset Pricing Model (CAPM) where the required return is a function of the project's covariation with the market portfolio (Black, 1972; Lintner, 1969; Sharpe, 1964). By employing stock price instead of return on investment as the focus of managerial activity the market value model overcomes the flaws inherent in the latter accounting-based economic model that were outlined before. In addition, by focusing on free cash flows instead of profits the market value model takes into account timing, duration and risk differences between returns and it does not create the behavioral problems that might occur when ROI is used as a control device (Hopwood, 1976).

The main line of criticism against the market value model questions whether managers are really motivated to pursue value maximization. Even if management rewards were directly correlated to the value of companies it is unlikely that management will solely act to maximize shareholder value. Managers were shown to take into account a wide variety of other factors in formulating strategy (Donaldson & Lorsch, 1983; Doyle & Hooley, 1992; Kaplan & Norton, 2000; Pickering & Cockerill, 1984). In addition, if a company is not traded publicly, the assessment of its value is a questionable task. Obviously, by applying the market value theory one assumes efficient capital markets, i.e. one believes that the current market value of a company adequately reflects all available information. Another argument questions the appropriateness of the CAPM to capture the risk inherent in product investments. It may be relevant to assess the risk inherent in stock investments, but doubts exist whether it is able to adequately adjust for business risk (Devinney, Stewart & Shocker, 1985). In addition, CAPM has been shown to have serious theoretical and practical weaknesses even in the context of stock market investments (Roll, 1977; Ross, 1976).

Adopting the organizational view outlined by the market-value model of the firm makes effectiveness measurement an obvious task: Market value is the one indicator of organizational success. The effectiveness of organizations that are not publicly traded cannot be measured if this organizational model is employed. Whether market value is a good indicator of effectiveness or whether the market is able to value firms correctly can be questioned, but these doubts are irrelevant if the market value theory of the firm is adopted.

1.2.3. Behavioral Model

Both the economic and the market value theory are in essence prescriptive. Their description of organizational decision-making provoked the criticism of researchers pointing towards the respective lack of convergence with business reality. The major proponents of this development questioned the assumption of single objective maximization and aimed for a more realistic description of corporate activity (Cyert & March, 1963; March & Simon, 1958; Simon, 1957): the behavioral theory of the firm. This model was the starting point for a number of different theories that stress behavioral factors in explaining the policies and decision-making of firms (Anderson, 1982; Mintzberg, 1979; Pfeffer & Salancik, 1978). In explaining the functioning of perfectly competitive markets the economic theory does a reasonable job. However, hardly any markets are perfectly competitive and the theory is not successful in explaining the decision making of organizations in imperfect markets where firms operate under uncertainty.

Instead of portraying the firm as a profit- or value-maximizing entity the behavioral theory views the firm as a coalition of individuals. Each coalition member wants the organization to pursue certain goals some (or all) of which are likely to differ. Organizational goals then evolve through a process of conflict resolution whereby individual goals are seen as "a series of independent aspiration-level constraints imposed on the organization by the members of the organizational coalition" (Cyert & March, 1963). Behavioral model proponents emphasize that decisions have to satisfy a number of constraints and singling out one of them – like profit or market value – is essentially arbitrary. Simon (1964) believes that often the constraints imposed on the organization will have more influence on a decision than some arbitrary goal may have and that the entire set of constraints should be treated as the organizational goal. Instead of prescribing certain goals the behavioral theory of the firm sees goals emerging through a bargaining process among potential coalition members. Therefore it is unlikely that a single goal prevails over an extended period of time since coalition members join and leave the organization and bargaining positions change. Organizations therefore will not pursue some given goal as claimed in the economic or market value theory of the firm since goals change over time. Additionally, the idea of maximization lacks realism. Since information is imperfect aspiration levels depending on the organization's past goals, its past performance and the performance of comparable organizations substitute maximization levels. Firm behavior concerning the various goals therefore becomes satisficing rather than maximizing (Simon, 1964).

Since this theory of the firm does not take a normative, but descriptive view of organizations, it has never been subject to criticism to an extent as the two other models. However, many ideas implicit in this model (like the existence of multiple constituencies that have claims on the organization or the possibly desirable presence of organizational slack) are still subject of major discussion (Friedman, 1971; Stewart, 1991).

1.2.3.1. Resource dependence theory. The key to organizational survival is the ability to acquire and maintain resources (Pfeffer & Salancik, 1978). If organizations were in complete control of these resources the problem would be a simple one. However, no organization has control over all the resources required for survival and therefore has to establish relationships with various elements in its environment to acquire these resources. The fact that an organization depends on the environment is a necessary but not a sufficient condition that survival is a legitimate organizational problem. If this environment were fully dependable – i.e. if resources were available at any time at foreseeable conditions – the problem would not exist. However, the organizational environment is not dependable. It changes, new entities continuously enter and exit the environment, hence making certain resources scarce and their supply difficult. Such a case requires the organization to adjust its activities to guarantee a continued supply of resources. To secure the ability to obtain resources and support from its external coalitions the organization has to offer various inducements (Barnard, 1938; March & Simon, 1958; Simon, 1964). The level of inducements offered depends on the value of the contribution of the specific exchange partner. Since sub coalitions within the organization specialize in providing inducements for certain external coalitions, those internal coalitions serving the most valued external ones by providing "behaviors, resources and capabilities that are most needed or desired by . . . (these external) . . . organizational participants come to have more influence and control over the organization" (Pfeffer & Salancik, 1978, p. 27). The fact that the demands of external coalitions change and may even be in conflict makes continuous adjustment of corporate activity necessary. Organizational behavior could therefore be described as a response to the constraints imposed by competing and changing demands of constituencies.

1.2.3.2. Constituency-based theory of the firm. Anderson's (1982) theory adopts the coalition model of organizations present in the behaviorally oriented theories. Organizations – in an attempt to better serve the demands of the various coalitions – have developed specialized internal units (functional areas) to negotiate with each one of them: Marketing to negotiate with customers, finance to negotiate with creditors, personnel to negotiate with employees or purchasing with suppliers, etc. Extending the resource dependence framework where external coalitions that control vital resources will have greater power over the organization, Anderson (1982) asserts that the functional areas negotiating these vital resources will have greater power as well. The rise and fall in the importance of certain functions within the organization can therefore be attributed to changes in the environment that made the resources these functions negotiated for more or less important for the organization, respectively. Hall (1984) described the policies of a Canadian newspaper over a period of 20 years and identified the critical problems faced by the company during distinct phases. In the beginning the company had to cope with lack of supplies of both paper and printing capacity. Hall (1984) assumes that the internal coalition responsible to negotiate for these resources would rise in status and power and hence be able to divert resources into its attainment. Not only did this prediction prove true, but also the coalition in question was interested to maintain this status. It extended the shortage of these supplies by allowing the number of pages to grow substantially. Through the acquisition of paper mills and timberland the production coalition rose to high status within the company. However, the extensive spending towards production lead to a reduction in promotion expenses and after some time to a lack of readers, also caused by strong promotion activities of competitors. Suddenly another external coalition – readers – became the most critical one for the company's survival and the internal coalition best fitted to negotiate with this group – circulation promotion – rose in importance. By reducing production costs, increasing advertising expenditures and the acquisition of a company that handled subscription sales and circulation promotion this department also tried to sustain its dominant position.

1.2.4. Discussion and Evaluation

Different theories of the firm explain some of the opposing prescriptions for organizational effectiveness found in the literature. The economic model and the market value model include these prescriptions in their respective definition: Management should either follow a strategy that maximizes the profits of this period, or maximizes the market value of the company. While these objectives are probably the most frequently used indicators of effectiveness (which is legitimate even when assuming the behavioral paradigm), they do not provide management with decision-making guidelines. Both profit and market value are the result of certain strategic actions. But they are not very helpful in making a choice between strategic actions. Management is aware neither of exact demand curves (which would be needed to choose the most profitable action) nor of exact future cash flows and the correct risk-adjusted rate of return of an investment (which would be needed to choose the action offering the highest net present value). Therefore these models may provide a rationale why organizations exist (i.e. to earn a profit or to generate shareholder value) and the respective purposes may still serve as indicators of success in organizational performance studies, but the adoption of these models does not make sense when assessing the impact of certain organizational behavior.

The behavioral theory as well as its extensions view the organization as an open system consisting of a number of constituencies where the boundary between the organization and its constituencies is fluid. The proponents of the theory claim that organizational behavior is constrained by the demands brought forward by these constituencies. Goals are not one-dimensional (the one "organizational goal"), but multidimensional (the many goals of the organization's constituencies) with possibly conflicting dimensions. Hence effectiveness assessment becomes a delicate task. One could argue that by taking such a view organizations themselves have no goal and therefore simply cannot be effective. Thus any attempt to measure effectiveness is inconsistent with the behavioral paradigm. While such opinions exist, different approaches to effectiveness measurement seem feasible (the next chapter contains a detailed discussion of approaches to effectiveness measurement). One possibility would be to define more than one goal and measure effectiveness as the simultaneous attainment of all of these goals. While these goals are not the ones of the abstract entity "organization," they are the ones of the people making up this organization, namely all constituencies. Obviously, the importance of these individual goals is likely to differ and the goals and their respective importance will change over time as the demands of constituencies and their impact on the organization change. In case of conflicting goal dimensions maximization of performance on any one of these dimensions would cease to be a sign of effectiveness. Organizations should then strive for the attainment of a certain level in every goal dimension. Another possibility would be to adopt the goal of the most important constituency as the one organizational goal (Pfeffer & Salancik, 1978) although maximization of this goal might well conflict with the expectations of other stakeholders. The basic question driving the development of the resource-dependence theory of the firm – why do organizations survive? – would call for survival as the ultimate indicator of effectiveness.

Given the research question of this work – responsiveness to which constituency (constituencies) matters most for performance – this study

- has to assume an open systems perspective of organization;
- has to consider the demands of various constituencies in an organization;
- has to question the assumption that organizations exist only to benefit their shareholders (as the market value model does);
- has to question the assumption that a single period perspective is appropriate (as the economic model does);

and must therefore take a behavioral view of organizations. Given the difficulties this paradigm poses for effectiveness measurement, the next chapter will deal with the construct effectiveness. The discussion to follow should highlight some of the problems in conceptualizing the term, explain some widely differing research results in spite of similar research traditions and provide a basis for the choice of an effectiveness approach most suitable for this study.

1.3. Different Views on the Effectiveness of Organizations

1.3.1. Introduction

Although organizational effectiveness (OE) is the ultimate dependent variable in organizational research, so much ambiguity and conceptual disarray may surround no other term in the discipline. Hundreds of books and articles written on the subject fail to produce a consistent definition of the term or synonyms used like performance, success, ability or improvement. More than 30 years ago Katz and Kahn (1966, p. 149) wrote: "There is no lack of material on criteria for organizational success . . . Most of what has been written on the meaning of these criteria and on their inter-relatedness, however, is judgmental and open to question. What is worse, it is filled with advice that seems sagacious but is tautological and contradictory." This statement still applies today. The writing on the construct of effectiveness has been so confusing that researchers postulated to abandon the search for a definition of the construct at all (see Goodman, 1979). Still, all theories of organizations build on notions of effective strategies, include explanations of drivers of effectiveness (Cameron & Whetten, 1983), and differentiate between effective and ineffective organizations. In empirical organizational research effectiveness is the most widely used criterion variable. If the aim is to show that some type of strategy, compensation plan or leadership style is superior to another, effectiveness becomes the central empirical issue. Effectiveness is the time test of any strategy (Schendel & Hofer, 1979). In practice, effectiveness is even less likely to disappear. Whenever judgments are made about organizational success, stock purchases or employment opportunities, the respective individual will use some definition of effectiveness to compare various entities. One of the reasons for the ambiguity of the term "effectiveness," lies in the fact that scholars have conceptualized organizations in a number of different ways, for example as coalitions of powerful constituencies (Pfeffer & Salancik, 1978), open systems (Thompson, 1967), collegiums (Millett, 1962), meaning-producing systems (Pondy & Mitroff, 1979), rational goal-pursuing entities (Perrow, 1970), machines (Taylor, 1911), coalitions of multiple, conflicting interests (Cyert & March, 1963), social contracts (Keeley, 1978), psychic prisons (Morgan, 1980) and information-processing units (Galbraith, 1973). Each of these views differs with respect to the core phenomena studied, the interrelationships among these phenomena and hence judges effectiveness differently.

If the construct space of OE is ambiguous, its measurement will also be unclear. Steers (1975) mention some of these shortcomings. Certain criteria may be a sign of effective management in certain time periods, but they may reflect an organization at risk in other periods. For example, investment may be an indicator of effectiveness during good economic conditions while it may be a liability in bad times. Another problem is the time perspective one chooses in evaluating effectiveness. While most researchers realize that different criteria need to be employed in the short, medium and long run assessments of organizational effectiveness, they rarely, if ever, consider that these criteria may be in conflict. By maximizing production efficiency – a criterion with short-term effects – at the expense of research and development – a criterion with long-term effects – an organization may threaten its survival. Researchers should be explicit about the time horizon of their models and the resulting limitations. The same is true if some of the applied measures are in conflict. If an improvement on one measure negatively affects another measure organizations cannot show stellar performance on all accounts. Finally, relevant criteria may differ for organizations of different size, for private, non-profit or public organizations. Katz, Werner and Brouthers (1999) show that performance has different meanings in different cultures. Therefore cross-national studies superimposing certain performance criteria may distort results if organizations in one culture may actively pursue these criteria while those in another culture may care little about them. Today most studies employing some type of effectiveness measure define it – unconsciously or not – using one of four frameworks: the goal approach, the systems resource approach, the internal approach or the strategic constituency approach (Dess & Robinson, 1984; Ford & Schellenberg, 1982).

1.3.2. The Goal Approach to Organizational Effectiveness

A majority of studies in the field of organizations assumes that organizations have ultimate and identifiable goals (Etzioni, 1964). In fact, many definitions of organizations rest on this very characteristic of orientation towards a specific goal. While some researchers derive goals by thoroughly studying the organization in question – be it through an examination of the organizational charter or through interviewing organization members – others derive goals coherent with their own organization theories. They hence may arrive at goals independent of the intentions or even awareness of organizational members. The latter procedure was termed functional approach (Yuchtman & Seashore, 1967) and is essentially normative – in contrast to the descriptive approach taken by the former procedure. In the functional approach the investigator reports what the goals of an organization should be as dictated by the logical consistency of his theory. While avoiding potential subjectivity of organizational members the researcher him(her)self superimposes the goals the organization should be striving for. It seems questionable to judge a company's performance through criteria set by others – maybe with no relevance to its own objectives. The prescribed goal approach assumes that the information given by organizational members or written down in company missions and statements is the most informative set of clues when it comes to deriving goals. Starbuck (1965) warns of the hazard to infer organizational goals from the behavior of its members: "To distinguish goal from effect is all but impossible. Environmental effects pollute the relation between goals and results, and people learn to pursue realistic goals. If growth is difficult, the organization will tend to pursue goals that are not growth oriented; if growth is easy, the organization will learn to pursue goals which are growth oriented. What one observes are the learned goals. Do these goals produce growth, or does growth produce these goals (p. 465)?"

Even if certain organizational goals are assumed to truly drive organizational action it is not clear whether an organization that attains these goals can be considered effective or whether an organization not attaining these goals can automatically be judged ineffective. This assessment may strongly depend on the point of view of the assessor. The reliance on internal criteria becomes even more complicated when considering the wide variety of individuals and groups within an organization that might hold different claims on the organization (Cyert & March, 1959). It is questionable to ask one layer in the organization – in most studies the management – in order to understand the goals of the whole organization: "The confusion arises because ultimately it makes only slightly more sense to say that the goal of the business organization is to maximize profit than it does to say that its goal is to maximize the salary of Sam Smith, Assistant to the Janitor (Cyert & March, 1963, p. 34)." Rhenman (1968) summarizes this argument by stating that "the firm itself simply does not have goals."

Typically organizational researchers superimpose single performance criteria that are either reported by or collected from top management are employed without taking into account potentially conflicting indicators or lagged effects of the chosen independent variables on performance. Usually scholars defend this decision by pointing towards the difficulty in operationalizing other conceptualizations. The shortcomings of their own perspective are accepted because no more widely accepted operationalization has yet been established. Most studies do not even mention possible problems in the identification of certain goals as Ambler and Kokkinaki (1997) do: "We therefore assume that firms pursue goals, whether consciously or not and whether set by themselves or others, and that the extent of "success" is defined by the proximity of achievement of those goals, whatever they may be (p. 666)."

Another argument in favor of the goal approach is the widespread use of goals in our daily lives. Everybody is clear about the meaning of goals, goals provide obvious criteria to aim for and the achievement of goals is relatively easy to measure. None of these points is true for the other approaches to effectiveness. It seems that both scholars and managers accept – knowingly or not – the deficiencies of the criteria chosen as long as they have criteria that are simple to understand and communicate.

1.3.3. The Systems Resource Approach to Organizational Effectiveness

According to Katz and Kahn (1966) most of the interrelations between an organization and its environment take place in the form of an exchange of scarce and valued resources. These resources have value insofar as they are a means for carrying out organizational activity rather than from their attachment to certain goals. These resources are also the major focus of competition between organizations. Competition is a continuous process, which leads to hierarchical differentiation among organizations. This hierarchy is an excellent yardstick against which to assess effectiveness: Organizations that achieve a better "bargaining position" in competition for scarce and valued resources are more effective than those with a worse bargaining position (Yuchtman & Seashore, 1967). The organization itself does not have goals but by enhancing its bargaining position it may be more capable to attain the different and often contradictory goals of its members.

While Katz and Kahn (1966) equate organizational effectiveness with the maximization of return to the organization in any form, the systems resource approach avoids the concept of maximization. It is even considered destructive from the viewpoint of the organization. Effectiveness is defined as the maximum ability to exploit the organization's environment not the actual use of this ability. By fully exploiting its environment an organization risks its own survival since an exploited environment may be unable to produce further resources. It also may induce strong opposition in pursuit of limiting this organization's bargaining ability. A recent example is the troubles Microsoft experienced after forcing its customers to install its Internet Browser together with its operating system. Both competition and jurisdiction accused Microsoft of over-stretching its bargaining position and steps were taken to limit the company's position both through a ban of the outlined practice as well as fines. The optimum bargaining position thus is that point beyond which an organization endangers itself, either through devaluing the resource, depleting the environment or inducing opposing forces within the environment.

1.3.4. The Internal Approach to Organizational Effectiveness

Some researchers also mention an internal approach to organizational effectiveness (Cameron, 1980) that defines effectiveness as the absence of internal strain. Effective organizations are those that function smoothly, where information flows freely and where trust and benevolence are omnipresent (Katz & Kahn, 1966; Likert, 1967). This approach will not be considered hereafter since its theoretical reasoning did not catch much attention and is irrelevant to the research question discussed here. The internal approach could also be integrated into the goal approach with the overriding goal of minimizing internal strain. To the author's knowledge no studies have yet claimed to measure organizational success employing the internal approach to organizational effectiveness. The approach was applied to assess the performance of organizational sub-units, especially personnel (McFarlin & Sweeney, 1992), but not for performance assessment on an organization-wide level. However, concepts like job rotation or employee involvement programs suggest that the central idea of this approach is seen by many as a possible driver for success rather than an indicator of success.

1.3.5. Strategic Constituency Approach to Organizational Effectiveness

This approach views the organization as existing to benefit numerous constituencies – both inside and outside of the organization – and assesses effectiveness by the organization's ability to fulfill these constituencies' needs. Adopting an open systems perspective the organization induces coalition members to stay in the coalition by meeting their expectations (Friedlander & Pickle, 1968; Thompson, 1967). While it may be unnecessary to strive for maximization in the fulfillment of these expectations, fulfillment must at least exceed comparable alternative (Keeley, 1978; Miles, 1980; Thibaut & Kelley, 1978). Because each constituent has its own aspirations, performance is defined as the constituent's evaluation of the organizational activities. Managers therefore must focus on achieving a balance among the often-conflicting demands of different constituents. Excellence is achieved if all constituencies are satisfied. Pushing satisfaction of any one group to extremes may harm the overall effectiveness since these actions may run counter to the satisfaction of another constituency.

This approach not only considers the existence of multiple and potentially conflicting performance criteria (Hage, 1980), but emphasizes that some of them may be more salient than others. Pfeffer and Salancik (1978) indicate that the salience is a function of the constituent's control over the resources on which the organization depends and the degree of this dependency. The constituent approach also recognizes that performance criteria are dynamic since expectations change with experience (March & Simon, 1958). While the constituent approach disagrees with the goal approach in its claim that organizations have ultimate identifiable goals, it explicitly considers the goals of constituents that have to be attended to. The goal approach suggests comparative evaluations of multiple organizations from a single perspective while the constituency approach suggests evaluations of a single company from multiple perspectives.

Not surprisingly, the difficulties in estimating preferences of all stakeholders limited the number of empirical studies applying this approach to organizational effectiveness. Researchers investigated whether companies really take into account constituencies in their decision-making, how these constituencies assess the effectiveness of organizations and how these assessments are reached (Hitt & Middlemist, 1979; Tsui, 1990) or focused on the effectiveness of organizational subunits and not-for-profit institutions (Cameron, 1984; Jobson & Schneck, 1982). To my knowledge no study has yet used the approach to arrive at an overall assessment of an organization's performance which – through an aggregation of effectiveness assessments by all constituencies – seems theoretically feasible. Two difficulties arise which complicate such an aggregation (Zammuto, 1984): First, researchers need to rate preferences of individual constituencies, second, constituency preferences are constantly changing and are changed through organizational actions that are difficult to account for.

1.3.6. Discussion and Evaluation

The domain of effectiveness for this project will be discussed by answering six questions Cameron (1980) wants every evaluator of OE to consider. At this stage the respective answers will not take into account whether the information needed for such an evaluation is available, which method will be employed or which type of organization will be included in the analysis.

- *What domain of activity should be the focus of the evaluation?*

Answering this question is similar to deciding which approach to organizational effectiveness to take. In the context of evaluating balanced versus focused responsiveness to multiple organizational constituencies the internal approach to effectiveness is the least relevant. Whether processes in the organization function more smoothly by attending particularly to certain stakeholders or balancing attention to all of them is of comparably little relevance to management scholars and practitioners. Adopting the strategic constituency approach would turn the proposed research question tautological. Given that the predictor variables are intended to grasp organizational behavior towards constituencies employing the fulfillment of their expectations as the dependent variable is unreasonable. The conceptualization of effectiveness in the systems-resource approach is again closely related to the idea of stakeholder management, but it does not turn the research question into a tautology like the strategic constituency approach does. Therefore, the availability and use of resources appears to be a useful measure to study the effect of attention to various stakeholders on OE. So far, only Chakravarthy (1986) has attempted the operationalization of this approach to OE. Most goals employed as dependent variables in empirical effectiveness studies are to be classified into the goal approach to OE. Although most goals used benefit certain constituencies more than others they are widely accepted as sensible goals to pursue and are frequently employed in popular and scientific effectiveness reports (see Ambler & Kokkinaki, 1997 for an overview). Market share, growth, profitability, market value and customer satisfaction are among the most prominent examples, and in spite of the criticism available against each, are widely used by organizations as yardsticks for performance. Managers and other organizational members more easily understand them than, for example, the availability and use of resources.

- *Whose perspective, or which organization's point of view should be considered?*

The criteria selected in organizational effectiveness studies always reflect the values of the assessor. Many effectiveness studies use indicators that reflect the expectations of the – in the evaluator's view – most powerful constituency. Such an approach is not suitable for this study. Ideally, the success measure used should be independent of any constituency's view. However, all constituencies within and outside of the organization (no matter where the organizational boundary is drawn) consist of human beings and they are the only ones that are interested in assessing and actually assess (consciously or not) effectiveness. Performance measures in the context of a study on stakeholder theory should be chosen to be as independent of a single stake holder's expectations as possible. Ideally, effectiveness assessments using these measures should result in similar evaluations no matter which constituency the evaluator belongs to.

- *What level of analysis should be used?*

The appropriate level of analysis for this study is the overall organization: The focus is on the effect of attending differently to stakeholders on organizational performance. However, this level of analysis is not appropriate for all types of organizations. When the organization consists of loosely coupled units (not from a functional, but from a geographical or product-market perspective) that differ in their behavior towards certain stakeholders, this level of analysis becomes inappropriate or one should treat these units as individual organizations. Therefore the analysis should be restricted to organizations (or, if necessary sub-units in the latter definition) that act consistently with regard to the responsiveness given to certain stakeholders.

- *What time frame should be employed?*

Cameron (1980) suggests for researchers to consider this question to enhance their awareness about the trade-off between effectiveness criteria that are used in assessing short-term and long-term performance (i.e. efficiency and adaptability). This distinction may be misleading since all performance measures can only reflect some current state of the organization. Adaptability reflects the current potential to react to future changes, efficiency the current relation of resource input to resource output and market value the current market estimate of all future free cash flows discounted at a risk-adjusted rate. The only difference is whether the measure reflects past certain information or future uncertain information. What is of equal relevance in the context of time perspective is the lag one assumes between certain organizational behavior and their effect on performance. Some organizational actions will show (not only, but also) immediate results on certain performance measures (i.e. price increases on sales or layoffs on market value) while the effect of others may take some time to be detectable (like advertising expenditures on brand awareness or research and development cut-backs on new product success). Therefore all studies on organizational effectiveness should consider the likely time lag between the drivers of performance studied and their impact on the chosen performance criteria.

- *What type of data should be used?*

This question is concerned with the decision whether to use information published in official documents or whether to collect information from organizational members. The question is a very critical one because studies showed that the information stored in official documents often differs from the information provided by organizational members (e.g. Cavusgil & Zou, 1994). Arguments could be brought forward in favor of both approaches but given the research question the former approach seems preferable.

- Since long-term data availability is crucial for the research question, secondary sources are more likely to contain the necessary information. In most organizations nobody may be identified to provide the necessary data over more than a couple of years.
- Organizational performance for the same company was found to differ when either organizational members are surveyed or it is assessed using data from external sources. While these differences may not be substantial (Dess & Robinson, 1984; Venkatraman & Ramanujam, 1986) biases cannot be ruled out, especially when using a key informant approach (Huber & Power, 1985; Phillips, 1981). Relying on more than one informant per organization to reduce the key informant bias seems unrealistic for a longitudinal study.
- With regard to the predictor variables to be used, the above arguments become even more critical. It seems difficult to identify enough members of all constituency groups of all organizations studied to obtain a satisfyingly unbiased estimator of firm behavior towards all these groups. It is also unlikely that management can adequately assess an organization's behavior towards its stakeholders. They may lack the necessary knowledge both in scope and in time to provide this information. Even in a cross-sectional study an approach relying on information from management does not determine actual behavior towards stakeholders, but rather orientation towards stakeholders (Greenley & Foxall, 1998).

These arguments favor the use of data from external sources where one should also be aware of potential drawbacks. By definition the measures to be used are all "official" criteria of effectiveness. They have been published and stored purposefully by the organization in question. Thus, the data is usually narrow in scope. Only behavior and results of behavior that can be quantified is present in such records. Quantitativeness has to be traded off against comprehensiveness. While the use of such data safeguards the researcher against criticism concerning honesty, accuracy and knowledge of the informant, it limits the broadness of usable criteria. Another criticism may focus on potential biases in the data arguing that the accounting system of companies may render data subject to the creativity of the organization's accountant.

- *What referent should be employed?*

Cameron (1980) mentions five possibilities how effectiveness can be evaluated once the appropriate indicators have been chosen:

- *Comparative Evaluation*: Performance on certain indicators is compared with the performance of other organizations on the same indicators.
- *Normative Evaluation*: Performance on certain indicators is compared with a predefined standard or ideal performance on these indicators.

- *Goal-Centered Evaluation*: Performance on certain indicators is compared with goals set by the organization on these indicators.
- *Improvement Evaluation*: Performance on certain indicators is compared with the performance of the same organization on these indicators in earlier periods.
- *Trait Evaluation*: Performance on certain indicators is related to certain desirable organizational characteristics independent of the performance indicators.

In a study among British companies Doyle and Hooley (1992) found that 28% of companies use some absolute performance measure in assessing their performance, 44% employ a goal-centered approach, 19% an improvement evaluation and only 7% a comparative evaluation. Some of these approaches cannot be used for answering the given research question: Being descriptive/instrumental in nature this study does not introduce ideal performance standards that ought to be achieved. Therefore normative evaluation is no useful avenue. Since the study will rely on data from external sources irrespective of goals set by the organization, the goal-centered evaluation cannot be used either. Trait evaluation requires the definition of characteristics of effective organizations. The definition of these characteristics is as complex and as contingent upon the assessor's values as is the choice of performance indicators themselves and will therefore not be attempted. The quality of comparative and improvement evaluations depends a good deal on the indicators chosen. Only when these indicators are meaningful in an improvement over the last period or better performance than other organizations is it a sign of effectiveness. If possible (depending on the availability of long-term data), both of these approaches will be employed: Whether behavior towards various stakeholders explains performance differences in comparison with other organizations, and whether performance levels over time change because of a change in behavior towards certain stakeholders are both relevant questions for this study.

However, some limitations for comparative evaluation studies have to be considered: Traditional economics assumes that performance differences are only transitory (Scherer & Ross, 1990; Weiss, 1974) and treated industries or markets as the unit of analysis (Bain, 1956). As this research analyzes the effect of the attention to certain constituencies on firm performance the unit of analysis must be the individual company. Industry effects in some studies account for much of the variance in performance measures like profitability (Schmalensee, 1985) or Tobin's q (Wernerfelt & Montgomery, 1988), in other studies their effect is less substantial (Rumelt, 1991). However, industries also differ with respect to other performance measures – growth, market value changes or customer satisfaction – and comparative evaluations should therefore be restricted to firms within a single industry. Another reason why only companies within one industry can be compared is that stakeholders are likely to be of different relevance for different industries.

Suppliers may be of greater relevance for industries where resources needed as input in the production process are scarce or concentrated among a small number of suppliers as opposed to markets where supply is vast and easy to obtain. In addition, since countries differ with respect to their understanding of effectiveness (Katz, Werner & Brouthers, 1999), comparative evaluation should be restricted to a single country.

2. DRIVERS OF PERFORMANCE

2.1. Introduction

It is probably impossible to find a single number of any business journals without a piece of research claiming to uncover some performance-enhancing aspect. Scholars in economics, management, business policy, marketing, accounting, finance, human resources, international business, sociology or management science strive to identify performance drivers. The emerging drivers are even more diverse than the operationalizations of performance themselves but usually come as (a combination of) elements of organizational environment, organizational strategy and organizational characteristics. As diverse as the relationships between drivers and performance proposed, are the methods employed to assess these relationships ranging from case study research to econometrics, from analysis of variance to causal analyses. Given this wide variety of backgrounds, phenomena studied and methods employed it is no wonder that no "grand theory" of organizational performance has yet emerged. With the exception of a handfull meta-analyses no comprehensive attempts have been made to integrate at least a portion of these studies in order to detect patterns or uncover contradicting results. Usually scholars only take into account those studies that support or contradict their own findings with little concern to the wider impact on the research tradition of which they are a part.

While no "official" classification scheme for organizational effectiveness studies exists they all have either an economic or organizational background. Studies in the economic tradition assume that external market factors are important in explaining performance differences between companies whereas those in the organizational tradition assume that the behavior of organizations is the major determinant of performance. While there is not enough room to report only a fraction of findings in either tradition a short summary of the respective assumptions and most important findings is attempted.

The majority of studies in the economic tradition is concerned with business profitability. One of the main assumptions of economic models is the existence of perfectly competitive markets that make it impossible for firms to persistently earn abnormal profits (returns that differ from the industry average). Thus, superior business performance is a disequilibrium phenomenon resulting from shocks like innovation or changes in supply or demand. As resources are directed into areas that earn such abnormal profits, these returns will sooner or later be driven back to competitive levels. The major determinants of profitability in economic models are industry characteristics (e.g. concentration or growth), a firm's position relative to its competitors (e.g. market share) or the quality and quantity of a firm's resources (e.g. firm size). The respective effect of these variables on performance is ambiguous: Schmalensee (1985) shows that industry differences measured by average industry return on assets account for almost all of the explained variance in business unit performance. Rumelt (1991) disagrees with these findings and states that both industry and corporate effects are negligible and that the most important sources of economic rents are business-specific. In contrast, Brush, Bromiley and Hendrickx (1999) claim that both corporate and industry effects significantly impact business performance. Hansen and Wernerfelt (1989) take a more cautious stand by stating that the typical economic model of firm performance explains between 15 and 40% of profit variance among firms. The remaining variance can be attributed to one of three possible factors:

- the existence of economic variables that cannot or have not yet been measured;
- a "true" model of economic performance that differs from case to case and hinders aggregate analysis;
- the possibility that all-remaining variance could be due to organizational factors disregarded in the economic literature.

Organizational researchers have developed an even wider variety of models to explain performance. These studies are more difficult to integrate than the economic ones not only because of a larger number of independent variables proposed but because the dependent variable "effectiveness" is subject to discussion as well. While the economic models unanimously consider (accounting or economic) return to be the appropriate measure of performance, organizational researchers have used performance measures ranging from employee satisfaction to market value, from customer loyalty to market share. Thus, both comparing and attempting to integrate results and their respective implications provided by different pieces of research becomes a thorny and arguable task.

2.2. *A Case for a Constituency-Based Performance Assessment*

This study wants to integrate some earlier research present in the organizational performance tradition and provide a basis for future research. Investigating the consequences of organizational behavior that affects the relationship to stakeholders is a useful framework for studying organizational effectiveness, both from a theoretical and a practical view. While it may be argued that all organizational behavior to some extent must affect stakeholders (given the definition of stakeholders), the following section is intended to provide a rationale for undertaking organizational performance research from a stakeholder perspective. Providing such a rationale is not only lacking in most other performance studies (which often seem to deliberately pick some driver of performance depending on the availability of data), it should also provide the basis for further research concentrating on the relationships between a company and its constituencies and how these relationships might affect performance.

A basis for studying relationships to stakeholders as sources of superior performance can be found in a number of publications. The opinions and assertions present in the literature on stakeholder management, marketing, human resources or corporate social responsibility form a broad basis since they all emphasize the necessity to attend to, collect and interpret certain environmental information. However, they are not useful in this context since such reasoning would be tautological: One cannot defend an approach by producing arguments calling for this approach in a normative way. These positions will be weighed against each other later when it comes to comparing approaches that differ with respect to the importance they attach to certain constituencies. But the meaningfulness of adopting a stakeholder perspective to study performance can also be derived from other theoretical backgrounds: The integration of the industry analysis framework and the resource view of the firm (Amit & Shoemaker, 1993), the concept of environmental orientation of managers (Lines & Gronhaug, 1993) or the organization-constituency interaction framework proposed by Rindova and Fombrun (1999) all support such a perspective to organizational effectiveness.

2.2.1. The Integration of Industry Analysis and Resource View of the Firm

The industry analysis framework (Porter, 1980; Schmalensee, 1985) views the sources of performance to be the characteristics of an industry and the firm's position within that industry. Since such a perspective might incorrectly overemphasize industry effects in explaining performance differences, Amit and Shoemaker (1993) attempt to integrate the industry analysis framework with the resource view of the firm (Barney, 1986a, 1991; Wernerfelt, 1984), the conceptualization of the firm as a collection of resources and capabilities. Amit and Shoemaker (1993) term the set of resources and capabilities that guarantees a firm's competitiveness its strategic assets. Whether the set of capabilities and resources a company has currently available are strategic assets depends on the strategic industry factors, those resources and capabilities that, at a given point in time, are the prime determinants of performance in an industry. Therefore, in order to generate a competitive advantage, a company has to identify current and future strategic industry factors and develop the respective strategic assets, that is, the set of resources and capabilities consistent with the strategic industry factors.

Not all players in the industry are aware of the current composition and future development of these strategic industry factors, otherwise these factors would lose relevance. Companies better in anticipating and reacting to these factors should, by definition, perform better than the rest. Interactions between key players in an industry – competitors, customers, suppliers, regulators and other stakeholders – determine strategic industry factors (Amit & Shoemaker, 1993). Therefore companies that are able to first understand the development and meaning of these interactions should be in a position to most effectively act on this knowledge. Thus, the ability to cultivate interactions with key constituencies must be the most important strategic asset at any time – without it a company is unable to establish a competitive advantage.

2.2.2. Environmental Orientation of Organizations

Lines and Gronhaug (1993) derive the concept of environmental orientation along the following line of argumentation:

- Organizations are open systems (Katz & Kahn, 1966).
- To be effective, they must create outcomes for the parts of the environment they depend on (Pfeffer & Salancik, 1978).
- Management matters as suggested in the managerial literature, as believed by practitioners and as demonstrated in research (Thomas, 1988).
- Organizational failure/success depends on choice, execution and monitoring of activities.
- Managers can only react to what they detect and interpret. While it has been shown that they are able to process vast amounts of data (Lord & Maher, 1990), it is also true that their attention is selective and biased (Dearborn & Simon, 1958).

Given the above assumptions, how managers attend to their environments is of crucial importance. Lines and Gronhaug (1993, p. 6) term the pattern of this attention the environmental orientation of managers: It is a "set of beliefs held by managers reflecting the relative importance of different environmental sectors (elements) for the goal achievement of their firms." Comprehensive attendance to the entire environment is not feasible: Managers were found to disregard important information because of departmental bias or wrong assessment of the relevancy of information (Dearborn & Simon, 1958; Levitt, 1960). Porac and Thomas (1990) even claim that it is impossible to completely monitor competitors, one single environmental sub-segment. Only a fraction of the vast amount of environmental data is noticed and even less reacted upon (Kiesler & Sproull, 1982), also in areas highly relevant for businesses which has caused a number of severe organizational crises (Starbuck & Hedberg, 1977; Zajac & Bazerman, 1991). Thus, whatever form the environmental orientation of managers takes, the orientation adopted results in incomplete and possibly unbalanced monitoring of environmental sub segments. The fact that different researchers as well as practitioners call for focused attention to specific environmental sub-segments is a sign that they hold different views about the relative importance of these segments in achieving their goals. Managers may attend most intensively to those parts in their environment they consider most relevant for improving OE.

2.2.3. Firm-Constituent Interactions as Sources of Performance

Rindova and Fombrun (1999) argue that four sources of a competitive advantage exist all of which are linked by processes connecting an organization and its environment. One source is an organization's differential market power. Such power allows the organization to control prices and earn monopoly rents (Bain, 1956; Mason, 1957; Porter, 1980; Scherer & Ross, 1990). Another source is control over a bundle of unique resources (in Rindova & Fombrun's, 1999 conceptualization: material resources) that allows exploitation of above-average economic rents (Barney, 1991; Penrose, 1959; Peteraf, 1993). These two economic sources need to be complemented by interpretational sources: cognitive research emphasizes the importance of useful interpretation of economic conditions guided by knowledge, values and beliefs. Within an organization these rare and difficult to imitate intangible resources and "sensemaking" skills are the third source of competitive advantage (Fiol, 1991; Spender & Grant, 1996; Weick, 1979). Superior sensemaking skills allow superior evaluations of rent-earning potential (Barney, 1986b; Penrose, 1959). In line with this organizational micro-culture (Rindova & Fombrun, 1999), external parties also exchange information, form opinions and create preferences (Fombrun, 1996; Hill & Jones, 1992) which can lead to competitive advantage. Six processes relating an organization to its environment nurture these sources of competitive advantage (Rindova & Fombrun, 1999).

(1) Investments build a competitive advantage by satisfying and creating needs of constituents. Organizations invest differently depending on their perceived opportunities for generating rents. The focus of these investments will affect an organization's performance by influencing its market position, resource endowment as well as internal and external interpretations.

(2) Strategic projections are explicit communication about firm characteristics through advertising, logo development, press releases or financial reports (Salancik & Meindl, 1984) to impress desirable symbols in constituents' minds. They not only serve to influence interpretations about investments, but contribute to the formation of firm-related schemata, such as reputation (Fombrun, 1996; Rindova, 1997).

(3) A strategic plot provides the long-term context allowing constituents to attribute meaning to investments and projections. Both the current resource endowment and the existing micro-culture determine an organization's strategic plot. If a plot is missing, resources and micro-culture are inconsistent: Investments not supported by projections may not realize their value-creating potential, projections not supported by investments may cause loss of credibility.

(4) Constituents will allocate their resources to those firms where they generate the highest value. Their assessment of alternatives will depend not only on their own objectives, but also on strategic investments and projections made by firms. Cognitive limitations render such assessments difficult (Schwenk, 1984) and often constituents will rely on macro-cultures to facilitate sense-making. For example, reputation schemata help investors, employees and customers alike to choose between competing alternatives (Wartick, 1992).

(5) Constituents assess firms not only through allocating resources but regularly make explicit statements about organizational success. Employees discuss fulfillment of their expectations in private as do investors, communities and customers. Institutionalized discussions like stock reports, consumer magazines or rankings (e.g. *Fortune* for business performance, *Business Week* for business schools) will bear on an organization's micro-culture and resources by defining what performance means in an industry (Dutton & Dukerich, 1991).

(6) Constituents over time develop interpretative frameworks to understand the meaning of organizational actions (Weick, 1995). By interpreting strategic projections key constituents (lead users, bank analysts, environmental groups) will lead the development of a shared understanding – an industry paradigm – by constituents (Rindova & Fombrun, 1999).

By taking into account both material and interpretational processes reinforcing each other (Porac, Thomas & Baden-Fuller, 1989) and conceptualizing competitive advantage as a systemic outcome, Rindova and Fombrun (1999) provide a strong argument that performance is the result of processes between an organization and its constituencies. How an organization manages its respective relationships becomes a critical factor in achieving performance.

The concept of environmental orientation, the idea of strategic assets (Amit & Shoemaker, 1993) as well as the Rindova and Fombrun (1999) model of organizational effectiveness provide a strong theoretical basis for a constituency-based framework to performance assessment. All emphasize the direct link between organizational performance and the relationships with constituencies providing the information necessary for organizational interpretation, action and, finally, competitiveness. Organizations that are able to maintain relationships to those environmental sub-segments (constituencies) providing the most relevant information should therefore be more effective.

In addition, such a perspective is also of major relevance to managerial practice. Given limited resources, attention and cognitive capacity, managers daily face the decision of how to respond to the demands of constituencies. Suggestions from colleagues, consultants or academia come in such a diversity and frequency that even full attention to these suggestions becomes difficult: One should enrich employees, be market oriented, maximize shareholder value, be customer-focused, etc. It is difficult to think of any managerial decision that does not impact on the relationship with some constituency and will not bear on the effectiveness of the organization.

All of the following approaches claiming that constituencies differ in their importance for achieving organizational success fit the above model (Rindova & Fombrun, 1999) but hold different assumptions where strategic investments from organizations pay off most. The term "responsiveness" used in the headlines of these approaches may encompass a certain organizational culture, a value system or certain types of behavior all of which are hypothesized to bear on the relationship to constituencies and finally to organizational performance. For each of these views, the author will provide a summary of the arguments brought forward by the proponents, report about empirical studies trying to establish the hypothesized link between responsiveness to some constituency and business performance, summarize the measures used to operationalize the relationship with the respective constituency and discuss the significance of this view for the stated research question.

An extensive debate will be limited to the most prominent approaches, namely customer responsiveness, shareholder responsiveness, employee responsiveness and stakeholder responsiveness. In addition, the author will shortly discuss the likely impact of the relationship with other constituencies on performance.

2.3. Customer-Responsive Strategies

2.3.1. Theoretical Foundation

The line of reasoning supporting the dominant importance of customers for organizational effectiveness is easy to follow: Companies by definition produce some output in the form of goods and/or services. In order to receive resources necessary for sustaining the production of these goods and services they have to identify somebody to sell them to. Due to the market changes in the second half of the 20th century, customers had the possibility of choosing their consumptive activities between a number of competing firms. Therefore, companies had to find ways to offer something more attractive than competition in order to being able to sell their output. This task calls for a close relationship with customers to guarantee alignment between their demands and the company's offers. Since customers are necessary for a company's survival, all business activities should be planned and implemented from the customer's point of view (Borch, 1964; Drucker, 1954; Levitt, 1960; McKitterick, 1957). To members of this research community, financial performance cannot be a sensible business objective, but should be considered a reward for creating satisfied customers (Webster, 1992).

Today's conceptualization of marketing has essentially grown from this perspective and – in its original definition – incorporates three elements, namely customer orientation, an integrated marketing effort and profit direction (Kotler, 1984). One could argue that this worldview explicitly benefits two constituencies, namely customers and shareholders. Customers should benefit because organizations show an orientation towards them and shareholders should benefit because profit directs company activities (although profit may not be the best indicator for shareholder wealth). However, the significance of customers is more prominent than that of shareholders. Organizations are suggested to craft products and services that fulfill the needs of customers, whereas profit is seen as a consequence of such activities and a choice criterion in case different options to fulfill customers' needs are available (Kotler, 1991, p. 71).

The idea that success comes from attending to customers is prominent in the management literature and among management consultants. Peters and Waterman (1982), two McKinsey consultants, state in their best selling business book for practitioners *In Search of Excellence* that a customer focus distinguishes the excellent companies from the ordinary ones: "The excellent companies really are close to their customers. That's it. Other companies talk about it, the excellent companies do it" (Peters & Waterman, 1982, p. 156).

The claim that satisfied customers are a sufficient precondition for organizational success has been criticized extensively: Anderson (1991) suggests that marketing should shift its focus away from reactively monitoring and meeting customers' needs. Instead it should proactively control the environment by communicating with key people in an industry that control the sources and flows of resources and information. Management (Donaldson & Lorsch, 1983; Freeman, 1984; Rhenman, 1968) as well as marketing (Day & Wensley, 1983; Doyle, 1992; Webster, 1992) scholars also criticizes the narrow focus of marketing researchers: "The marketing concept relies on inappropriate neoclassical economic premises and should be grounded in a more relevant constituency-based theory of the firm" (Day & Wensley, 1983, p. 81).

However, the narrow view of marketing towards customers outlined above has been broadened by a number of authors. The most prominent stream of research evolved around the idea of a "market orientation" (Dreher, 1996; Kohli & Jaworski, 1990; McCarthy & Perreault, 1984; Narver & Slater, 1990; Payne, 1988; Ruekert, 1992; Shapiro, 1988). Instead of treating marketing as a philosophy or a policy statement like the marketing concept does (Barksdale & Darden, 1971; McNamara, 1972), the "market orientation" of organizations was introduced as an indicator for the quality of marketing implementation (Kohli & Jaworski, 1990; Narver & Slater, 1990; Ruekert, 1992) or the extent to which orientations are aware about the importance of relationships for business survival and the possession of relevant knowledge about successfully maintaining these relationships (Dreher, 1996). The term is originally defined as the "organization-wide generation of market intelligence pertaining to current and future customer needs dissemination of the intelligence across departments and organization-wide responsiveness to it" (Kohli & Jaworski, 1990, p. 6). Narver and Slater (1990) operationalize this definition to consist of three behavioral components – customer orientation, competitor orientation, and interfunctional coordination – and two decision criteria – long-term focus and profitability. Another less focused definition is offered by Deshpande, Farley and Webster (1993, p. 27) who perceive customer orientation as a "set of beliefs that puts the customer's interest first while not excluding those of all other stakeholders such as owners, managers and employees in order to develop a long-term profitable enterprise." A positive relationship between market orientation and a number of other organizational variables is proposed, among them esprit de corps, job satisfaction, organizational commitment of employees, customer satisfaction, repeat business of customers and business performance (Kohli & Jaworski, 1990).

It is obvious that most marketing scholars today do not perceive outstanding customer responsiveness as a necessary and(!) sufficient precondition for organizational effectiveness. They emphasize that by ignoring other constituencies because of a sole focus on customers organizations may perform badly since the environment shaping an organization's performance does not consist of customers alone. However, one can still state that researchers in the field of marketing would not demand a balance between attention to constituencies, but call for an unbalanced view of the environment in favor of customers.

2.3.2. Empirical Support and Measurement

One would expect a large number of empirical studies on the effect of customer responsiveness on company performance to be reported in the marketing literature.

However, hardly any quantitative empirical research on the relationship between some facet of the marketing concept (orientation or actual behavior) and performance can be found until the operationalization of market orientation (Narver & Slater, 1990). As far as actual organizational behavior toward customers is concerned, this statement is still true today. One of the most publicized (and criticized) pieces of research on the effect of a customer focus on performance is the study by Peters and Waterman (1982) that claims to uncover a highly positive effect of an organization's focus towards customers on performance. The authors draw from their experience as management consultants, interviews with executives and their intuition to propose eight attributes that lead to company excellence: bias for action, staying close to the customer, autonomy and entrepreneurship, productivity through people, hands-on and value-driven, stick to the knitting, simple form and lean staff, simultaneous loose-tight values. Their suggestions are especially concerned with the promotion of soft management skills focusing on customers and employees in a time where many U.S. companies supposedly lost ground to their Japanese competitors because of an emphasis on "number-crunching." This love for numbers should be substituted by a love for products, customers and entrepreneurship. Peters and Waterman (1982) conclude that an obsession with customers is a prerequisite for organizational success. Carroll (1983) criticizes this study for not specifying how the authors derived their eight attributes of excellence. Other points of criticism focus on the chosen performance measures that only include accounting-based indices (Johnson, Natarajan & Rappaport, 1985), the usefulness of these eight attributes for any type of organization under any circumstance and the short-term nature of excellence of most companies in the sample (Hitt & Ireland, 1987; Krueger, 1989). Hitt and Ireland (1987) compare the performance of the Peters and Waterman (1982) sample with a sample representative of the *Fortune 1000* and find the latter to perform better on most performance measures applied. They conclude that ". . . the excellent firms identified by Peters and Waterman may not have been excellent performers, and they may not have applied the excellence principles to any greater extent than the general population" (Hitt & Ireland, 1987, p. 95).

Narver and Slater (1990) develop measures for an organization's market orientation and assess its effect on business performance. The measure consists of an item battery for each of the three behavioral components of market orientation – customer orientation, competitor orientation and interfunctional coordination – the answers to which are summed up to arrive at a score for an organization's market orientation. Narver and Slater (1990) uncover a positive relationship between market orientation and profitability. The suggested moderating role of competitive environment on this relationship (Day & Wensley, 1988; Kohli & Jaworski, 1990) was not strongly supported (Jaworski & Kohli, 1993; Slater

& Narver, 1994), suggesting that market-orientation is always a cost-effective strategic posture. Ruekert (1992) uses a different measure to assess a company's market orientation (not surprisingly given his differing definition), but identifies a positive effect on long-run financial performance as well. He also finds that the implementation of a market-oriented strategy is the best variable to discriminate between high and low-performing organizations.

Summing up, most empirical research studying the relationship between customer responsiveness and performance identifies a positive effect, but frequently shows that the strength of the effect is contingent on other macro-and micro-environmental factors. However, most studies rely on information obtained from management – thus they study the relationship between customer orientation of top management and performance. It is surprising to notice that the object of this research field and the method by which it is studied appear not to be highly compatible: While the research field claims that the customer should take the most prominent position among organizational constituencies, research focuses on the organization (or rather, organizational entities) as the most appropriate unit of inquiry. One could argue: to learn about the extent and quality of those organizational activities that are targeted towards customers, these customers should be the ones to provide the information.

2.4. Shareholder-Responsive Strategies

2.4.1. Theoretical Foundation

Strategies placing the constituency of stockholders in the center of organizational attention differ from the ones emphasizing other constituencies. While the latter ones claim that foremost responsiveness to their target group will maximize some – often-unspecified – measure of organizational effectiveness, most shareholder-centered strategies call for maximization of shareholder value as the means and end of organizational activity. Obviously, this claim renders a comparison with the other outlined strategies a difficult task: If one argues for shareholder value as the ultimate criterion of organizational performance and proposes the implementation of strategies with the highest NPVs as the only road to success, this line of reasoning undoubtedly makes sense. But it is also tautological. Therefore the discussion around the idea of shareholder value maximization must be presented in a more differentiated way: Two arguments are prevailing, only one of which is compatible with the purpose of this study.

First are the arguments of scholars working in the tradition of the market value theory of the firm: Market value maximization is considered the raison d'être of organizations. The advice that all activities with a positive net present value (or in the case of limited resources the ones with the highest net present values) should be carried out – since they add to an organization's market value – is difficult to contest. However, the normative nature of the claim turns this argument into a simple equation (as which it was originally formulated) that tells you how to "mechanically" maximize market value: the sum of a given number of values will – everything else equal – turn out bigger the more summands are in the equation or the bigger the individual summands are. Some scholars focus on the sources of value creation, namely these cash flows, and introduce techniques how to maximize them (Arzac, 1986; Rappaport, 1986). All of the variables in the market value equation – leverage, growth, cost of equity and other financial indicators – are analyzed for their potential to increase market value. Since this argument is not concerned with the attention a specific stakeholder group should receive in comparison to other constituencies it falls out of the domain of this study and need not be discussed further.

The second argument takes a more managerial approach and focuses on strategic actions generating free cash flows. By taking the "right" actions cash flows will be higher than by taking the "wrong" actions. In its 1990 company statement, George W. Merck, a family member of the pharmaceutical giant, stated: "We try never to forget that medicine is for the people. It is not for the profits. The profits follow, and if we have remembered that, they never fail to appear." Shareholder value is not the result of simply adding up cash flows, it is the result of organizational actions that produce these cash-flows. Thus, proponents of this view might even be found among scholars or business people arguing for disproportionately high customer or employee responsiveness but believe shareholder value to be the most appropriate indicator of organizational effectiveness. Only a subset of this argument falls into the research domain of this study. It claims that organizations will be most effective when attending more to shareholders than to other constituencies. As stated before, such an argument – when taken to extremes and applying shareholder value as the criterion of organizational effectiveness – may be tautological: (Correctly) Choosing actions that benefit shareholders will maximize shareholder value – but only if capital markets are efficient and if shareholder value is the only criterion in investment decisions. When these two assumptions are not met, this argument ceases to be tautological. Then organizations in their efforts to respond to shareholders may be forced to take actions besides analyzing cash-flows in order to increase value for their shareholders – depending on the very measure that is chosen to assess effectiveness. Even if shareholder value is the chosen measure, other types of actions in the relationship with stockholders than simply considering future cash-flows may become necessary.

Admittedly, the edge between cause and effect in this discussion is hard to determine. When a company claims that the "top priority continues to be the building of shareholder value" (Quaker Oat's Quarterly report for the period ending December 31, 1990) one may conclude that this company evaluates all its actions with respect to the likely effect on the stock price. Thus, one might argue, this company attends more to stockholders than other constituencies. The critical question in this context is whether we perceive the alignment with stock price as a filter that guides this organization in its choice for certain actions or whether we perceive it as the driver behind the design and implementation of all organizational actions. If we take the former perspective, the company may or may not hold some other type of orientation but perceives shareholder value as the yardstick for evaluating organizational actions. Only if we take the latter perspective would we be facing an organization with a behavior consistent with our definition of shareholder responsiveness. However, this view is not a common one in the business literature.

2.4.2. Empirical Support and Measurement

Studies applying the first two perspectives outlined above are nearly countless. Especially the 1990s have experienced an enormous increase of research papers promoting shareholder value as the ultimate organizational goal and introducing techniques how to calculate it (Arzac, 1986; Rappaport, 1986; Scott, 1998). The second perspective is a very common one, too. Numerous studies have investigated the effect of organizational structures or actions on shareholder value (e.g. Baliga, Moyer & Rao, 1996; Fruhan, 1979; Grundy, 1995; Lubatkin & Chatterjee, 1991; Srivastava, Shervani & Fahey, 1998; Varaiya, Kerin & Weeks, 1987). However, the perspective applicable for this study – organizations that respond more to shareholders than other constituencies will outperform other companies – has to the author's knowledge never been explicitly proposed and tested. While not explicitly taking such a view, a number of publications promoting the idea of shareholder value creation come close. For example, Arzac (1986) calls for a threshold value of ROI and margin on sales to be achieved for all projects. Each strategic discussion should be evaluated bearing this threshold in mind. Therefore responsiveness to stockholders seems to outdo responsiveness to other stakeholders. Bughin and Copeland (1997) claim that companies creating more shareholder value are also more productive and grow employment faster than other players, which is supported by a large-scale study among some 3000 companies in 20 countries. They identify a virtuous circle that by creating value releases more disposable income. This in turn increases consumption and thus spurs growth, higher employment and generates more shareholder value. Bughin and Copeland (1997) therefore demand a "shifting to a shareholder value mindset (p. 165)" which would not come at the expense, but to the benefit of other stakeholders. The concept of shareholder value – in the fourth perspective outlined above – can also be observed in the business community. After years of disappointing results Kodak in 1993 declared a shift in emphasis and promised to move shareholders from the "most underserved" constituency to the focus of organizational attention. Right after this announcement the stock price increased by some 17%. Once again, it is difficult to discern between cause and effect. When assessing organizational effectiveness by measuring shareholder value, an announcement that responsiveness to shareholders will be increased and efforts to boost shareholder value will be implemented should result in better performance.

Among others, return on equity, shareholder return, economic value added, market value added and share of new wealth creation are among the most popular measures to capture the construct of shareholder value. By far the most widely used yardstick of financial performance among managers and investors is return on equity (ROE), defined as Net Income divided by Shareholders' Equity. Its importance stems from the fact that it is a measure of the efficiency with which a company makes use of its owners' capital, i.e. how much is earned on every dollar invested by owners. Shareholder return shows the wealth increase of shareholders during the latest period in percentage terms. This measure is conceptually different from return on equity. While ROE shows accounting returns to the book value of equity, the latter shows the actual wealth increase of equity holders during the latest period from dividends and stock price movements. It is therefore a more useful measure to assess the benefits an organization provides to the constituency of shareholders. Economic Value Added applies the well-known demand that any investment should earn more than its cost of capital to performance evaluation. EVA can be calculated by subtracting the capital charge of an organization from its after-tax income. EVA therefore indicates how well a company makes use of the capital employed, i.e. how much value it creates after taking care of the cost of capital. Market value added – like stockholder return – is a measure directly related to the market's valuation of an organization. It is the market value of a company's debt and equity minus all what capital lenders and shareholders have contributed over time. In other words, it is the difference between cash in (what investors have contributed) and cash out (what they could sell their claims for today). Hamel (1997) spells doubt on the long-term ability of stock prices to reflect competitiveness; he argues that what really matters is the percentage a company can capture of the entire wealth increase in an industry.

While each of the above measures has its merits – be it the ease of measurement, the ease with which it is understood, the theoretical foundation provided, the competitive position communicated – all of them have been criticized extensively. Ferguson (1986) states an interesting, somewhat ironic message about attempts to measure investment performance:

> Nobody knows how to measure investment performance; nobody will ever know how to measure investment performance and nobody would ever want to know how to measure investment performance (p. 4).

By pointing at the inherent flaws of most market measures (like alphas and betas), the unknown and unpredictable psychology of investors, the number of sample periods required to detect significant performance differences between portfolios, and the knowledge needed to understand the characteristics of all portfolios available, he concludes that performance measurement is something providing jobs, but making no sense. Still, as discussed frequently in other parts of this study, everybody, especially investors, is keen on measuring some type of performance, whether it makes sense or not.

Bughin and Copeland (1997) claim that striving for shareholder value will benefit all stakeholders and call for a shareholder value mindset for managers. What exactly constitutes such a mindset is not elaborated on. Most other writers embracing the concept of shareholder value creation do not use it in the context of this research project. It is either conceptualized as a goal in itself, a screening mechanism to assess strategic and operational actions or as an indicator of organizational effectiveness (which can result from any form of strategic orientation, be it employee-, customer- or supplier-oriented). Hardly ever is a shareholder-focused strategy introduced to achieve some organizational purpose other than shareholder value itself. When applied in this latter form, shareholder value is implicitly considered an end in itself. Therefore it seems that these researchers often emphasize certain – positively acclaimed – relationships to other variables (like employment, income growth) without debating their causal connection (Bughin & Copeland, 1997).

2.5. Employee-Responsive Strategies

2.5.1. Theoretical Foundation

Like for most of the other constituencies, a case for an employee-responsive behavior of organizations is highly plausible. After all, employees – even in some industries dominated by a high degree of technology – are the ones in charge of creating (often) producing and selling the goods and services of a company. To be able to perform these tasks at the highest possible level the quality of employees becomes a critical success factor. To attract and retain such employees a company must aim to take care of its employees so that the benefits arising out of their efforts are maximized. In addition, a number of management practices implemented throughout the last decades, like the quality movement, efficient consumer response, supply chain management are highly dependent on the commitment of employees to support such movements. This dependence on employee commitment has spurred the development of a new model of personnel management in organization theory, the high commitment model (Walton, 1985). Such management is characterized by the use of personnel practices such as information dissemination, problem-solving groups, minimal status differences, job flexibility, and teamwork. This in turn should lead to relations within the organization characterized by mutual trust (Wood & Albanese, 1995), high involvement with the organization and its goals and consistently high performance.

A similar approach is taken by research on organizational climate that is interested in the effect of individual-organizational interactions on individual behavior (Field & Abelson, 1982; Glick, 1985; Litwin & Stringer, 1968). Organizational climate is determined by a wide variety of structures and processes including decision making practices, communication flow, group processes and job conditions (Hansen & Wernerfelt, 1989). A number of studies showed that a positive organizational climate through motivation of people (Litwin & Stringer, 1968) has a direct effect on individual (Pritchard & Karasick, 1973) and organizational performance (Barney, 1986b; Hansen & Wernerfelt, 1989; Lawler, Hall & Oldham, 1974).

The critical importance of employees for firm performance may be most obvious for service businesses (Naisbitt & Aburdene, 1985; Parasuraman, 1987). This industry is commonly cited as the one where an employee orientation and a customer orientation jointly affect performance since satisfaction of customers is closely linked to employee behavior. Many mission statements address a firm's commitment to both employees and customers – termed people orientation (Beatty, 1988). Such an orientation will provide employees with more purpose than a financial-goal orientation that does little to inspire employees (Peters & Waterman, 1982). Lack of commitment to employees, in turn, leads to low employee satisfaction, morale and commitment and negatively affects employee performance, and finally, organizational performance (Hunt, Chonko & Wood 1985; Schneider & Bowen, 1985).

2.5.2. Empirical Support and Measurement

By referring to a large number of research projects, case studies and drawing from his experience as an organizational researcher, Pfeffer (1994) argues that foremost responsiveness to employees is a sure path to high performance. He develops a guidebook for human resource management leading to higher levels of productivity, innovation, quality, customer satisfaction and profits. While most of the claims brought forward are supported by empirical evidence building on case studies and are only sparsely supported by large-scale cross-sectional or longitudinal studies the arguments brought forward are intuitively persuasive. Nemeroff (1980) studied eighteen companies that at that time were generally considered successful (e.g. McDonald's, Disney, IBM), and found three strong common characteristics leading to high service quality, all of which might be summarized as being evidence of a high level of employee responsiveness: intensive, action involvement on the part of senior management, a remarkable people orientation, a high intensity of measurement and feedback. If the winners of the Malcolm Baldridge National Quality award are considered high performers the importance of human resource practices can also be inferred. Most of the winners – for example Motorola, IBM, or Xerox – emphasize the need to effectively work with employees (*Competing*, 1990; *Managing*, 1990, 1991). Such behavior should lead to higher customer satisfaction and organizational effectiveness due to the frequently cited importance of employee behavior for customer satisfaction (Bitner, 1990; Bowen, 1996; Gronroos, 1984). CEOs themselves keep emphasizing the importance of employees for organizational effectiveness, sometimes even pronouncing their foremost importance among organizational constituencies. Richard Branson, CEO of Virgin, stated his ideas in a speech to the Institute of Directors: "We know that the customer satisfaction which generates all important . . . recommendations and fosters a repeat purchase depends on high standards of service from our people. And we know that high standards of service depend on having staff who are proud of the company. This is why the interests of our people come first . . . In the end the long-term interests of shareholders are actually damaged by giving them superficial short-term priority" (quoted from an INSEAD Case, 1993).

It is interesting to note that a meta-analysis relating job satisfaction to job performance only identified weak relationships (Iaffaldano & Muchinsky, 1985). However, when employee satisfaction was related to other performance indicators, especially customer satisfaction, these correlations turned out to be significant: A study among Sears outlets showed a highly negative relationship between employee turnover and customer satisfaction (Schneider & Bowen, 1993), and research both at Macy's and Allstate showed positive correlations between employee and customer satisfaction (Ashworth, Higgs, Schneider & Shepherd, 1995; Rock & Cayer, 1995). Positioning strategies may only be implemented effectively through a dedicated workforce (Fiol, 1991; Lado, Boyd & Wright, 1992; Lado & Wilson, 1994; Teece, Pisano & Shuen, 1997), eventually leading to higher return on assets (Lee & Miller, 1999).

One obvious avenue for assessing an organization's employee responsiveness would be to measure job satisfaction of employees. A variety of job satisfaction scales have been developed throughout the last decades (Aldag & Brief, 1978;

Koeske & Kirk, 1994; Smith, Kendall & Hulin, 1969). The dominant methodology in the job satisfaction literature is the facet approach that assumes that satisfaction derives from different facets of a job to all of which employees might have different attitudes. Typical facets would be the challenges a job provides, the physical environment in which a job is conducted or the salary received. While one could think of an almost endless list of facets, the job satisfaction literature suggests that only a limited number of critical dimensions – work, pay, promotion, supervision and coworkers – account for overall job (dis)satisfaction (Smith, Kendall & Hulin, 1969, 1985). Another possible approach would be to measure organizational climate as an indicator of an organization's employee orientation. Defined as "the perceived, subjective effects of the formal system, the informal management style and other important environmental factors on the attitudes, beliefs, values and motivations of the people who work in a particular organization (Litwin & Stringer, 1968, p. 5)," it should reflect employees' perceptions of organizational attention towards them. The most widely accepted measure of organizational climate builds on the Survey of Organizations (SOO) instrument developed by Taylor and Bowers (1972) and includes factors like decision-making practices, communication flow, job design or goal emphasis: Existence of work groups with clear standards, linked together through effective communication are indicators of high-quality human resource practices and will result in better performance. The SOO has been widely accepted as a useful measure of organizational climate (Glick, 1985; Mossholder & Bedeian, 1983) and hypothesized effects on performance have been confirmed (Hansen & Wernerfelt, 1989).

The utmost importance of employees for organizational effectiveness is a widely uncontested proposition. Largely gone are the days when people were simply considered additional means of production besides capital. Even though an emphasis on employee welfare may be a socially desired trait of management, one can perceive indicators of such attitude in a variety of organizational actions. Increased emphasis on training, selection, team building may all be taken as signals of attention to employees for organizational effectiveness. Whether these observations warrant a prime status among organizational constituencies must be questioned. The number of both academics and managers who claim employees to be the most important constituency is still small. Most consider investments into employees as a means to increase customer satisfaction and/or shareholder returns. While this is a legitimate view and fits the research question at hand – which wants to explore the effect of organizational behavior towards constituencies on (some, possibly shareholder- or customer-oriented measure of) performance – it renders employees a less important position among constituencies. Trends towards downsizing, part-time and seasonal employment also run counter demands set forth by advocates of employee responsiveness. How much credibility does a claim that skills and knowledge of experienced employees are invaluable have when these are among the first to be laid off in times of crises? Globally operating companies, which outsource unskilled manual mass production to low-wage countries, may also serve as evidence that employees are not the constituency receiving most organizational attention.

2.6. *Other Stakeholders*

A number of other stakeholder groups – by definition – will impact an organization's performance. Theoretically, an organization's stakeholders may even be countless, but only a few additional groups are commonly mentioned as playing major roles for organizational effectiveness. In contrast to the other groups outlined before they are usually not proposed as *the* most relevant constituency for organizational success, but as one of the factors in the market and macro environment of organizations impacting performance. Still, a short discussion of the likely impact the relationship with these groups might have on performance and some empirical evidence on the effect of attention towards them seems warranted.

2.6.1. The Public

"The public" is usually one of the constituencies mentioned in most publications dealing with the role of various groups in strategy development. But: What exactly is "the public?" Everybody (Everything) outside the organizational boundary – wherever it is drawn? The government? The community? The population? The environment? Does "the public" include all other constituencies? If it does, is a shareholder-responsive strategy also a public-responsive one? Is a company that attends to all stakeholders automatically a public-responsive strategy?

While it is obvious which persons make up the constituency of employees (at least from a legal point of view), who purchases from a business organization and who owns it, the public may be almost anything. It may be defined as anybody not part of another constituency or *anybody* whether part of another constituency or not. Given the research question the public will be defined – similar to the definition of stakeholders provided by Freeman (1984) – as anybody who affects or can be affected by an organization but not in his role as part of another constituency. Therefore an investor may also be part of the public when he is affected by philanthropic or environmental activities of an organization, but dividend payouts to investors are not constituting behavior in terms of a public orientation of organizations. Obviously, such activities are strongly interrelated – higher environmental standards may lower dividends while high dividends may prevent organizations from implementing such standards – and almost impossible to cleanly delineate from each other. As shown before, almost all organizational actions have some – often contrary – effects on different constituencies. Public responsiveness typically includes behavior targeted at society as a whole and affecting the environment, the social standard or the well-being of the community as opposed to the other orientations discussed which are usually assessed at an individual level. The only research stream, which made an attempt to provide a classification of such activities, is the field of corporate social responsibility.

The social responsibility of businesses has been a widely discussed issue for decades. In 1938 Barnard characterized the executive process as "involving a sense of fitness, of appropriateness, of responsibility" (p. 257). In the middle of this century a number of books on the subject were published (Bowen, 1953; Mason, 1960; McGuire, 1963) and the topic was so present that Peter Drucker (1954) wrote: "You might wonder, if you were a conscientious newspaper reader, when the managers of American business had any time for business." The real debate started in the 1960s when Milton Friedman began his widely-cited attacks on the idea of the social responsibility of businesses: "Few trends could so thoroughly undermine the very foundations of our free society as the acceptance by corporate officials of a social responsibility other than to make as much money for their stockholders as possible" (Friedman, 1962, p.133). While this idea is generally accepted as a useful normative demand by the finance school, other scholars frequently point to corporate reality: "When making a profit conflicts with respecting the welfare of the community, corporations do not always choose profits as their only goal" (Goodpaster & Matthews, 1982, p. 132). The fact that, according to the Social Investment Forum, some hundred institutional investors now allocate funds using social screens, may reflect genuine interest by investors for CSR or at least a belief in a direct relationship between the extent of social activities and shareholder returns.

Given the difficulties of conceptualizing social responsibilities of organizations the variety of methods, operationalizations, and results in empirical studies is not surprising. These studies show an interesting caveat: The idea of CSR arose out of frustration with the results of a purely economic orientation of businesses and its negative effects on society. One would therefore assume that empirical studies assessing the effect of taking on such responsibility would concentrate on some performance indicator related to the effects on society. The majority of studies, however, concentrates on the impact of CSR on financial performance. Two explanations come to the author's mind: On the one hand, financial indicators are the most common performance measures and applying them might lead to higher acceptance in business journals and among practitioners. On the other hand, being able to show that business can be socially responsive *and* financially successful renders criticism against the concept of CSR more difficult.

The relationship between an organization's social record and financial performance has been studied by a variety of authors with a variety of methods producing a variety of results. Moskowitz (1972) chose 14 firms that possessed what he believed good social responsibility credentials and compared their stock return with the Dow Jones index. The fact that his stocks had appreciated more than the index was taken as support of his hypothesis. Alexander and Buchholz (1978) as well as Aupperle (1984) found no relationship between CSR and a number of performance indicators like ROA, risk levels, or profitability. The existence of a CSR committee on board level or social forecasting methods did not impact performance either (Aupperle, Carroll & Hatfield, 1985). Greening (1995) studied the effect of investments into environmental conservation by utilities and found no impact on financial measures. Sharma and Vredenburg (1998) show that a proactive stance towards environmental issues leads to the development of unique organizational capabilities like stakeholder integration, continuous innovation or higher-order learning and positively impacts organizational performance. McGuire, Sundgren and Schneeweiss (1988) turned the common argument around: CSR may not only impact firm financial performance but only a strong financial record, and the existence of slack resources might allow a company to effectively take on CSR activities (Cyert & March, 1963; Ullmann, 1985). Their results imply that prior financial performance is generally a better predictor of CSR than is CSR for subsequent performance. Pava and Krausz (1996) provide a summary of twenty-one studies relating CSR and financial performance nearly all of which conclude ". . . that firms which are perceived as having met some social responsibility criteria, either outperform or perform as well as other firms which are not – necessarily – socially responsible" (p. 322). In detail, twelve studies find a positive, eight a negligible and only one a negative association. Although different approaches in sampling, statistical procedures or operationalizations of the variables were chosen, the results show such a uniform tendency that Pava and Krausz (1996) term their finding "the paradox of social costs." Among a variety of possible explanations given – like sampling or methodological shortcomings – they choose the following one to solve the paradox: "Sometimes, a *conscious* (emphasis added) pursuit of corporate social responsibility goals causes better financial performance" (p. 335).

2.6.2. Suppliers

Suppliers, like employees, customers or shareholders, provide an organization with resources necessary for the production process. The amount of resources needed and the dependence on them varies between industries; i.e. car manufacturers require a considerable amount of physical supplies, while a consulting business needs to be less concerned with securing access to this type of resources, but may exhibit a stronger demand for university-educated human resources. The resulting patterns of reaction to such forms of dependence also vary considerably: While car manufacturers engage in building very close relationships to their suppliers or even integrate vertically, consulting businesses do the same thing for their most valuable resource, i.e. professionals. This section considers suppliers of physical resources.

The way a business manages its relationships with such suppliers can be a source of competitive advantage (Richardson, 1993; Woodside, 1987). Supplier quality may have a significant impact on product quality and securing this quality may therefore be critical for organizations. Opinions differ as to how this quality may be secured. Following classic strategic management literature (Porter, 1985), strong competition between suppliers should increase performance. Organizations should therefore strive for multiple sourcing to reduce dependence, gain access to information about suppliers' cost structures and performance potential (Demski, Sappington & Spiller, 1987) and increase the innovation potential of the supplier group. But multiple sourcing also increases costs of quality control (Walton, 1986), costs of implementing low inventory production processes (Deming, 1986), and destroys the benefits gained through commitment in long-term relationships (Heide & John, 1990; Smitka, 1991). Close cooperation with suppliers may result in sustainable competitive advantages because the associated benefits may be causally ambiguous (Reed & deFilippi, 1990), i.e. hard to identify for competitors. Benefits like mutual trust and commitment can only develop over time suggesting usefulness of long-term cooperative partnerships (Anderson, Hakansson & Johanson, 1994; Asanuma, 1985; Mudambi & Helper, 1998). Rather high levels of dependence often characterize these relationships because of asset specificity, exclusivity rights and information sharing. A trade-off seems to exist between the benefits associated with long-term partnership and the benefits associated with low levels of dependence. The way an organization attends to the constituency of suppliers may therefore have a significant impact on performance.

In contrast to other stakeholder groups outlined, no argument has yet been made that suppliers are the group deserving foremost attention. One might think of situations where this constituency plays the most significant role (e.g. during the oil crisis in the 1970s), but business literature is usually more concerned with other stakeholder groups. To the author's knowledge no normative claim has yet been made to place suppliers on top of the stakeholder list.

2.6.3. Managers

Lists of stakeholder groups sometimes include management, sometimes they do not. Three reasons for not including management come to mind. One can consider management to be an integral part of the stakeholder group "employees." They also provide human capabilities in exchange for compensation, promotion, etc. Viewing them as an individual group apart from employees would necessitate a distinction between employees and management that is a difficult task. Often type of compensation is used as discriminating between employees and management. The latter group acting as agents for the organization's owners usually puts a part of its compensation at risk whereas employees receive fixed salaries irrespective of performance (up to a certain threshold, bankruptcy). However, given the growing tendency of having "regular" employees participate in an organization's performance, this distinction is rendered useless. Another distinction would be the influence over the organization's direction. Managers are said to shape organizational decisions, actions, and consequently, performance (Dean & Sharfman, 1996; Hambrick, 1989). Thus, management would be the part of employees who make strategic decisions whereas employees would be the ones carrying them out without consideration of their usefulness. Even if one assumed that it was true that an organization's employees can be split into these two groups, the problem would be to find the border between the groups. Following the arguments by behavioral scholars that organizational decisions are often more influenced by external than internal constituencies (March & Simon, 1958; Pfeffer & Salancik, 1978; Rindova & Fombrun, 1999), this problem is an even more tricky one.

The second reason for excluding management from the list of stakeholders may be the idea that the entity management – however delineated – is the group in charge of the decision which stakeholders to attend to. Management then simply serves as a synonym for the focal organization that allocates responsiveness to different constituencies. Research interested in organizational attention to the environment usually surveys top managers as the ones in charge of this decision (Greenley & Foxall, 1996; Sutcliffe & Huber, 1998). Attention to management would be difficult to measure if one takes the management team as the constituent responsible for the allocation of this very attention.

The fact that managers are often both employees and shareholders might be another reason for excluding them in analyses. While a number of constituents may well be part of more than one group (especially in the consumer goods industry where a number of stakeholders will also be customers), the conflict of interest inherent in the relation employee-owner might be higher than that for other constellations. Asking managers for the importance of employees and owners respectively may already pose a difficult task, including management may render a useable assessment even more unlikely.

Agreement exists that management, at least to a certain degree (Hambrick & Finkelstein, 1987) matters and will affect performance and that – vice versa – performance will affect management well-being. Therefore management is a stakeholder along Freeman's (1984) definition. Principal-agent theory and the resulting executive compensation schemes would also suggest a strong positive relationship between responsiveness to management (in terms of the number of stock options and an attractive strike price) and performance (in terms of stock price). However, given the fact that the boundary between the stakeholder groups of management and employees are hard to draw and that management is usually in charge of allocating responsiveness to different stakeholder groups – through investments, organizational structures or processes – it may make sense to exclude them from empirical stakeholder analyses. Since their primary role is one of contracting on behalf of the firm with other stakeholders (Coff, 1999; Hill & Jones, 1992), management may be conceptualized as the focal organization of interest.

2.6.4. Distribution Partners

The importance of a functioning distribution channel for successful business strategies is uncontested (Buzzell & Ortmeyer, 1995; Stern, El-Ansary & Coughlin, 1996). Many organizations would not be able to create and deliver offerings without support from retail and other distribution partners. Often the relationship between manufacturers and their vertical partners in the value-creation system is considered to be a major source of organizational success. For example, Procter and Gamble enjoyed substantial gains in competitive advantage through their close collaboration with Wal-Mart (and vice versa), Caterpillar mentions its distribution partners as one of the major sources of organizational success (Fites, 1996) and McDonald's as well as other franchise-based businesses would not be able to sustain their impressive growth rates without support from their franchise partners. In contrast, confrontational relationships with these partners can cause severe business crises or at least negatively impact the position in the mind of the final consumer. Lack of influence on POS and service quality or product availability may be results of non-functioning relationships to distribution partners. However, this constituency group will not be discussed at length since it never is cited as (one of) the most influential one(s) and one could also claim that these partners are in fact customers of an organization. Then the way such relationships are managed by an organization would be the result of their customer responsiveness.

2.6.5. Competitors

The central question for strategy researchers is "Why do some firms perform better than others?" (Rumelt, Schendel & Teece, 1991). The usual explanation for a given advantage is a unique access to rare resources (Barney, 1991; Wernerfelt, 1984). Anything rare and valuable is usually the target of competition. Identifying and developing such resources better than the competition should result in a sustainable competitive advantage. Competitors therefore qualify as important organizational stakeholders. One may even argue that responsiveness to competition is as vital as to customers, employees or shareholders. While this may be true the character of such responsiveness differs. Proponents of a competitor orientation would hardly claim that learning about competitors' expectations towards oneself and fulfilling them will enhance effectiveness. If one's competitors aim for market share or high profitability, few businesses will try to fulfill these demands.

Still, a few researchers propose a positive effect of competition on long-term profitability and growth by taking a dynamic industry-level perspective. A number of competitors allow spillover effects of individual firms' actions and help the industry as a whole (Jaffe, 1986; Leone & Schultz, 1989), form the basis for technological developments unthinkable for individual companies (Dollinger, 1990; Huff, 1982), and strengthen the network of suppliers, the quality of the labor force and the knowledge available (Krugman, 1993; Porter, 1991). However, it is also emphasized that organizations within an industry have to show diversity in their competitive strategy to benefit from competition (Miles, Snow & Sharfman, 1993). While these observations suggest a positive impact of competition on performance on an industry level, no argument is made that high responsiveness to competition would cause high performance. Much of what is learned from a competitor orientation rather falls into the domain of one of the other types of responsiveness, since the area of interest will be the marketing, human resource or finance activities of competition, not the competitors' interests per se.

2.7. Stakeholder-Responsive Strategies

2.7.1. Theoretical Foundation

In the classical input-output model of the corporation, investors, suppliers and employees are depicted as providing inputs which the firm transforms into some output to the benefit of its customers or, in the pure capitalist model, of the investor. The provision of inputs is rewarded at market competitive rates. The stakeholder model opposes this traditional model: First, various other groups are considered (like the government, neighbors or competitors). Second, none of these groups is prima facie considered more important than others. A number of theories provide a theoretical basis for such a model. Pfeffer and Salancik (1978) criticize that most books on organizations describe how they operate, not how they manage to survive. In contrast their resource dependence theory concentrates on factors that guarantee organizational survival. The key to survival is the ability to acquire and maintain resources. As long as organizations are in complete control of these resources the problem is a simple one. However, no organization has control over all the resources required for survival, and therefore has to establish relationships with various elements in its environment to be able to acquire these resources. If this environment were fully dependable – i.e. if resources were available at any time at foreseeable conditions – this challenge would not exist. However, the organizational environment is not dependable. It changes, new entities enter and exit the environment, making certain resources scarce and their supply difficult. Therefore an organization needs to adjust its activities towards the environment to guarantee a continued supply of resources. To secure the ability to obtain resources and support from its external coalitions the organization has to offer various inducements (Barnard, 1938; March & Simon, 1958; Simon, 1964). The level of inducements offered depends on the value of the contribution of the specific exchange partner and their level of interest in the firm which might depend on their choice set and the specificity of assets devoted to this partnership (Williamson, 1984). The fact that the demands of external coalitions change and may even be in conflict makes continuous adjustment of corporate activity necessary.

By applying the stakeholder concept, a company takes into account the importance of a number of internal and external coalitions. It will try to discern the expectations of each of these groups and then satisfy them. Meeting the demands of the various stakeholder groups might be easy if these demands were compatible: The organization would find that increasing the satisfaction of one group would raise the satisfaction of others. However, stakeholder groups vary in terms of their needs, values and expectations (Friedlander & Pickle, 1968). This implies that numerous actions from the company's part are necessary to satisfy them all. More actions usually require more resources, but, as frequently pointed out, a company's resources are limited (Barney, 1991): Time and capital obviously are scarce resources, but so are attention capacity, rationality (Simon, 1955) and information-processing capability. Therefore, any attention given to one stakeholder group consumes resources that cannot be employed in attending to another stakeholder group. The company hence has to decide how to allocate its limited resources to assure the satisfaction of all stakeholder groups. By definition, a company that manages along the stakeholder concept must not pursue the maximization of responsiveness to one stakeholder group but try to strike a balance between the satisfaction levels of all stakeholders. Even if a company's resources were unlimited, maximizing responsiveness to all constituencies would not be appropriate since stakeholders do not only vary in terms of their needs, values and expectations, but some of their requirements may even be in conflict: Putting too much emphasis on one stakeholder group may therefore result in the disappointment of another one and may put strain on the latter relationship.

Given the above difficulties, two causes of disequilibrium in the relationship with each stakeholder exist: By not meeting the minimum requirements of a stakeholder group, the company faces either termination of the respective relationship or a threat to its very existence (depending on the specific stakeholder). By over performing in fulfilling the requirements of one stakeholder group, the company increases the likelihood of not meeting the minimum requirements of others. The zone between the two limits mentioned above is a zone of tolerance (Doyle, 1992): Performance within this zone will satisfy the respective stakeholder while not endangering the relationship with others. This zone can be conceptualized as an area around the expectation level of the stakeholder (see Fig. 1). Exceeding these expectations – obviously at additional costs to the company – may result in higher satisfaction of the specific stakeholder group. However, these resources may be employed more wisely in a relationship with another stakeholder whose expectations are not met satisfactorily. These relationships may approach the lower boundary of the tolerance zone and may sooner or later be considered unacceptable. To be successful hence requires constant attention to all stakeholder expectations.

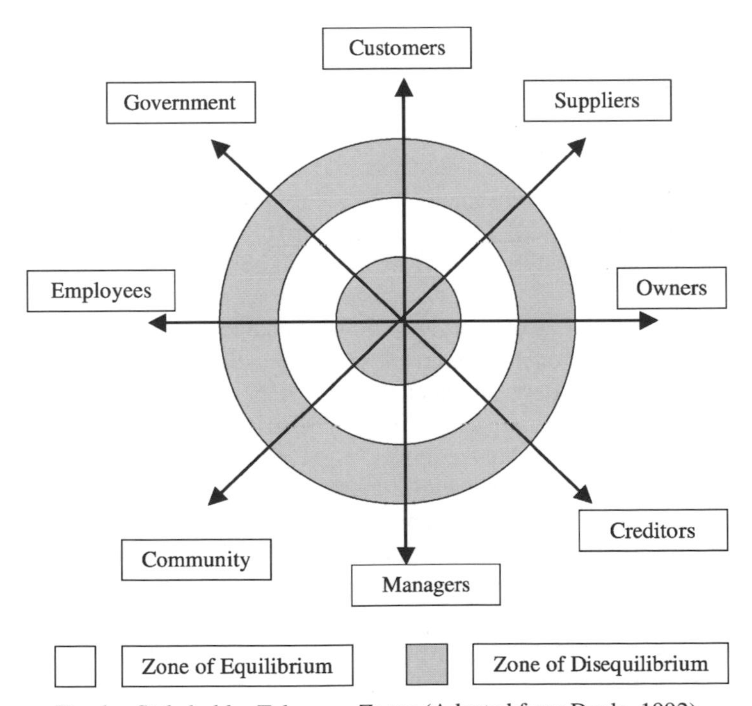

Fig. 1. Stakeholder Tolerance Zones (Adapted from Doyle, 1992).

The idea of stakeholder management has long been present in the marketing and strategy literature. Slater and Narver (1995) argue that "to be a powerful foundation for a learning organization and to provide the opportunity for generative learning, the scope of market orientation must include all stakeholders and constituencies..." (p. 68). Kimery and Rinehart (1998) call for a reconceptualization of the marketing concept "...based on a long-term multiple constituency approach...as a prescriptive model for more effective business orientation" (p. 117). Freeman (1983) developed a strategy formulation process applicable for each stakeholder group. By analyzing actual behavior, likely future behavior, commonalities in behavior generic strategies for dealing with stakeholder expectations can be formulated and implemented.

But even "classical" finance literature applies the idea of the stakeholder concept. Cornell and Shapiro (1987) argue that stakeholders play an important role in financial policy. They distinguish between explicit contractual claims which firms issue to non-financial stakeholders (wage contracts, product warranties or loan repayments) and implicit claims like job guarantees, continued service to customers or community care. The difference between these types of claims is their legal standing and riskiness. Defaulting on explicit claims will likely lead to bankruptcy while implicit claims need not necessarily be attended to. Explicit claims are usually senior to all claims by shareholders and bondholders and will only be endangered in times of financial distress. Implicit claims, however, are junior to financial claims by owners and lenders and carry a higher amount of risk. As these implicit claims cannot be unbundled and sold separately their risk is difficult to diversify and may negatively impact firm value. Since the value of a company depends on all claims issued, Cornell and Shapiro (1987) demand an extension of the original balance sheet to take into account all claims in the financial policy of organizations.

The observation of actual organizational behavior is another justification for the stakeholder model. A number of studies have shown that managers take into account a variety of stakeholders in their decision-making processes (Baumhart, 1968; Brenner & Molander, 1977; Clarkson, 1991; Kotter & Heskett, 1992). Therefore the stakeholder model might provide a more realistic picture of corporate behavior than any single-stakeholder view and thus deserves to be taken as the most useful model of corporate behavior. However, such reasoning connects a descriptive observation with a normative claim that does not provide a defensible case for the stakeholder concept. Just because people act in a certain manner does not guarantee its higher worthiness. Taking the step from *description* to *explanation* or from *is* to *ought* without intervening analysis poses the problem of so-called "natural fallacy" (Moore, 1959/1903).

All of the various definitions of the construct "stakeholder" contain at least two parts: the stakeholder and the entity to which the stakeholder is connected. Two of the most popular definitions come from Freeman (1984) – "any group or individual who can affect or is affected by the achievement of an organization's purpose" (p. 52) – and Carroll (1993) – "individuals or groups with which business interacts who have a stake, or vested interest, in the firm" (p. 22). The diverse character of the relationship between an organization and a stakeholder is obvious: While Carroll focuses on entities which have an "interest" or a "claim," Freeman's definition is broader in scope including everybody "affecting or being affected" without necessarily having a "claim" or any "right." Usually stakeholders are limited to a subset of living human beings as in Donaldson and Preston (1995) that argue that "stakeholders are persons or groups with legitimate interests in procedural and/or substantive aspects of corporate activity" (p. 67). Very few authors have extended this conceptualization to include non-human physical entities (Buchholz, 1993; Stead & Stead, 1992) arguing that, for example, air or raw materials can affect or are affected by carrying out a number of organizational activities. The results of hurricanes, droughts or energy shortage may have large impacts on a number of industries well comparable to or even surpassing the impact of certain human stakeholders. Where a useful boundary is drawn depends on a number of circumstances (industry, location, resource dependence, time), but including non-human beings might occasionally be necessary.

In contrast to the economic or financial theory of the firm, stakeholder theory (at least certain aspects of it) provides a value-free conceptualization of a business organization. In its simplest form it does not argue for maximizing the benefits of any given constituency but posits the idea that an organization is an aggregation of stakeholders who attempt to advance their interests (Wartick, 1994). However, the implications for managers and the practice of stakeholder management do not rest with this value-free conceptualization but are concerned with the central question "Who comes first?" In this sense some value structure must bear on the actual implementation of stakeholder thinking. If stakeholder theory is presented by outlining a certain business purpose like creating "increased wealth for its primary groups by using its resources efficiently" (Clarkson, 1995) or serving as a "vehicle for coordinating stakeholder interests" (Evan & Freeman, 1983), it ceases to be value-free but promotes a normative core. This distinction has been made explicit by Donaldson and Preston (1995) who criticize that the concepts stakeholder, stakeholder management, stakeholder model and stakeholder theory are used by various authors in different ways. They call for distinguishing between three different arguments present in stakeholder research, namely a descriptive, an instrumental and a normative one.

2.7.1.1. Descriptive stakeholder theory. In its descriptive version stakeholder theory is concerned with the description and explanation of corporate characteristics and behavior. As such, one might find studies investigating whether and how different stakeholders' interests affect corporate decisions, whether certain stakeholders receive more corporate attention than others or how the legal system deals with stakeholder demands. The descriptive stakeholder theory is therefore interested in the question whether corporations act *as if* stakeholders exist.

2.7.1.2. Normative stakeholder theory. Normative arguments are not concerned with correspondence between theory and observed corporate life. They aim to formulate the objectives of an organization on the basis of some underlying moral or philosophical principle. As such, normative theory prescribes that an organization should act in the interest of stakeholders because of its responsibility to society or because of certain ethical standards. It is concerned with the question why organizations *ought to act* in the interest of stakeholders. Kantian posture (Bowie, 1994; Evan & Freeman, 1983), feminist perspective (Wicks, Gilbert & Freeman, 1994), the fair contracts approach (Freeman, 1994; Philipps, 1997) or property rights (Becker, 1992; Coase, 1960) serve as moral principles supporting the normative foundation of stakeholder theory.

2.7.1.3. Instrumental stakeholder theory. In contrast, instrumental stakeholder theory assumes that corporations act as if stakeholders exist but aims to understand the relation of such behavior to corporate objectives. As such, it might investigate the effect of certain kinds of organizational behavior towards stakeholders or managerial orientations towards different stakeholder groups on profitability or some other performance measure. It is therefore interested in the question *what* happens *if* organizations do hold a stakeholder perspective.

This research project is mostly interested in the instrumental version of the stakeholder theory implicit even in the first traceable definition of stakeholders by the Stanford Research Institute that referred to "those groups without whose support an organization would cease to exist" (1963, quoted in Freeman, 1984, p. 31). Without establishing a clear causal link to performance, inducement of contributions from stakeholders is considered essential for organizational survival. Jones (1995) has formulated a more comprehensive instrumental argument for the stakeholder idea. He assumes that firms have relationships with many stakeholders, are run by professional managers and exist in markets where competitive pressures influence behavior but do not necessarily penalize inefficient behavior. The relationships to stakeholders can be described using the contract as a metaphor where contracts with different stakeholders vary in their extent of formality and specificity. Usually these contracts are used to reduce opportunism to an "efficient" level where the costs of further reduction outweigh the benefits. However, opportunistic behavior which is costly due to adverse selection (Akerlof, 1970), and contracting costs might also be reduced by the voluntary adoption of standards of behavior (e)limi(na)ting it. Jones (1995) provides evidence from game theory (Axelrod, 1984) and a number of arguments emphasizing the importance of emotions in business (Etzioni, 1988; Frank, 1988) to suggest that firms which contract on the basis of mutual trust and cooperation will have a competitive advantage over firms that do not, an argument which has been formulated by both strategy and marketing scholars (Barney & Hansen, 1994; Morgan & Hunt, 1994). Such firms will be more eligible to take part in certain types of relationships and transactions unavailable to opportunistic firms. As such, this instrumental stakeholder theory turns the classic economic theory upside down, since it implies that trusting, trustworthy and cooperative, not opportunistic behavior will give the firm a competitive advantage.

Hill and Jones (1992) draw on agency theory developed in financial economics (Jensen & Meckling, 1976; Ross, 1973) in supporting an instrumental stakeholder theory. By extending the relationships from a principal-agent view to all stakeholders, they view the firm as a nexus of implicit and explicit contracts with management in the center of the nexus acting as an agent for all other stakeholders. Since stakeholder preferences likely differ from management preferences, utility losses (the difference between the utility achieved if management acts in the stake holder's best interest and the utility achieved if management acts in its own interest) will occur. Minimizing these utility losses (through incentives, monitoring and enforcement) will increase management's access to stakeholder resources. For example, stock option plans, warranties, long-term contracts or tax-breaks for pollution equipment are means for reaching further alignment of interests between management and stockholders, consumers, suppliers and community respectively and should increase access to these stakeholders' resources. To keep the respective utility losses at acceptable levels close attention to all relationships seems warranted. One might also argue that incentive, monitoring and enforcing costs will decrease as the two parties establish closer relationships. Once utility losses exceed certain levels the respective stakeholder will cease to supply his resources which may negatively affect firm performance.

Freeman (1999) criticizes large parts of stakeholder theorizing for relying too strongly on the distinction between normative, instrumental and descriptive theory (Donaldson & Preston, 1995; Jones & Wicks, 1999). He argues that the idea of a value-free science implicit in this separation is impossible and that the term stakeholder itself carries a certain value structure as an "obvious literary device meant to call into question the emphasis on stockholders" (Freeman, 1999, p. 234). Therefore only instrumental research on stake holding provides an ideal synthesis of stakeholder theory, because any instrumental argument contains descriptive and normative statements at the same time. Purely normative arguments for the stakeholder idea without relation to real world consequences provide little avenue for progress whereas instrumental arguments may also further the normative core of stakeholder thinking. Gioia (1999) argues that all the theorizing on stakeholder theory will not foster progress as long as the pragmatic forces in the corporate world are ignored. By working from concept to concept without supporting data "theorizing ... easily ends up in the clouds, whereas a dose of data just might help ground us in practical reality" (Gioia, 1999, p. 230).

2.7.2. Empirical Support

The gap of empirical studies on the stakeholder management-performance association has been made explicit by Donaldson and Preston (1995), who demand that the "connection between stakeholder strategies and organizational performance should be examined. Consider, for example, the simple hypothesis that corporations whose managers adopt stakeholder principles and practices will perform better than those that do not. This hypothesis has never been tested directly, and its testing involves some formidable challenges" (p. 77). The central problem in this research effort is the lack of reliable indicators of the stakeholder management side of the equation, i.e. the independent variable. Comparable to the efforts in organizational effectiveness research, different researchers developed different operationalizations of the concept, used different data gathering methods and differ with respect to the methodology employed. This type of research is not only plagued with the major problem of all organizational performance research – how should the dependent variable be operationalized? – but with a similar difficulty for the independent variable – how should stakeholder orientation/management be operationalized? Comparability of different studies is therefore more limited than in traditional effectiveness research since both driver and effects usually differ substantially.

Most "research" on stakeholder management takes the form of anecdotal evidence related to macro-economic stereotypes or individual firm behavior. The stakeholder concept is usually presented as the natural opposition to shareholder capitalism where the former represents the European and Japanese economic systems and the latter the U.S. system. Usually the argument takes an evaluative stance promoting the superiority of one over the other (Bosshart, 1996; Economist, 1996) or compares the importance of different stakeholder groups in different economies (Steadman, Zimmerer & Green, 1995). The *Economist* (1996) argues that the shareholder-oriented U.S. economy recently performed better than the stakeholder-oriented U.K., Japanese or German economies and issues a warning against an adoption of a stakeholder mindset. This argument lacks rigor conceptually and methodically. No evidence is given how the difference in stakeholder attention between U.S. companies and their foreign counterparts is assessed. One-year GDP growth and current unemployment rates are then taken as evidence for the superiority of the shareholder system dismissing other firm-independent factors which may well influence these numbers (like the tax system, laws or interest rates). Since the stakeholder concept claims that companies with "healthy" relationships to their stakeholders may outperform those with less "healthy" ones, the relevant comparison set needs to be identified. Even in times of globalization, much competition for relationships to constituencies might still take place at a national more than at an international level (at least for certain constituencies) rendering such cross-national conclusions doubtful. Doyle (1992) argues that the sudden problems of many formerly profitable businesses are caused by their sole focus on financial goals. To remain successful, businesses need to be aware that trade-offs exist between different organizational goals. Since these are formulated by different stakeholder groups, businesses constantly need to monitor the expectations of their stakeholders in order to satisfy all their interests. A number of examples are cited for business failures due to ignorance of the diverse expectations of different stakeholders.

Empirical evidence for the stakeholder concept is often promoted in the form of executive statements. In 1950, the then-CEO of Sears, General Robert E. Wood, argued that if a business properly takes care of customers, employees and community, "...the stockholder will benefit in the long pull" (quoted in Worthy, 1984, p. 64). The chairman of Eastman Chemical claims that aligning the interests of all stakeholders to form a balanced, interdependent network can most effectively increase the value of a company. Eastman aims to tie the success of each of its primary stakeholder groups (customers, employees, suppliers, various publics and investors) to one another by aligning their interests. Different programs, like stock ownership, performance plans, quality programs, community services provide incentives to act in the interest of all major stakeholder groups without emphasizing a single one: "It's detrimental to all stakeholders when the interest of a single stakeholder group is allowed to become so strong that it actually consumes value" (Deavenport, 1996, p. 2). A number of books and articles illustrate the usefulness/dangers of stakeholder (mis-) management by providing descriptive evidence of organizations' corresponding activities. Ben and Jerry's efforts in supporting the community and local suppliers (Cohen & Greenfield, 1997), Basic Manufacturing Technology's commitment to a number of stakeholders (Maranville, 1989) and McDonald's or Mobil's environmental strategies (Lawrence, 1991; Polonsky, 1995), are cited as corporate examples where the creation of stakeholder value is deemed essential for organizational performance. In contrast, British Gas or Hoover's (Whysall, 2000) failures in addressing stakeholder concerns as well as Ben and Jerry's, then lackluster financial performance (Taylor, 1997) are among the examples provided for business problems due to ignorance of stakeholder expectations. Obviously, the Ben and Jerry's example could also serve as a case against stakeholder management and for an adoption of a purely shareholder-value driven strategy.

Only very few studies apply what commonly counts as scientific procedure in studying the stakeholder-performance relationship. Greenley and Foxall (1998) analyze the relationship between an organization's stakeholder orientation and (self-reported) performance and find no relationship between high levels of stakeholder orientation and performance. However, positive effects of orientation towards single stakeholders under certain environmental conditions are detected. These results suggest that a constant focus on single constituencies may pay off for certain performance measures while orientation towards others is worthwhile only under certain external circumstances. It also shows the sensitivity of this type of research towards the operationalization of the performance measure and the external conditions surrounding the investigation period. However, one should bear in mind that the measures chosen by Greenley and Foxall (1998) may not reflect actual organizational behavior but focus on managers' assessments of indicators reflecting prescriptive elements of stakeholder-oriented behavior. Preston and Sapienza (1990) also use perceptual, not behavioral data to assess an organization's stakeholder orientation, but make use of the yearly *Fortune* rankings on a number of organizational dimensions. They assess survey results related to four stakeholder groups, namely shareholders, employees, customers and community. Their findings show that the four dimensions of stakeholder reputation are strongly correlated and also are positively associated with company size, return and growth. Once again, no evidence for a trade-off among stakeholder objectives can be found.

The most comprehensive study to-date compares the strategic stakeholder model to the intrinsic commitment model (Berman, Wicks, Kotha & Kones, 1999). While the first assumes managerial concerns for stakeholders to exist solely because of their effect on financial performance, the second one views the firm as having commitments to stakeholders, which in turn shape strategy and impact performance indirectly. The authors assess five stakeholder relations (employee relations, natural environment, diversity, customers/product safety and community) and capture a firm's strategy along two dimensions, cost efficiency and differentiation. Performance is operationalized as return on assets. The authors use a sample of 486 firms and analyze six years of data. They find that two of the five stakeholder relationships, employee relations and product safety, are positively related to financial performance. When stakeholder relationships are modeled as moderators of the strategy-performance association, a number of interactions were identified. Specifically, 9 out of 20 possible interactions showed significant effects and all stakeholder relationships interacted with certain strategy variables. However, no significant effects of stakeholder relationships on strategy were identified, leading the authors to reject the intrinsic stakeholder commitment model (Berman et al., 1999). The authors conclude that careful attention to stakeholder groups and specifically to the interdependence between strategy and stakeholder relations pays off financially.

2.7.3. Measures of Stakeholder Responsiveness

Given that the stakeholder management process suggested by Freeman (1983) consists of several steps (observation, analysis, strategy development, and implementation) measures could be taken at different levels. One could assess the extent to which organizations observe their stakeholders, how much time is spent on analyzing them or how strongly they influence strategy development or their role in strategy implementation. Since actual observation of these steps would be prohibitively costly and difficult to administer (if one is not conducting case study research), research focusing on these factors is usually conducted in survey form asking organizational members about their perception of these activities. Greenley and Foxall (1996) take into account five elements as indicators of an organization's stakeholder orientation:

- importance of research on expectations of stakeholders carried out by an organization;
- importance of management judgement for understanding stakeholder interests;
- extent of planning of activities for each stakeholder group;
- extent to which stakeholders are accounted for in mission statements;
- impact of stakeholder groups when discussing corporate culture.

By aggregating management answers on seven point scales for each of these indicators and comparing the result of each stakeholder group (competitors, consumers, employees and shareholders) an organization's orientation towards these constituencies can be assessed.

Preston and Sapienza (1990) use the yearly *Fortune* rankings on different organizational issues as proxies for stakeholder orientation. For example, "ability to attract and keep talented people" is used as an indicator for employee relationships whereas "quality of products and services" applies to customers. The balanced scorecard approach by Kaplan and Norton (1996) takes into account different expectations by different stakeholders and assesses attention to each group by developing appropriate objectives and measuring attainment of these objectives. However, no specific measures are suggested rendering comparisons of various organizations' stakeholder practices difficult. Some studies in the field of stakeholder management borrow from the perspective of corporate social performance and use measures originally developed for the latter domain-like KLD ratings (Berman et al., 1999). While these measures are certainly related to stakeholder relationships, they lack both scope and depth by, for example, ignoring suppliers and limiting customer relationships to a five-point indicator of product safety.

Taking a "true" stakeholder perspective would imply inclusion of a number of factors besides financial return into the set of effectiveness measures. Taking into account a measure specifically related to the well-being of a single stakeholder group would contradict the central idea of the stakeholder approach. However, such a step seriously complicates empirical studies on the subject as noted by Berman et al. (1999): ". . . such an attempt can greatly complicate what constitutes the dependent and independent variable" (p. 503).

2.7.4. Discussion

Various aspects of stakeholder theory have gained considerable influence in organizational research. It may be argued over whether the stakeholder concept has attained status of a theory of the firm comparable to the economic or financial one, or whether it only covers certain aspects of a behavioral theory of the firm (Cyert & March, 1963). The idea that constituencies matter and shape an organization's strategy, however, has been well accepted by scholars in most business disciplines. One might even claim that it constitutes an umbrella encompassing the core ideas of marketing, human resources, finance or corporate social responsibility, since all of them call for the utmost importance of one single constituency in order to attain some end. No one working in these areas seriously demands total ignorance for other stakeholder groups but promotes supremacy of a certain one, irrespective of organization, industry, time and place. The stakeholder idea differs insofar as it pays attention to such circumstances. It does not demand equal responsiveness to all constituencies (as is often wrongly asserted and an idea far removed from corporate reality), but calls for such responsiveness that each stakeholder is satisfied and initiates/prolongs his relationship with the organization. Depending on circumstances, the most important constituency may even change over time. Someone embracing the stakeholder idea will adapt the responsiveness exhibited towards a stakeholder group if circumstances change and might, under certain conditions, at one time even hold the mindset of an employee-responsive strategy and at another time the one of a customer-responsive strategy.

To be able to act along the stakeholder concept close monitoring of the organizational environment is necessary. While some trade-off between responsiveness given to various parts of this environment seems likely due to limited capacities, organizations certainly differ in ability and interest to foster all relationships with constituencies. An organization spending its entire attention capacity will face a trade-off, but most organizations are probably not using their entire capacity or are not willing to learn how to increase this capacity. Empirical results showing positive correlations between attention/orientation to different stakeholders support this proposition. Greenley and Foxall (1996) find a positive association between orientation towards different stakeholder groups.

Scanning the literature on stakeholder management, one encounters a large number of theoretical pieces, but very little empirical research. Scholars demand more empirical studies to be carried out to increase acceptance of stakeholder management and to provide fresh input for conceptual work (Freeman, 1999; Gioia, 1999). The lack of studies can be attributed to a number of difficulties the empirical researcher faces. First, the meaning of the term stakeholder has to be specified. As discussed earlier, no widely accepted delineation exists and the delineation chosen might be influenced more by data availability or sampling considerations than due reasoning. Once such delineation is accepted, the researcher needs to decide which information on stakeholder activities are of interest since this decision will greatly influence the information, gathering process.

2.8. Summary

Each of the preceding approaches presents sensible arguments why certain stakeholder groups deserve special attention. Viewed in isolation, each makes a valid, well defensible, claim: Responding foremost to customers certainly makes sense because they are the ones who are paying the bills, responding foremost to employees is certainly necessary because they are to a large extent responsible for the quality of current and the development of future offerings, responding foremost to shareholders is more than sensible because they will otherwise withdraw the financial resources necessary to sustain the business, responding foremost to the public might be most useful because they are responsible for the overall image transferred which will then influence the perception of other stakeholders. However, these views do not exist in isolation, but do respectively affect the responsiveness to other stakeholder groups. Attending most to one stakeholder group implies attending less (or, unlikely, not at all) to other ones. But in spite of the exclusive nature of the different approaches, all enjoy empirical support and seem to co-exist. Subtle differences and lack of preciseness can explain such coexistence and render a comparative assessment of these approaches difficult:

- In their conceptual form these approaches propose the supremacy of a certain stakeholder group for attaining some organizational objective without specifying the exact effect of such management. While a tendency towards financial measures – which themselves are subject to discussion and come in many different forms as shown earlier – exists, most claims avoid explicit propositions which organizational objective is attained by following the advice outlined. Usually reference is made to a concept like competitiveness or performance which could translate in a large number of performance indicators respectively. It is easy to think of performance indicators maximized by responding most to employees and others being maximized by foremost attention to customers.

- In contrast to conceptual work empirical studies need to incorporate one or more performance indicators. But the diversity of causal relationships brought forward renders comparisons between these approaches very difficult. One of the few meta-analyses in effectiveness research (Capon, Farley & Honig, 1990) considered only financial dependent variables and reviewed studies using level, growth and variability of measures like profit, market share, assets, equity, cash flow, sales or market/book value. Including other non-financial dependent measures and combining them with the vast array of independent measures used to operationalize each of the orientations proposed produces a virtually unlimited set of possible causal relationships. While only a fraction of these has yet been tested, the variety of empirical research is "impressive," but prevents sensible comparisons between these studies.

Since these approaches derive from different research traditions, employ different measures and largely build on normative arguments difficult to counter, any comparison between them may be an impossible task. While a number of research projects investigate the validity of each individual approach, the likely impact on other organizational behavior – especially the responsiveness to other stakeholder groups – is ignored. The next section will suggest a methodology to investigate the joint effect of responsiveness to different stakeholder groups on performance since such a comparison seems valuable both for managerial research and practice. After all, it deals with the core question of organizational strategy research: Why are some organizations more competitive than others?

At the most basic level, balanced vs. focused strategies can be compared using the Lines and Gronhaug (1993) concept of environmental orientation. They argue that an unbalanced attention to the environment results in higher performance than some other configuration of attention. It is this imbalance which warrants further discussion to allow comparisons between these approaches. For illustration let us assume a business has to deal with only four stakeholder groups, that is, customers, employees, shareholders and suppliers. Then balancing attention to stakeholders could be conceptualized in two different ways (compare Fig. 2). One, each stakeholder group enjoys exactly the same level of responsiveness. This does not mean that they are treated the same way but that each of the stakeholder groups receives 25% of the organization's responsiveness capacity devoted to

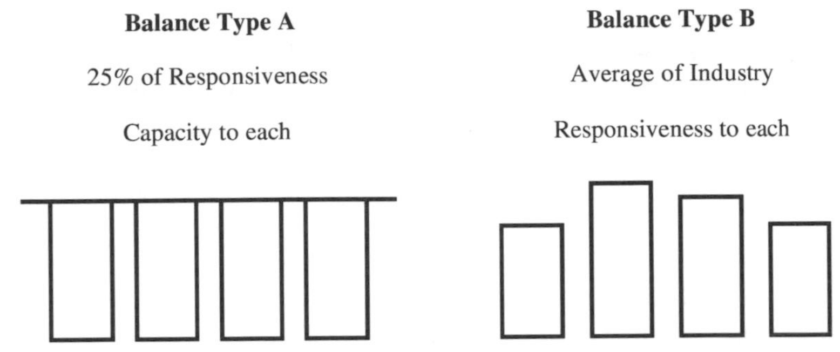

Fig. 2. Types of Operationalizations of Stakeholder Responsiveness.

constituencies. Two, each stakeholder group receives the average attention provided by the members of the industry the business is competing in. Industries are defined here as organizational categorizations by constituencies that act as mobility barriers between clusters of organizations (Porac & Thomas, 1990; Reger & Huff, 1993). Typically, they represent the peer groups along which an organization's performance is assessed and may therefore serve as standards for evaluations by constituencies.

This distinction is critical for understanding balance in responsiveness. For example, when employing the first type of argument, a number of organizations would show responsiveness towards two (or three), but at least towards one and never towards four constituencies (if one assumed only four relevant constituency groups). While this perception of balance may allow the most straightforward conceptualization of any type of stakeholder orientation it poses a number of problems. First, it is difficult to think of any business where a balanced allocation of responsiveness is likely. Some businesses may warrant above-average – here defined as full responsiveness capacity divided by the number of constituencies – responsiveness for certain constituencies simply to survive. Calling such behavior stakeholder x-responsive may then be inappropriate. For example, airline executives in Europe in the late 1980s and early 1990s were required to engage in extensive political lobbying to guarantee access to airports. Some executives then reported spending the majority of their time on lobbying (quoted in Lines & Gronhaug, 1993). This type of posture would by definition disallow companies to be customer-responsive or employee-responsive. The second problem would be a methodological one: Each stakeholder group expects different types of behavior from an organization and different resources are exchanged. Comparing these exchange relationships becomes a difficult, if not impossible task. How would one assess whether customers receive exactly the same level of responsiveness as employees? A third problem is the conceptualization of an organization's responsiveness capacity as either constant or variable across comparable units. In case of variations between organizations one may qualify as more employee-responsive than another one even though it attends less to employees. For example, organization A devotes 100 units of responsiveness capacity to its constituencies, 30 of which are allocated to customers and 20 to employees. Organization B only devotes 50 units and allocates ten to customers and 15 to employees. Organization B would qualify as employee-oriented (30% of responsiveness capacity) while organization A would not (20% of responsiveness capacity). The authors still assume that employees may prefer a relationship with organization A as the relevant comparison unit may not be customers of the same organization but employees of comparable organizations.

The second form of understanding "balance" therefore seems more useful (balance type B in Fig. 1) as the above problems are of little concern. First, above-average responsiveness to a stakeholder group would be captured at the industry or strategic group level where most constituencies will judge performance. No assumption is made that all constituencies "a priori" deserve the same level of responsiveness but that industry standards shape the expectations of constituencies. For example, if 50% of responsiveness capacity in an industry are devoted typically to customers, an organization devoting a higher level (or some threshold to be defined by the researcher) may be considered customer-oriented, ceteris paribus. Comparability between responsiveness to different constituencies would cease to be an issue: Since comparisons are now made among organizations and not within organizations, a certain set of measures applies for each constituency group. Finally, such a comparison need not be concerned with the issue of constant-sum. Given the results of empirical studies (Greenley & Foxall, 1996; Preston & Sapienza, 1990), it may turn out that some corporations are able to devote above-average responsiveness to a number of stakeholder groups while others attend less than average to all relevant constituencies.

Stakeholder theory is only consistent with the second type of balance outlined above. It makes no claim that all constituencies are equivalent in their importance for organizational effectiveness but that each group's expectations need to be fulfilled in order to receive continuous support. An organization therefore needs to take into account the expectations – shaped through industry standards – of each group and respond to them to prevent them from disrupting the relationship. It is well imaginable that some organizations can continuously over-fulfill all stakeholder expectations because of better practices in so doing, more available resources or higher commitment in doing so. Some limit of responsiveness may exist, but may be considerably higher than the sum of the industry averages for all constituencies.

3. THE MODEL AND ITS OPERATIONALIZATION

3.1. Measures of Responsiveness to Stakeholders

The relationship between an organization and its constituencies is characterized by the exchange of resources. To guarantee the continuous supply of these resources an organization has to offer inducements. These inducements can take a number of different forms, for example financial rewards, personal attention or emotional commitment. Capturing responsiveness to stakeholder groups requires the following steps: learning about the expectations of a certain stakeholder group, deriving all possible organizational behavior targeted at meeting these expectations, and measuring the extent to which organizations engage in these activities. Taking on these steps for all stakeholder groups and for a number of organizations represents an impressive task. Let us consider customers and for simplicity assume that all they are interested in is receiving a product offering high value at a low price. The activities an organization takes on for meeting this expectation may range from market research to sponsoring innovation competitions, from investing in manufacturing to demanding lower prices from suppliers, from advertising to changes in distribution. Dozens of other strategies are imaginable and measuring an organization's activity on each of them may present an impossible task, let alone for more than one company, for a number of periods and firms, and for activities related to other stakeholders. In order to compare competing strategies in terms of multiple constituency responsiveness, three possible research avenues addressing the above concerns come to mind. If the focus of the research were not comparison, an in-depth study of the administration of responsiveness within a single organization may also increase our knowledge on factors driving OE. Such an approach would correspond to the central idea of the resource-based view which claims that sources of sustainable competitive advantage are located within organizations and research should therefore not be done *on*, but *in* organizations (Rouse & Daellenbach, 1999).

The first two strategies for cross-sectional purposes rely on primary information sources whereas the last one uses secondary sources. Surveying management is one way of learning about an organization's posture to stakeholders. One could also survey the individual constituencies about the level of responsiveness they experience. A third option is to develop indicators for the organizational responsiveness towards its constituencies that need not be collected from organizational or constituency members but from secondary sources.

These three strategies differ with respect to the type of information collected, to the inferences drawn and to the relevant sample. Each actually constitutes a different research strategy for assessing an organization's stakeholder relationships. By gathering relevant information from managers, one assumes that a single (or possibly a number of) organizational member(s) can assess the responsiveness an organization devotes to different constituencies. Top management is usually considered most knowledgeable about an organization's various relationships, and the information provided is deemed suitable to learn about stakeholders' likely interpretation of and reaction to organizational behavior. This assumption is critical, but difficult to uphold. Management's knowledge on the diverse relationships to organizational stakeholders must be superficial. Given the cognitive limitations of humans it is unlikely that managers are informed about all organizational activities targeted towards certain constituencies. Thus, management surveys were mainly used to make intra-, not interorganizational assessments of the importance attached to certain stakeholders (Greenley & Foxall, 1996).

Collecting information from stakeholders may be the obvious way to learn about the organization-stakeholder interface since they are the ones affected by organizational behavior. While it may not be necessary to receive information from each constituency group of every organization included in the sample (if the conceptualization of "balanced responsiveness" refers to balance relative to an industry average), the methodological, organizational and financial demands are very high. Both the development of a suitable instrument for each stakeholder group, the identification of the relevant sampling frame and the data collection step might turn out to be overwhelming tasks. Some relevant respondents may have to provide more information than reasonable, for example if a supplier has a large number of customers in the industry under study. Information from other constituency groups would have to be collected from a possibly large number of individual sources (e.g. employees, customers or the local community).

The third strategy to learn about organizational responsiveness to stakeholders relies on information from secondary sources. Instead of questioning organizational or constituent representatives indicators for organizational attitudes and behavior targeted at stakeholder groups are developed. When cross-sectional research is carried out accounting or financial statements may serve as information sources. For example, overall sales, repurchase rates, and share-of-pocket may all indicate the fulfillment of customer expectations. As inherent in every indicator, debates concerning its validity may arise. While prone to managerial distortions bias may not be more present than in survey results. Data collection difficulties would depend on the indicators formulated and the availability and accessibility of the relevant material. No claim can be made that the respective expectations are fully covered nor that the indicators are able to fully capture how well these expectations are met.

3.2. Measures of Organizational Performance

Strategy research is interested in the development and sustainability of a competitive advantage. It was outlined before that relationships between an organization and its stakeholders may be the source of such advantage (Amit & Shoemaker, 1993; Lines & Gronhaug, 1993; Rindova & Fombrun, 1999). The way an organization manages these diverse relationships should therefore determine its competitive position and should translate into a certain level of performance. Since measurement of organizational effectiveness lies at the heart of strategy research and practice, the indicator used to assess performance must be cautiously chosen. No unanimously accepted conceptualization of organizational effectiveness, let alone a unanimously accepted measure exists. Especially when relationships with stakeholders are the object of study careful choice of this measure is necessary. Most measures usually applied show a tendency to measure satisfaction of certain stakeholder groups, and may be of limited use when the importance of the quality of relationships between an organization and its stakeholders is proposed to determine performance.

Any firm might enjoy a competitive advantage through development and management of strategic assets which does not fully translate into classical performance measures since parts of the rent are appropriated by non-shareholder nodes in the nexus. How much of the rent is appropriated depends on the bargaining power of these stakeholders. Coff (1999) gives a number of examples how internal stakeholders, especially the management team, may be in a position to appropriate large parts of organizational rents due to the limited bargaining power of shareholders. Because of separated ownership and control significant, rent generation may even go undetected as long as average returns are posted keeping investors from becoming alarmed. The idea of rent appropriation directly impacts on the research question of this study. A highly competitive firm may be less profitable or have lower shareholder returns than a less competitive one because other parts in the organizational nexus are able to appropriate rent. Thus, traditional performance measures may not be able to capture the competitive position of an organization because they confound rent generation with rent appropriation. It may even be argued that the difficulty of detecting rents in traditional performance measures may allow organizations to sustain competitive advantages for fairly long periods. After all, extra-high profits or shareholder returns may elicit the interest of competitors more than average ones.

Coff (1999) argues that survival may be the most sensible long-run performance measure. While this may be true, the discriminatory power of survival, its ability to show different levels of performance, is limited, at least in the short run. In spite of the frequently-made claim that short-run performance is irrelevant and the focus should be on long run measures, we usually focus on the short-run. The fact that "none of us will be around forever" (Ferguson, 1986, p. 4) prevents the employment of survival as the ultimate test of effectiveness. Using a variety of performance measures should allow detection of competitive advantages even in the case of significant stakeholder appropriation. It is unlikely that such an advantage can remain undetected when a number of different performance measures are jointly considered over multiple periods. A number of authors apply various performance measures in order to avoid easy criticism concerning the choice of a single performance criterion (Chakravarthy, 1986; Greenley & Foxall, 1998; Venkatraman & Grant, 1986). The balanced scorecard (Kaplan & Norton, 1996) takes a similar approach where minimum thresholds along various performance dimensions are set which have to be met by an organization. Another option would be to assess performance by calculating returns for each individual stakeholder taking into account the conceptualization of the firm as a nexus of interests (Coff, 1999). Overall performance may then be determined by taking the mean of these individual results or by applying some other algorithm.

Several options to assess organizational performance are available for this research project: By taking the "traditional" approach one could choose among a variety of performance measures which are well established in the management literature. While all of them are targets of criticism their frequent use in scholarly research, public discourse and managerial practice does not only provide support for using them but would allow comparison of results with other similar research studies. Measures developed by Chakravarthy (1986) to operationalize the systems resource approach to OE may offer additional insights. Organizational effectiveness may also be understood as a latent variable resulting from the performance each coalition involved can attain. An organization is therefore effective if all interests are well served and less effective if one (or more) of them is (are) badly served. By clustering organizations into groups attending disproportionately high to all, to some or to no constituencies, groups of differently effective organizations could be formed. Depending on one's understanding of organizations, qualitative interpretation of these groups (type and extent of news coverage, *Fortune* rankings, personnel fluctuation, customer satisfaction ratings, etc.) or checking for convergence with other performance measures could then be attempted. Table 2 provides a short overview of the respective possibilities which impact both the model-building process as well as data requirements. The return-to-constituencies approach represents a combination of the other two approaches. While it recognizes the difficulties related to typical success measures and thus aims to incorporate the interests of all organizational constituencies, it attempts to make a quantitative statement about an organization's performance.

Table 2. Options for the Operationalization of the Dependent Variable.

Approach to Organizational Effectiveness	Traditional Approach	Return-to-Constituencies Approach	Interpretative Approach
Operationalization of organizational effectiveness	*Goal approach*: Various single measures as proposed in the management literature *Systems resource approach*: Ability to secure and deploy resources	Various measures related to benefit-cost ratios of the individual constituency groups are combined	Success is a (latent) result of satisfaction of the individual constituency groups

3.3. A Stakeholder Model of Organizational Effectiveness

3.3.1. Model Requirements

Once the decision how to capture responsiveness to each stakeholder group is made the scholar needs to model their relationship with OE. Two broad options come to mind, one of which incorporates the hypothesized causal relationships into the statistical model in an attempt to assess the quantitative effect of responsiveness to each stakeholder group on performance. The second option takes an explorative approach in an attempt to assess the interrelationships among the respective behaviors to stakeholder groups and organizational performance. Causality is not built into the model but can be derived from the results obtained.

3.3.2. Causal Modeling

Strategy research usually relies on regression-based techniques to estimate effects of organizational behavior on performance. Bowen and Wiersema (1999) criticize that the method employed in strategy research is at odds with its central paradigm. While the latter assumes that firms differ in their responses to environmental factors and that responses may have different effects in different time periods cross-sectional regression techniques assume stable model parameters across firms and across time. If these parameters in fact vary across firms and over time, any inferences drawn from the results may be useless. Ignorance of firm- or time-specific effects cause most cross-sectional research to "suffer from an inability to determine the true causal relationships" (Hill & Hansen, 1991, p. 187). Single-period regression needs to assume that both the constant and the individual coefficients are constant across firms, otherwise the number of parameters to be estimated would exceed the number of observations. In addition, by claiming to uncover some stable relationship between dependent and independent variables, the researcher makes – consciously or not – the assumption that parameters are constant over time as well.

When model parameters are assumed to vary across firms, pooled time-series, cross-sectional data and corresponding statistical techniques need to be employed (Bass & Wittink, 1975; Greene, 1993). Several techniques – like least squares dummy variables – are available to allow individual coefficients to vary and test for stability of coefficients over time and across firms. In addition to these purely methodological concerns, multivariate statistical analysis tends to be "radically analytic because it breaks cases into parts – variables – that are difficult to reassemble into wholes" (Ragin, 1987, p. 10). By decomposing organizations and their behavior into variables that combine with corresponding variables from other organizations, the unit of analysis is being fragmented and cannot be united after running the analysis. Therefore specificity and insight are given up for generalizability. This results in shortcomings of traditional statistical techniques vs. the "ideal" technique of comparison, experimental design (Ragin, 1987), especially for testing multiple vs. focused responsiveness strategies.

The dependent variable cannot be examined under all combinations of the independent variables. This is obvious for continuous data, but even in the case of interval data four independent variables with four levels each would allow the independent effects to take on 64 different combinations all of which are unlikely to be present in the sample. Statistical techniques estimate non-present effects by assuming a certain continuous shape between dependent and independent variable. For example, the data set may include firms with high levels of customer responsiveness and firms with low levels, but none with medium levels. On average, firms with high levels of customer responsiveness also enjoy a better performance. Results may therefore suggest a monotonous positive effect of customer responsiveness on performance while in fact a medium orientation might have the exactly same effect as a high one.

The coefficients for the independent variables are assumed to be the same irrespective of the value of other independent variables. Context is assumed to have no effect. For example, high levels of employee responsiveness might only pay off for high levels of customer responsiveness but may be found to improve performance – ignoring context. Conjuncture causation may hence go undetected.

By collecting large samples, results are sensitive to the relative frequency of certain cases since statistical techniques estimate population parameters for a given set of observations. Assume a data set contains a large number of organizations with: (a) high levels of customer responsiveness and low levels of shareholder responsiveness; and (b) high levels of employee responsiveness and low levels of shareholder responsiveness. Both types of companies hold large market shares. The estimate of the effect of shareholder responsiveness on market share would be negative, although for other combinations of causes shareholder orientation may positively impact market share. When calculating the average effect in a population of observations statistical parameters are powerfully influenced by the relative frequency of cases.

Finally, the goal of estimating the independent contribution of each effect is inconsistent with the goal of determining the different combinations of conditions which cause a certain phenomenon. For example, a number of different combinations of responsiveness may result in high levels of performance, but may go undetected when the focus is on calculating the individual contribution of each type of responsiveness. The model of causation that is implicit in common additive statistical techniques contradicts the idea of multiple conjuncture causation.

The first problem outlined, plagues all non-experimental work and most multivariate techniques were actually developed to provide the social sciences with a technique circumventing the problem of data deficiencies resulting from the observation of real-world phenomena not present in the analysis of experiments. Applying more sophisticated statistical techniques can circumvent the second and fourth problem outlined. For example, interaction models avoid the problem of additive causality and allow investigation of effects of one variable on another within categories of other variables. However, testing for interaction effects requires the researcher to have some a priori hypotheses regarding possible interaction effects. For the purpose of this study a large variety of interaction effects could be formulated due to little knowledge on possible interactions between the independent variables in question. Trying different specifications of an argument until one fits the data set without having theory-backed arguments for these specification, corresponds with the notion of "data in search of a theory" (Ullmann, 1985). By splitting the sample and estimating different models for different sub-populations one could circumvent the third problem mentioned, the effect of the relative frequency of cases on the estimation of population parameters. The resulting parameters could then be tested for significant differences. In addition to sample size problems splitting should be the result of theoretical concerns and not be guided by the data at hand.

In summary, the statistical techniques commonly used in social sciences achieve rigor through statistical manipulation that allows to calculate partial effects and control other – possibly competing – explanatory variables. Thus, making simplifying assumptions about the world enables the aim of generalization (these assumptions are often built into the statistical procedures). Instead of excluding extreme cases from analysis it may even be worthwhile to take a close look because they may provide particularly pure examples of certain social phenomena.

3.3.3. Comparative Modeling

Social science advances through comparisons: "Thinking without comparison is unthinkable. And, in the absence of comparison, so is all scientific thought and research" (Swanson, 1971, p. 145). Lieberson (1985, p. 44) states that social research "in one form or other, is comparative research." Statistical techniques rely on comparisons, too, by comparing sub-populations or by comparing results against the null hypothesis. *The* comparative method, however, focuses on cases, not variables, by identifying comparable instances of a phenomenon and analyzing theoretically important similarities and differences among them (Easthope, 1974; Smelser, 1976). The strength of the approach is the ability to address questions about outcomes which result from multiple and conjuncture causes – where different conditions may combine to result in the same outcomes. The comparative method uses two of Mill's methods of inductive inquiry (1843): the method of agreement and the indirect method of difference. These methods use all available and pertinent data concerning the preconditions of a specific outcome and, by examining the similarities and differences among relevant instances, elucidate its causes.

The comparative researcher faces two challenges, the identification of the various types of cases and the identification of the various types of causes. The first problem is concerned with the construction of useful empirical typologies, and demands simplifying the complexity among combinations of characteristics of cases and then constructing a model of the types that exist. Basically, this step requires identification of variables that might help explain the phenomenon of interest. The challenge here comes "in trying to make sense of the diversity across cases in a way that unites similarities and differences in a single coherent framework" (Ragin, 1987, p. 20). The comparative method and traditional statistical techniques both offer opportunities, which combined, may improve the quality of comparative analysis. Multivariate statistical techniques allow analysis of a large number of cases and are ideal for estimating probabilistic relationships between variables over the widest possible population of observations. The comparative method, in contrast, is ideally suited to identify invariant patterns common to a small number of cases. The benefits of the variable-oriented approach are reduced by complex, conjuncture causal arguments that require estimation of a large number of interaction effects while the benefits of the case-oriented method cannot be enjoyed when the number of cases is large.

The technique suggested to combine these benefits is a Boolean approach to qualitative comparison. The basic methodology is explained using the

Table 3. A Simple Truth Table of Stakeholder Responsiveness.

Shareholder Responsiveness (S)	Customer Responsiveness (C)	Employee Responsiveness (E)	OE	Number of Instances
1	1	1	1	5
0	1	1	0	4
1	1	0	1	2
1	0	1	1	7
1	0	0	0	5
0	0	1	?	0
0	1	0	1	4
0	0	0	0	4

hypothetical example given in Table 3, where for illustrative purposes only responsiveness to three stakeholder groups is considered. Boolean techniques use binary data, that is the presence/absence of explanatory variables and of the outcome variable are coded with 1/0. For a more differentiated discrimination one can employ several binary variables for each explanatory factor. The data are presented in the form of a truth table where each logical combination of values on the independent variable is represented in one row. Thus, each row may contain more than one case or no case at all. The number of cell entries does not directly impact Boolean analysis which focuses on combinations of independent variables leading to certain outcomes. However, these frequencies provide information about typical, rare or non-existent configurations of independent factors. In addition they might be used to solve contradictory rows which include both successful and unsuccessful companies.

Once such a complete truth table is constructed, various procedures (Boolean addition, multiplication, minimization and implication) allow the determination of causal conditions that lead to the outcome of interest. In the above examples the following combinations lead to OE:

$$\mathrm{OE} = \mathrm{SCE} + \mathrm{SCe} + \mathrm{ScE} + \mathrm{sCe}$$

A capital letter implies presence of the explanatory factor and a lower case letter implies absence. These expressions may now be reduced using Boolean algebra. For example, SCE combines with SCe to SC – implying that presence/absence of employee responsiveness is irrelevant in case of a presence of responsiveness to the other two constituencies. The most parsimonious expression after minimizing the other elements is the following:

$$\mathrm{OE} = \mathrm{SE} + \mathrm{Ce}$$

This equation states that organizations are successful when responsiveness to shareholders AND to employees is present OR when responsiveness to customers is combined with no employee responsiveness.

Assessing the conditions under which an outcome does not occur may be equally informative. Learning when organizations are not successful may provide as much information as learning about conditions that lead to success. The various causes identified may be assessed in terms of necessity and sufficiency. A cause is necessary if it must be present to produce an outcome, and sufficient if it can produce an outcome by itself. Ragin (1987) also outlines extensions of the Boolean approach to deal with less clear-cut truth tables as the one shown above. For example, some rows may not allow identification of an output value because no cases exhibiting such a combination of explanatory factors exist. This condition may be a sign for a socially constructed order, that is, not all possible combinations of features of cases exist. In order to address such diversity one may also employ the Boolean approach. Then the output variable in our example would not be concerned with the presence/absence of success, but whether cases exhibiting such a combination of causes exist. The resulting equation then informs about the combinations of responsiveness present in a population and thus shows how much diversity among cases exists. Another – likely – problem may arise if the output variable varies between cases: i.e. if the same combination of responsiveness leads to different outcomes. Following the tradition of case-oriented research, one should investigate whether any other explanatory variables were omitted from the analysis. Boolean analysis can once again be used to investigate the combinations of explanatory variables which lead to contradictory results. This can be a starting point to clarify the meaning of certain explanatory variables or to identify additional variables omitted from the analysis.

In summary, the Boolean approach can handle a large number of cases. The number of cases is of little relevance, what matters more is the number of combinations of causal variables. The technique is ideally suited to address patterns of multiple conjuncture causation. It takes into account each possible combination of causes and simplifies this complexity through Boolean principles. It allows identification of parsimonious explanations by developing logically minimal statements of conditions producing each outcome. The Boolean approach is both analytic and holistic. It does not take into account each individual case since cases with identical combinations of values on causal variables are pooled. But by comparing different combinations of values of explanatory variables it does not decompose wholes into parts. Finally, different explanations for certain outcomes can be evaluated and may coexist: "The typical end product of Boolean analysis is a statement of the limits of the causal variables identified with different theories, not their mechanical rejection or acceptance" (Ragin, 1987, p. 123).

4. METHODOLOGY

4.1. Database

This study will employ data from the COMPUSTAT database, a secondary source widely used in strategy research (Ailawadi, Borin & Farris, 1995; Baker, Faulkner & Fisher, 1998; Bowen & Wiersema, 1999; Glueck & Willis, 1979; Miles, Snow & Sharfman, 1993). The Compustat database contains financial information for over 10,300 U.S. corporations and provides 20 years of annual data, 12 years of quarterly data plus business and geographic segment data. While organizations have a great deal of latitude regarding their reporting practices, Compustat claims to remove any reporting variability through standardized collection techniques.

4.2. Sampling

Although concerns about the quality of the data in the Compustat database may be unwarranted with regard to accuracy, such concerns are certainly warranted regarding comprehensiveness. It might be preferable to learn about the variety of organizational behavior towards stakeholders through observation and in-depth interviews with the parties involved. However, such a procedure seems unrealistic given the requirements for the number of observations and number of time periods to be studied. Therefore, this research project aims to depict the behavior of organizations towards key stakeholders by analyzing data coming from the financial system of organizations, which is the only available source providing information for a large number of organizations over a significant period of time. Well aware that such information can only provide limited insight into the complex nature of these relationships, the analyses to follow will rely on this type of information. The following questions need to be answered to render these explorative analyses as illustrative and generalizable as possible: Which data is to be included? Which firms are to be included? Which years are to be included?

4.2.1. Choice of Variables

4.2.1.1. Responsiveness to organizational constituencies. Due to the limitation to financial variables, Table 4 presents the data items available in the Compustat database which seem most suitable in capturing the responsiveness of an organization towards the five key stakeholder groups outlined.

Organizations providing customers with goods and services offering better benefits than the competition should be able to enjoy growth in sales (Kohli &

Table 4. Stakeholder Responsiveness Measures Available in Compustat.

Customers	Percentage change in sales, gross profit margin, R&D expenses, advertising expenses
Employees	Labor and related expenses/employee, pension expenses/employee, labor and pension/sales
Shareholders	Total financial return
Suppliers	Days to pay accounts payable
Public	Tax rate, change in the number of employees

Jaworski, 1990; Kotler, 1991; Ruekert, 1992), as the provision of these benefits becomes more widely known and should attract business from new customers in addition to constant or even increasing sales generated from current customers. Furthermore, these companies should ceteris paribus be able to charge comparably high margins as their offerings provide higher value to customers. However, these two factors may show some degree of interrelatedness. For example, when Toyota introduced the Lexus brand targeted at the upper segment of the car market, a decision was made to initially give up margins in order to attract customers, i.e. lure them away from high-priced competitors like Mercedes or BMW who enjoy comparatively high gross profit margins but only moderate sales increases. While the former approach (i.e. the one chosen by Lexus) may lead to an above average increase in sales, the same increase may be even more impressive when customers are willing to provide the manufacturer with a high gross profit margin, i.e. pay a comparatively high price. Therefore, a combination of these two indices might be more informative than employing one of them separately to assess an organization's responsiveness towards customers. Organizations able to increase their sales while still enjoying high gross profit margins might provide more valuable offerings to customers than organizations where both indicators are low. When one indicator is low and the other one is high, the organization may currently aim more for market share than economic profit, target a segment limited in market size or be at a transitory state moving towards one of the extremes. Table 5 gives an overview of combinations for these variables relating them to the degree of customer responsiveness exhibited by an organization.

Table 5. Combinations of Gross Profit Margin and Sales Increase.

	High Sales Increase	Low Sales Increase
High GPM	Exceptionally attractive offering High customer responsiveness	Skimming strategy Targets a limited market
Low GPM	Fight for market share	Unattractive offering Low customer responsiveness

It may also be proposed that responsiveness towards customers translates into higher R&D as well as advertising expenses. Organizations interested in providing customers with offerings which are superior to competitor's offerings may decide to spend a higher percentage of sales for R&D activities and may also decide to invest into advertising activities to inform customers about the efforts in providing superior offerings. Once again, these indicators must not be viewed in isolation. It may well be possible to provide superior offerings with average-only investments in R&D by, for example, closely and quickly following moves by lead competitors. An organization able to learn effectively from competitor decisions or from signals other stakeholders send – be it through superior market research capabilities or experienced personnel – may be able to generate above-average offerings even with low R&D investments. Superior offerings might even be in little need of heavy advertising but will be chosen by customers through positive communication by various constituencies (e.g. the media providing enthusiastic coverage, other customers raving about the product, competition adapting their offering to correspond closely to the attractive offering). In isolation each of these indicators is of limited value. However, an organization scoring high on gross profit margin, sales change, R&D as well as advertising expenses, can be assumed to be more responsive to its customers than a comparable organization scoring low on all these indicators.

While a number of "soft" factors will influence the satisfaction employees derive from their work (which are not captured in the Compustat database), financial benefits will play a prominent role in determining employee satisfaction (Pfeffer, 1998; Smith, Kendall & Hulin, 1985). Expenses for labor and pensions (related to sales or the number of employees) illustrate how well employees fare financially. Especially in the U.S. where the provision of pension benefits is largely at the organization's discretion, this figure also provides relevant information how responsive an organization behaves towards its employees. One could argue that labor and pension expenses should be related to sales indicating the percentage of revenues that human resources are able to appropriate (Coff, 1999). Such an approach might show a biased picture of employee responsiveness for various reasons. One could think of an organization which achieves impressive sales by employee because of a highly motivated work force. Even when this work force is paid above average, an indicator relating these payments to sales might convey low employee responsiveness while in fact the high level of responsiveness is the reason for the high level of sales. In contrast, an organization which is not able to attract above-average business or has just encountered plummeting sales may show an impressive figure for employee financial benefits by sales (given the organization has not yet set part of its work force free) while in fact the payment and pension benefits may well be below industry standards. Therefore, relating labor and pension expenses to employees may make more sense than relating them to sales. After all, this figure (i.e. salary and pension benefits) compared to an industry average or an industry maximum will affect employees' assessment how much an organization values their input.

Shareholders will mostly be interested in the financial return an investment in the respective company offers (Friedman, 1962; Rappaport, 1986). Therefore the annual percentage return (generated by an increase in share price and dividend payouts) should be a useful indicator to measure responsiveness to shareholders. As stated earlier, it may be argued that an organization's influence on share prices is limited, but when market return is not taken at an absolute level but in comparison to relevant competitors, this relative performance should correspond to the evaluation of responsiveness to shareholders by this constituency. If the whole market shows a negative performance but the shares of the focal organization fare relatively better (i.e. share prices fall to a lesser extent), owners will most likely judge the organization's performance more positively than when the whole market increases significantly and share prices of the focal organization fare poorly (i.e. share prices increase only moderately).

Suppliers may be interested in a number of factors few of which are derivable from the financial system of organizations. Actual days to pay accounts payable is one factor which may be of interest to suppliers. However, since this value is usually (and also in the Compustat database) determined by various accounting figures showing which percentage of cost of goods sold is made up by accounts payable, it may hardly correspond to the actual speed at which supplier bills are paid. The interest of the public must be restricted to its fiscal dimension since other activities are not captured by an organization's financial system (although the public may also be interested in the financial well-being of organizations because of usually positive effects on employment or national welfare). While employment figures may be of interest to the public, relating them to sales or assets may produce weird results as a higher number may also be a sign of inefficiency. However, change in employment numbers – when combined with other data which is not available in Compustat, like environmental records or community activities – might be a useful indicator in this respect.

Therefore, this study will restrict analysis to three organizational constituencies, namely employees, shareholders and customers. First, these groups are the only ones for which organizational researchers and practitioners frequently claim a dominant position among organizational stakeholders. The respective effect of responsiveness to these constituencies on organizational performance is of vast interest to business research and practice. These three stakeholder groups are also the ones usually assumed to have the largest impact on organizational behavior. Donaldson and Lorsch (1983) argue that three constituencies significantly constrain managerial action, namely the capital market, the product market and the organization which roughly correspond to shareholders, customers and employees respectively. Kaplan and Norton (1996) also highlight the importance of these three groups by including measures specifically related to their well-being in the generic model of the Balanced Scorecard.

After examining the database certain measures outlined above are not used in the analyses. Some of the indicators are not compulsory items for inclusion in an organization's financial system and are therefore reported infrequently. For example, advertising as well as research and development expenses are reported by a minority of firms as are employee and pension expenses. Data availability for the construct of employee responsiveness turns out to be a critical selection criterion for organizations to be included in the analysis. In addition, meaningful analysis requires availability of data for a substantial number of companies within single industries for an extended period of time. As no industry in the database showed enough depth in data availability advertising and R&D expenses will not be employed for further analysis. The final indicators employed are the following: gross profit margin and percentage change in sales for customer responsiveness, labor and pension expenses per employee for employee responsiveness and total financial return to owners for shareholder responsiveness.

4.2.1.2. Success measures. Choice of a suitable performance measure may be the most critical question in any OE study. Reconsidering the four different approaches present in performance research, we remember that the process approach did not meet the requirement of acceptance among organizational researchers and practitioners. The strategic constituency approach faces the problem of tautology as responsiveness to the various constituencies is regarded a sign of effectiveness by these stakeholder groups. This research project adopts the strategic constituency approach but in search of drivers of effectiveness not in its operationalization. This approach will therefore be employed in a slightly different form by clustering organizations with high, average or low levels of responsiveness to all three stakeholder groups and then describe these groups by examining their performance on other more conventional success measures. The goal approach incorporating certain predefined success measures will therefore be employed. Specifically, return on assets, one of the most widely used criteria of organizational effectiveness (Berman et al., 1999; Capon, Farley & Hoenig, 1990; Narver & Slater, 1990) will be employed as well as the *z*-score measure of bankruptcy (Altman, 1971; Argenti, 1976) which assesses the potential of an organization to survive. As such, this measure closely corresponds to theoretical foundations of the stakeholder idea like resource dependence theory (Pfeffer & Salancik, 1978) or open systems theory (Katz & Kahn, 1966), which emphasize the importance of constituencies for organizational survival. The systems resource approach, not yet employed in empirical research except for one study (Chakravarthy, 1986), seems to be an attractive approach from a conceptual viewpoint since it focuses on the resources necessary for organizational survival and how well organizations fare in securing these resources. However, a more elaborate operationalization of the approach seems warranted before making it useable for empirical research. The decision to include two dependent variables in the analysis follows the suggestion by Venkatraman and Ramanujam (1986) to use multiple effectiveness measures to check for consistency of results.

4.2.2. Choice of Organizations

Comparing responsiveness to stakeholders only makes sense within single industries as the impact of different constituencies on organizational behavior is likely to differ between industries. One may argue that the borders between industries are less clear-cut than they used to be due to increased cooperation between firms coming from different industries. Especially organizations which show a high degree of diversification cannot easily be allocated to a single industry and caution is warranted when their responsiveness to stakeholders is compared to an industry standard. It must be assumed that respective organizational behavior differs between the different markets this organization competes in. By assessing the responsiveness to constituencies employing the above measures derived from the financial system one would end up with an average level across these markets. As such, the conclusions drawn might rely on a distorted picture of actual organizational behavior. For example, an organization like General Electric competing in industries like household apparel, electronics or banking may not exhibit the same types of behavior in each of these markets. GE should therefore not be analyzed on an organizational level but on a business unit level to take into account the differences between the various markets GE competes in. Since most of the measures chosen to assess responsiveness are available on an organizational level only, analyses should be restricted to industries showing a low level of diversification. Choice of such industries is "facilitated" by data availability resulting in satisfactory sample sizes only for three different industries: oil industry (SIC code 2911), utilities (SIC code 4911) and airlines (SIC code 4512). Within these industries, firms were included only if at least one indicator per constituency group was available for every year of data analysis resulting in 17, 35, and 17 organizations in the final samples respectively (see Appendix for an overview of the firms included).

4.2.3. Choice of Time Periods

Two reasons support analysis for multiple periods for the cross-sectional samples chosen. One, the value of findings increases once certain associations can be

Table 6. Data and Sample for the Empirical Analysis.

Success Measures	Responsiveness Measures	Industries Studied	Period of Analysis
Return on assets	Customers: Relative sales change, gross profit margin	Airlines	1988–1997
Measure of bankruptcy (*z*-score)	Employees: Average salary, average pension	Oil	
	Shareholders: Total financial return	Utilities	

established for more than a single period. While most research in strategic management claims to identify effects which are stable over time, such findings would be more valuable if the results really were based on the analysis of multiple periods (Bowen & Wiersema, 1999) and not on a tacit assumption that interactions between cause and effect remain constant over time. In addition, some part of organizational behavior may not immediately translate into some measure of organizational effectiveness calling for an evaluation of time lags between behavior and result of such behavior. Therefore a ten-year period including data from 1988 to 1997 is chosen for this analysis. Table 6 provides an overview of the measures, companies and periods employed in the data analysis.

4.3. Analytical Procedures

Three assumptions guide the data analysis process:

- The average level of responsiveness may change over time as the importance of constituencies for organizational performance is not a constant but will depend on the relationship between external market forces. Certain periods may call for extremely high levels of responsiveness towards employees (e.g. times where qualified human resources are limited) while other periods may warrant a high degree of customer responsiveness (e.g. times of fierce competition).
- Organizations may show a certain degree of consistency in their behavior towards constituencies. However, within a decade the behavior of an organization may change from being above average customer-responsive towards above-average shareholder-responsive as management decides that the relative importance of constituencies has changed. Therefore it makes sense to assess the respective level of responsiveness on a yearly basis (and then check for consistency in behavior) instead of taking the mean over the ten years and use this figure as the one measure to evaluate organizational behavior towards the respective constituency.

- Constituencies assess organizational responsiveness towards themselves not by evaluating the absolute levels of certain indicators but by comparing them to the ones of relevant alternatives, i.e. comparable organizations. Customer responsiveness measured as a function of sales change and gross profit margin may be high compared to the former period, but if all competitors show an even higher level of responsiveness this organization may still fare poorly when judged from the customer perspective. Responsiveness to stakeholders should thus not be evaluated at an absolute level but in comparison to relevant competitors. Therefore, responsiveness towards each of the constituencies will be measured in relation to the relevant set of competitors, i.e. within the respective industry, for every single period. The following steps are taken to assess an organization's level of responsiveness towards the three stakeholder groups observed; i.e. customers, shareholders and employees.
- For every period the relevant measures are ranked for each of the three industries. Missing values are substituted by assuming a linear development for each measure for the relevant organization. For example, labor expenses of 50,000 per employee in 1989 and 60,000 in 1991 would suggest expenses of 55,000 for 1990. Organizations are included in the analysis only if at least one measure per constituency per period is available. Rank 1 is assigned to the lowest absolute value and, given the nature of the measures used, reflects poorest performance on this indicator for the given year within an industry.
- If there are two indicators for one stakeholder group (as for customers and employees) the respective ranks are multiplied. For these cases an organization is considered more stakeholders x-responsive if it scores high on both indicators (high gross profit margin and high positive sales change or high labor expenses and high pension expenses) than if it scores high on one and low on the other. The results of these multiplications are ranked again and taken as indicators for the responsiveness towards this stakeholder group.
- For every year organizations are allocated to one of three groups on each of the three resulting responsiveness scores (employee responsiveness, shareholder responsiveness and customer responsiveness). The three groups comprise 20%, 60% and 20% of the respective organizations in an industry. The first group is made up of the top responsive organizations, the second group of the medium responsive ones and the third group of the least responsive ones. The idea guiding this allocation is that a constituency may only recognize – positively or negatively – outstanding responsiveness, not average responsiveness. It is assumed that constituencies have a tolerance zone (Doyle, 1992; Strandvik, 1994) where small variations in responsiveness do not lead to increased support or withdrawal of support by this constituency. Only when organizations do significantly over- or underperform compared to the standard responsiveness within an industry

Table 7. Combinations of Responsiveness to Employees, Customers and Shareholders.

	High Shareholder Responsiveness	Medium Shareholder Responsiveness	Low Shareholder Responsiveness
High customer responsiveness			
High employee responsiveness	27	26	25
Medium employee responsiveness	24	23	22
Low employee responsiveness	21	20	19
Medium customer responsiveness			
High employee responsiveness	18	17	16
Medium employee responsiveness	15	14	13
Low employee responsiveness	12	11	10
Low customer responsiveness			
High employee responsiveness	9	8	7
Medium employee responsiveness	6	5	4
Low employee responsiveness	3	2	1

will constituencies be – positively or negatively – affected. The cut-off level is chosen somewhat arbitrarily but it may be argued that performance has to differ substantially from industry standards before likely affecting performance. Therefore rather small clusters of organizations are formed at the two extremes, and a rather large one in the middle instead of taking three groups which are identical in size.

- Each organization is then assigned membership to one of 27 groups depending on the responsiveness levels for each of the three constituencies (see Table 7). By definition the expected size of these 27 groups will differ since the likelihood of an organization to be among the top or bottom 20% of organizations is lower than to be among the middle 60%. Cluster 14 is therefore be expected to contain significantly more organizations than, say, clusters 1, 2 or 27. If group membership for each of the three responsiveness scores were independent of each other, likelihood to end up in cluster 14 would be 21.6% ($0.6 \times 0.6 \times 0.6$) while likelihood to end up in cluster 2 would only be 2.4% ($0.6 \times 0.2 \times 0.2$).

4.4. Research Questions

4.4.1. How Frequent are the Various Combinations of Responsiveness?

It can be assumed that the different types of responsiveness will not be completely independent of each other and that certain degrees of responsiveness for stakeholder x might typically coincide with certain degrees of responsiveness for stakeholder y. By comparing the observed number of entries per cell with the expected number such instances of a "socially constructed order" may be identified. An organization may also not be able to show high responsiveness to all constituencies, because showing higher degrees of responsiveness to one constituency group consumes resources that cannot be used in being responsive to another group (Dearborn & Simon, 1958; Doyle, 1992). In addition, expectations held by different groups may not be compatible so that higher responsiveness to one group actually means lower responsiveness to another one. Therefore, those combinations of responsiveness where more than one constituency groups is enjoying above average responsiveness warrant close examination.

4.4.2. How are the Various Combinations Related to Organizational Success?
The author expects to find substantial performance differences for the various combinations of responsiveness. Based on the various arguments that the relationships to constituencies matter for organizational performance (Lines & Gronhaug, 1993; Rindova & Fombrun, 1999), one should be able to identify performance differentials between the various groups. Whether one of the three stakeholder groups deserves special responsiveness and an imbalance of attention to stakeholder groups seems warranted or whether organizations should aim to strike a balance between the attention levels to stakeholder groups constitutes the central question of this research project. To answer this question one could compare the mean for the success measure for each of the combinations. In line with the idea of comparative analysis one might also discriminate between different levels of performance (for example, by splitting organizations in three groups, namely the top 20%, the medium 60% and the bottom 20%) and assess the distribution of group memberships for the different combinations of responsiveness. Proponents of customer responsiveness might expect clusters 19–27 to fare exceptionally well, proponents of employee responsiveness would expect clusters 7–9, 16–18 and 25–27 to be among the top performers while shareholder responsiveness believers would argue that clusters 3, 6, 9, 12, 15, 18, 21, 24 and 27 should show top performance. Following the idea of stakeholder management, especially those organizations which show low responsiveness to at least one stakeholder group should perform worse than organizations succeeding in striking a balance among the different forms of responsiveness. Obviously, these analyses should also examine lagged effects of behavior on performance since responsiveness to a stakeholder group may not instantaneously translate into performance.

4.5. Analytical Limitations

The limitations of the procedure outlined explain the lack of studies focusing on the effect of attention to constituencies on organizational performance. A comprehensive attempt to fully capture the domain of each relationship with a stakeholder group seems unrealistic given the data requirements. The complexity and uniqueness of each individual relationship an organization is part of may even question the appropriateness of clustering certain relationship partners into stakeholder groups. After all, the relationship with customer A may be very different from the relationship with customer B and employee A may be in a very different relationship than employee B. However, it seems unrealistic – and impractical for deriving managerial implications – to take into account all relationships an organization is part of. Clustering partners into groups of stakeholders therefore seems a reasonable step. The responsiveness of an organization towards these groups may be indicated by a variety of practices, policies or mindsets. Studying these variables may present an impossible task and evaluation of how a stakeholder group assesses these factors may therefore only be inferred by studying a variety of different indicators.

The approach chosen in this study may therefore be considered incomprehensive but might constitute the most adequate procedure available. Financial figures are the most widely reported and least intrusive indicators to assess organizational responsiveness. However, some additional – partly financial, partly non-financial – information would be very valuable to attain broader knowledge about an organization's responsiveness. In the case of customers, information on rebates, free products, costs to fulfill guarantees may all contain additional information. Employees may also benefit from a number of additional organizational actions not included in their base salaries. Collecting such information for a number of organizations over multiple periods might not be a big problem once their interest in the research question is gained.

Taking into consideration three organizational constituencies only and leaving out some other important groups may be a focus of criticism. After all, in some industries responsiveness towards the community or towards suppliers may affect organizational performance as much as behavior towards the constituencies studied. Two reasons for the focus chosen can be given: First, the data available would not allow useful assessment of behavior towards other constituencies. Most commonly used financial items (taxes for the public or days to pay accounts available for suppliers) are either out of organizational control (and would therefore not be a sign of responsiveness), or do not show any resemblance to actual organizational behavior since they are artifacts caused by accounting standards. Secondly, the three groups outlined are the ones usually cited as the most influential ones and most frequently are top-of-mind when allocation of organizational responsiveness to different constituencies is discussed.

The procedure chosen to assess responsiveness and allocating organizations to groups differing in their levels of responsiveness may be criticized for the loss of information involved. After all, whether an organization holds rank 8, 12 or rank 26 for a given year on some responsiveness measure in the utility industry is irrelevant, it will be allocated to the medium group even though the raw measures used to derive group membership might vary substantially. However, due to the limited extent to which the measures do capture the respective character of the relationship, this procedure may be more suitable than taking the raw measures. Comparative analysis is not used to claim that a certain change in one of the measures causes a certain change in organizational effectiveness, but uses group membership to derive any implications for managing effectiveness. Therefore the type of analysis used may limit the concerns due to choice of variables. The method chosen seems to provide a highly suitable instrument to answer the research question, namely how the allocation of responsiveness to stakeholder groups affects effectiveness. If one were interested in assessing the precise impact of increasing responsiveness to some stakeholder on performance concerns for data quality might be larger than when allocation to one of three groups matters. If results turn out to show interesting patterns, more elaborate causal analysis may be initiated.

5. RESULTS

5.1. Combinations of Stakeholder Responsiveness

The extent to which organizations can attend to their respective stakeholders is limited by two factors: First, the amount of time and attention capacity, while probably not fixed, may not be unlimited (Doyle, 1992; Simon, 1957). Thus, the degree of attention to various groups has to be weighed against each other. Secondly, some constituencies may perceive high levels of responsiveness to another group as negatively affecting the responsiveness they experience; e.g. when employees are able to appropriate large parts of organizational rent (Coff, 1999) and shareholders realize these activities they may react negatively. Therefore, some limit is put on the extent to which individual constituency groups are to be attended to.

The allocation of organizations to one of 27 groups depending on their extent of responsiveness (low, middle, high) to the three constituencies studied contains some information regarding the possibility of attending disproportionately

Table 8. Frequency of Stakeholder Responsiveness Combinations.

	Group			Airline		Oil		Utility		Overall	
	CUST	EMP	SHA	EXP	OBS	EXP	OBS	EXP	OBS	EXP	OBS
1	L	L	L	2.2	2	**2.2**	**7**	**2.8**	**9**	**7.2**	**18**
2	L	L	M	**5**	**0**	**5**	**12**	8.4	5	18.4	17
3	L	L	H	**2.2**	**0**	2.2	2	2.8	4	7.2	6
4	L	M	L	**5**	**12**	5	3	8.4	16	18.4	31
5	L	M	M	11.2	10	**11.2**	**4**	25.2	22	47.6	36
6	L	M	H	**5**	**2**	5	3	8.4	5	18.4	10
7	L	H	L	2.2	4	2.2	2	2.8	5	7.2	11
8	L	H	M	5	6	5	5	**8.4**	**2**	18.4	13
9	L	H	H	2.2	4	**2.2**	**1**	2.8	2	7.2	7
10	M	L	L	**5**	**1**	5	3	8.4	10	18.4	14
11	M	L	M	11.2	12	11.2	7	25.2	25	47.6	44
12	M	L	H	**5**	**2**	**5**	**2**	8.4	9	18.4	13
13	M	M	L	11.2	7	11.2	8	25.2	15	47.6	30
14	M	M	M	25.3	34	25.3	31	75.6	76	125.6	141
15	M	M	H	11.2	9	11.2	15	25.2	26	47.6	50
16	M	H	L	5	7	5	9	8.4	6	18.4	22
17	M	H	M	11.2	12	11.2	12	25.2	34	47.6	58
18	M	H	H	5	6	5	5	8.4	9	18.4	20
19	H	L	L	2.2	4	**2.2**	**1**	**2.8**	**1**	7.2	6
20	H	L	M	5	9	5	4	8.4	7	18.4	20
21	H	L	H	**2.2**	**9**	2.2	2	**2.8**	**0**	7.2	11
22	H	M	L	**5**	**2**	5	7	8.4	7	18.4	16
23	H	M	M	11.2	7	11.2	13	25.2	30	47.6	50
24	H	M	H	5	8	5	6	8.4	12	18.4	26
25	H	H	L	**2.2**	**1**	**2.2**	**0**	**2.8**	**1**	**7.2**	**2**
26	H	H	M	**5**	**0**	5	4	8.4	10	18.4	14
27	H	H	H	**2.2**	**0**	2.2	2	2.8	2	7.2	4

high to all stakeholder groups and the nature of the trade-offs between these responsiveness levels. If one assumes independence between the responsiveness level exhibited to each of the three groups studied the expected sizes regarding the combinations of responsiveness should roughly correspond to the results from multiplying the probabilities of belonging to each group. Table 8 contains the expected (EXP) and observed (OBS) number of organizations belonging to each of the 27 groups. The groups are described regarding the extent of responsiveness (low L, medium M or high H) exhibited to the three constituencies customers (CUST), employees (EMP), and shareholders (SHA). If the number of observed expectations more than doubles, the expected observations, or only reaches 50% of the expected observations, the respective numbers are shown in bold, if three times as many observations are found than would be expected or only one third the figures in these cells are shown in bold and the cells are shaded.

A number of considerable deviations from what could be expected if a random process determined group membership can be identified. The most significant deviation directly relates to the trade-off between responsiveness to different constituencies discussed before. Group 1 consisting of organizations which exhibit low responsiveness to all three constituencies contains substantially more organizations than would be expected while group 27 contains less. In fact, 4.5 times more organizations showed low responsiveness to all groups than high responsiveness to all groups while the expected number of observations would be equal. One may conclude that it is more difficult to be highly responsive to all three constituency groups observed than it is to be below average responsive to all of them. The other outstanding deviation for the combined samples shows an interesting pattern as well. The combination of above average responsiveness to employees and customers with below average responsiveness to shareholders is less frequent than expected. Attending disproportionately to the former two groups at the cost of the latter may be a difficult road to take since one would expect shareholders to penalize such behavior by reallocating their funds. Overall, comparatively few organizations manage to score high on responsiveness scores for both employees and customers, whereas comparatively many organizations score high on one and low on the other measure.

Looking at the above results from an industry perspective we find that the major source of the trade-off between employee responsiveness and customer responsiveness seems to be the airline industry. For all combinations of High and Low for either employees or customers we find a larger-than expected number of observations. In addition, only three observations exhibit a high-high or low-low combination for these two scores, further supporting the idea of a strong trade-off between the two responsiveness scores. While not entirely consistent with claims made in the service literature (Bitner, 1990; Parasuraman, 1987), we find close correspondence in the Austrian airline industry where one carrier is commonly claimed to have low employee responsiveness and high customer responsiveness (Tyrolean airways), and another one high employee responsiveness and rather low customer responsiveness (Austrian Airlines). The former is also the financially by far more successful one, an observation which needs to be investigated later on for this sample. In contrast, the oil industry contains comparably many organizations exhibiting both low employee and customer responsiveness.

The expected number of entries for cells with at least two responsiveness scores being high or at least two being low is 84, whereas the actual numbers are 84 and 103 respectively further supporting the idea that it is difficult to be highly responsive to more than one constituency. Whether or not it is even worthwhile

Table 9. Summary of Success Measures in the Three Industries.

Measure	Airline ($n = 170$)	Oil ($n = 170$)	Utilities ($n = 350$)
Average ROA	2.01%	4.08%	2.99%
Std. Dev. of ROA	11.6	3.14	2.67
Average z-score	2.06	2.52	1.17
Std. Dev. of z-score	1.47	0.61	0.35

to aim for such behavior and bother about being responsive to more than one (or even one) group is dealt with in the next section.

5.2. Impact on Organizational Effectiveness

Two measures are used to evaluate an organization's effectiveness, return on assets and the z-score measure of bankruptcy. Some of the analyses to follow will use the raw scores of these measures, while most will distinguish between three levels of performance, corresponding to the distinctions made for the independent variables, extent of responsiveness to constituencies. Three groups of performers will be distinguished, exceptional performers (top 20% of an industry), medium or average performers (60% of an industry) and low performers (bottom 20% of an industry for a given year). Table 9 contains some descriptive measures to support understanding of results to follow.

We see that the three industries show considerable differences in their average success measures with the airline industry showing by far the highest degree of variability. During the time of analysis, the airline industry was characterized by extremely strong economic cycles causing success rates to fluctuate wildly while the two other industries are showing a higher level of stability. It is also interesting to note that the threshold for the z-score measure of bankruptcy (Altman, 1971; Argenti, 1976) was originally set at 3; i.e. companies showing levels higher than three faced little to no danger of going bankrupt while lower scores were taken as a sign of financial distress. Both a higher degree of leverage during the time of this analysis as well as a lower level of retained earnings and fairly high asset intensity in the three industries studied may have caused this value to be significantly below 3.

5.2.1. Individual Constituency Responsiveness and Performance

5.2.1.1. Effects on absolute levels of performance. A number of scholars state that certain constituencies deserve high organizational responsiveness and that such behavior will be rewarded by higher organizational performance levels. Marketing

Table 10. Effect of Customer Responsiveness on Performance.

	Air		Oil		Utilities	
	ROA	*z*-score	ROA	*z*-score	ROA	*z*-score
Low CUST	−6.96%	0.89	2.79%	2.3	1.72%	1.04
Medium CUST	2.60%	1.94	4.62%	2.63	3.32%	1.22
High CUST	9.64%	3.48	4.08%	2.46	3.26%	1.17
F-Value	27.168	49.165	4.889	4.427	10.359	6.755
Sign	0.000	0.000	0.009	0.013	0.000	0.001

scholars most often focus on customers, finance scholars on shareholders and human resource scholars on personnel when making such claims. While this research project wants to investigate the joint effect of behavior towards different groups on performance, the following section will highlight how different levels of responsiveness to individual constituency groups relate to performance. Table 10 focuses on the relationship between different levels of responsiveness to customers (CUST) and average effectiveness, both in terms of return on assets (ROA) and the bankruptcy measure (*z*-score).

Low levels of customer responsiveness seem to be detrimental for success in all three industries with the most substantial impact in the airline industry. Here moving from one group to the next higher one in terms of responsiveness translates into an increase in return on assets of more than 7%. The differences between the groups are significant for all industries but being highly customer responsive may not pay off for the oil and utility industry where medium levels show the highest success rates. One may conclude that avoidance of low levels of customer responsiveness is generally helpful and aiming for high levels pays off in the airline industry.

Table 11 focuses on the relationship between different levels of responsiveness to employees (EMP) and effectiveness, both in terms of average ROA and the

Table 11. Effect of Employee Responsiveness on Performance.

	Air		Oil		Utilities	
	ROA	*z*-score	ROA	*z*-score	ROA	*z*-score
Low EMP	7.92%	3.56	2.92%	2.33	3.02%	1.23
Medium EMP	0.76%	1.76	4.08%	2.52	2.84%	1.14
High EMP	−0.92%	1.27	5.23%	2.69	3.4%	1.2
F-Value	7.39	40.86	5.702	3.542	1.185	1.721
Sign	0.001	0.000	0.004	0.031	0.307	0.18

Table 12. Effect of Shareholder Responsiveness on Performance.

	Air		Oil		Utilities	
	ROA	*z*-score	ROA	*z*-score	ROA	*z*-score
Low SHA	−4.2%	1.3	2.52%	2.33	1.43%	0.96
Medium SHA	3.52%	2.17	4.43%	2.57	3.46%	1.23
High SHA	4.82%	2.57	4.84%	2.58	3.13%	3.13
F-Value	8.347	8.605	7.134	2.417	16.537	17.754
Sign	0.000	0.000	0.001	0.092	0.000	0.000

bankruptcy measure (*z*-score). The results for employee responsiveness are less clear-cut. In fact, the relationship is contradictory for the industries studied. Organizations in the utility industry do hardly differ in performance whether employee responsiveness is comparatively high, medium or low. The oil industry benefits from higher levels of employee responsiveness with a steady increase both in terms of return on assets and *z*-score when moving from one of the groups to another. Interestingly enough, high levels of employee responsiveness are not rewarded in the airline industry where they result in negative return on assets. In addition, organizations showing low levels of responsiveness substantially outperform the rest.

Table 12 focuses on the relationship between different levels of responsiveness to shareholders (SHA) and average effectiveness, both in terms of ROA and the bankruptcy measure. Results for shareholders are comparable to results for customers with a generally positive effect of higher responsiveness on performance. Especially the differences between low and medium levels of responsiveness are significant with additional benefits resulting from high levels of shareholder responsiveness in the airline and oil industry and a small loss in the utility industry. Results suggest that comparatively low levels of shareholder responsiveness are not favoring high performance levels in terms of return or the *z*-score and aiming for high levels of shareholder responsiveness seems most beneficial in the airline industry.

5.2.1.2. Effects on group membership. To avoid potentially biasing effects of outliers on the interpretation of results and to illustrate findings more in line with the comparative approach to analysis (Ragin, 1987), a slightly different view is now offered. Organizations are allocated to one of three groups depending on their organizational performance (top 20, medium 60 and bottom 20%). Then group membership of the individual stakeholder responsiveness scores are related to group membership of performance to explore which combinations are typical and which hardly appear jointly. In addition to concurrent performance levels, lagged performance is examined as well. Results are given for return on assets only since the pattern is almost identical for *z*-scores. The number of observations is 10% lower for the lagged performance columns since only nine years of data can be used. For the airline and oil industry the split is not 20–60–20, but 23–53–23 because the number of observations per year (17) does not allow the allocation of organizations into three groups corresponding to the 20–60–20 split.

The level of customer responsiveness appears to affect group membership for both the concurrent and the lagged performance levels in all industries (see Table 13). Generally higher levels of customer responsiveness are more typically related to higher levels of performance, while low levels of customer responsiveness are more often found in combination with low levels of performance. Especially the airline industry shows an interesting pattern. While high customer responsiveness is not a necessary condition for high performance, it renders low performance very unlikely, and while low customer responsiveness does not necessarily lead to low performance achieving high performance becomes quite

Table 13. Effect of Customer Responsiveness on Membership in Performance Groups.

	Performance			Performance (1 Year Later)		
	Bottom	Medium	Top	Bottom	Medium	Top
Airline						
Low CUST	25 (63%)	14 (35%)	1 (2%)	20 (56%)	14 (39%)	2 (6%)
Medium CUST	14 (16%)	61 (67%)	15 (17%)	16 (20%)	51 (63%)	14 (17%)
High CUST	1 (2%)	15 (38%)	24 (60%)		16 (44%)	20 (56%)
Sum	40 (23%)	90 (53%)	40 (23%)	36 (23%)	81 (53%)	36 (23%)
χ^2 (sign)		75.675 (0.000)			50.003 (0.000)	
Oil						
Low CUST	19 (49%)	17 (44%)	3 (8%)	13 (37%)	19 (54%)	3 (9%)
Medium CUST	13 (14%)	53 (57%)	26 (28%)	15 (18%)	43 (52%)	25 (30%)
High CUST	8 (20%)	20 (51%)	11 (28%)	8 (23%)	19 (54%)	8 (23%)
Sum	40 (23%)	90 (53%)	40 (23%)	36 (23%)	81 (53%)	36 (23%)
χ^2 (sign)		20.599 (0.000)			8.725 (0.068)	
Utilities						
Low CUST	26 (37%)	34 (48%)	10 (14%)	24 (38%)	29 (46%)	10 (16%)
Medium CUST	30 (14%)	134 (64%)	46 (21%)	31 (16%)	116 (61%)	42 (22%)
High CUST	14 (20%)	42 (60%)	14 (20%)	8 (13%)	44 (70%)	11 (17%)
Sum	70 (20%)	210 (60%)	70 (20%)	63 (20%)	189 (60%)	63 (20%)
χ^2 (sign)		17.270 (0.002)			17.549 (0.002)	

difficult. In fact, only once is an organization with high customer responsiveness exhibiting low performance and only once is low customer responsiveness related to top performance. This relationship holds for the lagged performance level as well where we can observe that highly customer responsive organizations are never among the worst performers one year later. Medium customer responsiveness does not show such a poignant distribution, but does at least decrease the likelihood to be among the worst performers (but also among the top ones). More than half of the bottom group in terms of customer responsiveness also are among the bottom performers.

The patterns for the oil and utility industry are fairly similar with low levels of customer responsiveness not often related to high levels of performance. Moving from a medium to a high level of customer responsiveness does not seem to make much of a difference in terms of performance. The likelihood of being among the bottom level performers is even higher for concurrent high levels of customer responsiveness than for concurrent medium levels.

One of the reasons for the differences between the airline and the other two industries in terms of the effect of high customer responsiveness may be the respective level of competition. It may well be that both customers of utilities and oil companies show a certain level of loyalty which is not endangered by providing just average levels of responsiveness (this may be "true" loyalty as a result of satisfaction or high exit barriers for customers of oil or utility companies). However, the airline industry does not have extremely high exit barriers and high levels of customer responsiveness may therefore result in higher organizational effectiveness.

Employee responsiveness significantly affects the distribution of performance only in the airline and the oil industry (see Table 14). Organizational success levels in the utility industry do not appear to be related to employee responsiveness although high employee responsiveness tends to decrease the likelihood of low performance in the following year and low employee responsiveness seems to increase the likelihood of low performance one year later. Results for the oil industry suggest negative consequences for low employee responsiveness. No such organization achieves high performance in the same year or the next year. Similarly, high employee responsiveness is rewarded by above average likelihood to perform well, and below-average likelihood to perform poorly, both in the same and the following year. Medium levels of responsiveness tend to be related to above-average levels of performance.

Quite the opposite is true for the airline industry where low levels of employee responsiveness combine poignantly with high levels of performance. Specifically, among the top performers more than 50% of organizations boast low levels of employee responsiveness. High levels of employee responsiveness typically

Table 14. Effect of Employee Responsiveness on Membership in Performance Groups.

	Performance			Performance (1 Year Later)		
	Bottom	Medium	Top	Bottom	Medium	Top
Airline						
Low EMP	4 (10%)	13 (33%)	22 (56%)	2 (6%)	15 (43%)	18 (51%)
Medium EMP	21 (23%)	55 (60%)	15 (17%)	23 (28%)	43 (52%)	16 (20%)
High EMP	15 (38%)	22 (55%)	3 (8%)	11 (31%)	23 (64%)	2 (6%)
Sum	40 (23%)	90 (53%)	40 (23%)	36 (23%)	81 (53%)	36 (23%)
χ^2 (sign)		73.716 (0.000)			34.285 (0.000)	
Oil						
Low EMP	19 (48%)	21 (52%)		19 (53%)	17 (47%)	
Medium EMP	19 (21%)	47 (52%)	24 (27%)	12 (15%)	49 (61%)	20 (25%)
High EMP	2 (5%)	22 (55%)	16 (40%)	5 (14%)	15 (42%)	16 (44%)
Sum	40 (23%)	90 (53%)	40 (23%)	36 (23%)	81 (53%)	36 (23%)
χ^2 (sign)		30.271 (0.000)			34.294 (0.000)	
Utilities						
Low EMP	12 (17%)	48 (69%)	10 (14%)	13 (21%)	41 (65%)	9 (14%)
Medium EMP	43 (21%)	120 (57%)	46 (22%)	41 (22%)	105 (56%)	42 (22%)
High EMP	15 (21%)	42 (60%)	14 (20%)	9 (14%)	43 (67%)	12 (19%)
Sum	70 (20%)	210 (60%)	70 (20%)	63 (20%)	189 (60%)	63 (20%)
χ^2 (sign)		3.031 (0.553)			4.403 (0.354)	

combine with poor performance and even medium levels are rather related to low than high performance. Whether the findings support the hypothesized cause-effect relationship between responsiveness to a stakeholder group and performance is not obvious. In the oil industry, for example, organizations are either performing because of high levels of responsiveness to employees or only those organizations which are successful may be able to afford to be above-average responsive to their employees. Why performance and employee responsiveness are negatively related in the airline industry can only be guessed. It may well be that the airline industry during that time faced substantial cost pressure and an abundant pool of human resources that made attractive payment packages unnecessary and even rewarded organizations with "lousy" employee responsiveness for such behavior.

The general pattern of combinations for shareholder responsiveness levels and levels of organizational performance (see Table 15) are similar to the results found for customer responsiveness. All industries show a tendency towards a positive relationship between shareholder responsiveness and performance. When

Table 15. Effect of Shareholder Responsiveness on Membership in Performance Groups.

	Performance			Performance (1 Year Later)		
	Bottom	Medium	Top	Bottom	Medium	Top
Airline						
Low SHA	21 (53%)	16 (40%)	3 (8%)	17 (47%)	15 (42%)	4 (11%)
Medium SHA	15 (17%)	53 (59%)	22 (24%)	14 (17%)	48 (59%)	19 (24%)
High SHA	4 (10%)	21 (53%)	15 (38%)	5 (14%)	18 (50%)	13 (36%)
Sum	40 (23%)	90 (53%)	40 (23%)	36 (23%)	81 (53%)	36 (23%)
χ^2 (sign)		17.669 (0.001)			15.164 (0.004)	
Oil						
Low SHA	17 (43%)	19 (48%)	4 (10%)	13 (36%)	17 (47%)	6 (17%)
Medium SHA	15 (16%)	53 (58%)	24 (26%)	18 (22%)	44 (53%)	21 (25%)
High SHA	8 (21%)	18 (47%)	12 (32%)	5 (15%)	20 (59%)	9 (27%)
Sum	40 (23%)	90 (53%)	40 (23%)	36 (23%)	81 (53%)	36 (23%)
χ^2 (sign)		13.497 (0.009)			5.068 (0.280)	
Utilities						
Low SHA	27 (39%)	36 (51%)	7 (10%)	23 (37%)	32 (51%)	8 (13%)
Medium SHA	29 (14%)	133 (63%)	49 (23%)	29 (15%)	122 (64%)	39 (21%)
High SHA	14 (20%)	41 (59%)	14 (20%)	11 (18%)	35 (57%)	16 (26%)
Sum	70 (20%)	210 (60%)	70 (20%)	63 (20%)	189 (60%)	63 (20%)
χ^2 (sign)		21.986 (0.000)			15.206 (0.004)	

an organization belongs to the bottom group it infrequently enjoys high levels of performance. In contrast, medium and high levels of shareholder responsiveness show varying patterns for the three industries studied. While relatively many combinations of high performance which characterize the airline industry, high shareholder responsiveness results for the other two industries do not unanimously support striving for high shareholder responsiveness. Medium levels may even be preferable since high levels increase the likelihood of belonging to the bottom performers.

In sum, results for individual stakeholder relationships support most arguments on the importance of single constituency groups found in the literature: Typically higher levels of responsiveness are related to higher levels of performance with the notable exception of employee responsiveness in the airline industry. Here results would suggest to strive for low responsiveness. Given that competition between airline companies during the decade of analysis has strongly focused on price and other differentiation possibilities were not frequently attempted, the identified pattern is no big surprise. Targeting segments with lower levels of price sensitivity and focusing more on other features may warrant higher levels of employee responsiveness to be successful. Findings also support the basic hypothesis of the stakeholder idea, namely that there is a limit to what responsiveness to one single stakeholder group can do for organizational effectiveness: Most findings support the idea of both a minimum and (in most cases) a maximum level of responsiveness which is advisable. Faring both below and above this level causes either low performance or is (usually) not related to high performance, suggesting average levels of responsiveness to be a useful strategy. This argument will be investigated more thoroughly in the next chapter where the joint effect of different levels of responsiveness to the three stakeholder groups is studied.

5.2.2. Stakeholder Value and Performance

This section will concentrate on the relationship between organizational performance and group membership in terms of combinations of the respective responsiveness levels to customers, employees and shareholders. Thus, it focuses on the core question of instrumental stakeholder theory, namely whether behavior which is in line with the stakeholder idea benefits OE and whether behavior contrary to the stakeholder idea is detrimental to OE. By allocating organizations to one of 27 groups, depending on the level of responsiveness to three constituencies differences in the level of performance may be identified and some explorative implications as to the effects of stakeholder responsiveness derived. Two options to study the research question are available: One may either compare absolute performance levels for each of the 27 groups and hence evaluate strategic postures towards constituencies or one might take a more explorative avenue and discriminate between top, medium and bottom performance levels in the same way as was discriminated between responsiveness levels towards stakeholder groups. While the first avenue would provide precise figures, these results would warrant some caution. First, the effect of outliers may bias findings and lead to conclusions not supported by the remaining data. While outliers could be removed before analysis, this step always involves subjective decision making by the researcher and reduces the number of available observations even more. Second, absolute performance figures tend to fluctuate between periods. Then, overall findings may be biased towards periods which exceptional industry upswings or downturns. The impact of these fluctuations can be avoided by replacing absolute performance levels with standardized ones (where the average industry performance is scaled to 0 with a standard deviation of 1), but, once again, when averaging these values over a number of observations, periods with extreme outliers may strongly affect results. Finally, certain types of behavior may only occur in one or very few periods. If extremely high or low industry performance levels characterize such periods, conclusions may hinge on such periodical circumstances.

Therefore the analyses focus on group membership; i.e. it investigates the effect of the combination of responsiveness to constituencies on three different performance levels to avoid the problems discussed above. First, outliers cease to be a problem. Extremely high or low performance puts an organization into either the top or bottom group in terms of performance, but does not bias overall findings upwards or downwards since absolute performance levels are not taken into account for interpreting results (only for allocating organizations into groups). Second, industry-wide fluctuations in OE do not affect results since, irrespective of the average performance level within an industry, organizations are allocated to one of three groups, top, medium or low performers. While this procedure may even put an organization with negative ROA figures into the top performing group, it still provides information whether a certain behavior allows an organization to be less affected by industrial downturns. Finally, whether or not such a given type of behavior has only occurred in few periods or all periods does not matter since no absolute performance figures are taken into account, but group membership depending on the performance in relation to the rest of the industry is discussed.

Results will be presented for each industry, first showing how combinations of responsiveness to each of the three groups relate to the distribution of organizational performance. Thus, information is provided to what extent certain combinations relate to low, medium or high performance. Boolean analysis will be employed to summarize these findings in a more comprehensive way and possibly identify necessary and sufficient preconditions for organizational performance. These conditions can then be compared for all three industries to maybe identify combinations of responsiveness which are effective across industries. Such an ex-post approach to identify cross-industry relationships may be more sensible than analyzing all cases jointly since the results of the former chapters showed considerable differences between industries. It may therefore be anticipated that the most effective combinations differ and an overall analysis may well disguise certain combinations effective in any one industry. In addition, the 27 groups are then split into two responsiveness types where for one type the relationship to no organizational constituency is characterized by low responsiveness (groups 14, 15, 17, 18, 23, 24, 26, 27), and for the other type at least one relationship is characterized by low responsiveness (the remaining groups). These two groups may roughly correspond to organizations with and without a stakeholder orientation. Organizations with at least one relationship to a stakeholder group characterized by low responsiveness may be considered to not act along the stakeholder idea, while those in no relationship characterized by low responsiveness may be considered stakeholder-oriented. Performance can then be assessed for these two groups in the same way as for the 27 groups outlined before.

Table 16. Effect of Stakeholder Responsiveness on Performance in the Airline Industry.

Bool	CUST	EMP	SHA	Current ROA			Lagged ROA		
				Low	Med	Top	Low	Med	Top
1	L	L	L	2				1	
2	L	L	M						
3	L	L	H						
4	L	M	L	9	3		8	3	1
5	L	M	M	3	7		4	4	
6	L	M	H	2			1	1	
7	L	H	L	4			3	1	
8	L	H	M	4	2		3	3	
9	L	H	H	1	2	1	1	1	1
10	M	L	L		1		1		
11	M	L	M	1	7	4	1	5	5
12	M	L	H			2			1
13	M	M	L	4	3		3	2	
14	M	M	M	2	26	6	4	22	6
15	M	M	H	1	7	1	3	4	1
16	M	H	L	1	6		2	4	1
17	M	H	M	5	6	1	2	8	
18	M	H	H		5	1		6	
19	H	L	L	1	1	2		3	1
20	H	L	M		2	7		4	5
21	H	L	H		2	7		2	6
22	H	M	L		1	1		1	1
23	H	M	M		3	4		2	3
24	H	M	H		5	3		4	4
25	H	H	L		1				
26	H	H	M						
27	H	H	H						

5.2.2.1. Airline industry. The groups with the lowest absolute level of ROA are group 4 ($-13\%, n = 12$), 1 ($-10\%, n = 2$), and 6 ($-8\%, n = 2$), all characterized by low levels of customer responsiveness. The groups with the highest absolute performance levels are group 20 (18%, $n = 9$), 12 (11%, $n = 2$) and 23 (9%, $n = 7$), all characterized by at least medium responsiveness levels to shareholders and customers. Table 16 shows how different combinations of responsiveness are related to different performance levels, both concurrent ones and that one year later.

To belong to the top performers within the airline industry at least average responsiveness to customers is warranted, exhibiting a high level of responsiveness to customers makes bottom performance very unlikely. Combinations of medium responsiveness to customers with high or medium responsiveness to employees seem to be negatively related to performance. The, by definition, most typical combination of medium responsiveness to all three stakeholders appears to be a fruitful avenue to avoid low performance since the expected number of bottom performers, eight (= 34 × 0.235) is not reached.

In order to summarize the above table a comparative approach to analyzing these 170 cases is taken (Ragin, 1987). To construct a truth table with a smaller number of rows, only two types of responsiveness to a constituency are distinguished. Presence of responsiveness encompasses both the medium and top group. While the level of responsiveness exhibited may not be impressive, it means that an organization is not among the bottom fifth of the industry in a given year. Absence of responsiveness indicates an extremely low level of responsiveness with the cut-off chosen to be the bottom 20% of the industry in a given year (for the airline and oil industry 23.5% due to the number of observations available for each year, 17). With three constituencies studied, the truth table consists of eight rows with all possible combinations of responsiveness presence or absence (see Table 17). For example, row six in the truth table below consists of the combinations 13, 16, 22 and 25 from the above table each characterized by presence of responsiveness to customers and employees and absence of responsiveness to shareholders. The next column shows the number of observations exhibiting this very combination of presence/absence and the following three columns the performance levels of these observations. The "Outcome" column shows a 0 when there is a tendency towards low performance, a 1 when the combination of responsiveness typically leads to high performance, and a question mark when the combination leads to both high and low performance.

The following combinations of presence or absence of responsiveness to the three constituencies typically lead to low performance (LP): absence of responsiveness to all three stakeholder groups, absence of responsiveness to customers

Table 17. Truth Table for the Airline Industry.

	CUST	EMP	SHA	Instances	Low	Medium	High	Outcome
1	0	0	0	2	2	0	0	0
2	0	0	1	0	0	0	0	–
3	0	1	0	16	13	3	0	0
4	0	1	1	22	10	11	1	0
5	1	0	0	5	1	2	2	?
6	1	1	0	17	5	11	1	0
7	1	0	1	32	1	11	20	1
8	1	1	1	76	8	52	16	1

and shareholders and presence of responsiveness to employees, presence for both employees and shareholders combined with absence for customers, and absence of shareholder responsiveness with presence for employees and customers. High performance (HP) typically is achieved when all relationships are characterized by the presence of responsiveness or when it is absent only for employees. In Boolean terms the following expressions describe these observations:

$$LP = ces + cEs + cES + CEs$$

$$HP = CeS + CES$$

Combining ces and cEs to cs can further reduce the combinations leading to low performance, since employee responsiveness is irrelevant once both other responsiveness types are absent. In the same way cEs and CEs can be reduced to Es as well as cEs and cES to cE. The combinations related to high performance (CeS and CES) may accordingly be summarized as CS since neither presence or absence of employee responsiveness does lead to a different outcome.

$$LP = cs + cE + Es$$

$$HP = CS$$

In summary, the following combinations of responsiveness characterize performance in the airline industry. Responsiveness to both customers and shareholders is typically related to above average performance whereas low responsiveness to these two groups combines with below average performance. Responsiveness to employees combined with no responsiveness to either customers or shareholders may not be an effective strategy. Therefore being responsive to employees needs to occur in conjunction with responsiveness to the two other groups. If one assumed that row 2 does not exist (because it either drives organizations out of business or forces some other responsiveness combination on them), the Boolean expression for combinations of low performance might be reduced even further to

$$LP = c + Es$$

indicating that low responsiveness to customers is a strategy likely to fail. Organizations in the airline industry may therefore be forced to avoid absence of responsiveness to customers no matter how responsive they are to the other two constituencies, and presence of responsiveness to customers seems to be a useful strategy except when it combines with responsiveness to employees and no responsiveness to shareholders. Given the wide choice set in the industry it appears plausible that organizations with no attractive offering for their customers cannot thrive.

Table 18. Stakeholder vs. Non-Stakeholder Responsiveness in the Airline Industry.

	Performance			Performance (1 Year Later)		
	Low	Medium	High	Low	Medium	High
Stake	8 (11%)	52 (69%)	16 (21%)	9 (13%)	46 (67%)	14 (20%)
No stake	32 (34%)	38 (40%)	24 (26%)	27 (32%)	35 (42%)	22 (26%)
χ^2 (sign)		16.456 (0.000)			10.906 (0.004)	

Table 18 provides a comparison between two different types of behavior, one where at least one stakeholder group is responded to at the bottom of the industry level ("*No Stake*"-strategy), the other where each of the three groups is at least responded to at an average industry level ("*Stake*"-strategy). In comparison to the expected percentages (23%, 53% and 23%), the *Stake* strategy reduces the likelihood of being among the low performers (from 23.5 to 10.5%), but does not increase the likelihood of being among the top performers. In contrast, the *No Stake* strategy increases the likelihood of being among the extreme performers, both positively and negatively. While this tendency is stronger for weak performance, a *No Stake* strategy does not seem to prevent organizational success. However, this finding must be seen in light of the negative relationship between employee responsiveness and performance. In fact, 22 of the 24 times when this strategy lead to high performance, responsiveness to employees was low.

When comparing the means for these two strategies (see Table 19) we see that companies pursuing a *Stake* strategy enjoy a higher average level in terms of return on assets. The difference between the two groups is not significant given the high variation in performance for the *No Stake* strategy. This variation stems from the cases where responsiveness to employees is low which usually is related to above average performance.

Table 19. ANOVA Results for Stakeholder vs. Non-Stakeholder Responsiveness in the Airline Industry.

	Stake	No Stake	Between Groups	Within Groups	Total
Average ROA	3.58%	0.74%			
Standard deviation	4.97	14.86			
Sum of squares			339.17	22392.65	22731.82
Degrees of freedom			1	168	169
F-Value (sign)			2.545 (0.113)		

5.2.2.2. Oil industry. Scanning Table 20, one recognizes that low levels of responsiveness to all groups or any two of the three groups appear to be highly related to low levels of performance. In fact, such combinations never appear in conjunction with high performance (once for lagged performance). Interestingly, low levels of responsiveness to shareholders seem to be related to low levels of performance, irrespective of the organizational behavior towards the other two groups (except for high responsiveness to employees and medium level to customers). In contrast, high levels of responsiveness seem to be strongly related to high performance. Once an organization acts in a highly responsive manner to at least two groups low levels of performance are very unlikely. In addition, cases exhibiting average

Table 20. Effect of Stakeholder Responsiveness on Performance in the Oil Industry.

Bool	CUST	EMP	SHA	Current ROA			Lagged ROA		
				Low	Med	Top	Low	Med	Top
1	L	L	L	4	5		4	3	
2	L	L	M	7	5		5	6	
3	L	L	H	2			1		
4	L	M	L	2	1		1	1	1
5	L	M	M	1	2	1	1	2	
6	L	M	H	1	1	1	1	1	1
7	L	H	L	1	1			2	
8	L	H	M	1	3	1		3	1
9	L	H	H		1			1	
10	M	L	L	1	2		1	1	
11	M	L	M	2	5		5	2	
12	M	L	H	1	1		1	1	
13	M	M	L	5	2	1	4	2	1
14	M	M	M	1	20	10	1	18	10
15	M	M	H	3	8	4		11	2
16	M	H	L		7	2	1	4	3
17	M	H	M		6	6	1	4	6
18	M	H	H		2	3	1		3
19	H	L	L		1				
20	H	L	M	1	3		2	2	
21	H	L	H	1	1			2	
22	H	M	L	4	2	1	2	4	1
23	H	M	M	2	7	4	1	7	2
24	H	M	H		4	2	1	3	2
25	H	H	L						
26	H	H	M		2	2	2		2
27	H	H	H			2		1	1

levels on all stakeholder relationships hardly combine with low levels of performance and disproportionately relate to above average performance. A balance in attending to the three constituencies studied thus seems to be a fruitful strategy in the oil industry.

In absolute numbers, the best performers in the oil industry are groups 18 (ROA = 6.9%, $n = 5$), 27 (ROA = 6.1%, $n = 2$), 19 (ROA = 5.9%, $n = 1$) and 17 (ROA = 5.8%, $n = 12$). With the exception of group 19, high levels of responsiveness to employees characterize these groups, a result contrary to what was found for the airline industry. Closer inspection of group 19, consisting of just one case and characterized by low responsiveness to employees, supports this conclusion. It reveals that the high absolute performance level is less impressive when compared to the average industry performance in that year, 1997, where the 5.9% return on assets achieved by this company corresponds to average industry performance. Groups 3 (ROA = 0.4%, $n = 2$), 22 (ROA = 0.9%, $n = 7$), 13 (ROA = 1.3%, $n = 8$) and 5 (ROA = 1.8%, $n = 4$) are the low performers.

Transferring these observations into the truth table consisting of combinations of presence/absence of responsiveness for each stakeholder group supports the above findings (see Table 21). The typical outcomes are less clear-cut than for the airline industry, but show an interesting pattern as well. Rows 4, 5, and 6 shows a distribution that do not allow a unanimous decision but row 6 disproportionately leans towards low performance and is coded 0. The Boolean expressions for high (HP) and low performance (LP) thus turn out to be

$$\text{LP} = \text{ces} + \text{ceS} + \text{cEs} + \text{CeS} + \text{CEs}$$

$$\text{HP} = \text{CSE}.$$

High performance in the oil industry seems to warrant a minimum level of responsiveness to all constituencies. Organizations adopting such an approach are very likely to end up among at least average performers but also to be among

Table 21. Truth Table for the Oil Industry.

	CUST	EMP	SHA	Instances	Low	Medium	High	Outcome
1	0	0	0	7	4	3		0
2	0	0	1	14	9	5		0
3	0	1	0	5	3	2		0
4	0	1	1	13	3	7	3	?
5	1	0	0	4	1	3		?
6	1	1	0	24	9	11	4	0
7	1	0	1	15	5	10		0
8	1	1	1	88	6	49	33	1

the top performers. In fact, out of the 40 top performers within the oil industry, 83% showed a CSE combination of responsiveness (compared to less than 50% of the cases exhibiting this combination). Interestingly enough, the only other combinations where organizations achieved high performance are characterized by presence of responsiveness towards employees suggesting that employee responsiveness in contrast to the airline industry is a necessary precondition for high performance in the oil industry. The combinations typically leading to low performance can be simplified by combining ces and ceS into ce, ces and cEs into cs, CeS and ceS into eS, cEs and CEs into Es resulting in the following expression

$$\text{LP} = \text{ce} + \text{cs} + \text{eS} + \text{Es}.$$

Poor performance in the oil industry results from absence of responsiveness to customers combined with absence of responsiveness to either employees or shareholders or absence of responsiveness to employees with presence of responsiveness to shareholders and vice versa. Some balance in responsiveness between employees and shareholders seems to be necessary, otherwise performance appears to suffer. The outcomes for cES and Ces are not clear-cut. One may defer that absence of responsiveness to both shareholders and employees might be compensated by presence of responsiveness to customers. Similarly, when customers are responded to below average, presence of responsiveness to both employees and shareholders may still allow an organization to perform well.

The results for the dichotomous distinction between stakeholder-oriented cases and non-stakeholder-oriented cases show a highly significant distribution (see Table 22): Organizations which lack responsiveness to at least one group (No Stake group) are extremely unlikely to perform above average. Only 9% compared to an expected percentage of 23% rank among the top performers. In contrast, likelihood of belonging to the poor performers within the industry jumps from the expected 23% to some 41%. In contrast, companies exhibiting at least average levels of responsiveness to all stakeholder groups (Stake group) are unlikely to perform poorly (7% compared to the expected 23%) and significantly more likely to be among the top performers of the industry in a given year (38%

Table 22. Stakeholder vs. Non-Stakeholder Responsiveness in the Oil Industry.

	Performance			Performance (1 Year Later)		
	Low	Medium	High	Low	Medium	High
Stake	6 (7%)	49 (56%)	33 (37%)	7 (9%)	44 (56%)	28 (36%)
No stake	34 (42%)	41 (50%)	7 (9%)	29 (39%)	37 (50%)	8 (11%)
χ^2 (sign)		37.045 (0.000)			25.024 (0.000)	

Table 23. ANOVA Results for Stakeholder vs. Non-Stakeholder Responsiveness in the Oil Industry.

	Stake	No Stake	Between Groups	Within Groups	Total
Average ROA	5.16%	2.92%			
Standard deviation	2.05	3.67			
Sum of squares			213.17	1457.16	1670.34
Degrees of freedom			1	168	169
F-Value (sign)			24.577 (0.000)		

compared to the expected 23%). Acting along the stakeholder concept thus seems to be a highly reasonable strategy in the oil industry, which is also supported by the respective means for the *Stake* and the *No Stake* strategies (see Table 23).

5.2.2.3. Utility industry. At first sight, the utility industry seems to be the one industry least affected by different levels of responsiveness towards the three stakeholder groups. When scanning Table 24, the distribution of performance is spread across the three predetermined levels. Some rows with contradictory organizational behavior show similar performance distributions (for example, rows 4 and 26), whereas rows with similar combinations differ widely in performance (rows 4/5 or 18/26). The highest return on assets belongs to group 27 (ROA = 4.5%, $n = 2$), 18 (ROA = 4.1%, $n = 9$) and 9 (ROA = 4%, $n = 2$). All these groups show at least average responsiveness to both shareholders and customers. Groups 4 (ROA = −2.3%, $n = 7$), 10 (ROA = 1.5%, $n = 10$), 7 (ROA = 1.5%, $n = 5$) and 3 (ROA = 1.6%, $n = 4$) show the lowest performance numbers. Low levels of responsiveness to shareholders and customers respectively characterize three of these groups.

The truth table for the utility industry (see Table 25) does not show a clear-cut pattern. Rows 2, 4 and 6 exhibit quite uniform distributions between low, medium and high levels of performance. Row 7 might be allocated to outcome level 1 since such a combinations at least significantly lowers the likelihood of low success. Translating the contents of the truth table into Boolean expressions the following statements for high (HP) and low performance (LP) result:

$$\mathrm{HP} = \mathrm{CSE} + \mathrm{CSe}$$

$$\mathrm{LP} = \mathrm{ces} + \mathrm{cEs} + \mathrm{Ces}.$$

The expression for high performance can be reduced to CS implying that an organization responding to both customers and shareholders performs above average. In the case of low performance, ces and cEs combine to cs, and ces and

Table 24. Effect of Stakeholder Responsiveness on Performance in the Utility Industry.

Bool	CUST	EMP	SHA	Current ROA			Lagged ROA		
				Low	Med	Top	Low	Med	Top
1	L	L	L	3	5	1	3	5	
2	L	L	M		4	1	1	3	1
3	L	L	H	1	3		1	1	1
4	L	M	L	10	6		9	5	1
5	L	M	M	3	14	5	5	12	4
6	L	M	H	4		1	2		1
7	L	H	L	4	1		3	2	
8	L	H	M	1		1		1	1
9	L	H	H		1	1			1
10	M	L	L	4	4	2	2	5	2
11	M	L	M	2	19	4	4	18	1
12	M	L	H	1	7	1	2	4	3
13	M	M	L	2	12	1	4	8	2
14	M	M	M	12	42	22	10	37	19
15	M	M	H	4	17	5	5	14	6
16	M	H	L	1	3	2		4	2
17	M	H	M	4	23	7	4	20	6
18	M	H	H		7	2		6	1
19	H	L	L	1				1	
20	H	L	M		6	1		4	1
21	H	L	H						
22	H	M	L	2	4	1	2	1	1
23	H	M	M	2	20	8	3	19	6
24	H	M	H	4	5	3	1	9	2
25	H	H	L		1			1	
26	H	H	M	5	5		2	8	
27	H	H	H		1	1		1	1

Table 25. Truth Table for the Utility Industry.

	CUST	EMP	SHA	Instances	Low	Medium	High	Outcome
1	0	0	0	9	3	5	1	0
2	0	0	1	9	1	7	1	?
3	0	1	0	21	14	7		0
4	0	1	1	31	8	15	8	?
5	1	0	0	11	5	4	2	0
6	1	1	0	29	5	20	4	?
7	1	0	1	41	3	32	6	1
8	1	1	1	199	31	120	48	1

Table 26. Stakeholder vs. Non-Stakeholder Responsiveness in the Utility Industry.

	Performance			Performance (1 Year Later)		
	Low	Medium	High	Low	Medium	High
Stake	31 (16%)	120 (60%)	48 (24%)	25 (14%)	114 (63%)	41 (23%)
No stake	39 (26%)	90 (60%)	22 (15%)	38 (28%)	75 (56%)	22 (16%)
χ^2 (sign)		8.433 (0.015)			10.241 (0.006)	

Ces combine to es. Therefore, an absence of either both customer and shareholder responsiveness or both shareholder and employee responsiveness result in below average performance. The simplified expressions therefore are

$$\mathrm{HP} = \mathrm{CS}$$

$$\mathrm{LP} = \mathrm{cs} + \mathrm{es}.$$

The effect of presence of shareholder responsiveness with absence for the other constituencies (ceS), presence for employees and shareholders with absence for customers (cES) and presence for customers and employees with absence for shareholders (CEs) are related to both negative and positive outcomes. When summarizing these terms, one ends up with cS and CEs, suggesting that a lack of customer responsiveness may be compensated by shareholder responsiveness and a lack of shareholder responsiveness only by both customer and employee responsiveness.

The allocation of all combinations to one of two groups distinguished by whether at least one constituency group experiences low responsiveness results in significant differences, albeit less substantial than for the other stakeholder groups (see Table 26). Combinations characterized by a stakeholder approach decrease the probability of being a low performer and increase the probability of being a

Table 27. ANOVA Results for Stakeholder vs. Non-Stakeholder Responsiveness in the Utility Industry.

	Stake	No Stake	Between Groups	Within Groups	Total
Average ROA	3.38%	2.48%			
Standard deviation	1.57	3.58			
Sum of squares			69.26	2412.78	2482.05
Degrees of freedom			1	348	349
F-Value (sign)			9.990 (0.002)		

high performer, the opposite is true for "No Stake"-strategies. When comparing the average return on assets of the utility companies following a *Stake* strategy with average return on assets of companies which show low responsiveness to at least one of the three stakeholder groups studied, one encounters significant performance differences (Table 27). On average, the former group enjoys 0.9% higher return on assets than the latter group.

6. CONCLUSIONS AND IMPLICATIONS

Table 28 summarizes the findings for the airline, oil and utility industry in terms of the Boolean expressions derived for observations of high (HP), low (LP) and inconclusive performance. While the results differ between the three industries studied, some combinations of responsiveness are related to the same level of performance for all industries. For example, presence of responsiveness for both customers and shareholders is typically related to high performance in the airline and oil industry, and, when combined with employee responsiveness, also for the oil industry. In contrast, absence of responsiveness to all groups always combines with a tendency towards low performance. Even a lack of responsiveness for only customers and shareholders warrants low performance levels. While none of the three causes studied is sufficient for organizations to perform above average, both customer and shareholder responsiveness are necessary conditions for above-average effectiveness.

Figure 3 summarizes these findings and offers an overview how combinations of responsiveness absence or presence relate to organizational performance. The area within the circles corresponds to the presence of such a responsiveness type whereas the area outside a circle corresponds to its absence. It is easy to recognize that certain strategic postures towards constituencies never seem to be useful (for example, responsiveness towards employees without a minimum level of responsiveness to the other two groups or no responsiveness to each of the groups), whereas others are sensible for each of the industries (presence of responsiveness to all groups). In addition, strategies focusing on single constituencies and

Table 28. Comparison of the Boolean Expressions Across Industries.

	Airline	Oil	Utility
High performance	CS	CES	CS
Low performance	cs + cE + Es	ce + cs + eS + Es	cs + es
Inconclusive	Ces	cES + Ces	cS + CEs

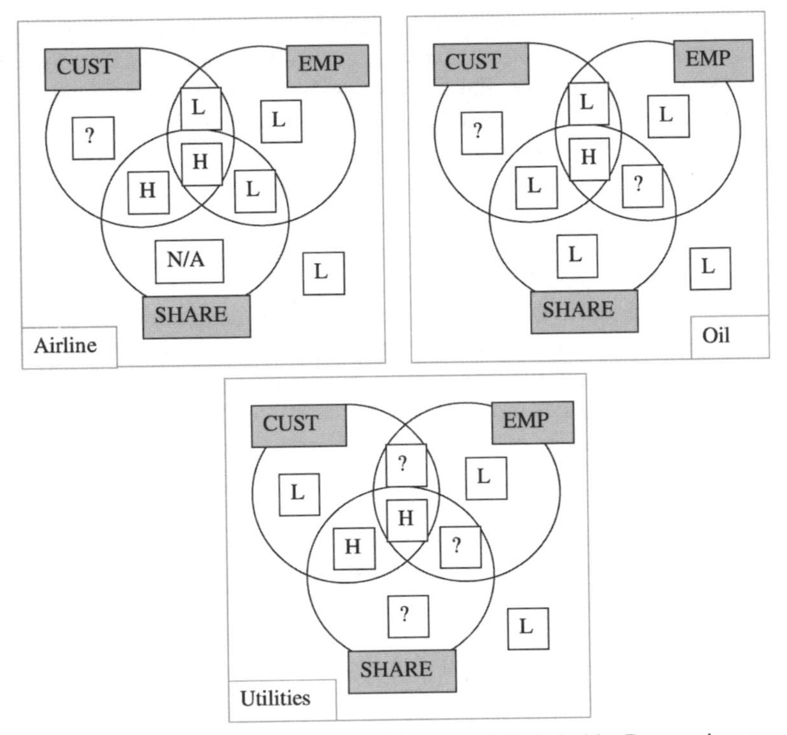

Fig. 3. Effective vs. Ineffective Combinations of Stakeholder Responsiveness.

disregarding other ones usually lead to low performance or inconclusive results at best. Presence of responsiveness to all groups is often rewarded with above-average performance and hardly ever related to below-average performance.

While the number of similarities between the industries clearly outnumbers the differences, some of these partly contradictory findings warrant some closer analysis. The most notable is the negative relationship between employee responsiveness and performance in the airline industry which is not present in the other two industries. There, especially in the oil industry, a positive relationship between responsiveness to employees and organizational effectiveness was identified. Some explanations for these results are feasible. The airline industry is characterized by comparatively high labor costs (not on average, but compared to sales) which may more strongly affect return on assets than labor costs in industries where they do not impact the bottom line as strongly. In addition, a higher percentage of employees in the airline industry might choose to pursue such a career in pursuit of benefits apart from salary. A substantial percentage of employees may enter such a relationship for a limited time only and aim for benefits which may matter as much as the average salary (like extended stays at certain locations, vouchers for family and friends, free periods that allow private travels, working climate, etc.). Low salary levels may therefore not automatically have a negative effect on employee performance as usually suggested (Pfeffer, 1998). In contrast, for people working in the oil or utility industry the absolute salary levels they receive may motivate them relatively more than airline employees. The positive effect of low employee responsiveness on organizational effectiveness still warrants further discussion. Going back to the original data one learns that the big airline companies like United, USAir, KLM, or Delta which usually ranked among the top performers were also the ones with the lowest average salary levels. An explanation for the negative relationship between employee responsiveness (which in this study only captures the average salary) and performance come to mind: As mentioned before, these airlines may be able to offer some benefits not provided to the same extent by other airlines (attractive destinations, non-salary social benefits, career opportunities), and are therefore able to afford paying lower salaries. In combination with the competitive pressure during that period which lead to immense price competition, low salary levels might have been a necessity to achieve above average return on assets.

While both the oil and airline industry showed a fairly distinct, though different, pattern of performance outcomes for different combinations of responsiveness to shareholders the picture for the utility industry was less clear-cut. The fact that the explanatory factors chosen for this research project seem to somewhat better explain effectiveness levels in the airline and oil industry than in the utility industry may be the result of different competitive circumstances in these industries. The author assumes that the regional concentration of the players in this industry and their historical development affect the relationships to stakeholders by limiting their respective choice set. For example, customers may not easily be able (or willing due to high exit costs) to switch their electricity suppliers. To some degree, shareholders often are state- or municipality-related bodies which may be less taxing on the return their investment provides. Even employees, especially when they possess highly specialized abilities may not easily be able to terminate their relationship with an organization and pursue a different professional career. These factors may be less influential for the other two industries where investors usually have a more financially-oriented relationship, employees may more easily be able to switch employers and customers have a broad choice set to satisfy their demands.

In light of the findings of this research project one may conclude that managers should strive to aim for a minimum level of responsiveness to all stakeholder groups. Given the thresholds chosen for this analysis any organization should aim to belong to the top 80% of an industry in terms of responsiveness to employees, shareholders and customers. Such a strategy makes low performance very unlikely and is disproportionately related to high levels of performance. A trade-off between the responsiveness to each of the three groups studied may be derived from the fact that the number of organizations able to respond above average to more than one constituency is lower than the expected number of organizations given the distribution chosen to allocate the organization into different groups. Therefore, it may be argued that managers should not aim for extremely high levels of responsiveness to any one group if such behavior results in below-average responsiveness to another group.

Even with the modest demands (be among the top 80% of an industry for a given year) put on the examined organizations to rate their responsiveness as "present," only some 57% of observations in the utility industry, 52% in the oil industry and 45% in the airline industry manage to avoid a bottom 20% rank on all three responsiveness scores. While this number is higher or equal than the statistical likelihood of 51% (46% for the airline and oil industry) one wonders whether it would really be difficult to be among the top 80% performers on each responsiveness score (which, given the data used, would not automatically be sensible for airlines). The fact that more than the expected number of organizations are able to avoid low responsibility to any of the three groups might be a sign of a positive correlation between the measures chosen or show that some organizations simply are more able in responding to all their stakeholders. The data suggests that some organizations simply are more apt to act in line with the stakeholder idea, since a number of organizations manage to be responsive for a majority of the periods studied while as many manage to hardly ever be responsive to all three constituencies. Out of the 69 organizations only four manage to avoid low responsiveness to any of the three groups for all ten periods. Southwest Airlines in the airline industry, GPU in the utility sector and Royal Dutch PET and Royal Dutch/Shell Inc. in the oil industry never show low levels of responsiveness to any one group. On the other end of the spectrum, seven organizations "manage" to never avoid low responsiveness to any of the three groups. In all ten periods Mesaba Holdings, Skywest and TWA in the airline industry, Murphy Oil and USX in the oil sector and Central Vermont Public Service among the utilities are never able to be responsive to all three groups at the same time.

Out of the 69 corporations 38 qualify as stakeholder-oriented for at least six periods (23 of the utilities, 7 in the airline and 8 in the oil industry) whereas 22 are stakeholder-oriented for a maximum of three periods. One might conclude that some organizations simply are more capable/willing of responding to their constituencies whereas others show a certain level of ignorance to at least one of these groups. Given the results obtained (with the exception of the airline industry), such a "stakeholder-ignorant" posture may well be less smart than one incorporating to some degree the interests of all stakeholder groups.

NOTE

1. Organizational performance, organizational success and organizational effectiveness will be used interchangeably throughout this paper.

REFERENCES

Ailawadi, K. L., Borin, N., & Farris, P. W. (1995). Market power and performance: A cross-industry analysis of manufacturers and retailers. *Journal of Retailing*, *71*(3), 211–250.

Akerlof, G. A. (1970). The market for lemons: Quality, uncertainty and the market mechanism. *Quarterly Journal of Economics*, *84*, 488–500.

Aldag, R. J., & Brief, A. (1978). Examination of alternative models of job satisfaction. *Human Relations*, *31*(1), 91–98.

Alexander, G. J., & Buchholz, R. A. (1978). Corporate social responsibility and stock market performance. *Academy of Management Journal*, *21*(3), 479–486.

Altman, E. I. (1971). *Corporate bankruptcy in America*. Lexington, MA: Heath Lexington Books.

Ambler, T., & Kokkinaki, F. (1997). Measures of marketing success. *Journal of Marketing Management*, *13*, 665–678.

Amit, R., & Shoemaker, P. J. H. (1993). Strategic assets and organizational rent. *Strategic Management Journal*, *14*, 33–46.

Anderson, P. F. (1982). Marketing, strategic planning and the theory of the firm. *Journal of Marketing*, *46*(Spring), 15–26.

Anderson, W. T. (1991). Is the purpose of the organization to create satisfied customers? No! *Marketing and Research Today* (August), 127–141.

Anderson, J. C., Hakansson, H., & Johanson, J. (1994). Dyadic business relationships within a business network context. *Journal of Marketing*, *58*(4), 1–15.

Argenti, J. (1976). *Corporate collapse: The causes and symptoms*. New York: John Wiley.

Arzac, E. R. (1986). Do your business units create shareholder value? *Harvard Business Review* (January/February), 121–126.

Asanuma, B. (1985). The organization of parts supply in the Japanese automotive industry. *Japanese Economic Studies*, *15*, 32–53.

Ashworth, S. D., Higgs, C., Schneider, B., & Shepherd, W. (1995). The linkage between customer satisfaction data and employee-based measures of a company's strategic intent. Paper presented at the Annual Meeting of the Society for Industrial and Organizational Psychology, Orlando, Florida.

Aupperle, K. E. (1984). An empirical measure of corporate social orientation. In: J. E. Post (Ed.), *Research in Corporate Social Performance and Policy* (Vol. 6, pp. 27–54). Greenwich, CT: JAI Press.

Aupperle, K. E., Carroll, A. B., & Hatfield, J. D. (1985). An empirical examination of the relationship between corporate social responsibility and profitability. *Academy of Management Journal*, *28*(2), 446–463.

Axelrod, R. (1984). *The evolution of cooperation*. New York: Basic Books.

Bain, J. S. (1956). *Barriers to new competition*. Cambridge, MA: Harvard University Press.

Baker, W. E., Faulkner, R. R., & Fisher, G. A. (1998). Hazards of the market: The continuity and dissolution of interorganizational market relationships. *American Sociological Review, 63*, 147–177.

Baliga, B. R., Moyer, C. R., & Rao, R. S. (1996). CEO duality and firm performance: What's the fuss? *Strategic Management Journal, 17*(1), 41–53.

Barksdale, H. C., & Darden, B. (1971). Marketers' attitude towards the marketing concept. *Journal of Marketing, 35*, 29–36.

Barnard, C. I. (1938). *The functions of the executive*. Cambridge: Harvard University Press.

Barney, J. B. (1986a). Organizational culture: Can it be a source of sustained competitive advantage? *Academy of Management Review, 11*(3), 656–665.

Barney, J. B. (1986b). Strategic factor markets: Expectations, luck and business strategy. *Management Science, 32*, 1231–1241.

Barney, J. B. (1991). Firm resources and sustained competitive advantage. *Journal of Management, 17*, 99–120.

Barney, J. B., & Hansen, M. H. (1994). Trustworthiness as a source of competitive advantage. *Strategic Management Journal, 15*, 715–7190.

Bass, F. M., & Wittink, D. R. (1975). Pooling issues and methods in regression analysis with examples in marketing research. *Journal of Marketing Research, 12*(November), 414–425.

Baumhart, R. (1968). *An honest profit: What businessmen say about ethics in business*. New York: Holt, Rinehart and Winston.

Beatty, S. E. (1988). An exploratory study of organizational values with a focus on people orientation. *Journal of Retailing, 53*(4), 405–425.

Becker, L. C. (1992). Too much property. *Philosophy and Public Affairs, 21*, 196–206.

Berman, S. L., Wicks, A. C., Kotha, S., & Jones, T. M. (1999). Does stakeholder orientation matter? The relationship between stakeholder management models and firm financial performance. *Academy of Management Journal, 42*(55), 488–506.

Bitner, M. J. (1990). Evaluating service encounters: The effects of physical surroundings and employee responses. *Journal of Marketing, 54*(April), 69–82.

Black, F. (1972). Capital market equilibrium with restricted borrowing. *Journal of Business* (July), 444–455.

Borch, F. J. (1964). The marketing philosophy as a way of business life. In: H. C. Barksdale (Ed.), *Marketing in Process*. New York: Holt, Rinehart and Winston.

Bosshart, D. (1996). Shareholder capitalism and stakeholder capitalism. *GDI, 2*, 26–38.

Bowen, H. R. (1953). *Social responsibilities of the businessman*. New York: Harper & Row.

Bowen, D. E. (1996). Market-focused HRM in service organizations: Satisfying internal and external customers. *Journal of Market-Focused Management, 1*, 31–47.

Bowen, H. P., & Wiersema, M. F. (1999). Matching method to paradigm in strategy research: Limitations of cross-sectional analysis and some methodological alternatives. *Strategic Management Journal, 20*, 625–636.

Bowie, N. E. (1994). A Kantian theory of capitalism. In: *Ruffin Lectures in Business Ethics* (pp. 1–40). Charlottesville, VA: Darden School, University of Virginia.

Brenner, S. N., & Molander, E. A. (1977). Is the ethics of business changing? *Harvard Business Review, 58*(1), 54–65.

Brush, T. H., Bromiley, P., & Hendrickx, M. (1999). The relative influence of industry and corporation on business segment performance: An alternative estimate. *Strategic Management Journal, 20*, 519–547.

Buchholz, R. A. (1993). *Principles of environmental management: The greening of business*. Englewood Cliffs, NJ: Prentice-Hall.

Bughin, J., & Copeland, T. E. (1997). The virtuous cycle of shareholder value creation. *The McKinsey Quarterly*, *2*, 156–167.

Buzzell, R. D., & Ortmeyer, G. (1995). Channel partnerships streamline distribution. *Sloan Management Review* (Spring), 85–96.

Cameron, K. S. (1980). Critical questions in assessing organizational effectiveness. *Organizational Dynamics*, *9*, 66–80.

Cameron, K. S. (1984). An empirical investigation of the multiple constituency model of organizational effectiveness. Working Paper, National Center for Higher Education Management Systems, Boulder, CO.

Cameron, K. S., & Whetten, D. A. (1983). Organizational effectiveness: One model or several? In: K. S. Cameron & D. A. Whetten (Eds), *Organizational Effectiveness: A Comparison of Multiple Models* (pp. 1–27). Orlando: Academic Press.

Capon, N., Farley, J. U., & Hoenig, S. (1990). Determinants of financial performance: A meta-analysis. *Management Science*, *36*(10), 1143–1159.

Carlsmith, J. M., Ellsworth, C., & Aronson, E. (1976). *Methods of research in social psychology*. Reading, MA: Addison-Wesley.

Carroll, A. B. (1993). *Business and society: Ethics and stakeholder management*. Cincinnati: South-Western.

Carroll, D. T. (1983). A disappointing search for excellence. *Harvard Business Review*, *61*(6), 78–88.

Cavusgil, S. T., & Zou, S. (1994). Marketing strategy-performance relationship: An investigation of the empirical link in export market ventures. *Journal of Marketing*, *58*(1), 1–21.

Chakravarthy, B. S. (1986). Measuring strategic performance. *Strategic Management Journal*, *7*, 437–458.

Clarkson, M. E. (1991). Defining, evaluating and managing corporate social performance. In: J. E. Post (Ed.), *Research in Corporate Social Performance and Policy* (Vol. 12, pp. 331–358). Greenwich, CT: JAI Press.

Clarkson, M. E. (1995). A stakeholder framework for analyzing and evaluating corporate social performance. *Academy of Management Review*, *20*(1), 92–117.

Coase, R. H. (1960). The problem of social cost. *Journal of Law and Economics*, *3*, 1–44.

Coff, R. W. (1999). When competitive advantage doesn't lead to performance: The resource-based view and stakeholder bargaining power. *Organization Science*, *10*(2), 119–133.

Cohen, B., & Greenfield, J. (1997). *Ben & Jerry's double-dip: Lead with your values and make money, too*. New York: Simon & Schuster.

Copeland, T., & Weston, F. (1992). *Financial theory and corporate policy* (3rd ed.). Reading, MA: Addison-Wesley.

Cornell, B., & Shapiro, A. C. (1987). Corporate stakeholders and corporate finance. *Financial Management* (Spring), 5–14.

Cyert, R. M., & March, J. G. (1959). A behavioral theory of organizational objectives. In: M. Haire (Ed.), *Modern Organization Theory*. New York: Wiley.

Cyert, R. M., & March, J. G. (1963). *A behavioral theory of the firm*. Englewood Cliffs, NJ: Prentice-Hall.

Day, G. S., & Wensley, R. (1983). Marketing theory with a strategic orientation. *Journal of Marketing*, *47*(Fall), 79–89.

Day, G. S., & Wensley, R. (1988). Assessing advantage: A framework for diagnosing competitive superiority. *Journal of Marketing*, *52*(2), 1–20.

Dean, J. W., & Sharfman, M. P. (1996). Does decision process matter? A study of strategic decision-making effectiveness. *Academy of Management Journal*, *39*, 368–396.
Dearborn, D., & Simon, H. A. (1958). Selective perception: A note on the departmental identification of executives. *Sociometry*, *35*, 38–48.
Deavenport, E. (1996). Walking the high wire. *Executive Speeches* (February/March), 1–3.
Deming, W. E. (1986). *Out of the crisis*. Cambridge, MA: MIT Center for Advanced Engineering Study.
Demski, J. S., Sappington, D. E. M., & Spiller, P. T. (1987). Managing supplier switching. *Rand Journal of Economics*, *18*, 77–97.
Deshpande, R., Farley, J. U., & Webster, F. E. (1993). Corporate culture, customer orientation, and innovativeness in Japanese firms: A quadrad analysis. *Journal of Marketing*, *57*(1), 23–37.
Dess, G. G., & Robinson, R. B. (1984). Measuring organizational performance in the absence of objective measures: The case of the privately-held firm and conglomerate business unit. *Strategic Management Journal*, *5*, 265–273.
Devinney, T. M., Stewart, D. W., & Shocker, A. D. (1985). A note on product portfolio theory: A rejoinder to Cardozo and Smith. *Journal of Marketing*, *49*, 107–112.
Dibb, S., Simkin, L., Pride, W. M., & Ferrell, O. C. (1997). *Marketing: Concepts and strategies*. Boston, MA: Houghton Mifflin.
Dollinger, M. J. (1990). The evolution of collective strategies in fragmented markets. *Academy of Management Review*, *15*, 266–285.
Donaldson, G., & Lorsch, J. W. (1983). *Decision making at the top: The shaping of strategic direction*. New York, NY: Basic Books.
Donaldson, T., & Preston, L. E. (1995). The stakeholder theory of the corporation: Concepts, evidence and implications. *Academy of Management Review*, *20*(1), 65–91.
Doyle, P. (1992). What are the excellent companies? *Journal of Marketing Management*, *8*, 101–116.
Doyle, P., & Hooley, G. J. (1992). Strategic orientation and corporate performance. *International Journal of Research in Marketing*, *9*, 59–73.
Dreher, A. (1996). *Marketingorientierung als Unternehmensphilosophie*. Wiesbaden: Deutscher Universitätsverlag.
Drucker, P. (1954). *The practice of management*. New York: Harper and Row.
Dutton, J. E., & Dukerich, J. M. (1991). Keeping an eye on the mirror: Image and identity in organizational adaptation. *Academy of Management Journal*, *34*, 517–554.
Easthope, G. (1974). *A history of social research methods*. London: Longman.
Economist (1996). Stakeholder capitalism: Unhappy families (February 10), 21–25.
Etzioni, A. (1964). *Modern organizations*. Englewood Cliffs, NJ: Prentice-Hall.
Etzioni, A. (1988). *The moral dimension*. New York: Basic Books.
Evan, W. M., & Freeman, E. R. (1983). A stakeholder theory of the modern corporation: Kantian capitalism. In: T. Beauchamp & N. Bowie (Eds), *Ethical Theory and Business* (pp. 75–93). Englewood Cliffs, NJ: Prentice-Hall.
Fama, E. F., & Miller, M. H. (1972). *The theory of finance*. New York: Holt, Rinehart and Winston.
Ferguson, R. (1986). The trouble with performance measurement. *The Journal of Portfolio Management* (Spring), 4–9.
Field, G. R. H., & Abelson, M. A. (1982). Climate: A reconceptualization and proposed model. *Human Relations*, *35*(3), 181–201.
Fiol, M. (1991). Managing culture as a competitive resource: An identity-based view of sustainable competitive advantage. *Journal of Management*, *17*, 191–211.

Fisher, I. (1930). *The theory of interest: As determined by impatience to spend income and opportunity to invest it*. New York: Macmillan.
Fisher, F. M., & McGowan, J. J. (1983). On the misuse of accounting rates of return to infer monopoly profits. *American Economic Review, 73*(1), 82–97.
Fites, D. V. (1996). Making your dealers your partners. *Harvard Business Review* (March/April), 84–95.
Fombrun, C. (1996). *Reputation: Realizing value from the corporate image*. Cambridge, MA: Harvard Business School Press.
Ford, J. D., & Schellenberg, D. A. (1982). Conceptual issues of linkage in the assessment of organizational performance. *Academy of Management Review, 7*(1), 49–58.
Frank, R. H. (1988). *Passions within reason: The strategic role of emotions*. New York: Norton.
Freeman, R. E. (1983). Strategic management: A stakeholder approach. In: R. Lamb (Ed.), *Advances in Strategic Management* (Vol. 1, pp. 31–60). Greenwich, CT: JAI Press.
Freeman, R. E. (1984). *Strategic management: A stakeholder approach*. Boston: Pitman.
Freeman, R. E. (1994). The politics of stakeholder theory: Some future directions. *Business Ethics Quarterly, 4*, 409–422.
Freeman, R. E. (1999). Divergent stakeholder theory. *Academy of Management Review, 24*(2), 233–236.
Friedlander, F., & Pickle, H. (1968). Components of effectiveness in small organizations. *Administrative Science Quarterly, 13*, 289–304.
Friedman, M. (1962). *Capitalism and freedom*. Chicago: University of Chicago Press.
Friedman, M. (1971). Does business have a social responsibility? *Bank Administration* (April), 13–14.
Fruhan, W. E. (1979). *Financial strategy: Studies in the creation, transfer, and destruction of shareholder value*. Homewood, IL: Richard D. Irwin.
Galbraith, J. (1973). *Designing complex organisations*. Reading, MA: Addison-Wesley.
Gioia, D. A. (1999). Practicability, paradigms, and problems in stakeholder theorizing. *Academy of Management Review, 24*(2), 228–232.
Glick, W. H. (1985). Conceptualizing and measuring organizational and psychological climate: Pitfalls in multi-level research. *Academy of Management Review, 10*(3), 601–616.
Glueck, W. F., & Willis, R. (1979). Documentary sources and strategic management research. *Academy of Management Review, 4*, 95–101.
Goodman, P. S. (1979). Organizational effectiveness as a decision-making process. Paper presented at the 39th Annual Meeting of the Academy of Management. Atlanta.
Goodpaster, K. E., & Matthews, J. B., Jr. (1982). Can a corporation have a conscience. *Harvard Business Review* (January–February), 132–141.
Gordon, R. A. (1948). Short-period price determination. *American Economic Review, 38*, 265–288.
Greene, W. F. (1993). *Econometric analysis*. New York: MacMillan.
Greening, D. W. (1995). Conservation strategies, firm performance, and corporate reputation in the U.S. electric utility industry. In: J. E. Post (Ed.), *Research in Corporate Social Performance and Policy* (Suppl. 1, pp. 345–368). Greenwich, CT: JAI Press.
Greenley, G. E., & Foxall, G. R. (1996). Consumer and non-consumer stakeholder orientation in U.K. companies. *Journal of Business Research, 35*, 105–116.
Greenley, G. E., & Foxall, G. R. (1998). The external moderation of associations among stakeholder orientations and company performance. *International Journal of Research in Marketing, 15*(1), 51–70.
Gronroos, C. (1984). A service quality model and its marketing implications. *European Journal of Marketing, 18*(4), 36–44.
Grundy, T. (1995). Destroying shareholder value: Ten easy ways. *Long Range Planning, 28*(3), 76–82.

Hage, J. (1980). *Theories of organizations: Form, process, and transformation*. New York: Wiley.
Hall, R. I. (1984). The natural logic of management policy making: Its implications for the survival of an organization. *Management Science*, *30*(8), 905–927.
Hambrick, D. C. (1989). Guest editor's introduction: Putting top managers back into the strategy picture. *Strategic Management Journal*, *10*(Summer Special Issue), 5–16.
Hambrick, D. C., & Finkelstein, S. (1987). Managerial discretion: A bridge between polar views on organizations. In: B. Staw & L. L. Cummings (Eds), *Research in Organizational Behavior* (Vol. 9, pp. 369–406). Greenwich, CT: JAI Press.
Hamel, G. (1997). How killers count. *Fortune*, *135*(12), 74.
Hansen, G. S., & Wernerfelt, B. (1989). Determinants of firm performance: The relative importance of economic and organizational factors. *Strategic Management Journal*, *10*, 399–411.
Heide, J. B., & John, G. (1990). Alliances in industrial purchasing: The determinants of joint action in buyer-supplier relationships. *Journal of Marketing Research*, *27*, 24–36.
Hill, C. W. L., & Hansen, G. S. (1991). A longitudinal study of the cause and consequences of changes in diversification in the U.S. pharmaceutical industry 1977–1986. *Strategic Management Journal*, *12*(3), 187–199.
Hill, C. W. L., & Jones, T. M. (1992). Stakeholder-agency theory. *Journal of Management Studies*, *29*(2), 131–154.
Hitt, M. A., & Middlemist, D. R. (1979). A methodology to develop the criteria and criteria weightings for assessing subunit effectiveness in organizations. *Academy of Management Journal*, *22*, 356–374.
Hitt, M. A., & Ireland, R. D. (1987). Peters and Waterman revisited: The unended quest for excellence. *Academy of Management Executive*, *1*(2), 91–98.
Hopwood, A. G. (1976). *Accounting and human behavior*. Englewood Cliffs, NJ: Prentice-Hall.
Huber, G. P., & Power, D. J. (1985). Retrospective reports of strategic-level managers: Guidelines for increasing their accuracy. *Strategic Management Journal*, *6*, 171–180.
Huff, A. S. (1982). Industry influences on strategy formulation. *Strategic Management Journal*, *3*, 119–131.
Hunt, S. D., Chonko, L. B., & Wood, V. R. (1985). Organizational commitment and marketing. *Journal of Marketing*, *49*(Winter), 112–126.
Iaffaldano, M., & Muchinsky, P. (1985). Job satisfaction and job performance.
Jaffe, A. B. (1986). Technological opportunities and spillovers of R&D: Evidence from firms' patents, profits and market value. *The American Economic Review*, *76*, 984–1001.
Jaworski, B. J., & Kohli, A. (1993). Marketing orientation: Antecedents and consequences. *Journal of Marketing*, *57*(July), 53–70.
Jensen, M. C., & Meckling, W. H. (1976). Theory of the firm: Managerial behavior, agency costs, and ownership structure. *Journal of Financial Economics*, *3*, 305–360.
Johnson, B., Natarajan, A., & Rappaport, A. (1985). Shareholder returns and corporate excellence. *Journal of Business Strategy* (Fall), 52–62.
Jobson, J. D., & Schneck, R. (1982). Constituent view of organizational effectiveness: Evidence from police organizations. *Academy of Management Journal*, *25*, 25–46.
Jones, T. M. (1995). Instrumental stakeholder theory: A synthesis of ethics and economics. *Academy of Management Review*, *20*(2), 404–437.
Jones, T. M., & Wicks, A. C. (1999). Convergent stakeholder theory. *Academy of Management Review*, *24*(2), 206–221.
Kaplan, R. S., & Norton, D. P. (1996). Using the balanced scorecard as a strategic management system. *Harvard Business Review*, *74*(January–February), 75–85.

Kaplan, R. S., & Norton, D. P. (2000). *The strategy-focused organization*. Boston, MA: Harvard Business School Press

Katona, G. (1951). *Psychological analysis of economic behavior*. New York: McGraw-Hill.

Katz, D., & Kahn, R. L. (1966). *The social psychology of organizations*. New York: Wiley.

Katz, J. P., Werner, S., & Brouthers, L. (1999). Does winning mean the same thing around the world? National ideology and the performance of global competitors. *Journal of Business Research*, *44*(2), 117–126.

Keeley, M. A. (1978). A social justice approach to organizational evaluation. *Administrative Science Quarterly*, *22*, 272–292.

Kiesler, S., & Sproull, L. (1982). Managerial response to changing environments. *Administrative Science Quarterly*, *27*, 548–570.

Kimery, K. M., & Rinehart, S. M. (1998). Markets and constituencies: An alternative view of the marketing concept. *Journal of Business Research*, *43*, 117–124.

Koeske, G. F., & Kirk, S. A. (1994). Measuring the Monday blues: Validation of a job satisfaction scale for the human services. *Social Work Research*, *18*(1), 27–35.

Kohli, A. K., & Jaworski, B. J. (1990). Market orientation: The construct, research propositions and managerial implications. *Journal of Marketing*, *54*, 1–18.

Kotler, P. (1984). *Marketing management: Analysis, planning, and control* (5th ed.). Englewood Cliffs, NJ: Prentice-Hall.

Kotler, P. (1991). *Marketing management: Analysis, planning and control* (7th ed.). Englewood Cliffs, NJ: Prentice-Hall.

Kotter, J., & Heskett, J. (1992). *Corporate culture and performance*. New York: Free Press.

Krueger, W. (1989). Hier irrten Peters and Waterman. *Harvard Manager*, *1*, 13–18.

Krugman, P. A. (1993). Myths and realities of U.S. competitiveness. *Science*, *254*(5033), 811–815.

Lado, A. A., & Wilson, M. (1994). Human resource systems and sustained competitive advantage. *Academy of Management Review*, *19*, 699–727.

Lado, A. A., Boyd, N. G., & Wright, P. (1992). A competency model of sustained competitive advantage. *Journal of Management*, *18*, 77–91.

Lawler, E. E., Hall, D. T., & Oldham, G. R. (1974). Organizational climate, relationship to organizational structure, process and performance. *Organizational Behavior and Human Performance*, *11*, 139–155.

Lawrence, J. (1991). The green revolution: Mobil. *Advertising Age*, *62*(5), 12–13.

Lee, J., & Miller, D. (1999). People matter: Commitment to employees, strategy and performance in Korean firms. *Strategic Management Journal*, *20*, 579–593.

Leone, R. P., & Schultz, R. L. (1989). A study of marketing generalizations. *Journal of Marketing*, *44*(Winter), 10–18.

Levitt, T. (1960). Marketing myopia. *Harvard Business Review*, *38*, 24–47.

Lieberson, S. (1985). *Making it count: The improvement of social research and theory*. Berkeley: University of California Press.

Likert, R. (1967). *The human organization*. New York: McGraw-Hill.

Lines, R., & Gronhaug, K. (1993). Environmental orientation of managers: The construct, its antecedents and consequences. Working paper.

Lintner, J. (1969). The aggregation of investors' diverse judgments and preferences in purely competitive security markets. *Journal of Financial and Quantitative Analysis* (December), 347–400.

Litwin, G. H., & Stringer, R. A. (1968). *Motivation and organizational climate*. Cambridge, MA: Harvard University Press.

Lord, R. G., & Maher, K. J. (1990). Alternative information-processing models and their implications for theory, research and practice. *Academy of Management Review*, *15*(1), 9–28.

Lubatkin, M., & Chatterjee, S. (1991). The strategy-shareholder value relationship: Testing temporal stability across market cycles. *Strategic Management Journal*, *12*, 251–270.

Maranville, S. J. (1989). You can't make steel without having some smoke: A case study in stakeholder analysis. *Journal of Business Ethics*, *8*, 57–63.

March, J.G., & Simon, H. A. (1958). *Organizations*. New York: Wiley.

Margolis, J. (1958). The analysis of the firm: Rationalism, conventionalism, and behaviorism. *Journal of Business*, *31*, 187–199.

Mason, E. S. (1957). *Economic concentration and the monopoly problem*. Cambridge, MA: Harvard University Press.

Mason, E. S. (1960). *The corporation in modern society*. Cambridge, MA: Harvard University Press.

McCarthy, J. E., & Perreault, W. D. (1984). *Basic marketing* (8th ed.). Homewood: Richard D. Irwin.

McCrory, F. V., & Gerstberger, P. G. (1992). The new math of performance measurement. *The Journal of Business Strategy* (March/April), 33–38.

McFarlin, D. B., & Sweeney, P. D. (1992). Distributive and procedural justice as predictors of satisfaction with personal and organizational outcomes. *Academy of Management Journal*, *35*, 626–637.

McGuire, J. W. (1963). *Business and society*. New York: McGraw-Hill.

McGuire, J. B., Sundgren, A., & Schneeweis, T. (1988). Corporate social responsibility and firm financial performance. *Academy of Management Journal*, *31*(4), 854–872.

McKitterick, J. B. (1957). What is the marketing management concept? In: F. M. Bass (Ed.), *The Frontiers of Marketing Thought* (pp. 65–79). Chicago: American Marketing Association.

McNamara, C. P. (1972). The present status of the marketing concept. *Journal of Marketing*, *36*, 50–57.

Miles, G., Snow, C. C., & Sharfman, M. P. (1993). Industry variety and performance. *Strategic Management Journal*, *14*, 163–177.

Miles, R. E. (1980). *Macro-organizational-behavior*. Glenview, IL: Scott.

Mill, J. S. (1843, 1967). *A system of logic: Ratiocinative and inductive*. Toronto, University of Toronto Press.

Millett, J. A. (1962). *The academic community*. New York: McGraw-Hill.

Mintzberg, H. D. (1979). Organizational power and goals: A skeletal theory. In: D. E. Schendel & C. W. Hofer (Eds), *Strategic Management*. Boston, MA: Little, Brown and Company.

Moore, G. E. (1959). *Principia ethica*. Cambridge, UK: Cambridge University Press (original published in 1903).

Morgan, G. (1980). Paradigms, metaphors, and puzzle solving in organizational theory. *Administrative Science Quarterly*, *25*, 605–622.

Morgan, R. M., & Hunt, S. D. (1994). The commitment-trust theory of relationship marketing. *Journal of Marketing*, *58*(3), 20–38.

Moskowitz, M. (1972). Choosing socially responsible stocks. *Business and Society Review*, *1*, 71–75.

Mossholder, K. W., & Bedeian, A. G. (1983). Cross-level inference and organizational research: Perspectives of interpretation and application. *Academy of Management Review*, *8*(4), 547–558.

Mudambi, R., & Helper, S. (1998). The close but adversarial model of supplier relationships in the U.S. auto industry. *Strategic Management Journal*, *19*, 775–792.

Naisbitt, J., & Aburdene, P. (1985). *Reinventing the corporation*. New York: Warner.

Narver, J. C., & Slater, S. F. (1990). The effect of a market orientation on business profitability. *Journal of Marketing*, *54*, 20–36.

Nemeroff, D. (1980). *Service delivery practices and issues in leading consumer service businesses: A report to participating companies*. New York, NY: Citibank.

Parasuraman, A. (1987). Customer-oriented corporate cultures are crucial to services marketing success. *Journal of Services Marketing*, *1*(Summer), 39–46.

Pava, M. L., & Krausz, J. (1996). The association between corporate social-responsibility and financial performance: The paradox of social cost. *Journal of Business Ethics*, *15*, 321–357.

Payne, A. F. (1988). Developing a market-oriented organization. *Business Horizons*, *31*(3), 46–53.

Penrose, E. T. (1959). *The theory of the growth of the firm*. New York: Wiley.

Perrow, C. (1970). Departmental power and perspectives in industrial firms. In: M. N. Zald (Ed.), *Power in Organizations* (pp. 59–89). Nashville: Vanderbilt University Press.

Peteraf, M. A. (1993). The cornerstones of competitive advantage: A resource-based view. *Strategic Management Journal*, *14*(2), 179–191.

Peters, T. J., & Waterman, R. H. (1982). *In search of excellence: Lessons form America's best run companies*. New York: Harper and Row.

Pfeffer, J. (1994). *Competitive advantage through people: Unleashing the power of the work force*. Boston: Harvard Business School Press.

Pfeffer, J. (1998). *The human equation: Building profits by putting people first*. Boston: Harvard Business School Press.

Pfeffer, J., & Salancik, G. R. (1978). *The external control of organizations: A resource dependence perspective*. New York: Harper Row.

Philipps, R. A. (1997). Stakeholder theory and a principle of fairness. *Business Ethics Quarterly*, *7*, 51–66.

Phillips, L. W. (1981). Assessing measurement error in key informant reports: A methodological note on organizational analysis in marketing. *Journal of Marketing Research*, *18*, 395–415.

Pickering, J. F., & Cockerill, T. A. J. (1984). *The economic management of the firm*. Oxford: Philip Allen.

Polonsky, M. J. (1995). A stakeholder theory approach to designing environmental marketing strategy. *Journal of Business and Industrial Marketing*, *10*(3), 29–46.

Pondy, L. R., & Mitroff, I. (1979). Beyond open systems models of organization. In: B. Staw (Ed.), *Research in Organizational Behavior* (pp. 13–40). Greenwood, CN: JAI Press.

Porac, J. F., Thomas, H., & Baden-Fuller, C. (1989). Competitive groups as cognitive communities: The case of the Scottish knitwear industry. *Journal of Management Studies*, *26*, 397–416.

Porac, J. F., & Thomas, H. (1990). Taxonomic mental models in competitor definition. *Academy of Management Review*, *15*, 224–240.

Porter, M. E. (1980). *Competitive strategy*. New York: Free Press.

Porter, M. E. (1985). *Competitive advantage*. New York: Free Press.

Porter, M. E. (1991). Towards a dynamic theory of strategy. *Strategic Management Journal* (Winter Special Issue), 95–117.

Preston, L. E., & Sapienza, H. J. (1990). Stakeholder management and corporate performance. *The Journal of Behavioral Economics*, *19*(4), 361–375.

Pritchard, R. D., & Karasick, B. W. (1973). The effects of organizational climate on managerial job performance and job satisfaction. *Organizational Behavior and Human Performance*, *9*, 126–146.

Ragin, C. (1987). *The comparative method – Moving beyond qualitative and quantitative strategies*. Berkeley: University of California Press.

Rappaport, A. (1986). *Creating shareholder value: The new standard for business performance*. New York, NY: Free Press.

Reed, R., & deFilippi, R. J. (1990). Causal ambiguity, barriers to imitation and sustainable competitive advantage. *Academy of Management Review*, *15*, 88–102.

Reger, R., & Huff, A. S. (1993). Strategic groups: A cognitive perspective. *Strategic Management Journal*, *14*(2), 103–123.

Rhenman, E. (1968). *Industrial democracy and industrial management*. London: Tavistock.

Richardson, J. (1993). Parallel sourcing and supplier performance in the Japanese automobile industry. *Strategic Management Journal*, *14*, 339–350.

Rindova, V. P. (1997). The image cascade and the formation of corporate reputations. *Corporate Reputation Review*, *1*, 189–194.

Rindova, V. P., & Fombrun, C. J. (1999). Constructing competitive advantage: The role of firm-constituent interactions. *Strategic Management Journal*, *20*, 691–710.

Rock, J., & Cayer, M. (1995). Who knows best about customer satisfaction, management, employees or customers themselves? Paper presented at the Annual Meeting of the Society for Industrial and Organizational Psychology. Orlando, Florida.

Roll, R. (1977). A critique of the asset pricing theory's tests. *Journal of Financial Economics*, 129–176.

Ross, S. (1973). The economic theory of agency: The principal's problem. *American Economic Review*, *63*, 134–139.

Ross, S. (1976). The arbitrage theory of capital asset pricing. *Journal of Economic Theory*, 343–362.

Rouse, M. J., & Daellenbach, U. S. (1999). Rethinking research methods for the resource-based perspective: Isolating sources of sustainable competitive advantage. *Strategic Management Journal*, *20*, 487–494.

Ruekert, R. W. (1992). Developing a market orientation: An organizational strategy perspective. *International Journal of Research in Marketing*, *9*, 225–245.

Rumelt, R. P. (1991). How much does industry matter? *Strategic Management Journal*, *12*(3), 167–185.

Rumelt, R. P., Schendel, D., & Teece, D. J. (1991). Strategic management and economics. *Strategic Management Journal*, *12*, 5–29.

Salancik, G., & Meindl, J. (1984). Corporate attributions as strategic illusions of management control. *Administrative Science Quarterly*, *29*, 238–254.

Schendel, D., & Hofer, C. (1979). *Strategic management: A new view of business policy and planning*. Boston: Little, Brown & Co.

Scherer, F., & Ross, D. (1990). *Industrial market structure and economic performance*. Boston, MA: Houghton-Mifflin.

Schmalensee, R. (1985). Entry deterrence in the ready-to-eat breakfast industry. *Bell Journal of Economics*, *9*, 305–327.

Schneider, B., & Bowen, D. E. (1985). Employee and customer perceptions of service in banks: Replication and extension. *Journal of Applied Psychology*, *70*(August), 423–433.

Schneider, B., & Bowen, D. E. (1993). The service organization: Human resource management is crucial. *Organizational Dynamics*, *21*, 39–52.

Schwenk, C. R. (1984). Cognitive simplification processes in strategic decision-making. *Strategic Management Journal*, *5*(2), 111–128.

Scott, M. C. (1998). *Value drivers*. Chichester, UK: John Wiley.

Shapiro, B. P. (1988). What the hell is market-oriented? *Harvard Business Review* (Nov–Dec), 119–125.

Sharma, S., & Vredenburg, H. (1998). Proactive corporate environmental strategy and the development of competitively valuable organizational capabilities. *Strategic Management Journal*, *19*, 729–753.

Sharpe, W. F. (1964). Capital asset prices: A theory of market equilibrium under conditions of risk. *Journal of Finance*, 425–442.

Simon, H. A. (1955). A behavioral model of rational choice. *Quarterly Journal of Economics*, *69*(February), 99–118.

Simon, H. A. (1957). *Administrative behavior*. New York: Macmillan.

Simon, H. A. (1964). On the concept of organizational goal. *Administrative Science Quarterly*, *9*(June), 253–283.

Slater, S. F., & Narver, J. C. (1994). Does competitive environment moderate the market orientation-performance relationship. *Journal of Marketing*, *58*, 46–55.

Slater, S. F., & Narver, J. C. (1995). Market orientation and the learning organization. *Journal of Marketing*, *59*, 63–74.

Smelser, N. (1976). *Comparative methods in the social sciences*. Englewood Cliffs, NJ: Prentice-Hall.

Smith, A. (1776). *The wealth of nations*. New York: Penguin.

Smith, P. C., Kendall, L. M., & Hulin, C. L. (1969). *The measurement of satisfaction in work and retirement*. Chicago: Rand McNally.

Smith, P. C., Kendall, L. M., & Hulin, C. L. (1985). *The job descriptive index*. Bowling Green, OH: Department of Psychology, Bowling Green State University.

Smitka, M. J. (1991). *Competitive ties: Subcontracting in the Japanese auto industry*. New York: Columbia University Press.

Solomon, E. (1969). *The theory of financial management* (6th ed.). New York, Columbia University Press.

Spender, J. C., & Grant, R. M. (1996). Knowledge and the firm: Overview. *Strategic Management Journal*, *17*(Winter), 5–9.

Srivastava, R., Shervani, T. A., & Fahey, L. (1998). Market-based assets and shareholder value, a framework for analysis. *Journal of Marketing*, *62*(January), 2–18.

Starbuck, W. H. (1965). Organizational growth and development. In: J. G. March (Ed.), *Handbook of Organizations*. Chicago: Rand McNally.

Starbuck, W. H., & Hedberg, B. L. T. (1977). Saving an organization from a stagnating environment. In: H. B. Thorelli (Ed.), *Strategy + Structure = Performance*. Bloomington, IN: Indiana University Press.

Stead, W. E., & Stead, J. G. (1992). *Management for a small planet*. Newbury Park, CA: Sage.

Steadman, M. E., Zimmerer, T. W., & Green, R. F. (1995). Pressures from stakeholders hit Japanese companies. *Long Range Planning*, *28*(6), 29–37.

Steers, R. M. (1975). Problems in the measurement of organizational effectiveness. *Administrative Science Quarterly*, *20*, 546–558.

Stern, L., El-Ansary, A. I., & Coughlan, A. T. (1996). *Marketing channels*. Upper Saddle River, NJ: Prentice-Hall.

Stewart, G. B., III (1991). *The quest for value*. New York: Harper Business.

Strandvik, T. (1994). *Tolerance zones in perceived service quality*. Swedish School of Economics and Business Administration, Helsinki.

Sutcliffe, K. M., & Huber, G. P. (1998). Firm and industry as determinants of executive perceptions of the environment. *Strategic Management Journal*, *19*, 793–807.

Swanson, G. (1971). Frameworks for comparative research: Structural anthropology and the theory of action. In: I. Vallier (Ed.), *Comparative Methods in Sociology: Essays on Trends and Applications* (pp. 141–202). Berkeley: University of California Press.

Taylor, A. (1997). Yo, Ben! Yo, Jerry! It's just ice cream! *Fortune* (April 28).

Taylor, F. W. (1911). *Principles of scientific management*. New York: Harper and Row.

Taylor, J., & Bowers, D. G. (1972). *Survey of organizations: A machine-scored standardized questionnaire investment*. Ann Arbor, MI: University of Michigan.
Teece, D. J., Pisano, G., & Shuen, A. (1997). Dynamic capabilities and strategic management. *Strategic Management Journal, 18*(7), 509–533.
Thibaut, J. W., & Kelley, H. H. (1978). *The social psychology of groups*. New York: John Wiley.
Thomas, A. B. (1988). Does leadership make a difference for organizational performance? *Administrative Science Quarterly, 31*, 439–465.
Thompson, J. D. (1967). *Organization in action*. New York: McGraw-Hill.
Tsui, A. S. (1990). A multiple constituency model of effectiveness: An empirical examination at the human resource subunit level. *Administrative Science Quarterly, 35*, 458–483.
Ullmann, A. H. (1985). Data in search of a theory: A critical examination of the relationships among social performance, social disclosure, and economic performance of U.S. firms. *Academy of Management Review, 10*(3), 540–557.
Varaiya, N., Kerin, R. A., & Weeks, D. (1987). The relationship between growth, profitability and market value. *Strategic Management Journal, 8*, 487–497.
Venkatraman, N., & Grant, J. H. (1986). Construct measurement in organizational strategy research: A critique and proposal. *Academy of Management Review, 11*(1), 71–87.
Venkatraman, N., & Ramanujam, V. (1986). Measurement of business performance in strategy research: A comparison of approaches. *Academy of Management Review, 11*(4), 801–814.
Walton, R. (1985). From "control" to "commitment" in the workplace. *Havard Business Review, 63*(2), 77–84.
Walton, M. (1986). *The Deming management method*. New York: Dodd & Mead.
Wartick, S. L. (1992). The relationship between intense media exposure and change in corporate reputation. *Business and Society, 31*, 33–49.
Wartick, S. L. (1994). The Toronto conference: Reflections on stakeholder theory. *Business and Society* (April), 110–117.
Webster, F. (1992). The changing role of marketing in the corporation. *Journal of Marketing, 56*(October), 1–17.
Weick, C. E. (1979). Cognitive processes in organizations. In: B. Staw (Ed.), *Research in Organizational Behavior* (Vol. 1, pp. 41–74). Greenwich, CT: JAI Press.
Weick, C. E. (1995). *Sensemaking in organizations*. Thousand Oaks, CA: Sage.
Weiss, L. W. (1974). The concentration-profits relationship and antitrust. In: H. J. Goldschmidt et al. (Eds), *Industrial Concentration: The New Learning*. Boston: Little, Brown.
Wernerfelt, B. (1984). A resource-based view of the firm. *Strategic Management Journal, 5*, 171–180.
Wernerfelt, B., & Montgomery, C. A. (1988). Tobin's q and the importance of focus in firm performance. *American Economic Review, 78*, 246–251.
Whysall, P. (2000). Stakeholder mismanagement in retailing: A British perspective. *Journal of Business Ethics, 23*(2), 19–28.
Wicks, A. C., Gilbert, D. R., & Freeman, E. R. (1994). A feminist reinterpretation of the stakeholder concept. *Business Ethics Quarterly, 4*, 475–498.
Williamson, O. E. (1984). Corporate governance. *Yale Law Review, 93*, 1197–1230.
Wood, S., & Albanese, M. (1995). Can we speak of a high commitment management on the shop floor? *Journal of Management Studies, 32*(2), 215–247.
Woodside, A. G. (1987). Measuring customer awareness and share-of-requirements awarded to competing industrial distributors. *Industrial Marketing & Purchasing, 2*(2), 47–68.
Worthy, J. C. (1984). *Shaping an American institution: Robert E. Wood and Sears, Roebuck*. Urbana: University of Illinois.

Yuchtman, E., & Seashore, S. E. (1967). A system resource approach to organizational effectiveness. *American Sociological Review*, *32*, 891–903.

Zajac, E. J., & Bazerman, M. H. (1991). Blind spots in industry and competitor analysis. *Academy of Management Review*, *16*, 37–56.

Zammuto, R. F. (1984). A comparison of multiple constituency models of organizational effectiveness. *Academy of Management Review*, *9*(4), 606–616.

APPENDIX

Sampled Organizations

Utilities

- ALLEGHENY ENERAMEREN CORP
- AMERICAN ELECTRIC POWER
- ATLANTIC ENERGY
- BEC ENERGY
- CAROLINA POWER & LIGHT
- CENTRAL & SOUTH WEST
- CENTRAL VERMONT PUBLIC SERVICE
- CLECO CORP
- CMP GROUP INC
- COMMONWEALTH EDISON CO
- DTE ENERGY CO
- DUKE ENERGY CORP
- EASTERN UTILITIES ASSOC
- EL PASO ELECTRIC CO
- EMPIRE DISTRICT ELECTRIC CO
- ENTERGY CORP
- FIRSTENERGY CORP
- GPU INC
  IDACORP INC
- KANSAS CITY POWER & LIGHT
- NEVADA POWER CO
- NEW ENGLAND ELECTRIC SYSTEM
- NORTHEAST UTILITIES
- OGE ENERGY CORP
- OTTER TAIL POWER
- PECO ENERGY CO
- PP&L RESOURCES INC

- SOUTHERN CO
- TEXAS UTILITIES CO
- TNP ENTERPRISES INC
- UNICOM CORP
- UNISOURCE ENERGY
- UNITED ILLUMINATING
- UPPER PENINSULA ENERGY CORP

Airlines

- ALASKA AIR GROUP INC
- AMERICA WEST HLDG
- AMR CORPORATION
- ASA HOLDINGS INC
- BRITISH AIRWAYS PLC
- COMAIR HOLDINGS INC
- CONTINENTAL AIRLS
- DELTA AIR LINES INC
- KLM ROYAL DUTCH AIR
- MESA AIR GROUP INC
- MESABA HOLDINGS INC
- NORTHWEST AIRLINES
- SKYWEST INC
- SOUTHWEST AIRLINES
- TRANS WORLD AIRLINES
- UNITED AIRINES CORP
- US AIRWAYS GROUP INC

Oil Companies

- AMOCO CORP
- ASHLAND INC
- ATLANTIC RICHFIELD CO
- BRITISH PETROLEUM PLC
- CHEVRON CORP
- EXXON CORP
- IMPERIAL OIL LTD
- MOBIL CORP
- MURPHY OIL CORP
- PENNZOIL CO
- PHILLIPS PETROLEUM CO
- ROYAL DUTCH PET

- ROYAL DUTCH/SHELL
- SHELL TRAN&TRADE
- SUNCOR ENERGY INC
- USX CORP

BUILDING EFFECTIVE BUYER-SELLER DYADIC RELATIONSHIPS

Michael W. Preis, Salvatore F. Divita and Amy K. Smith

INTRODUCTION

Missing in most of the research on selling has been an examination of the process from the point of view of the customer. When satisfaction in selling has been considered, researchers have focused on the satisfaction of the salesperson with his job and/or the impact of this job satisfaction on performance (e.g. Bluen, Barling & Burns, 1990; Churchill, Ford & Walker, 1979; Pruden & Peterson, 1971). To concentrate on salesperson performance while neglecting customers is to ignore the most important half of the relationship between buyers and sellers and entirely disregards the marketing concept and the streams of research in customer satisfaction. This research takes a different approach and examines customers' satisfaction with salespeople.

Importance of the Study

In the highly competitive markets of the 21st century, traditional forms of product differentiation are declining rapidly. Markets are overcrowded. Marketers and their organizations quickly match competitive products' features. Instantaneous communication and flexible manufacturing methods mean that proprietary manufacturing processes are soon copied throughout industries (Coyne, 1989). Styling is quickly copied, making a distinctive appearance more difficult to achieve.

Evaluating Marketing Actions and Outcomes
Advances in Business Marketing and Purchasing, Volume 12, 263–358

ISSN: 1069-0964/doi:10.1016/S1069-0964(03)12005-4

Product reliability and warranties improve in order to remain competitive. As a consequence, substantive differences between product offerings decrease.

Meanwhile, pricing, advertising and promotion strategies, and distribution policies are also quickly copied. In this competitive and highly charged environment, it becomes more and more difficult to achieve a sustainable competitive advantage; customer satisfaction is thought to offer such an advantage (Oliver, 1997). Products and services that provide high levels of overall customer satisfaction are less vulnerable to competition, have higher proportions of repeat business and have higher gross margins. Zeithaml, Berry and Parasuraman (1996) contend that high levels of customer satisfaction serve both offensive and defensive purposes for marketers. Used as an offensive tool, customer satisfaction may attract new customers, while used as a defensive tool it may create brand loyalty and thus keep present customers from defecting. In short, high customer satisfaction is an indispensable means of creating a lasting competitive advantage (e.g. Fornell, 1992; Patterson, Johnson & Spreng, 1997).

"Personal selling is the most important element in marketing communications to most business concerns" (Weitz, 1978). Personal selling represents what is often the only personal contact between customers and the selling organization. This customer contact represents a unique opportunity for marketers to adapt their messages to the particular requirements of each individual customer and to influence how that customer feels about the product, the manufacturer, the retailer, and the salesperson. If organizations could prospectively select and hire salespeople who would increase customer satisfaction, this would be an important advance in the field of selling and sales management. In addition it would offer manufacturers the opportunity to create a distinct competitive advantage (Dwyer, Schurr & Oh, 1987).

Several researchers suggest that interpersonal satisfaction is an important component of overall satisfaction (e.g. Bitner, Booms & Tetreault, 1990; Hempel, 1977; Oliver & Swan, 1989a). Because interpersonal satisfaction is so important, a question that may be asked is, what psychological attributes of customers and salespeople are important in determining the level of interpersonal satisfaction that is attained? The research described here examines the influence of two aspects of the personalities (referred to here as temperament and value system) of both customers and salespeople on the creation of interpersonal satisfaction and on repurchase intentions. (In order to narrow the scope of the research, repurchase rather than initial purchase intentions are examined.) Temperament describes behavioral patterns or "characteristic ways of coping with the environment" (Kragness & Rening, 1996) and value systems relate to motives (Vinson, Scott & Lamont, 1977) and how people use what they know (Divita, 1993). Both terms are described more fully in subsequent sections.

Purpose of the Study

This study investigates the effects of temperament-type and value system-type on interpersonal satisfaction and the effect of interpersonal satisfaction on repurchase intentions. Specifically, the study serves the following purposes:

(1) To introduce the concept of value systems to the selling literature.
(2) To determine the impact of value systems of salespeople and customers on customers' interpersonal satisfaction.
(3) To introduce Marston's typology of temperaments to the marketing literature.
(4) To determine the impact of temperaments of salespeople and customers on customers' interpersonal satisfaction.
(5) To determine the impact of interpersonal satisfaction on intention to repurchase from the same salesperson.
(6) To provide managers with criteria useful in the selection and training of salespeople in order to achieve high levels of customer interpersonal satisfaction.
(7) To further the knowledge of dyadic interactions in selling relationships.

Temperament

Two aspects of personality, temperaments and value systems, remain relatively unexplored in the marketing literature. Based on the work of Harvard psychologist William Moulton Marston (1928), Thomas Hendrickson developed a typology of behavioral patterns called the Personal Profile Analysis (PPA) which is sold as a diagnostic tool for self-improvement and to improve interpersonal relationships (TIMS, 1993). The aspects of personality that Hendrickson classified are referred to as *temperament* in this study. Temperament can be thought of as patterns of behavior, such as being introverted or extroverted, that individuals use in relating to one another.

The Marston model, on which the PPA is based, classifies behavior into four archetypes called Dominance, Influence, Supportive, and Critical Thinking (also referred to as D-, I-, S-, and C-type temperaments, respectively, in this study). Behaviors typical of the Dominance archetype include task orientation; people with this pattern like to take control and get things done. They also have high energy levels, appear restless, have erect postures, rapid speech, and speak with authority and emphasis. The Influence pattern is typified by extroverted individuals who establish rapport instantly and who relish social interaction and relationships. People with this style are expressive and animated – some part of

Table 1. Characteristics of the Four Temperament Archetypes.

Temperament Archetypes	Temperament Traits	Temperament Identifiers	Communication Style
Dominance	Task oriented, take control, high energy, think strategically	Speak with authority and emphasis	Direct, quick, forceful, interrupt others
Influence	Need social interaction and relationships, extroverted	Large vocabularies, articulate, many gestures	Like to talk, may stray from topic, talk feelings and ideas
Supportive	Supportive of others, deliberate behaviors, shy, team player	Speak slowly, hesitantly, deliberately,	Take time to open up, suggest or imply, quiet
Critical Thinking	Detail thinker, high standards, picky, orderly	Slow and soft pattern of speech	Distant, question details, need time for processing

their bodies always seems to be in motion. They have large vocabularies and speak articulately. The Supportive pattern is typified by behavior supportive of other people. Individuals with this style are often shy and uncomfortable with strangers, avoid attention and speak slowly and hesitantly. They have deliberate work patterns and often communicate indirectly, often by implication. Overt behaviors of people with the Critical Thinking style include slow and soft patterns of speech with measured gestures. Individuals exhibiting this pattern are generally seen as reserved, cold, and formal, and as paying great attention to detail and accuracy (Divita, 1993). Characteristics of the four temperament archetypes are summarized in Table 1.

Hendrickson claims that all individuals display some aspects of each archetype in their behaviors but that one type is preferred or dominant; the dominant archetype is the strongest influence on the behavioral patterns of the individual and controls more of that individual's behavior than the other types. If the intensity of the dominant archetype is low, then the behavioral patterns characteristic of that archetype will still dominate the individual's behavior, though with less force than if the intensity were stronger.

Temperament is innate and is very influential in determining non-verbal overt behaviors such as speech patterns (volume, fluency, and speed), extroversion or introversion, gait, and body language. In addition, temperament controls such behaviors as levels of patience, supportiveness, procrastination, and self-discipline. Non-verbal language, which consists of such characteristics as tone of voice, facial expressions, body language, rapidity of speech, and eye contact, is an important component in interpersonal communication. People with similar patterns of non-verbal communication understand each other intuitively and easily; hence, they communicate better. Increased understanding is likely to increase interpersonal satisfaction. People with dissimilar patterns of non-verbal language may talk past each other, leaving both parties with the feeling that he or she has not been understood. This lack of communication increases the stress level of both parties and leads to diminished interpersonal satisfaction (Divita, 1993, 1995).

Value System

The second aspect of personality examined in the research described here was value systems. Value systems have been studied and classified in the fields of psychology, sociology, and anthropology but the construct has received little attention or recognition in the field of marketing. Value systems are sets of standards, principles, and values which influence decisions we make (O'Connor, 1986; Vinson, Scott & Lamont, 1977) and volitional aspects of behavior that reflect how we use our knowledge (Divita, 1993, 1995).

Values determine the goals that provide a positive drive toward personal satisfaction and the perspective that one has on society and the world in general. One's values determine whether material goods, fulfillment of personal goals and satisfaction with a job well done, approval of others, or respect of others motivate actions (O'Connor, 1986).

Values and Lifestyles (VALS) is a widely used approach to market segmentation (Engel, Blackwell & Miniard, 1995). VALS divides the population of the United States into nine segments, each with unique identifying characteristics. The system is based, in part, on conformity of different segments to values, such as machismo, competitiveness, societal responsibility, and materialism. In addition to values, however, VALS also uses demographic data (such as age and income) and lifestyle data (such as whether the individual lives in the city or the country) to discriminate between segments. Another approach used for market segmentation is the list of values (LOV) developed by Kahle (1983). In this approach marketers assign consumers to segments based on the values that the consumers ranked highest from among a list of values. Both of these tools are used primarily for market segmentation and not to predict individuals' behaviors or communication styles.

The research described here used the framework for classifying value systems developed by Massey (1979). As with Hendrickson's behavioral patterns, Massey posits there are four archetypes and that each individual's value system profile combines them all. While people exhibit elements of each of the four types, one of the four is usually stronger than the other three. The strongest value system archetype will be more salient than the others and will exert a greater influence on the individual's value-system-driven behavior than the remaining three types. The strongest type is referred to as dominant and is more visible to observers than the remaining types.

The four archetypal value systems in Massey's framework are Traditionalist, Challenger, In-betweener, and Synthesizer (also referred to as T-, C-, I-, and S-type value systems). Briefly, people holding the Traditionalist value system accept authority that comes from rank, are loyal, and keep their work and personal relationships relatively formal. Challengers, on the other hand, question authority. A person with this value system may appear selfish – never being satisfied with his or her present condition but always wanting more. Individuals with the In-betweener value system appear to switch back and forth between Traditionalist and Challenger value system types – sometimes accepting authority, sometimes questioning it. A person with the Synthesizer value system will work in the interests of others, even if it means sacrificing his or her time, effort, and possibly well being. While material rewards are most significant for Challengers, the satisfaction of a job well done and the fulfillment of personal objectives are most significant to a Synthesizer. Characteristics of the four value system archetypes are summarized in Table 2.

Theoretical Model

Customer satisfaction has multiple components (e.g. Howard & Sheth, 1968; Westbrook & Reilly, 1983) and one of the components of customer satisfaction is interpersonal satisfaction, the satisfaction a customer feels with the salesperson (Oliver & Swan, 1989b). In order to create interpersonal satisfaction, Divita (1993, 1995) and Webster (1965) suggest that customers' human needs (such as love, desire for personal recognition or respect, trust, or avoidance of interpersonal stress) must be met by salespeople. This view is an interesting parallel to Maslow's (1954) hierarchy of needs in which higher level needs, such as self-actualization, motivate individuals once lower level needs, such as food and shelter, are met. Once the basic need of purchasing the good or service is met, the more the salesperson is able to fulfill each customer's human needs, the more satisfying the relationship will be to that customer. Relationship selling is predicated on the idea that it is crucial to build a continuing long-term personal relationship between customers and salespeople (Clabaugh & Forbes, 1992). Temperaments, which drive non-verbal overt behavioral patterns, and value systems, which represent what is important to each individual and how that individual will use his or her

Table 2. Characteristics of the Four Value System Archetypes.

Value System Archetype	Primary Motivation	Key Value	Characteristics	Communication Style
Traditionalist	Social respect	Loyalty	Accept authority, formal relationships, maintain status quo	Speak of "the way things are"
In-betweener (transitional stage between Challenger and Traditionalist)	Social approval	Equality	Influenced by peer pressure, lack commitment, avoid decisions	Wait to commit until highest authority has spoken and then echo
Challenger	Material rewards such as money, power, pleasure	Freedom	Question authority, Selfish, unethical, manipulative	Subtle complaints, how bad things are
Synthesizer	Personal fulfillment, compulsion to make the world a better place, operates in future timeframe	Justice	Work in interests of others, idealistic, initiate change which benefits others	Passionate about beliefs, specific ideas to benefit organization

knowledge, are important aspects of satisfying human needs (Divita, 1993, 1995; McIntyre, 1991).

The theoretical model used in this research incorporates the relationships suggested by Divita and is also consistent with and incorporates the expectancy disconfirmation model of customer satisfaction/dissatisfaction as proposed by Oliver (1997). As depicted in Fig. 1, each customer's unique temperament and value system determine his or her human needs. The salesperson's unique temperament, value system, and knowledge of the product being sold, determine his or her behavior; each aspect's (i.e. temperament, value system and knowledge base) influence is additive. The salesperson's behavior is also influenced by the customer's human needs. How well that salesperson meets the customer's human needs will determine that customer's level of interpersonal satisfaction. Interpersonal satisfaction, since it is a component of overall customer satisfaction, affects that evaluation and also affects repurchase intentions (Grewal & Sharma, 1991; Woodside, Frey & Daly, 1989).

Customers' expectations of products are formed from customers' knowledge of those products (both brand knowledge and category knowledge) and from product needs. Salespeople's behaviors, in the form of promises and portrayal of products will further influence customers' product expectations. By definition, expectations

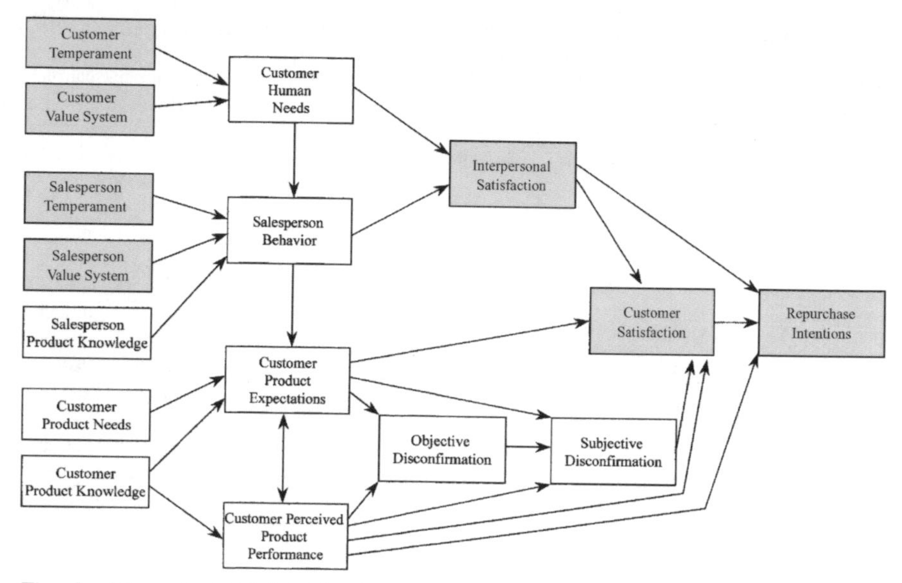

Fig. 1. Theoretical Model Showing Impact of Temperament and Value System on Customer Satisfaction. *Note:* Shaded areas represent constructs examined in this study.

directly influence both objective and subjective disconfirmation and have been shown to impact perceived product performance (Churchill & Surprenant, 1982; Oliver, 1980; Tse & Wilton, 1988). Objective disconfirmation is based on objective (actual) product performance compared to expectations of performance; subjective disconfirmation is based on subjective (perceived) performance compared to expected performance.

Product expectations, perceived performance, and subjective disconfirmation influence customer satisfaction. Repurchase intentions (i.e. customer intention to repurchase the product) are directly influenced not only by customer satisfaction, of which interpersonal satisfaction is a component, but also by perceived product performance. The shaded boxes in Fig. 1 represent the components of the model that were examined in the research described here.

Organization

This section has provided an overview of the research presented here and has provided background information to put the issues being studied into context. The theoretical model of the study was presented to provide a framework for assistance in understanding the relationships between the various concepts discussed. The next section presents the theoretical foundations of the research and includes a review of relevant literature on customer satisfaction as well as temperament and value systems. The third section details the operational model and the methodology used, including the sample selection method, survey technique, questionnaires, hypotheses, statistical techniques used to test the hypothesized relationships, and the decision rules used in determining the significance of the results. The fourth section presents the analysis of data and results of the analyses. The final section provides a summary of the study and discusses the significance, managerial implications, and limitations of the findings, and ends with a discussion of the implications of the study for future research.

REVIEW OF THE LITERATURE

Heretofore, personality has been considered in only a superficial manner as an explanatory variable in buyer-seller relationships. Certain temperament characteristics, including forcefulness, sociability, self-esteem, and self-confidence have been examined for relationships to sales performance with very mixed results (see Table 3). But temperament can explain the need for different contents and styles of sales messages (non-verbal communication). And value systems have not yet been examined in interpersonal marketing relationships. Together, these constructs explain many of the motivations for sales and purchasing actions and clarify why some sellers are more or less satisfying to buyers than are others.

Temperament and value systems were considered independent variables and interpersonal satisfaction was the dependent variable examined in the study described here. These independent variables were shown to influence interpersonal satisfaction, helping explain some of the disparate findings of earlier studies, but more importantly, providing guidance to managers on both personnel selection for sales managers and on appropriate adaptive behaviors for salespeople.

This section presents the conceptual framework guiding the research described in this study, details the relationships between the independent and dependent variables, and specifies the effects that were anticipated. In addition, this section reviews the relevant literature supporting the underlying model.

"Personal selling is the most important element in marketing communications to most business concerns" (Weitz, 1978). Personal selling represents what is often the only personal contact between customers and the selling organization. This customer contact represents a unique opportunity for marketers to adapt their messages to the particular requirements of each individual customer and to influence how that customer feels about the product, the manufacturer, the retailer,

Table 3. Relationships of Selected Variables to Salesperson Performance in Customer Interactions as Reported for Selected Research Studies.[a]

Significantly Related to Performance	Not Significantly Related to Performance
Age	
Industrial (Kirchner et al., 1960)	Life insurance (Merenda & Clarke, 1959; Tanofsky et al., 1969)
Retail (Mosel, 1952)	Retail (French, 1960)
	Industrial (Lamont & Lundstrom, 1977)
Education	
Life insurance (Merenda & Clarke, 1959)	Specialty food (Baehr & Williams, 1968)
Retail (Mosel, 1952)	Life insurance (Schultz, 1935; Tanofsky et al., 1969)
Industrial (Lamont & Lundstrom, 1977)	Retail (French, 1960; Mosel, 1952)
Sales Related Knowledge	
Sales Experience, Product Knowledege, Training	
Meta-analysis (Churchill et al., 1985)	Life insurance (Merenda & Clarke, 1959; Tanofsky et al., 1969)
Telemarketing (Sujan et al., 1988)	Specialty food (Baehr & Williams, 1968)
Retail (Brock, 1965; Woodside & Davenport, 1974)	Retail (French, 1960)
Intelligence	
Oil company (Miner, 1962)	Oil company (Harrell, 1960)
Industrial (Bagozzi, 1978)[b]	Meta-analysis (Churchill et al., 1985)
Stockbrokers (Ghiselli, 1973, 1969)	Industrial (Dunnette & Kirchner, 1960)
Job Satisfaction	
Specialty foods (Baehr & Williams, 1968)[c]	Appliances (Cotham, 1968)
Industrial (Bagozzi, 1978)	Retail (French, 1960)
Self-Esteem/Self-Confidence	
Industrial (Bagozzi, 1978)	
Stockbrokers (Ghiselli, 1969)	
Forcefulness	
Oil company (Harrell, 1960)	Oil company (Miner, 1962)
Life insurance (Merenda & Clarke, 1959)	Retail (Howells, 1968)
Trade (Dunnette & Kirchner, 1960)	Stockbroker (Ghiselli, 1973)
Industrial (Dunnette & Kirchner, 1960)	
Retail (Howells, 1968)	
Technical rep (Howells, 1968)	
Commodities (Howells, 1968)	
Stockbroker (Ghiselli, 1973)	

Table 3. *(Continued)*

Significantly Related to Performance	Not Significantly Related to Performance
Sociability	
Life insurance (Merenda & Clarke, 1959)	Oil company (Harrell, 1960; Miner, 1962)
Technical representative (Howells, 1968)	Industrial (Bagozzi, 1978; Pruden & Peterson, 1971)
Retail (Howells, 1968)	
Food (Howells, 1968)	
Expertise	
Retail (Brock, 1965; Woodside & Davenport, 1974)	
Life insurance (Taylor & Woodside, 1981)	

[a] After Weitz (1979, 1981b).
[b] Statistically significant but negatively correlated.
[c] Either positively or negatively correlated depending on measure of performance.

and the salesperson (Weitz, Sujan & Sujan, 1986). Presently, sales managers hire salespeople based on subjective assessments of potential salespeople's abilities to perform, including the ability to satisfy customers (Manning & Reece, 2001). If sales managers could use an objective process to prospectively select and hire salespeople who would increase customer satisfaction, this would be an important contribution to the field of selling and sales management. In addition it would offer manufacturers opportunities to create distinct competitive advantages (Dwyer, Schurr & Oh, 1987). This was one of the primary purposes in undertaking this research project. The following subsections discuss the fundamental constructs upon which the research was based; these are the constructs of interpersonal satisfaction, temperaments, value systems, and repurchase intentions.

Interpersonal Satisfaction

In the multi-attribute model, overall customer satisfaction consists of many components, including, of course, satisfaction with the product itself (e.g. Crosby & Stephens, 1987; Oliver, 1993; Woodruff, Cadotte & Jenkins, 1983). But, as products become more similar, manufacturers find it increasingly difficult to differentiate their offerings from those of competitors (Coyne, 1989). Because of the similarity of offerings it is likewise difficult to create differential levels of customer satisfaction.

One component of overall satisfaction, interpersonal satisfaction (I/S) (Crosby, Evans & Cowles, 1990; Crosby & Stephens, 1987) represents an important facet of the overall satisfaction construct. It is also a facet that manufacturers might constructively employ to differentiate themselves while increasing the satisfaction of their customers.

Satisfaction with relationships between customers and salespeople has been examined without explicitly defining the concept (e.g. Crosby, Evans & Cowles, 1990). Only one reference to the term interpersonal satisfaction has been found in the literature (Oliver & Swan, 1989a) yet the authors did not define the construct. Because the construct is so central to this research and because the term is so potentially useful, interpersonal satisfaction is defined as follows:

> interpersonal satisfaction is an emotional response to a salesperson resulting from evaluation of one's interactions with that salesperson during the shopping, negotiating, payment, delivery, and post-delivery usage stages of acquisition of a good or service.

Several aspects of this definition are noteworthy. First, the construct deals with the customer's response to the person and not the offering. That response is emotional and not analytical; feelings toward the salesperson could involve liking, trusting, warmth, friendliness, repulsion, or any other affective state or states. Second, it is a response *to* the salesperson. Generally, with respect to the salesperson from any particular company, customers do not choose the salesperson with whom they deal, the salesperson is thrust upon them by the seller; hence the customer reacts to the salesperson that the seller has thrust on him or her. The reaction is to the salesperson only, i.e. to his or her behavior, actions, attitudes, demeanor, and to what he or she says; the reaction is not to the product, the firm, or other qualities or characteristics not directly related to the salesperson. Third, any of the steps in the sales process may be omitted without affecting the definition or the emotion. Even the acquisition could be absent; for example, an item might not be purchased or might be purchased and returned, yet the customer could experience high interpersonal satisfaction from his or her interactions with the salesperson. Fourth, the acquisition need not involve purchase of a physical product, but could be a lease or the performance of a service. Finally, interpersonal satisfaction is relevant to salespeople providing both goods and services. Since interpersonal satisfaction is relevant to personal selling situations, attention is next turned to the literature of this subject.

Personal Selling

Unlike customer satisfaction and dissatisfaction, personal selling is not a construct; it is an activity that takes place every day, in many formats, and under many circumstances. Over 16 million people in the United States (more than 12% of the labor force) are engaged in selling and selling activities (U.S. Census Bureau, 2001). "Executives in industrial, consumer durable, and consumer non-durable companies indicated personal selling and sales management were respectively 5.2, 1.8, and 1.1 times more important than advertising in their company's marketing efforts" (Weitz, 1978).

Despite the importance that selling has to marketing and to the profitability of companies large and small, relatively little research has been conducted in this area of marketing. Bagozzi (1976) noted that only 17 of 325 (5%) articles published in the *Journal of Marketing* and the *Journal of Marketing Research* in the two-year period from 1973 to 1975 "dealt with personal selling problems (and about half of these were concerned with normative, managerial issues)" (p. 32). The situation appears to have deteriorated in the intervening twenty-plus years; only five articles (out of 137 published) related to selling were counted in those same two journals in the years 1995 and 1996, and all of these dealt with normative or sales management issues.

Stimulus-Response Model

In historical terms and from a conceptual point of view, most of the research into salesperson performance has been limited in scope and orientation. The model or paradigm that has been used most often is the stimulus-response (S-R) model of behavior. This model suggests that a person's behavior is dependent upon either the situation in which he or she is operating (i.e. his or her environment) or upon his or her situation and personal attributes (Bagozzi, 1976). Thus, in this model, salespeople can elicit desired responses from customers if the salespeople provide the correct stimuli. This approach treats customers in mechanistic terms and places all of the responsibility for sales on salespeople (Bagozzi, 1976). This view contrasts to the dyadic model in which relationships between buyers and sellers are dependent upon both of those people (Evans, 1963). The dyadic model will be discussed later.

Personal Characteristics

The first research into salesperson performance sought to identify universal traits and characteristics associated with high performers. Thus, researchers looked at physical, demographic, and personal traits such as age, height, appearance, intelligence, and education to explain differences in performance. Results of these studies were mixed; while some studies found positive correlations between characteristics, such as age, and performance (e.g. Kirchner et al., 1960; Mosel, 1952), others found no significant correlation with those same characteristics (e.g. French, 1960; Tanofsky et al., 1969).

Churchill et al. (1985) performed a meta-analysis of 116 prior studies. They found that personal attributes, such as age, height, gender, weight, race, appearance, education, marital status, number of dependents, etc. accounted for slightly less than 3% of the variance in salesperson performance. Aptitude accounted for slightly less than 2% of the variation and motivation accounted for just over 3% of the variance. Role perceptions accounted for the highest proportion of any of the categories of characteristics but still accounted for less than 9% of the variance in performance. Organizational and environmental factors had the lowest correlation with performance, accounting for only 1% of the variance. Thus, the relationship of personal characteristics to salesperson performance has been found to be weak, at best.

Peterson, Cannito and Brown (1995) report that speech attributes have a significant impact on customer perceptions of plausibility. More rapid rates of speech and decreasing fundamental frequencies (i.e. declarative sentences as opposed to questions) correlate to successful salespeople. These voice characteristics are an example of a behavioral trait typical of temperaments classified by the Marston typology and measured by the PPA instrument. Rapid speech and falling fundamental frequencies seem to fit with descriptions of the I-type temperament archetype. The research described here may explain, in part, the findings of the Peterson et al. study.

The power of individual traits is low in explaining significant levels of variance in sales performance. These variables do not capture the dynamic properties of dyadic interactions sufficiently to explain dyadic relational constructs such as trust, satisfaction, and reciprocity of personal disclosures (Iacobucci & Hopkins, 1992). The study described here examines personality traits (temperament and value system) of both buyers and sellers to attempt to capture some of these dynamics and thus overcome the weaknesses of earlier studies.

In one of the few studies involving both salespeople and customers, McIntyre (1991) examined the effects of cognitive style, as determined by the MBTI, on customer preference. He showed that matching marketing efforts by modifying the interaction style of contact personnel, including salespeople, modifying audio-visual aids, or utilizing compatible seller contact personnel, can increase buyer preference for a seller. While the McIntyre study was a quasi-experimental design based on MBTI personality types, it clearly supports the concept of matched personality types in dyads. The research described here differs in that it utilized a different instrument for determining similarity, used cross-sectional data from working dyads, and used interpersonal satisfaction (as opposed to preference) as the dependent variable.

Dion, Easterling and DiLorenzo-Aiss (1994) also used the MBTI to assess the influence of personality types on sales performance. In contrast to the McIntyre study, Dion et al. found that similar personality types were not associated with increased buyer trust or salesperson performance. However, perceived similarity, rather than actual similarity, was linked to buyer trust and sales performance. That study differs from the research described here in that *perceived* salesperson performance, rather than interpersonal satisfaction, was examined as a dependent variable. The Dion et al. study should more properly have called salesperson performance perceived salesperson performance because the measures used to determine performance included such items as "Overall, this salesperson represents his/her company well." Similarly, perceived similarity was measured with items such as, "This salesperson is the same type of person as I am." Both of these sample items are worded vaguely given the specific nature of the conclusions reached. Nevertheless, the lack of effect on perceived performance resulting from personality similarity is at odds with the results that were expected in the research described here. However, note that a different personality instrument was used and a different dependent variable was examined in the prior research than were used in this study.

Dispositional Characteristics

The scientific management movement, beginning in the early part of the twentieth century, led to the study of tests for the selection and evaluation of job applicants. Substantial amounts of information regarding the validities of various tests was accumulated and analyzed. This approach to explaining salesperson performance focused on personality, ability, and aptitude tests. Many of the studies were atheoretical and anticipated relationships were rarely specified prior to performance of the research (Ghiselli, 1973; Weitz, 1979). Table 4 shows some of the tests that have been used to predict sales performance.

Ghiselli (1973) examines the literature published between 1920 and 1971 relating aptitude test scores to personnel selection. Success in different jobs was measured by criteria relevant to the respective jobs. Sixteen test types were found to be correlated with performance of salespeople. Using a weighted average correlation coefficient, Ghiselli determined that intelligence tests had the highest validity coefficients, 0.34, for salesmen, while personality tests had the highest coefficients, 0.36, for sales clerks. Tests of personality traits had the second highest correlations, 0.30, for salesmen, while intelligence tests were negatively and weakly correlated (−0.03) to sales clerk performance. Thus, even the strongest tests were able to account for less than 13% of the variance in salesperson performance. After surveying the extant literature on the relationship between personality testing and job performance, Guion and Gottier (1965) conclude that while such tests achieve better results than mere chance, they are not very successful in specific situations. The research described here differs from these earlier studies in that it used a different instrument for determining personality type and the dependent variable was interpersonal satisfaction.

Table 4. Correlation of Selected Test Results with Sales Performance.

	Study	Significant	Not Significant
Capability Tests			
Terman Concept Mastery	Miner (1962)		X
Vocabulary Test G-T	Miner (1962)		X
Wesman Personnel Classification Form A	Miner (1962)		X
Survey of Mechanical Insight	Miner (1962)		X
Wechsler Adult Intelligence Scale-Arithmetic	Miner (1962)	X	
Borgatta's Word Association Form	Bagozzi (1978)		X
Collins "Other Directedness"	Bagozzi (1978)		X
Otis Test of Mental Ability	Harrell (1960)	X[a]	
Canfield Sales Sense	Harrell (1960)		X
Personality and Interest Tests			
Kuder Preference Record	Miner (1962)		X
Tomkins-Horn Picture Arrangement Test	Miner (1962)	X	
Sentence Completion	Miner (1962)		X
Myers-Briggs Type Indicator	McIntyre (1991)	X	
Myers-Briggs Type Indicator	Dion et al. (1994)		X
Activity Vector Analysis	Merenda and Clarke (1959)	X	
Bernreuter Personality Inventory	Harrell (1960)	4 scales[a]	3 scales
Moss-Hunt Social Intelligence	Harrell (1960)	1 scale[a]	2 scales
Strong Vocational Interest Blank	Harrell (1960)	1 scale[a]	8 scales
Strong Vocational Interest Blank	Dunnette and Kirchner (1960)	18 scales	30 scales
Washburne Social Adjustment Inventory	Harrell (1960)		X
Edwards Personal Preference Schedule	Dunnette and Kirchner (1960)	2 scales	13 scales

[a] Significant at the 1% level.

A particularly interesting "scientific" study was performed by Chapple and Donald (1947) using the Interaction Chronograph. This electro-mechanical device was used to record the Activity or Energy Curve (the relative frequency of the subject's speech and gestures) and the Initiative-Dominance Curve (the relative frequency of the subject's initiative and domination of conversation in comparison to the interviewer's). Output of the device was highly correlated with sales performance of department store sales clerks with rank order correlations of 0.85–1.0 (though the sample sizes were small). When tested prospectively (i.e. on new hires), predictions based on outputs of the device were also highly correlated with sales performance. No validations of the study appear to have been performed. Judging from the descriptions of behavioral traits recorded by the instrument, it measures I-type temperament characteristics. The results of that study lend support to the concept tested in the research described here, namely that I-type temperaments are important characteristics for salespeople. Unlike the Chapple and Donald (1947) research, the present study was based on a theory of the importance of personality to interpersonal satisfaction.

Evaluative Comments on the Stimulus-Response Model

The S-R model ignores the complexities of human behavior as displayed in purposeful behavior, information seeking, cognitive processes, and social relationships. In addition, it does not explain why characteristics cause certain effects but, rather, is limited to describing behaviors and predicting outcomes. Finally, the internal and external validities have been challenged as the result of certain anomalies (Bagozzi, 1976). As Bagozzi (1976) states, the S-R approach focuses "primarily on the behavior of single actors and rel[ies] on unilateral influence principles." The equivocal nature of the findings relating personal and personality characteristics to performance was shown in Table 3. Note that different researchers have found conflicting results, even in the same industry (e.g. for life insurance agents, Merenda & Clarke, 1959, found education significantly related to performance while Tanofsky et al., 1969; and Schultz, 1935, did not). The studies that examined forcefulness, sociability, and self-esteem/self-confidence are of particular interest because these are personality traits that are associated with the Dominance and Influence archetypes in the Marston typology. In this research investigation, salespeople with dominant I-type and secondary D-type temperaments were expected to outperform salespeople with other dominant temperament-types. Too much forcefulness without the moderating effects of I-type sociability, or too much sociability without the drive of the D-archetype to close sales, could account for the findings not being significantly related to performance.

Performance was measured by many different methods in the studies presented in Table 4. For example, Harrell (1960) used supervisor interviews and quotas, Howells (1968) found it impossible to develop inter-company scales and so divided the salespeople into two categories, more- and less-successful, Bagozzi (1978) used workload, Baehr and Williams (1968) used sales volume, Cotham (1968) used sales volume and commission earnings, and Dunnette and Kirchner (1960) used overall selling effectiveness, as measures of salesperson performance. While performance is not the same as customer interpersonal satisfaction, the two could be expected to be correlated since satisfied customers are expected to purchase greater quantities or more frequently than dissatisfied customers (see subsection on repurchase intention). However, note that the business environment in the 1960s when many of these studies were conducted was less competitive than it is in the 21st century. A fundamental difference between measures of salesperson performance and measures of customer satisfaction is that the former is based on the perspective of the seller and the latter is based on the perspective of the buyer. In spite of the conflicting evidence presented by these prior studies, none of them has used the same independent and dependent variables as were used in the study described here.

In the face of these shortcomings of the S-R model and of the studies undertaken, the search continued for better explanations of selling effectiveness. The next approach to be discussed considers the roles of buyers, sellers, and their interactions.

Dyadic Model

Evans (1963) was the first to suggest that earlier theories of personal selling did not take into account the human element of the customer and that interactions of customers with salespeople influence outcomes of sales calls. Evans stated, "The sale is a product of the particular dyadic interaction of a given salesman and prospect rather than a result of individual qualities of either alone" (p. 76).

The importance of the customer-salesperson dyad varies depending upon the nature of the exchange relationship, growing progressively more important as exchanges move along the continuum from discrete to relational. A discrete transaction is one that stands alone, characterized by limited communications, narrow content, and void of intrusion of the identities of the parties. While such an extreme may be impossible, it would be approximated by a one-time purchase of an unbranded consumable from an independent vendor and paid for with cash (Dwyer, Schurr & Oh, 1987); a traveler purchasing gasoline from an independent unbranded service station for cash is an example of a such a transaction. A one-time on-line purchase of an unbranded product from a website previously unknown to the buyer might be a somewhat more contemporary example.

In contrast, "relational exchange participants can be expected to derive complex, personal, non-economic satisfactions and engage in *social* exchange" (Dwyer, Schurr & Oh, 1987, p. 12, emphasis in original). Relational exchanges are more apt to be involved with expensive, complex, highly technical products, or with products having long lead times, such as capital goods (e.g. machine tools) or complex components used in manufacturing such as wiring harnesses for automobiles.

Static Characteristics

Many researchers have explored sales performance within dyads by examining characteristics that are static in nature and that do not change as part of the social relationship (Weitz, 1979). Thus, the emphasis changed from studying the characteristics of salespeople alone, to examining the characteristics of both buyers and sellers.

Similarities

Similarities of many types have been examined to determine their impacts on sales within dyads. Following are some representative examples of similarities that have been researched. Churchill, Collins and Strang (1975) found that demographic factors such as age, height, race, and gender were correlated with the size of the purchase made and with whether or not a purchase was made in four departments of a department store. The correlation was weak, however, explaining just 2% of the variance in size of purchase and less than 0.2% of the variance in whether a purchase was made. Similarity in taste in music between the salesperson and customer was found to increase sales of a tape head cleaner in a music store (Woodside & Davenport, 1974). Likewise, when paint salesmen's experience levels were manipulated to be similar to customers' consumption needs, salesmen were more successful at influencing customers than when experience was much different (Brock, 1965). Riordan, Oliver and Donnelly (1977) found greater similarity in attitudes between sold and unsold prospects and the life insurance agents that called on them.

Weitz (1981b) reviewed empirical research and suggested that dyadic similarities (operationalized as physical characteristics, demographics, and irrelevant attitudes) are, at best, weakly correlated to performance. In a study of persuasion, when similarities were relevant to influence attempts, greater opinion change occurred than when similarities were not relevant to the persuasion attempts (Berscheid, 1966). Wilson and Ghingold (1981) reviewed the literature of the similarity-dissimilarity concept and concluded that in spite of being intuitively appealing, "similarity has not proven to be a powerful concept despite all the attention it has received" (p. 88). Many of the similarities tested for in the studies examined by Wilson and Ghingold were demographic variables, expertise, "attitudes," and product usage, so it is not surprising that their conclusions were similar to Weitz's. Nevertheless, they concluded that similarity, as a concept, should be tested "within the context of more elegant models of buyer-seller interaction" (p. 98). The study described here did that by examining temperaments and value systems and their effects on interpersonal satisfaction within dyadic relationships.

Social Exchange Paradigm

The social exchange paradigm provides that "the essential feature of social behavior is that each of the persons in face-to-face interaction influences the behavior of the other" (Webster, 1968, p. 8). Bagozzi (1976) found that seven characteristics differentiate social exchange models from single-party models. First, the parties in an exchange relationship have some mutual or shared interest or value. Each may want what the other has or they are in some other manner dependent upon each other. That they have a mutual interest does not imply that the parties will necessarily cooperate, nor that they will be in conflict; rather, the implication is that each finds it in its own interests to engage in the relationship. Second, the parties will engage in efforts of social influence. Such efforts may use threats, promises, warnings, deception, flattery, or other strategies. Third, forces outside the dyad may constrain or shape the relationship. Such factors could be a powerful third party or socio-economic conditions, or perhaps legal restrictions such as anti-trust regulations. Fourth, characteristics of the parties may influence the relationship through such factors as organizational structures, personality characteristics, or individual or organizational needs. Fifth, conceptions of how one ought to behave (norms) can play an important role in dyadic relationships. Policies and standards of conduct, rituals, rules of etiquette, demeanor, and deference can be strong influences on behavior or expectations in social relationships. "Courtesy calls" by industrial salespeople are a common ritual that can have deep meaning and significance in a relationship. Sixth, purposive behavior such as plans, goals, and intentions influence relationships by establishing bounds on what is acceptable and they determine some of the subjects of discussion. Lastly, the course and duration of the relationship, whether it is cooperative or competitive, and even the eventual outcome, are all uncertain. The creation and resolution of such tensions depends on the actions, interpretations, evaluations and decisions of the parties.

Johnston (1981) shows the importance of the social process within buying organizations. Using the "snowball" interview technique, Johnston examined 62 cases and concluded that purchasing is a "socially negotiated outcome" involving many people in both the buying and selling organizations. There was not a single situation in which a lone individual made all the decisions on his or her own.

Dynamic adjustment of the message and the offering in response to the factors enumerated by Bagozzi is one of the most compelling arguments for the social exchange paradigm. Salespeople use feedback from customers to adjust their communications and sales strategies to meet the unique requirements of each customer (see below). While we are primarily concerned here with salesperson behavior, customer behavior can also change in dyadic relationships and adjust to each situation and salesperson.

Conceptual Frameworks

Several conceptual models of exchange and selling processes are relevant to the discussion here. The models vary in their levels of abstraction, the variables identified, and in the anticipated relationships between them.

Both content and style of communications were proposed as determinants of buyer-seller interactions by Sheth (1976). His framework consisted of four major elements. In the first element, product, organizational, and personal factors set the tone for buyer-seller communications. While the effect of personality on salesperson effectiveness has empirically been found to be weak (Kassarjian, 1971, 1979; Kassarjian & Sheffet, 1975, 1981), these characteristics were included in this element of the model. The second element was seller communication, consisting of both content and style. Content was defined as the words and sentences used to describe product-related features, benefits, novelty, and the like, while style referred to non-verbal communication such as "the format, ritual or mannerism which the buyer and the seller adopt in their interaction." Note that this non-verbal communication included many of the characteristics determined by Marston's temperament-types. The third element in Sheth's framework was the interaction of communication styles of both buyer and seller. The fourth element in Sheth's framework was the outcome of the transaction.

Bagozzi (1979) proposes a formal theory of exchange in the marketplace in which four factors are considered determinants of exchange. These factors were *situational contingencies*, *characteristics of social actors*, *third parties*, and *social influence between actors*; each of these was thought to place limitations on the actions of the parties to an exchange. Situational contingencies consisted of the physical environment, the psychological climate, and the legal setting. Characteristics of social actors could influence exchanges through such factors as expertise, credibility, self-confidence, background, cognitive style, personality traits, and interpersonal orientation (cooperation, competitiveness, rigidity, altruism, and the like). Social parties outside of exchange relationships (i.e. third parties) afforded constraints or opportunities to the actors and, to the extent that what a third party offered was more satisfying to one of the actors, he or she might have left the relationship for the more satisfying alternative. Finally, parties to exchanges had abilities to satisfy individuals' needs and could have accomplished this through social negotiations involving offers, adjustments, threats, promises, and warnings.

Believing that communications play a central and vital role in personal selling, Williams, Spiro and Fine (1990) modeled the customer-salesperson interaction process as consisting of three parts – a customer component, a salesperson component, and an interaction component. The customer and salesperson components were modeled as symmetrical, with role specific processes and information processing segments. The binding force between the salesperson and the customer was the interaction component, which consisted of interpersonal communications. This model is especially relevant to the research described here.

All of these models reflect the interactive relational nature of dyads. Only the Sheth (1976) and Williams, Spiro and Fine (1990) frameworks model the interaction process per se. The Williams et al. model adds information processing, communication coding and communication rules and so includes more factors as being determinants of relationship success. In addition, the latter framework is considerably richer in the interactions and processes specified. As noted previously, this model corresponds in many respects to the research described here.

Much of the discussion about personal selling has revolved around communication between buyers and sellers. Since this is the crucial aspect of personal selling, the discussion now turns to communication.

Communication

Communication is a dynamic process in which one person attempts to change the cognitions of another (Andersen, 1972). Because personal selling is the only form of communication that allows marketers to adapt their messages to the specific needs, desires, style, and beliefs of each customer (Spiro & Weitz, 1990), salespeople have unique opportunities to influence customers and customer satisfaction (Grewal & Sharma, 1991).

Communication has four aspects: content, code, rules, and style (Andersen, 1972; Williams, Spiro & Fine, 1990). Content is the ideational material to be conveyed in the message. Code is the symbolism used to convey the content, and consists of the verbal aspects including grammar and syntax and non-verbal aspects such as smiles, tears, gestures, rate, volume and posture. The rules of communication deal with what to say (i.e. content), how and when to speak, interpersonal space, and touching. Style refers to the pattern of communication including type of arguments used, and voice patterns and represents a synthesis of content, code, and rules (Williams & Spiro, 1985). In spite of the importance of communication, it has been ignored in most conceptualizations of the selling process; when communication has been considered, consideration has usually been limited to content (Williams, Spiro & Fine, 1990).

In one study of the importance of style in sales communications, Williams and Spiro (1985) found that the communication styles of salespeople and customers accounted for 17% of the variance in sales performance. They also found that when salespeople use a "self-oriented style," in which salespeople are preoccupied with themselves and their own welfare, performance is hindered.

Several researchers have found that similarity of communication style is crucial to maximizing sales performance (Kale & Barnes, 1992; Miles, Arnold & Nash, 1990; Mitroff & Mitroff, 1979). The current research investigation builds on these results and explores whether temperaments and value systems of buyers and sellers are correlated with interpersonal satisfaction. The expectation was that the research results would help explain the prior findings.

Steinberg and Miller (1975) suggest that people who understand communications know that non-verbal behavior is at least as important as the words people are mouthing. Effective salespeople were found to detect more cues, especially non-verbal cues, among their customers, than less effective salespeople (Grikscheit, 1971 cited in Weitz, Sujan & Sujan, 1986). These findings are consistent with and provide support for the hypotheses being tested in the research described here.In a rather surprising finding, Cronin (1994) reports no support for the hypothesis that buyers see high-performing salespeople as especially good communicators. Similarly, a precise communication style alone is not perceived as well as when it is combined with a friendly style (Dion & Notarantonio, 1992). These findings further suggest that the open friendly communication style of the dominant I-type temperament is an important element in the personalities of successful salespeople and that the precise style of the dominant C-type temperament is not an important element in the personalities of successful salespeople. These relationships were examined in the research described here.

Cognitive Style

Cognitive style (or the manner in which people understand material that is presented to them) relates closely to communication. A satisfactory interaction between a buyer and a seller, including satisfaction with the negotiation process, requires buyer-seller compatibility with respect to content and style (Spiro & Weitz, 1990; Weitz, 1981a, b; Williams & Spiro, 1985). Content and style within a dyadic interaction are shaped by the personalities of the individuals involved (Kale & Barnes, 1992).

Dion, Easterling and DiLorenzo-Aiss (1994) find that most industrial buyers and salespeople fit a small number of personality types as determined by the Myers-Briggs Type Indicator (MBTI):

- 43% of the salespeople were classified into three of the 16 possible types,
- 53% of the buyers were classified into two personality types.

And the most common buyer personality type (ESTJ) was also the most common salesperson personality type (32 and 23% respectively). The researchers conclude that perceived personality similarity, not actual personality similarity, was correlated with buyer trust and sales performance.

In an approach that utilizes the concept of adaptive behavior, Mitroff and Mitroff (1979) suggest that likes (i.e. people with the same cognitive styles) speak the same language and deal best with likes. The implication for selling is that salespeople should know each customer's style and use that style to make sales presentations. McIntyre (1991), using the MBTI, affirmed the value of this approach; he found that buyers preferred sellers of the same cognitive style more than sellers of the opposite cognitive style. This seems to contradict the stated findings of the Dion et al. study, although the dependent variables – performance, trust, and preference – are different. Nevertheless, Dion et al. found that in their sample, both salespeople and customers were tightly concentrated in a small number of personality types. That this is so seems to confirm McIntyre's findings, especially since the customers in the Dion et al. study selected the salespeople to participate in the study.

These results suggest intriguing links between personality types and interpersonal satisfaction. Since Marston's temperament types are based on behavioral traits which, in turn, are related to non-verbal communication styles, it seems likely that temperaments are related to interpersonal satisfaction. This relationship was examined in the study described here.

Adaptive Selling

The core concept behind the dyadic model is adaptive selling (Weitz, 1981b; Weitz, Sujan & Sujan, 1986), which is usually described as the ability to fashion sales messages and presentations to the unique characteristics and requirements of each customer (e.g. Sujan, Weitz & Sujan, 1988; Weitz, 1981b; Weitz, Sujan & Sujan, 1986). The ability of a salesperson to customize his or her presentations to the requirements of each individual customer is a unique and crucial advantage over canned presentations (Jolson, 1975).

While adaptive selling is recognized for its inherent strengths, few attempts have been made to measure the degree to which salespeople engage in adaptive behavior. Though he did not quantify the extent to which they do so, Cronin (1994) claimed that salespeople do engage in adaptive behavior; Spiro and Weitz (1990) developed a 16-item instrument to measure adaptive behavior. Called ADAPTS, this paper-and-pencil instrument, when completed by salespeople, scores self-responses to statements designed to determine the degree to which a salesperson varies his or her presentation from one situation to another (e.g. "I treat all of my buyers pretty much the same").

Several textbooks on selling suggest that understanding communication styles associated with various personality types is key for salespeople (e.g. Clabaugh & Forbes, 1992; Manning & Reece, 2001). For example, Clabaugh and Forbes (1992) describe Jung's *perception mode* and *informational mode*. The former (intuitor/sensor) relates to how people learn and become aware of objects and ideas while the latter (thinker/feeler) deals with how people process the information gathered in the perception mode. The authors then build a scheme for classifying customers into one of four types based on these scales. In order to utilize such a system, one must learn to both classify or diagnose styles and then to adapt his or her communication style to that of the prospect. The Marston archetypes provide a framework for salespeople, first, to classify the behaviors of their customers and, second, to adapt their behaviors to match those of their customers. The research described here has the potential to demonstrate the value of Marston's framework.

Needs of the Customer

Each customer has two kinds of needs: *personal* needs which motivate what he or she does, and *social* needs which represent activities which are acceptable to relevant others (whether in an organization, a family unit, or to the individual). For example, in an organization, personal needs might involve the need for recognition and advancement while social needs involve the desire to satisfy the requisitioner of the item being purchased (Webster, 1968).

Marketers and researchers have concentrated on satisfying customers' product needs and have neglected personal and social needs (Webster, 1968).

> Specifically, customers have desires and wants in relation to each aspect of the *selling process*... Failure to consider and evaluate the customer's *sales process* needs lessens the probability that a sale will take place. Only by satisfying all of the client's requirements – those related to both the product and the sales process – can the salesperson maximize the chances for a sale. In an environment where competition is keen and many goods can satisfy the client's product requirements, a competitive advantage may accrue to the salesperson who additionally satisfies the client's sales process requirements (Szymanski, 1988, p. 65, emphasis in original).

According to Divita (1995), whether a customer is satisfied (i.e. experiences interpersonal satisfaction) with a salesperson depends upon how well that salesperson meets the customer's human needs and works in the customer's interests. This differs from fairness (equity theory) which suggests that satisfaction is dependent upon outcomes for each party being equitable and that the ratios of inputs to outcomes for each party should be approximately equal.

> In short, customer satisfaction will no longer be achievable by the vendor's commitment to technical excellence or quality of product offering; customer satisfaction will require the satisfaction of both the human and functional needs (Divita, 1995, p. 21).

Each individual's human needs can be met by appropriate verbal and non-verbal behaviors on the part of a salesperson.

> For example, a customer who has a need, e.g. compulsion, to think, talk, and act rapidly also has a need to have other people think, talk and act at the same speed. If a salesperson thought, talked, and acted at a slower speed, he/she would not satisfy the customer's human need for processing speed and thus would impose stress on that customer (Divita, n.d., p. 2).

Such non-verbal behavior is important because when the seller uses the same non-verbal patterns of communication (e.g. code, rules, and style) as the buyer, the buyer can "hear" more of the message and attend to it better (this subject is dealt with in more detail in a subsequent section). Similarly, value system-driven communication is also important for creating understanding and for satisfaction of human needs. Thus, the temperaments and value systems of both customers and salespeople are determinants of interpersonal satisfaction (Divita, n.d.). This relationship is expressed as:

$$IS_{Customer} = f(T_{Salesperson}, VS_{Salesperson}, I_{Salesperson\ VS, Customer\ VS}, I_{Salesperson\ T, Customer\ T})$$

where, IS represents interpersonal satisfaction of the customer, T represents temperament, VS represents value system, and the I terms represent the interactions of the customer's and salesperson's value systems and temperaments.

This relationship is consistent with others' findings. For example, Kale and Barnes (1992) suggest that a satisfactory interaction between a buyer and seller requires buyer-seller compatibility with respect to content (substantive aspects such as product attributes) and style (rituals, format, mannerisms, and ground-rules) (cf. Spiro & Weitz, 1990; Weitz, 1981a; Williams & Spiro, 1985). Williams, Spiro and Fine (1990) say that non-verbal communication is more important than is generally realized; it is less formalized than verbal communication and yet it can enhance or detract from the verbal message. As discussed below, both content and style of the communicator are shaped by the communicator's temperament and value system; the listener's temperament and value system shape that part of the content that can be attended to by the listener and determine whether the communicator's style helps or hinders that listener's understanding.

The concept of interpersonal satisfaction is relevant to all manner of transactions including sales or leases of products and services whether at retail or business-to-business. In short, interpersonal satisfaction relates to virtually all selling activities (excepting, of course, ATM (automated teller machines) transactions, vending machine purchases, Internet sales, and other similar automated transactions in which no personal contact between buyer and seller is made).

Repurchase Intention

Research into customer satisfaction and dissatisfaction would be of little relevance to marketers unless satisfaction had a significant impact on profitability or firm performance. An increase of one percentage point in customer satisfaction "has a net present value of $7.48 million over five years for a typical firm in Sweden" (Anderson, Fornell & Lehmann, 1994). The same authors conclude that the impact on the *Business Week* 1000 would be worth $94 million or 11.4% of current ROI, for a similar increase in overall satisfaction.

Satisfied customers are important because they represent "an indispensable means of creating a sustainable advantage in the competitive environment of the 1990s" (Patterson, Johnson & Spreng, 1997). This advantage manifests itself in positive word-of-mouth, in readier acceptance of other products in the product line, and as brand loyalty or increased intentions to repurchase (Grewal & Sharma, 1991; Rogers, Peyton & Berl, 1992). A strong link between customer satisfaction and repurchase intention, with customer satisfaction explaining 78% of the variance in repurchase intention, was found by Patterson, Johnson and Spreng (1997).

Johnson and Fornell (1991) state, "In a competitive environment, people generally do not continue to purchase products toward which they are ambivalent or hold negative evaluations." Conversely, because repurchase intention is correlated with overall customer satisfaction (Cooper, Cooper & Duhan, 1989), achievement of greater customer satisfaction is desired. The greater the customer satisfaction with past buying or consuming experiences, the lower the probability of searching for external information in future buying situations (Sheth & Parvatiyar, 1995). This, in turn, makes it more likely that customers will repurchase in future buying situations.

"The best predictor of a customer's likelihood of seeking future contact with a salesperson is the quality of the relationship to date" (Crosby, Evans & Cowles, 1990). Bitner, Booms and Tetreault (1990) find that for services, the quality of customers' contacts with salespeople is critical to overall satisfaction. This evidence shows that interpersonal satisfaction is an important component of overall satisfaction and that overall satisfaction is an important factor with respect to repurchase decisions.

Thus far, the emphasis in this review of the literature has been on the creation of interpersonal satisfaction and the role that interpersonal satisfaction plays in repurchase intentions. Communication, both verbal and non-verbal, was seen to be important in the selling process in conveying content and in meeting social needs. Adaptive selling was explored as a method for improving communication in dyads. Now attention turns to examination of the aspects of personality, temperament and value system that are the independent variables in the model that was tested in this research.

Temperament

The temperament construct describes innate fixed behavioral patterns. Thus, introversion or extroversion, submissiveness or aggressiveness, being demanding or accepting, fearful or fearless, and trusting or distrusting are specific traits characteristic of behavioral patterns. How people learn or acquire knowledge is another aspect of temperament. Generally, temperament can be said to account for how we relate to other people.

The American psychologist William Moulton Marston, working in the early part of the 20th century and at about the same time as Carl Jung, developed a scheme for classifying behavioral styles or patterns. While Marston and Jung worked independently of each other, they both developed typologies of human behavior; Marston's framework involved four prototypical types (Marston, 1928) while Jung's scheme involved three (later expanded to four by Myers and Briggs) bi-polar dimensions (Performax, 1987).

Marston (1928) believed that human behavior is a response to the nature of the environment in which the individual finds himself or herself. Environments occur along a continuum ranging from unfavorable (antagonistic) at one extreme to favorable at the other, with unfavorable and favorable environments generating different forms of responses from individuals. The preferred or habitual behaviors exhibited by individuals in response to their environments can be classified as either active or passive.

From the interactions of favorable and unfavorable environments and active and passive responses, Marston postulated four basic survival strategies. Irvine (1989) described the circumstances leading to each:

> Marston argued that an active reaction in an antagonistic environment would foster qualities of dominance, designed to overcome obstacles by forceful personal actions. On the opposite pole, a disposition to passive responses to environments perceived as hostile would induce cautious and calculative compliance. Both strategies are designed to reduce the harmful effects of the environment but they are differentiated by dispositional emotional tendencies in individuals.
>
> In a favourable climate of work or of social organisation, an active emotive response would allow, and foster inducements or persuading and influencing others rather than compelling them by dominance. On the other hand, a habitual passive response to favourable situations

> would produce submission, a form of easy-going acceptance of routine in a low emotional key (pp. 24–25).

Marston called these four basic patterns of behavior *Dominance*, *Inducement*, *Submission*, and *Compliance*. (Earlier it was noted that the PPA instrument, which was used in this study, classifies behavior into archetypes called *Dominance*, *Influence*, *Supportive*, and *Compliance*. Hendrickson, the developer of the PPA, chose to use Influence and Supportive rather than Inducement and Submission because the latter terms have more negative connotations than those of the former. To conform to the usage of the instrument used to operationalize the temperament construct, either Hendrickson's terms or the shorthand notations of D-, I-, S-, and C-type, will be used throughout the remainder of this article.)

The Personal Profile Analysis instrument was used in this research because it classifies temperaments into four archetypes, unlike the Myers-Briggs Type Indicator, which uses 16 archetypes. The smaller number of archetypes makes it easier for practitioners to learn the Marston typology. Additionally, the PPA, is a measure of surface traits and is not designed to describe characteristics that are not readily observed (Kragness & Rening, 1996).

While most people exhibit characteristics of each of the four behavioral archetypes at times, most individuals gradually develop a behavioral pattern which emphasizes only one archetype (TIMS, n.d.). Thus, one archetype outweighs the others and hence dominates behavior (Performax, 1987). This natural and most comfortable pattern of behavior determines what is called the dominant archetype. (The dominant archetype can be any one of Marston's four archetypes; it is important that the term "dominant archetype" not be confused with the "Dominance" archetype, which describes a specific pattern of behavior, described in the next subsection.) The dominant archetype also represents the behavioral pattern of the other people with whom the individual will be most comfortable and to whom the individual can respond with the least amount of effort and stress. Thus, if a salesperson attempts to relate to a customer with a profile similar to his or her own, each person will be more comfortable with the relationship than if they possessed dissimilar profiles. In the latter case, stress will be experienced by each person and the relationship will not be as comfortable for either person as it would be if they possessed similar profiles.

The characteristics of each archetype are described in the following subsections. Readers may "look for themselves" in the following descriptions and may find that multiple archetypes seem applicable. This frequently occurs because several archetypes are commonly detectable in a given individual's behavior; an individual's behavior is typically, however, *best* characterized by only one (dominant) archetype. Furthermore, some behaviors characteristic of a given archetype may seem to fit, while other behaviors of that same archetype do not seem to fit. This occurs because most people's behaviors combine characteristics of several archetypes and are rarely drawn purely from one.

Dominance

Marston reasoned that the Dominance archetype is developed by people taking active roles in what they perceive to be a hostile environment (Marston, 1928). Adjectives describing behavior typical of people with high Dominance elements of their temperaments include: driving, competitive, self-indulging, egocentric, daring, forceful, aggressive, venturesome, decisive, inquisitive, self-assured, dominating, direct, demanding, self-starting, blunt, overbearing, and assertive (Irvine, 1989). An individual with a dominant D-type or Dominance temperament in Marston's framework would be called a "Senser" in the Jungian system (Performax, 1987).

People with dominant D-type elements are task oriented and are driven by accomplishment. They like to take control and get things done. At the same time they tend to be perceived as insensitive to the feelings of others. Other overt non-verbal behaviors include high energy levels, restlessness, erect posture, rapid speech, speaking with authority and emphasis, and impatience. People with this dominant behavioral pattern think strategically (Performax, 1987).

Influence

Like the Dominance pattern, Marston reasoned that individuals with dominant I-type or Influence temperaments would take active roles in their environments; unlike dominant D-types, however, dominant I-types operate in favorable environments (Marston, 1928). Adjectives descriptive of the dominant I-type temperament are: charismatic, optimistic, self-promoting, outgoing, effusive, gregarious, sympathetic, generous, influential, persuasive, affable, friendly, confident, trusting, poised, charming, verbal, communicative, participative, and positive (Irvine, 1989). Jung called individuals with dominant I-type temperaments Feelers (Performax, 1987).

The I-type pattern is typified by an extroverted individual who establishes rapport instantly and relishes social interaction and relationships. People with this style are expressive and animated – some parts of their bodies always seems to be in motion. They have large vocabularies and speak articulately. In addition, they procrastinate, are often late for meetings, and show little respect for process or procedure (Divita, 1993).

Supportive

The Supportive style is the result of passivity in a favorable environment (Marston, 1928). Irvine (1989) says that dependable, self-controlled, easy-going, serene, relaxed, non-demonstrative, predictable, patient, deliberate, amiable, steady, even-tempered, persistent, passive, good-listener, kind, lenient, and accommodating are descriptive of people with dominant S-type temperaments. In the Jungian framework such people would be called Thinkers (Performax, 1987).

People with dominant S-type temperaments, while often shy and uncomfortable with strangers, are supportive of other people. They avoid attention and speak slowly and hesitantly, often making weak eye contact. Other characteristics include deliberate work patterns and communication may be indirect, often by implication or feeling (Divita, 1993).

Compliance

Marston (1928) said that the Compliance (also known as Critical Thinker) pattern results from passivity in antagonistic or unfavorable environments. The Compliant pattern is typified as: disciplined, compliant, self-effacing, evasive, overly-dependent, worrisome, careful, systematic, precise, diplomatic, accurate, conventional, cautious, conservative, perfection-seeking, and logical (Irvine, 1989). Intuitor is the name given by Jung to this pattern of behavior (Performax, 1987).

Overt behaviors of people with dominant C-type temperaments include slow and soft patterns of speech and measured gestures. Characteristically, Compliance individuals are seen as reserved, cold, and formal, and paying great attention to detail and accuracy. Individuals with this style are systematic thinkers, tend to distrust others, and set high performance standards for themselves and for others (Divita, 1993).

Influence of Temperament on Communication

Communication is "a dynamic process in which man consciously or unconsciously affects the cognitions of another through materials or agencies used in symbolic ways" (Andersen, 1972, p. 5); communication has four aspects: content, organization, language and style, and delivery. Content is the message; it contains the ideational material, e.g. ideas and thoughts that the communicator intends to convey. Organization is the order in which the content is presented. Style is the way in which the thoughts are expressed and includes a person's characteristic behavior patterns, such as bodily actions, organizational patterns, characteristic forms of support and mannerisms. Thus, style includes the choice of words and word order. Delivery includes voice quality, pitch, inflections, and rate. Face-to-face communication, then, has both verbal and non-verbal components covering all aspects of communication (Andersen, 1972).

As seen from the preceding discussion of Marston's typology, each archetype is characterized by a different style of communicating. Each seller's temperament determines the organization of his or her presentation, the language and style, including body language, used in the presentation, and the delivery. The seller's temperament influences the content (e.g. the message) that is communicated; this influence will be manifested primarily in the degree of detail. Similarly, the buyer's temperament determines what the buyer "hears" or can attend to. If the buyer and seller have the same dominant behavioral styles they will understand each other intuitively and easily since they each communicate in similar characteristic patterns (McIntyre, 1991). However, if they have different dominant behavioral styles, they will cause stress for each other; this stress may interfere with communication, for example, the salesperson may not understand the buyer's needs. Thus, recognition of these differences in the manner of communicating is crucial to successful communication (Divita, 1993, 1995).

A person with a dominant I-type temperament is naturally extroverted, friendly, and trusting. A salesperson with a dominant I-type temperament will instinctively and easily begin conversations and will find it easier to establish rapport with customers than will salespeople with other temperament-types. (A buyer with a dominant I-type temperament will need no help in opening up to a salesperson; a seller with a dominant I-type temperament will experience even greater ease in establishing rapport when dealing with buyers with dominant I-type temperaments than with buyers with other dominant temperament-types.)

Dion and Notarantonio (1992) found that a precise communication style is not perceived as positively as when it is combined with a friendly style. The I-component of temperament is responsible for such a friendly style. Thus, salesperson temperament is important for establishing communication with customers. For these reasons, people with I-type temperaments may make the best salespeople.

Differences in communication patterns, if not corrected, will strain relationships. Differences in content, organization, language and style, and delivery may all cause stress for buyers. Through the technique of adaptive behavior (e.g. Weitz, 1981b) a salesperson can adapt his or her overt behavior to closely match that of the listener. Shyness can be overcome (Toftoy & Shakalova, 1995). Content (for example, level of detail) and skillful sellers to match the behavioral patterns (temperaments) of customers can vary style (for example, rate, volume, and body language). This adaptive process can be learned with training, though most people are untrained (Divita, 1993, 1995).

Adaptive behavior is not mere mimicry; it is presentation of sales messages using the verbal and non-verbal language most easily understood by each customer. Note that the goal is not the use of adaptive behavior but meeting buyers' human needs.

As powerful as adaptive behavior is as a technique for improving communication, it does not fully compensate for differences in temperament-types. For example, an introverted salesperson might adopt a more extroverted behavioral pattern than is that person's norm in order to adapt to the naturally extroverted behavior of a customer. As a consequence, the salesperson would be able to establish better communication with the customer than if no adaptive behavior were engaged in. Nevertheless, communication would not be as natural as if he or she were a natural extrovert (Williams, Spiro & Fine, 1990).

VALUE SYSTEMS

While temperaments relate to behavioral traits or characteristics which are innate or inherent, value systems relate to motives or how people use what they know. For example, two people with different value systems might both volunteer at a social agency; one might do so because he or she gets satisfaction out of helping others while the other might do so because he or she expects some kind of recognition or reward. Thus, identical conscious choices and decisions may be made based on different value systems (England, 1967).

Definition of Value Systems

Just as related behavioral traits or characteristics can be grouped into patterns of behavior or temperaments, values can be grouped into value systems. "Value systems consist of an organized framework or mental set within which various values are prioritized, centralized and/or discarded as part of the process of perceiving and responding to the life stimuli we encounter" (O'Connor, 1986, p. 20). Value systems have also been said to consist of "your standards/principles upon which you base your choices and actions" (Performax, 1985, pp. 4–6) and "closely held personal values which are of high salience in important evaluations and choices" (Vinson, Scott & Lamont, 1977).

Five attributes are characteristic of all value systems:

(1) Values are arranged in a hierarchical structure and are rank-ordered or prioritized.
(2) Values are either central or peripheral. Central values are more important than peripheral values.
(3) Values are not random; values are selected purposely and serve functional goals.

(4) Values are interrelated; they fit (or do not fit) together with other values held by the individual.
(5) Values cluster together and are not independent of each other (Performax, 1985).

The value system construct is universal and the number of different value systems is limited. Furthermore, all value systems "are present in all societies at all times but are differentially preferred" (Kluckhohn & Strodtbeck, p. 10). Thus, for example, during the 1960s there was widespread acceptance of the counterculture movement and its emphasis on social justice. Similarly, during the 1980s there was general acceptance of Yuppies and their emphasis on material rewards. Both value systems were present during both decades, yet each was more widely accepted during its respective period.

Rokeach (1973) noted that values and value systems are both dependent and independent variables. They are dependent in that the development of any given individual's value system will be the result of the cultural, institutional and personal forces that that individual has experienced, and they are independent in that they influence all areas of an individual's actions.

Value System Classification

Value systems have been measured and classified in the fields of psychology and sociology (e.g. Kahle, 1986; Rokeach, 1971) but the construct has received little attention or recognition in the field of marketing. This subsection reviews the model of value systems that was used in the study described here.

Massey (1979) developed a classification system for value systems involving four archetypes. As with Marston's temperament scheme, Massey posits that each individual's value system profile combines elements of each of the four archetypes. While individuals exhibit elements of each of the four patterns, one of the four is usually stronger than the other three. The strongest value system archetype will be referred to as the dominant archetype and will be more prominent and will exert a greater influence on the individual's values-driven behavior than the other three types. The four archetypal value systems are called T (Traditionalist), I (In-betweener), C (Challenger), and S (Synthesizer).

Traditionalist Value System

Traditionalists accept authority and concur with the value systems of their bosses or other authority figures. They are loyal and prefer to keep their personal and work relationships formal. Respect of others provides a sense of personal worth and motivation for people with this value system.

Challenger Value System
Challengers believe in individual rights over group rights or institutional considerations. This leads Challengers to question authority and traditional values. They are never satisfied with what they have and always want more. Believing that the world has been unfair to them, they may be considered by others as being unethical in pursuing what they see as rightfully theirs. Others usually dislike people with this value system because Challengers are seen as always wanting or taking, manipulative, and conniving. Material rewards, such as money, power, and pleasure, are motivational for Challengers and such rewards give Challengers a sense of self-worth.

In-Betweener Value System
In-betweeners find it difficult to commit themselves to anything, hence they appear to switch back and forth between Traditionalist and Challenger value system types; they do not like to be forced to make decisions but prefer to keep their options open. They accept the value system of the highest authority figure and are influenced by trends and fads. In-betweeners frequently quote others and borrow words or phrases from others. In-betweeners obtain a sense of personal worth and motivation from the approval of other people. The In-betweener value system is a transitional stage as individuals change from Challengers to Traditionalists; typically, the stage does not last for a long period of time.

Synthesizer Value System
Fulfillment of personal goals and the satisfaction of a job well done are motivating and rewarding for Synthesizers. Synthesizers want to make the world a better place for others – even if others do not want it; they will work in the interests of others, even at the expense of themselves and their families. Peers tend to dislike Synthesizers because Synthesizers appear to be out of synch, pessimistic (about today, though optimistic about the future), and because they are change agents, always striving to make things the best they can be. While Challengers may bend or break rules to accomplish their ends, Synthesizers maintain the highest ethical standards.

Progression Through Value Systems

Massey (1979) says that value systems are not necessarily permanent the way that temperaments are. Significant emotional events or personal crises can trigger changes in an individual's value system. Typically, young children are Traditionalists but become Challengers as they become teenagers. Many then progress to In-betweeners, then Traditionalists (again), and finally Synthesizers. It is unusual for adults under age 30 to be synthesizers. Likewise, it is unusual for an individual older than 30 and who still has a dominant challenger value system to change dominant value system types.

Influence of Value Systems on Relationships and Communication

As with temperaments, each value system type has a somewhat different style of communication. Value systems influence primarily the message being communicated as opposed to style which is influenced primarily by temperament. For example, Traditionalists will speak in terms of the way things are and will prefer to maintain the status quo while Challengers will couch their statements in terms of complaints, such as how bad things are or how unfairly they (the Challengers) have been treated. In-betweeners lack commitment and will wait to take a stand on an issue until a higher authority has spoken – and then will chorus that position without taking ownership of the message. Synthesizers want to talk passionately about specific ideas which will benefit others or their organizations.

Communication style is important in customer-salesperson relationships, particularly in ones with moderate to long-term outlooks. Such relationships require collaboration and dedication to mutually beneficial gains; high levels of commitment "presume a shared value system between trading partners" (Frazier, Spekman & O'Neal, 1988, p. 58).

England (1967) examined the personal value systems of American managers. He asserts that "personal value systems influence the way in which a manager looks at other individuals and groups of individuals; thus they influence interpersonal relationships" (p. 54). The research described in this report examined interpersonal relationships between buyers and sellers for support for this assertion.

Thus far, this section describes the theoretical foundations of the model used in this study including interpersonal satisfaction and the personality constructs of temperaments and value systems. Attention is now turned to reviewing the relevant marketing literature on these subjects.

Value Systems and Marketing

Several studies attempt to draw a connection between values and marketing. The most common connection has been to use values as a method of market segmentation (Kahle, 1986; Prakash & Munson, 1985; Sukhdial, Chakraborty & Steger, 1995). In this vein, Beatty et al. (1985) stated, "values underlie the consumption behavior and are thus more inherently useful than demographics in understanding attitudes and behaviors" (p. 184). In addition, it has been suggested that values could be used to enhance the marketing mix (Prakash & Munson, 1985; Sukhdial, Chakroborty & Steger, 1995) or to enhance promotional strategy (Vinson, Scott & Lamont, 1977). In the only study that was found relating values to customer satisfaction, Prakash and Munson (1985) propose that a consumer's personal values may be linked to customer satisfaction.

Too many individual values (as opposed to value systems) exist for the values construct to be of significant help in segmenting markets or differentiating customers. Furthermore, it is not simply a matter of which values are important to an individual that affects behavior, the priorities of those values are also important (Kluckhohn & Strodtbeck, 1961). Value systems provide a means of classifying values that accounts for which values are important and for the priorities of those values (Performax, 1985). Additionally, the impact of a specific value on behavior is less apparent than the impact of a value system. For example, one might hold the value of accepting the authority of those in a hierarchy. When included in the Traditionalist value system, this value is typically associated with acceptance of traditional work practices, loyalty to the work group, and (relatively) formal work and personal relationships (Massey & O'Connor, 1989). Thus, one's behavior is likely to be impacted more by the value system than by a single value. For these reasons, value systems were used as an independent variable in the study described here.

This section has reviewed relevant literature on personal selling, communication, adaptive selling, temperament, and value systems. Next, attention is turned to the development of the hypotheses that were tested, the operational model of the research, and the research design including the sampling methodology, survey questionnaire, and the statistical tests and decision rules employed in the analysis.

HYPOTHESIS DEVELOPMENT AND RESEARCH METHODOLOGY

This study focuses primarily on the relationships between the personality traits of temperament and value system and interpersonal satisfaction and repurchases intention. Because the essential feature of the social exchange paradigm is that the behavior of each individual affects the behavior of the other, the hypotheses were developed specifically to test the strength and statistical significance of relationships in customer-salesperson dyads.

One of the variables of interest was the temperament-types of customers and salespeople in organizational (often called business-to-business as opposed to retail or business-to-consumer) transactions. As previously discussed, each person typically has a dominant temperament-type. Further, people with the same dominant temperament-type understand each other more readily and easily than people with dissimilar temperament-types. In the language of psychometrics, when two people have similar dominant types they are referred to as being "reciprocal and compatible." It was hypothesized that when customers' and salespeople's dominant personality types matched (i.e. they have the same dominant temperament type), interpersonal satisfaction felt by customers would be higher than when the dominant temperament-types did not match. The formal hypothesis state:

H1. When customer and salesperson dominant temperament-types match, interpersonal satisfaction of purchasing agents is higher than when the dominant temperament-types do not match.

One of the characteristics of people with dominant I-type temperaments is that they are naturally extroverted. These people "know no strangers" and make friends and establish rapport easily. This ability was thought to be important in helping salespeople establish open communications with and gain the confidence of their customers. Thus, it was expected that salespeople with dominant I-type temperaments would create higher interpersonal satisfaction with their customers than would salespeople with other dominant temperament-types. Formally, the hypothesis was stated:

H2. Salespeople with dominant I-type temperaments generate higher levels of interpersonal satisfaction among purchasing agents than do salespeople with other dominant temperament-types.

The I-type temperament, in addition to being extroverted and compelled to talk, finds it difficult to focus on tasks and to accomplish goals, especially within short or limited time frames. People with this temperament-type are particularly prone to procrastination. For salespeople this may mean never asking for the order or never closing the sale. Thus, while salespeople with this dominant temperament-type may easily establish rapport with their customers, they may not achieve high sales performance because they are so busy talking and participating in the relationship. Conversely, individuals with dominant D-type temperaments are extremely task-oriented but are not able to establish rapport easily; thus, while able to ask for the order, salespeople with this type of dominant temperament are not likely to create high levels of interpersonal satisfaction. It was theorized that people with dominant I-type temperaments with the second strongest temperament component (secondary temperament type) being the D-type, would be the most productive and would achieve the highest levels of interpersonal satisfaction among salespeople with I-type temperaments. This is because those salespeople combine the ability to establish rapport with the drive to ask for orders and close sales. The research hypothesis was stated thus:

> **H2A.** Salespeople with dominant I-type temperaments and secondary D-type temperaments generate higher levels of interpersonal satisfaction among purchasing agents than do salespeople with dominant I-type temperaments and other secondary-type temperaments.

The second variable of interest in this research was the value systems of the parties to a transaction. As previously discussed, "Shared values directly influence both commitment and trust" (Morgan & Hunt, 1994, p. 24). Shared values imply similar value systems. Thus, people with the same dominant value system-types would intuitively tend to accept each other more than people with different dominant value system-types. Because of their shared basic beliefs they would have greater commitment to each other and to their respective relationships. It was expected that this would lead to higher levels of interpersonal satisfaction. The relevant research hypothesis was stated as follows:

> **H3.** When customer and salesperson dominant value system-types match, interpersonal satisfaction among purchasing agents is higher than when the dominant value system-types do not match.

People with S-type value systems work "in the interest of others out of self-denial" (Divita, 1995, p. 16). Salespeople with dominant S-type value systems, by their natures, work in the interests of their customers – even if it means forgoing some sales or commissions. Salespeople with S-type value systems "tend to use the product the company has packaged to develop solutions that remedy customers' problems" (Divita, 1995, p. 19). It was expected that customers sense this caring on the part of the salesperson and that this would result in higher interpersonal satisfaction. The relevant research hypothesis was stated:

> **H4.** Salespeople with dominant S-type value systems generate higher levels of interpersonal satisfaction among purchasing agents than do salespeople with other dominant value system-types.

The greater the intensity with which salespeople hold their value systems, the more those value systems influence salespeople's behaviors. It was expected that customers sense that salespeople with S-type value systems genuinely care about solving customers' problems and that the more the salespeople cared, the greater would be the interpersonal satisfaction of customers. This leads to an important related hypothesis:

H4A. Among salespeople with dominant S-type value systems, the intensity with which the value system is held is positively correlated with interpersonal satisfaction of purchasing agents.

As was described in the literature review of adaptive selling, salespeople can learn to adapt their sales messages (content) and deliveries (styles) (Andersen, 1972; Williams, Spiro & Fine, 1990) to each customer's unique requirements (Clabaugh & Forbes, 1992; Sujan, Weitz & Sujan, 1988; Weitz, 1981b; Weitz, Sujan & Sujan, 1986). If a salesperson adapts his or her behavior to that of a customer it would tend to mask that salesperson's own temperament-type. While an intensive literature search found no studies supporting the idea, if a salesperson adapts his or her behavior, communication is not likely to be as successful as if the salesperson and customer have the same temperament type. It should, however, be better than if no attempt at adaptation is made. Thus, for example, if a naturally extroverted salesperson (dominant I-type temperament) tried to adapt his or her behavior to a quiet, introverted, and detail-oriented customer (dominant C-type temperament), communication would not be as effective as if the salesperson had a dominant C-type temperament.

Changing overt behavior to mimic different temperament-types can be accomplished but adapting to a different value system would mean changing one's values and the way one thinks. This is not possible (Divita, n.d.). "When the value judgments of the salesperson and the prospect are not alike (incongruent value levels), the message will be screened out and ignored" (Clabaugh & Forbes, 1992, p. 110). Salespeople can, however, change the contents of their sales presentations to be coincident with the value systems of customers and thus have the messages attended to. For example, if a customer has a challenger value system (motivated by material rewards), the salesperson should state the benefits in terms of what they mean to the purchaser personally. Alternatively, if the customer has a synthesizer value system (works in the interests of others) the benefits to the organization should be stressed.

This research investigation did not take adaptive behavior into account. While the 16-item ADAPTS scale (Spiro & Weitz, 1990) measures the degree to which salespeople practice adaptive behavior, it neither measures the degree to which salespeople modify their behaviors in the direction of any of the temperament archetypes nor in the direction of any of the value system archetypes. If the salespeople that participated in this study did engage in adaptive behavior it would have tended to hide the main effects being sought. For example, some salespeople called on customers with different dominant temperament-types than their own. If those salespeople used adaptive behavior to mask their true temperament-types and thus created interpersonal satisfaction, the effect would have been to show high interpersonal satisfaction for people with *mismatched* rather than matched dominant temperament-types. This would have masked the hypothesized effect since customers in mismatched dyads would still have experienced high levels of interpersonal satisfaction. As a result of this effect, the measured difference in interpersonal satisfaction between matched and mismatched dyads may have been lower than it would have been had the salespeople not engaged in adaptive behavior. The implications of this for the analysis of the data are that it would be more difficult to detect evidence of greater interpersonal satisfaction in matched dyads than if no adaptive behavior were practiced.

Customers' experiences with goods and services are of concern to marketers because those experiences influence customers' future choices and repurchase intentions (Oliver & Swan, 1989a). Oliver and Swan found that customer satisfaction after a purchase was related to customers' intentions to repurchase the same product. While a similar relationship was anticipated, it is important to marketers to know if interpersonal satisfaction is related to repurchase intentions. Thus, it was hypothesized:

> **H5.** Among purchasing agents, interpersonal satisfaction is positively related to repurchase intentions.

The hypotheses developed above tested the relationships expressed in the theoretical framework. Having stated the hypotheses, we now discuss the design of the research in which they were tested.

Operational Model

Figure 2 depicts the operational model for this study. This figure represents characteristics and behaviors of buyers and sellers on the extreme left and right, respectively.

Each relationship in the model that was tested in this research investigation is labeled with the number of the corresponding hypothesis developed above. If the buyer's and seller's dominant temperament-types matched (H1), the interpersonal satisfaction of the buyer was expected to be higher than for buyers in dyads in which temperaments were not matched. As previously explained, this is because people with similar dominant temperament-types have similar verbal and non-verbal styles of communications and hence understand each other more easily than people with mismatched dominant temperament-types.

For dyads in which dominant temperament-types were unmatched, if the seller had a dominant I-type-temperament (H2) interpersonal satisfaction was expected to be higher than if the seller's dominant temperament-type was one of the other

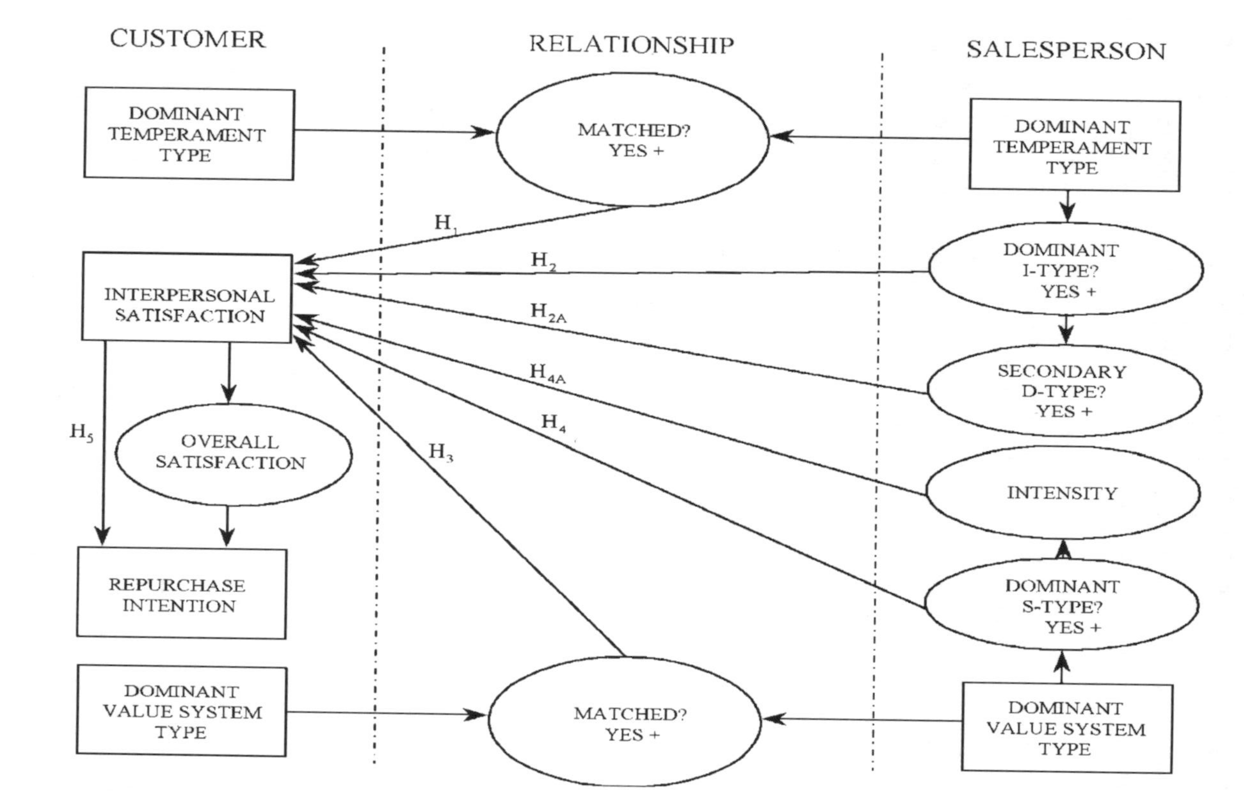

Fig. 2. Operational Model: Influence of Temperament and Value System on Interpersonal Satisfaction. (After Williams, Spiro & Fine, 1990.)

three types. This effect was expected to occur because people with dominant I-type temperaments verbalize easily and they open and maintain communications with others more easily than people with other dominant temperament-types (Irvine, 1989; TIMS, 1993). People with dominant I-type temperaments talk easily and establish rapport instantly. In addition, they are trusting of others (Divita, n.d.; TIMS, 1993) and people who exhibit trust are more likely to be trusted in return (Hawes, Mast & Swan, 1989).

If matched temperament-types lead to higher interpersonal satisfaction, one might ask why a dominant I-type personality would be beneficial for a salesperson calling on a customer that has other than a dominant I-type temperament. There could be two reasons. First, a salesperson cannot know prior to meeting a customer what the customer's dominant temperament-type is; therefore it would be most beneficial for the salesperson to have the dominant temperament-type most likely to succeed with the greatest number of people. Second, and more importantly, salespeople are frequently required to establish relationships with many people in buying organizations. For example, using the snowball interview technique in a sample of 62 companies purchasing both capital goods and industrial services, Johnston (1981) found that in no case was only one person involved in purchasing decisions; the smallest number of people involved was three and the number ranged up to over 25 people. For capital goods and services the purchasing department was involved in 97 and 90%, respectively, of the cases. Typically, the purchasing department had "the role of interfacing with the vendors" (Johnston, 1981, p. 18). Thus, salespeople will have to deal with many people within the buying organization and it is likely those people will have many different dominant temperament-types. Salespeople with dominant I-type temperaments are expected to find it easier to establish rapport with the many different people involved in the decision-making process than are salespeople with other dominant temperament-types. The implication of this for managers would be that salespeople should have dominant I-type temperaments in order to achieve the highest levels of interpersonal satisfaction. An additional implication would be that adaptive behavior could be based on learning and adapting to the four temperament archetypes.

As with temperament-types, if the dominant value systems of the buyer and seller are the same type, the buyer's interpersonal satisfaction will be higher than if the dominant value system-types are mismatched (H3). People with matched value system-types may find it easier to establish feelings of shared commitment and trust – important elements in long-term relationships – because they have shared beliefs. Further, if the seller's dominant value system is the S-type, the buyer's interpersonal satisfaction will be higher than if the seller's dominant value system-type is one of the other 3 types (H4). People with S-type value systems are, by their natures, concerned about other people and are compelled to work for the benefits of others because they "realize that failure to attend to the needs of others is also likely to heighten conflict and dissatisfaction for all" (O'Connor, 1986). Thus, it is also hypothesized that the greater the intensity of the S-type value system, the greater the resulting interpersonal satisfaction (H4a). As with temperaments, the implication for managers is that there is one dominant value system-type that leads to higher levels of interpersonal satisfaction than the others.

Interpersonal satisfaction is thought to influence overall customer satisfaction and customer satisfaction has been correlated with repurchase intentions (e.g. Oliver & Swan, 1989a). Interpersonal satisfaction is also thought to influence repurchase intentions; if a customer is highly satisfied with the relationship with a salesperson, then it is likely that the customer will want to continue to deal with that salesperson rather than risk a new relationship that might be less satisfactory (H5). It should be noted that Fig. 2 does not include the relationship of interpersonal satisfaction to overall customer satisfaction. Nor does Fig. 2 depict other factors, such as perceived product performance and product satisfaction which also influence repurchase intention. Overall satisfaction, perceived product performance, and product satisfaction were not included in this research investigation in order to simplify the model and because the purpose of the research was to examine the effect of temperaments and value systems on interpersonal satisfaction and the effect of interpersonal satisfaction on repurchase intention. The theoretical model, including these constructs, was presented in Fig. 1.

Research Design

A review of the literature found no prior studies of the relationships being examined here; it appears that no researchers have investigated the relationships between the Marston typology of behavioral patterns or any typology of value systems to customer satisfaction or repurchase intentions. The study described here sought to demonstrate whether or not relationships exist between these variables.

Operationalization of Constructs

Temperament- and value system-types are constructs developed to explain human behavior. The measurement systems used to operationalize these constructs are described in this section.

Operationalization of the Temperament Construct

The instrument designed by Thomas Hendrickson, called the "Personal Profile Analysis" (PPA), to assess temperaments according to Marston's theory was used in this study to determine the dominant temperament-types of the subjects. The PPA instrument consists of twenty-four tetrads of descriptive words. The original words used by Marston (1928) in the development of his theory were used to the extent that they were statistically validated as discriminating between the four dimensions described by Marston. Subjects select the word from each tetrad that is "most like" themselves and the word that is "least like" themselves. Scoring is then based on which words were selected. The survey questionnaire used in this study maintained the exact wording and order of presentation of the PPA instrument.

Substantial work has been performed on the validity and reliability of the PPA and no simple measure can convey the complexity of this information. Readers desiring additional information on this instrument are referred to Kaplan (1983), TIMS (n.d.), and Irvine (1989).

Determination of Dominant Temperament Type

Instructions for the PPA direct how responses are to be tallied and scored. Results are then plotted on a series of three graphs. Each graph has a separate scale for each of the four behavioral archetypes. For each graph, the archetype that is plotted at the highest point above the horizontal axis in the center of the graph is called the dominant temperament-type. Plotting is necessary because each of the scales is nonlinear and all of the scales differ from each other; thus, simply comparing scores is not sufficient to determine the dominant archetype.

The words the respondent selected as "most like" are tallied and plotted on graph one; this graph indicates how others see the individual in the work environment. The second graph plots the scores for the words the respondent selected as "least like" and indicates the individual's behavior under pressure. The arithmetic difference between the scores for graphs one and two are plotted on the third graph; this graph depicts how the individual sees himself or herself (TIMS, 1993). (For example, if the D-component on graph one is 12 and the D-component on graph two is 3, the D-component on graph three will plotted as 12 minus 3 or 9.) The effects of the third graph are permanent and are reflective of the individual's comfort zone or preferred behavioral mode (Performax, 1987).

It is the first graph that represents how the individual behaves and is perceived in the work environment (Performax, 1987). In this research, the purchasers and the sellers were interacting in the work environment. They formed impressions of each other and reacted to each other based on their behaviors in this environment. Hence, for purposes of this study, the results of the first graph were used for analysis. The second graph represents the behavior of the individual at home or at play and the third graph represents the way that the individual sees himself or herself. The third graph is permanent while the first two are temporary, representing the individual's adaptation to the respective environments.

In the commercial version of the instrument, subjects self-score their own responses and are then led through the interpretation of the results. For the purposes of this study, self-scoring was neither appropriate nor desired. Respondents did not have instructions for self-scoring and interpretations were not made available to subjects.

Matching of dominant temperament-types of salespeople and customers in dyads was determined by the dominant temperament-type of each without regard to the strength or intensity (height of the bar on the bar graph) of that temperament type. For example, if the dominant temperament-type of the salesperson was "challenger," then the dyad was considered matched if the customer's dominant temperament-type was also "challenger," regardless of the intensity of that temperament-type for either party. Otherwise, the dyad was considered unmatched with respect to temperaments.

Operationalization of the Value System Construct

An instrument similar to the instrument designed by Morris E. Massey and Michael J. O'Connor to assess value system-types was used in this study. The instrument consists of the 40 statements written by Massey and O'Connor representing various values followed by a parenthetical statement clarifying the first statement. For example, the statement "haste makes waste" is followed by the statement "when I do things quickly I make careless errors" enclosed in parentheses. Thus, each item contains the exact wording of the original instrument followed by a parenthetical statement of similar meaning. The parenthetical statements were written and tested in research by Salvatore Divita, D. B. A, Professor of Marketing at The George Washington University. Subjects responded to each set of statements by marking one of the five following responses: Strongly Agree/Very Important; Agree/Important; Undecided; Disagree/Unimportant; Strongly Disagree/Very Unimportant. Scoring was then conducted in accordance with the instructions for the instrument.

In the commercial version of the instrument, subjects self-score their own responses and are then led through the interpretation of the results. As with the temperament instrument, for the purposes of this study, self-scoring was neither appropriate nor desired.

Determination of Dominant Value System Type

As with the temperament instrument, instructions direct how responses are to be scored and tallied; the tallies are then plotted on a bar graph with one bar for each of the four value system archetypes. The archetype corresponding to the highest bar on the graph was considered the dominant value system-type and the archetype corresponding to the second highest bar was the secondary value system-type. As with the temperament graphs, the scales for the four bars are nonlinear and different. For example, a raw Traditionalist score of 32 is plotted twice as far above the axis as a raw score of 22 (the axis is 0). A raw score of 32 on the Traditionalist scale is plotted at the same distance above the axis as a raw score of 26 on the Challenger scale. Thus, simply comparing raw scores is not sufficient to determine the dominant archetype. The determination of the dominant value system-type used in the research described here was consistent with the instructions for the instrument.

Matching of dominant value system-types of salespeople and customers in dyads was determined by the dominant value system-type of each without regard to the strength or intensity (height of the bar on the bar graph) of that value system-type. For example, if the dominant value system-type of the salesperson was "synthesizer," then the dyad was considered matched if the customer's dominant value system-type was also "synthesizer," regardless of the intensity of that value system-type for either party. Otherwise, the dyad was considered unmatched with respect to value systems.

The intensity with which the value system was held was the independent variable in hypothesis H4a. This was determined as described above; the higher the bar on the bar graphs the greater the intensity. This interpretation is consistent with the instructions for the instrument.

Operationalizing Interpersonal Satisfaction

In the interests of parsimony and consistent with other researchers (e.g. Bitner & Hubbert, 1994; Drolet & Morrison, 2001; Smith, Bolton & Wagner, 1999), a single item was used to measure interpersonal satisfaction. Purchasing agents were asked, "How satisfied or dissatisfied are you with the *relationship* you have with this salesperson?" As with Oliver and Swan (1989a), a 7-point scale with the ends labeled "satisfied" and "dissatisfied" and the center labeled "neither" was used. The score on this item was used to indicate the level of interpersonal satisfaction.

Operationalization of Repurchase Intention

Considerable confusion exists in the literature between intention and expectation (see Oliver & Swan, 1989a) and the terms have sometimes been used interchangeably (Warshaw & Davis, 1985). The distinction between intention (conscious plan to perform a specific action) and expectation is that the latter includes a probability or likelihood that the plan will actually be carried out. Thus, when marketers express interest in repurchase intention, they are (or should be) interested in repurchase expectation. However, to use language consistent with the marketing literature, the term repurchase intention is retained here. As with interpersonal satisfaction, repurchase intention was measured using a single item, worded, "Assume that in the near future you have to buy or rebuy the product or service this salesperson sells. How likely would you be to buy from this salesperson?" A 7-point scale with the ends labeled "certain to use same salesperson" and "certain to use different salesperson" and the center labeled "neutral (not sure)" was used. Next we turn to the development of the questionnaire.

Questionnaire Design

Since cross-sectional data are required to test the hypotheses, it was decided to use self-administered survey questionnaires to collect data from a sample of purchasing agents and salespeople that call on them. Because some of the information required from purchasing agents was different from that required from salespeople, two different questionnaires were used. For example, purchasing agents were asked how satisfied or dissatisfied they were with the relationships with the salespeople; this is a question that does not apply to salespeople.

Each questionnaire consisted of three sections. The first and second sections of each type of questionnaire were identical; the first sections assessed temperament, utilizing the PPA, and the second sections assessed value systems, utilizing the Values Systems Profile. The third section of the customer questionnaire related to the buying experience and interpersonal satisfaction experienced during the purchase of the product. The third section of the salesperson questionnaire related to selling experience and the nature of the goods or services sold. Demographic data such as age, gender, education, and years of experience were also gathered in the third sections of both types of questionnaire.

Questionnaire Pretesting

The questionnaires were pretested using the concurrent and retrospective think-aloud interview methods (Blair & Presser, 1993; Bolton, 1993; Presser & Blair, 1994). In the concurrent think-aloud process, respondents are directed to say aloud everything that comes into their minds as they are completing the questionnaire. This technique uncovers difficulties in understanding and following directions as well as difficulties in evaluating survey items. In the retrospective think-aloud technique, respondents are stopped at various points throughout the questionnaire (e.g. at the end of each page or section) and asked to describe their thoughts.

Such pretesting helped clarify instructions and wordings and the questionnaires were revised to reflect changes and correct problems identified during the pretest. Separate pretests were conducted for the on-line versions of the questionnaires (used by respondents with Internet access, discussed below) and for the paper versions of the questionnaires for use by respondents without Internet access.

Unit of Analysis

In order to measure interpersonal satisfaction of buyers it was important to select a sample frame for which interpersonal satisfaction was likely to be important. Most retail transactions are so short in duration and low in involvement that little opportunity exists for customers to develop any sense of interpersonal satisfaction. Consequently, for most retail transactions it seemed that interpersonal satisfaction was not likely to be a factor in repurchase decisions. For these reasons, business-to-business marketing relationships were selected for examination.

Many business-to-business relationships endure for years, as vendors serve as regular sources of supply, fulfilling ongoing needs for raw materials, consumables, components (i.e. used in the assembly of a product), or services. Such relationships provide ample opportunity for personal relationships to develop and for interpersonal satisfaction to become a factor in purchase decisions. Other lengthy relationships develop as the result of purchases of capital items with long decision cycles or long lead times. Additionally, while purchases of capital items may be infrequent and made irregularly, there may be continuing relationships due to the need for assistance with installation, purchase of service or maintenance contracts, spare and replacement parts, and consumables. Since relationships are involved with purchases of many types of goods and services, this study was not restricted to any one type of purchase.

In this investigation dyads were the unit of analysis. In dyadic studies it is important that the data from both parties be collected and matched. Since customer/salesperson dyads were the unit of analysis, a questionnaire had to be obtained from each customer and each salesperson participating in the study. Then each customer-questionnaire had to be matched with the questionnaire from the salesperson that called on that customer. Because of the difficulty of finding dyads and then matching the respective questionnaires, few dyadic studies have been undertaken. Crosby, Evans and Cowles (1990) undertook a study of 59 purchasing agents and "124 salespeople who sell to them" (p. 396). This resulted in 59 independent observations rather than 124. McIntyre (1991) conducted a quasi-experiment in which college students performed role-playing to classify the preferred cognitive styles of "sellers." Earlier dyadic studies were conducted by Riordan, Oliver and Donnelly (1977), Woodside and Davenport (1974), Mathews, Wilson and Monoky (1972), Tosi (1966), Brock (1965), and McIntyre (1991).

Questionnaire Distribution

Since a large sample was desired, use of e-mail for contacting the sample frame and inviting participation was selected. This method had several advantages over mail contact (distribution by fax was not considered because of cost and the length of the questionnaires). First, the costs and logistics of contacting a large sample frame would have been expensive and difficult. Second, a faster response speed was expected (Weible & Wallace, 1998). Third, in the event that the response rate was low, reminder messages could be sent at no cost. These considerations outweighed the disadvantage that e-mail-based surveys have lower response rates than mail surveys (Tse, 1998).

Three types of e-mail surveys are possible. If the questionnaire is relatively short and simple, it can be included in the message of the e-mail. Longer and more involved questionnaires can be sent as attachments to the covering e-mail. The third type is "URL embedded"; respondents click on a hypertext link which then invokes their web browsers which retrieve a web-based questionnaire (Bradley, 1999). Because of the length and complexity of the questionnaires and because of the need to have both salespeople and purchasing agents complete separate questionnaires, the URL embedded design was selected.

Sample Frame

Since dyads were the unit of study in this research, it was first necessary to identify customer/salesperson pairs willing to participate. It was expected that salespeople would participate in the study and complete questionnaires if requested to do so by buyers they called upon; it was considered far less likely that buyers would participate at the request of salespeople that call upon them. For this reason it was decided to contact buyers and have each buyer, in turn, select one salesperson that called upon him or her, thus completing the dyad.

The National Association of Purchasing Managers (NAPM) permitted use of their membership role for this research. (The NAPM is now known as the Institute for Supply Management, however, all references to the organization retain the name of the organization at the time of data collection.) Of the approximately 40,000 members of the organization, 7,240 had e-mail addresses on file with the organization. For this study, the entire list of 7,240 members was contacted. The NAPM provided the e-mail addresses for the sample in electronic form.

This sample might not be representative of the entire NAPM membership. Some members with e-mail addresses might not be sufficiently computer-literate to find the website and complete the interview. For these reasons, provision was made to send hardcopy questionnaires to any respondents requesting them. While use of a web-savvy sample frame presented a possible threat to external validity, computer-literacy and access to the Internet are both growing rapidly and will become more and more prevalent. Furthermore, many members that had not listed e-mail addresses with the NAPM (and hence were not surveyed in this research) may nevertheless be computer-literate. To the extent that this is the case, the sample is more representative of the entire membership than is indicated by the percentage of members that had e-mail addresses on file with the NAPM (18.1%).

E-mail Message

An e-mail message was sent to each person in the sample frame (each NAPM member with an e-mail address) explaining that a study was being conducted and requesting that individual's participation. The e-mail message (discussed below) contained complete instructions on how to select a salesperson to participate and how to access the questionnaires.

In order to demonstrate the legitimacy of the study and to show the NAPM's support, the executive vice president and chief operating officer of the NAPM, Mr. Paul Novak, was asked to write a brief message to members urging their participation. This message was included as part of the invitation to participate. Some of the people contacted for pretesting of the survey questionnaires were members of the NAPM and commented that the letter from Mr. Novak would be helpful in eliciting participation from other NAPM members. The pretesters also expressed interest in the results of the study, when they became available. Accordingly, the e-mail message soliciting participation was changed and promised that results of the study would be summarized on the same website as was established for the purpose of collecting data. In addition, it was decided to offer an incentive in order to achieve as high a response rate as possible. The incentive was two drawings, each for one Palm IIIe® personal digital assistant; the first drawing was limited to purchasing agent respondents and the second drawing was limited to salespeople that responded to the survey. The invitation to participate explained the incentives. Personal digital assistants were selected as the incentives because they are popular and would potentially be of interest to business people. It was thought that they would be more of an incentive than the gift of a nominal amount to a charity selected by the respondent or other commonly used items, such as cash. One drawing was limited to purchasing agents and one was limited to salespeople to assure each group that they had a reasonable prospect of "winning." Anecdotal evidence indicated that while the incentives induced some people to participate, others felt that an incentive offered to purchasing agents was unethical and may have kept some from participating.

The procedure to be followed by the participants, including how they were to select the salespeople they asked to participate, was also detailed in the e-mail message. The criteria to be used by purchasing agents in selecting salespeople were: (1) purchasing agents were to select the next salesperson that called on them; (2) purchasing agents were to have decision-making authority about whether to purchase the goods or services the salespeople were selling; and (3) the salespeople had to be willing to participate. The selection of the next salesperson that called on the purchasing agent was stipulated in order to randomize the types of relationships sampled in terms of the degree of satisfaction experienced, the types and natures of the products purchased, and in order to include short- and long-term relationships. The directions also stated that if either party (i.e. either the salesperson or the purchasing agent) did not have Internet access, that questionnaires would be mailed or faxed upon request.

It was necessary to obtain questionnaires from salespeople and purchasing agents and to be able to identify each individual dyad. Thus, a unique identifier was needed for each dyad; the telephone number of the salesperson participant in each dyad was used for this purpose. Telephone numbers are unique; each participant was already familiar with and had ready access to the number, making loss of the number unlikely. An added benefit was that by using the salesperson's telephone number, purchasing agents could not complete questionnaires prior to selection of the salesperson participant. This latter point was important because items in section three of the questionnaire related to the specific salesperson selected to participate.

Confidentiality was promised all participants. Anonymity, however, could not be promised since use of telephone numbers as dyadic identifiers would have made it possible to identify participants. To prevent lack of anonymity from being an issue with respondents, all participants were assured that they would not be contacted unless they gave specific permission.

Questionnaire Access

The questionnaires were established on web pages hosted on The George Washington University's website (it was considered important that the URL for the research site reflect the university's connection with the research). Respondents first went to a greeting page containing the University's logo and explanatory comments about the research. This page also described the incentives (including a color picture of the item being offered) and set forth general instructions for completion of the survey. Two hyperlinks were provided; purchasing agents clicked on one and salespeople clicked on the other. By clicking on their respective links, purchasing agents and salespeople were taken to their respective questionnaires. For those unable to access the Internet or those experiencing difficulty downloading the web pages, an offer was made to either mail or fax copies of the appropriate questionnaires. The e-mail address of the researcher was also provided in the event that respondents desired to ask questions or make comments.

Statistical Tests

One of the benefits of using the embedded URL design is that respondents perform their own data entry. When completed questionnaires were submitted, the data were captured in one of two database files (one for purchasing agent data and one for salesperson data). Dyads were matched manually using the telephone numbers, as previously described, prior to conducting statistical analyses.

Hypotheses H1, H2, H2a, and H3 were tested using the *t*-test for the difference between means. Churchill (1991) recommends (p. 790) using the *z*-test for sample sizes large enough that the Central Limit Theorem applies to the individual sample means and when those samples are drawn from asymmetrical distributions of variables in parent populations. The *t*-test is similar to the *z*-test but the *t*-test uses the sample standard deviation as an estimator of the unknown standard error (Hamburg, 1991). The level of significance was selected as 0.10 for all tests. Because the direction of the difference between the sample means was predicted in the hypotheses, one-tailed tests were used. In all instances, results were in the direction hypothesized.

The correlation coefficient was used to test H5. Had sufficient data been collected, the correlation coefficient would also have been used to test H4a while the *t*-test would have been used to test H4.

The next section will present the findings and analyses of the research data.

ANALYSIS AND FINDINGS

Responses and Response Rate

Of the 7298 invitations to participate that were sent, 1787 (representing 24.7% of the NAPM sample frame) messages were returned due to incorrect or invalid e-addresses. Of those who received the invitation to participate, 141 responded that they were ineligible (e.g. had no salesperson contact, were consultants, etc.). Thus, the sample frame consisted of 5370 qualified individuals (Table 5).

Completed questionnaires were received from 247 purchasing agents and from 216 salespeople. Three questionnaires were discarded for failure to

Table 5. Determination of Sample Frame Size.

7240	NAPM e-addresses in proper format
+58	Convenience sample e-addresses
7298	Total invitations sent
−1787	Invalid/improper e-addresses
5511	Recipients of invitation
−141	Ineligible recipients
5370	Sample frame

follow directions or for incomplete data. The completed questionnaires formed 189 matched dyads (54 purchasing agents responded but no corresponding responses were received from their salespeople and 23 salespeople completed questionnaires but no matching purchasing agent questionnaires were received). Since only complete dyadic data could be used for analysis, 189 was considered the number of completed responses. This represents a response rate of 3.5% (189/5370).

Weible and Wallace (1998) report response rates for e-mail surveys as ranging from 6 to 73% compared to response rates for mail ranging from 27 to 56%. While the response rate actually achieved in this study is lower than those cited, there are several factors that may account for this. Among such factors are the length of the survey (over 100 items on both salesperson and purchasing agent surveys), participants' possible lack of familiarity with navigating the Internet, the requirement that a salesperson be recruited to participate, and possible concerns about anonymity and/or confidentiality of responses.

Of the 378 completed questionnaires comprising the sample (189 dyads × 2 questionnaires per dyad), 361 or 96% were completed on-line; of the paper questionnaires in the sample, 7 were received from purchasing agents and 18 from salespeople. The response rate from the hardcopy surveys was 69% (25 returned of 36 sent). This higher response rate is probably due to the request these respondents had to make in order to receive the hardcopy. In effect, such a request was a commitment to complete the survey. Some of the hardcopy responses were received from respondents who had tried, unsuccessfully, to complete questionnaires on-line. The number of people who sent e-mails stating that they were experiencing difficulty either accessing the website or submitting completed questionnaires was quite small. Others experiencing difficulty may have simply given up. No clear pattern emerged as to the cause of these difficulties. A few respondents were inexperienced at navigating the web and at least one misread the URL and consequently was entering it incorrectly. Some appeared to have

Table 6. Sample and NAPM Membership by Gender.

	Sample		NAPM
	n	Percent	
Male	101	53.4%	66.4%
Female	88	46.6%	33.6%
Total	189	100.0%	100.0%

encountered firewalls on their computer systems that they were unable to penetrate and others had difficulties that were never fully explained.

Sample Demographics

Demographic data were collected from respondents and compared to the 1998 NAPM membership profile (NAPM, 1998) to determine how representative the sample was of the entire organization. Four categories of information were collected and compared: gender, age, education (highest level attained), and years worked in the purchasing profession.

Gender

Of the 189 purchasing agent respondents, 88 or 46% were women compared with 34% of the NAPM general membership (Table 6). Differences in the gender profiles are probably not accounted for by changes in the makeup of the NAPM membership; the proportion of males and females within the organization was constant from 1995 to 1998 (NAPM, 1998).

Age

With respect to age, the makeup of the sample closely approximated the NAPM profile, with a tendency toward a slightly older sample than the NAPM membership (Table 7). Because of the reporting method used by the NAPM (age brackets rather than age in years), the mean age of respondents of 45.0 years cannot be compared with a comparable figure for the organization's membership. Since age is sensitive to many people, age was made an optional field on the questionnaire (all other fields were required). Of the 189 purchasing agent respondents, 180 (97.6%) provided their ages; three men and six women elected not to provide this information. This is consistent with findings by Basi (1999) that "when individuals make the decision to complete a web survey, they complete most items" (p. 400).

Table 7. Sample and NAPM Membership by Age.

Age	Sample		NAPM
	n	Percent	
<26	3	0.5%	2.2%
<36	24	13.6%	14.2%
<46	59	33.7%	35.7%
<56	72	40.2%	38.1%
>55	22	12.0%	9.9%
Total	180	100.0%	100.1%

Note: Age not required on questionnaire; respondents not providing age: 3 male, 6 female.

It seems likely that at least a part of the difference in the age profiles between the sample and the NAPM is due to the aging of the NAPM membership; the sample was gathered approximately 21 months after the most recent demographic profile of the organization was compiled (NAPM, 1998). Aging of the membership was observed between 1995 and 1998, when the two most recent profiles of the organization were completed (NAPM, 1998). Overall, the differences between the sample and the NAPM membership do not appear to be significant.

Educational Level

The educational level attained by the sample appears to be slightly lower than the educational level of the NAPM membership-at-large (Table 8). It is possible that this lower educational level is due to disqualification of some of the more highly educated people who would otherwise have completed questionnaires but who responded by e-mail that they were consultants (8) or educators (13). Consultants and educators would be expected to have at least bachelor's or master's degrees and by eliminating these people, those remaining in the sample would appear to have a somewhat lower level of educational attainment. The differences between

Table 8. Sample and NAPM Membership by Education.

Education	Sample		NAPM
	n	Percent	
HS and HS with some college	68	36.0%	30.8%
BA	89	47.1%	48.0%
Grad degree	32	16.9%	21.3%
Total	189	100.0%	100.1%

Table 9. Sample and NAPM Membership by Years in Purchasing Profession.

Years in Purchasing	Sample		NAPM
	n	Percent	
<3	21	11.1%	8.1%
3–8	51	27.0%	21.2%
9–15	46	24.3%	28.3%
16–24	46	24.3%	31.3%
>24	25	13.2%	11.0%
Total	189	99.9%	99.9%

the educational levels of the sample and the NAPM membership appear to be minor.

Professional Purchasing Experience

The mean number of years of professional purchasing experience of the sample respondents was 13.0. As with age, the NAPM gathered this information in categories rather than in years, hence a comparison of means is not possible. Nevertheless, the sample data were categorized in the same way as the NAPM data for a detailed comparison. As can be seen in Table 9, the proportions in each category for the two groups are quite similar.

Overall, the demographics of the sample are quite similar to those of the NAPM membership. The minor differences in the demographics of the two groups are readily explained.

Geographic Diversity

The telephone numbers of salespeople were used for identification of dyads. Collection of this information had the additional benefit of providing data on the geographic diversity of the sample. Respondents came from regions represented by at least 98 different area codes in at least 41 different states within the United States, including Alaska and Hawaii; in addition, 2 toll-free area codes were used. The greatest number of responses from any single area code was 7. All respondents were from the United States.

Industry Diversity

Salespeople were asked to classify the items sold to the purchaser into one of five categories: services, components, raw materials, consumables, or capital equipment. A separate open-ended item asked for a generic description of the offerings. The types of offerings of the vendors represented by the salespeople were not limited to any single category, as shown in Table 10. The types of

Table 10. Types of Offerings by Vendors.

Type of Offering	*n*	Percent
Services	31	16.1%
Components	57	30.1%
Raw materials	22	11.9%
Consumables	51	26.9%
Capital equipment	28	15.0%
Total	189	100.0%

services included staffing, consulting, banking services, and training. Goods ranged from low-tech items such as printing, food and vending, and drywall to such high-tech items as computers, copiers, and avionics, and such exotic items as niobium cans and titanium discs.

Independent and Dependent Variables

Before examining results of tests of the hypotheses we review the characteristics of the sample as they relate to the independent and dependent variables. The independent variables involved in the study included dominant temperament-type and dominant value system-type.

Independent Variables

Two independent variables were of interest in this study. This subsection presents data about the sample with respect to these variables.

Value System Types of the Sample

The value system of each respondent was evaluated in accordance with the procedures described earlier. The predominant value system in the sample was Traditionalist, with 142 or 75.1% of the purchasing agents and 144 or 76.2% of the salespeople exhibiting this value system. Table 11 shows the number of respondents with each dominant value system type for purchasing agents and for salespeople.

Matched Value System Dyads

Among the sample of 189 dyads, 120 consisted of salespeople and purchasing agents that had the same dominant value system types (matched dominant value

Table 11. Dominant Value System Types for Purchasing Agents and Salespeople.

Dominant Value System Type	Purch. Agents		Salespeople	
	n	Percent	*n*	Percent
T	142	75.1%	144	76.2%
I	14	7.4%	17	9.0%
C	20	10.6%	20	10.6%
S	2	1.1%	1	0.5%
Dual-dominant types	11	5.8%	7	3.7%
Total	189	100.0%	189	100.0%

system types) and 69 that had different (mismatched) dominant value system types. Of the 69 dyads with mismatched dominant value system types, 17 dyads contained at least one respondent with a value system profile with two value system types of equal height on the bar graphs used to determine the dominant value system type; for convenience, such respondents will be referred to as having dual-dominant value system types. Table 12 details the numbers of each dominant value system type among the matched and mismatched dyads. Since no single, dominant, value system type can be determined for individuals with dual-dominant value system types, they are not included in this tabulation. Only profiles that have exactly the same dominant value system type(s) can be considered to be matched. Since profiles with dual-dominant value system types cannot match profiles with single-dominant value system types, they were considered to be mismatched for analytical purposes. (Results for H3, the only hypothesis for which this was relevant, did not change when dyads with dual-dominant value system types were dropped from the analysis.)

Table 12. Dominant Value System Types Among Matched and Mismatched Dyads.

	Matched	Mismatched	
	All	Purch. Agents	Salespeople
T	114	28	30
I	1	13	16
C	5	15	15
S	0	2	1
Total[a]	120	58	62

[a] Does not include 17 dyads with respondents with dual-dominant value system types.

Temperament Types of the Sample

Whereas the value system instrument uses only one graph to interpret raw scores, the PPA (the instrument used to determine temperament types) uses three graphs to interpret raw scores. Graph 1 represents the individual's behavior at work, Graph 2 represents the individual's behavior at home or at play, and Graph 3 represents the way the individual sees him or herself. It was assumed that the output of Graph 1 would provide the most important information for analyzing relationships within dyads, since the dyads represent people in working situations. Output based on Graph 2 has no theoretical basis for being included in the analysis and discussion; in order to prevent confusion and to avoid reporting unnecessary information, findings and discussion of results based on the output of Graph 2 are not presented in the balance of this article. (Note that results based on the output of Graph 2 were tested and were found to be not statistically significant for all hypotheses involving temperaments.) Results based on the output of Graph 3 could be important since Graph 3 reflects the way individuals see themselves which could influence behavior in working situations. Thus, temperaments as determined only by Graphs 1 and 3 are reported, analyzed, and discussed below.

Table 13 shows the number and percentage of respondents with each dominant temperament type among purchasing agents and salespeople as determined by Graphs 1 and 3 of the PPA. As this table shows, more respondents among both purchasing agents and salespeople exhibit dominant I-type temperaments than any other single type. This is true for both graphs.

For convenience, respondents with two temperament types of equal height on the bar graphs used to determine the dominant temperament type will be referred to as dual-dominant temperament types. Since no single dominant temperament type

Table 13. Dominant Temperament Types Among Purchasing Agents and Salespeople.

Temp. Type	Purchasing Agents				Salespeople			
	Graph I		Graph II		Graph I		Graph III	
	n	Percent	*n*	Percent	*n*	Percent	*n*	Percent
D	39	20.5	30	15.8	39	20.4	25	13.1
I	60	31.6	72	37.9	80	41.9	96	50.3
S	32	16.8	29	15.3	35	18.3	36	18.8
C	59	31.1	59	31.1	37	19.4	34	17.8
Total[a]	190	100.0	190	100.0	191	100.0	191	100.0

[a] Does not include respondents with dual dominant temperament types.

can be determined for these individuals, they are not included in the tabulation. As with value systems, only profiles that have exactly the same dominant temperament type(s) can be considered to be matched. Since profiles with dual-dominant temperament types cannot match profiles with single-dominant temperament types, they were considered to be mismatched for analytical purposes. (Results for H1, H2, and H2a, the only hypotheses for which this was relevant, did not change when dyads with dual-dominant temperament types were dropped from the analyses.)

Matched Temperament Type Dyads

For Graph 1, among the sample of 189 dyads, 49 consisted of salespeople and purchasing agents that had matched dominant temperament types and 140 dyads that had mismatched dominant temperament types. For Graph 3 52 dyads had matched dominant temperament types and 137 had mismatched dominant temperament types. Table 14 details the numbers of respondents with each dominant temperament type among the matched and mismatched dyads as determined by Graphs 1 and 3 of the PPA. There were 31 dyads matched on dominant temperament types as determined by both Graphs 1 and 3.

Since no single dominant temperament type can be determined for individuals with dual-dominant temperament types they are not included in the tabulation. No dyads had matched dual-dominant temperament types. As with dyads containing individuals with dual-dominant value system types, dyads containing respondents with dual-dominant temperament types were included in the mismatched group for analytical purposes.

Table 14. Dominant Temperament Types Among Matched and Mismatched Dyads.

	PPA Graph I			PPA Graph III		
	Matched	Mismatched		Matched	Mismatched	
	All	PA	S	All	PA	S
D	8	32	31	4	27	21
I	21	36	56	33	36	60
S	8	22	25	7	20	27
C	12	47	26	8	51	27
Total[a]	49	137	138	52	134	135

PA = Purchasing agents, S = Salespeople.

[a] Does not include respondents with dual-dominant temperament types.

Dependent Variables

Intention to repurchase and interpersonal satisfaction were the dependent variables in the hypotheses that were proposed. This section discusses repurchase intention, the characteristics of the interpersonal satisfaction experienced by the sample and the relationships of interpersonal satisfaction to other measures.

Interpersonal Satisfaction

Each participant's level of interpersonal satisfaction was measured with a single item using a semantic differential scale with extremes of "dissatisfied" and "satisfied" and with the middle position labeled "neither." As Table 15 shows, the number of respondents reporting each level of the variable is highly concentrated at the "satisfied" end of the scale. This pattern, called restriction of range, is true for both purchasing agents and for salespeople. The mean level of interpersonal satisfaction experienced by purchasing agents was 6.50.

Restriction of range or skewness of satisfaction ratings in other studies was observed and commented on by Peterson and Wilson (1992).

> Virtually all self-reports of customer satisfaction possess a distribution in which a majority of the responses indicate [*sic*] that customers are satisfied and the distribution itself is negatively skewed (p. 62).

These researchers note four possible explanations for the phenomenon. First, the distribution could reflect actual satisfaction with goods and services. Second, "satisfaction is caused by factors such as expectations and requires considerable cognitive effort, [and] its antecedents may influence the shape and level of the

Table 15. Interpersonal Satisfaction of Purchasing Agents.

Interpersonal Satisfaction Scale	Purchasing Agents	
	n	Percentage
"1 Dissatisfied"	0	0.0
"2"	0	0.0
"3"	3	1.6
"4 Neither"	5	2.6
"5"	7	3.7
"6"	53	28.0
"7 Satisfied"	121	64.0
Total	189	99.9

observed distributions." Third, satisfaction may not be normally distributed. Fourth, satisfaction ratings may be artifacts of research methodologies employed and inherent characteristics of customers. Of these possible causes, only researchers can control the last. For example, if the survey question is phrased "how satisfied are you with ______?" responses may be different than when the question is phrased "how dissatisfied are you with ______?" (Sudman & Bradburn, 1991). In this research, in order to avoid such bias and in accordance with the recommendations of Sudman and Bradburn, the question was phrased "how satisfied or dissatisfied are you with ______?"

The restriction of range of interpersonal satisfaction had one important potential implication for this research: the evidence of differences between groups would be more difficult to detect than if the characteristic had displayed a wider range of values. As will be seen subsequently, statistically significant differences were detectable, despite the restriction of range.

Intention to Repurchase

Each purchasing agent's intention to repurchase was measured with a single item worded as follows: "Assume that in the near future you have to buy or repurchase the product or service this salesperson sells. How likely would you be to buy from this salesperson?" Responses were recorded using a 7-point semantic differential scale with extremes of "certain to use different salesperson" (1 on the scale) and "certain to use same salesperson" (7) and with the middle position (4) labeled "neutral (not sure)." As Table 16 shows, the number of respondents reporting each level of the variable is highly concentrated at the "certain to repurchase" end of the scale. The mean level of intention to repurchase in all dyads was 6.4 and no purchasing agent responded with a response lower than "neutral."

Table 16. Purchasing Agents' Intention to Repurchase.

Intention to Repurchase	*n*	Percentage
"1" Certain not to Repurchase	0	0.0
"2"	0	0.0
"3"	0	0.0
"4 Neutral"	8	4.2
"5"	11	5.8
"6"	68	36.0
"7 Certain to Repurchase"	102	54.0
Total	189	100.0

Table 17. Significance of Covariates on Interpersonal Satisfaction.

Covariate	p
PA age	0.42
S age	0.85
PA gender	0.20
S gender	0.55
PA years of experience	0.33
S years of experience	0.11
PA educational level	0.78
S educational level	0.13[a]
Length of relationship	0.12
PA age × S age	0.09
PA gender × S gender	0.59
PA educ. × S educ.	0.26
Nature of purchase	0.63

PA = Purchasing agent, S = Salesperson.
[a] After elimination of one cell with only two entries.

Potential Covariates

Data on gender, age, years of experience in their profession, and educational level were obtained for purchasing agents and for salespeople. In addition, the length of the relationship between each salesperson and purchasing agent, in years, was obtained. ANOVAs were used to test the effect of each of these items on the level of interpersonal satisfaction experienced by purchasing agents. The nature of the salesperson's offering (nature of purchase) was tested using the chi-square test for homogeneity to determine whether purchases of services, raw materials, consumables, components, or capital equipment were correlated with differing levels of interpersonal satisfaction. Only the interaction of purchasing agents' and salespeople's ages was found to be related to interpersonal satisfaction to a statistically significant ($p < 0.10$) extent, as shown in Table 17. The combined effect of this interaction and the effects of ages of sellers and buyers on interpersonal satisfaction was very small, with $R^2 = 0.02$.

TESTS OF HYPOTHESES

So far, this section has examined the demographics of the sample and, where possible, compared the demographics of the sample to the demographics of the NAPM membership. Other characteristics of the sample have also been analyzed in order to obtain an understanding of the respondents. This is important since judgments about the representativeness of the sample and the external validity of the results will be based upon this understanding. Now the examination turns to the tests of hypotheses and whether the null hypotheses can be rejected.

The first hypothesis was written to test the effects of matched dominant temperament types on interpersonal satisfaction in dyads and was written:

> **H1.** When customer and salesperson dominant temperament-types match, interpersonal satisfaction of purchasing agents is higher than when the dominant temperament-types do not match.

Three graphs are used to interpret the raw scores of the PPA. Since Graph 1 depicts individuals' profiles in work settings, it was proposed that matching of profiles on Graph 1 of the PPA should be used to identify dyads with matched and mismatched profiles. It was decided to test the hypothesis for dyads matched on Graph 1 and then, independently of that matching, to test the hypothesis for dyads matched on Graph 3. (As previously discussed, use of data based on output from Graph 2 had no theoretical basis but were tested and found to be not statistically significant for any hypotheses.) Such testing would yield greater information than testing output from Graph 1 only and would verify use of Graph 1 as being appropriate. Table 18 shows the numbers of matched and mismatched dyads based on the dominant temperament types of each of the participants as determined by Graphs 1 and 3 of the PPA.

Utilizing the t-test for the difference between means with a one-tailed test, the null hypothesis could not be rejected. This was true regardless of which graph was used; the p-values were 0.32 and 0.16 for Graphs 1 and 3, respectively. Thus, it cannot be stated that purchasing agents in dyads with matched temperament types experience greater interpersonal satisfaction than purchasing agents in dyads with mismatched temperament types. It should be noted that the difference between the means is in the hypothesized direction.

Table 18. Results of Test of H1: Matched and Mismatched Dyads Determined by Graphs I and III of the PPA.

	Graph I		Graph III	
	n	Percent	*n*	Percent
Matched dyads	49	25.9	52	27.5
Mismatched dyads	140	74.1	137	72.5
Total	189	100.0	189	100.0

Note: 31 dyads matched on both Graphs I and III, 119 dyads mismatched on both Graphs I and III.

Table 19. Numbers of Dyads in Which Salespeople's Dominant Temperament Types were I-types or not I-types.

Salesperson's Dominant Temperament Type	Graph I	Graph III
I-type temperament	77	93
Not I-type temperament	112	96
Total	189	189

The second hypothesis tests the effects of dominant I-type temperaments of salespeople and is stated formally as:

> **H2.** Salespeople with dominant I-type temperaments generate higher levels of interpersonal satisfaction among purchasing agents than do salespeople with other dominant temperament-types.

As with the first hypothesis, it was proposed that the test be performed on mean customer interpersonal satisfaction for groups of dyads in which salespeople's dominant temperament types were determined by Graph 1. Again, the test was performed for dyads in which salespeople's dominant temperament types were determined by Graph 3, as well. The number of dyads within each group is shown in Table 19.

As predicted, the null hypothesis was rejected for dyads based on the results of Graph 1 ($p = 0.002$); the null hypothesis was also rejected for dyads based on the results of Graph 3 ($p = 0.007$). The one-tailed t-test has shown that the mean interpersonal satisfaction of purchasing agents in the two groups (those dyads with salespeople with dominant I-type temperaments and those dyads with salespeople with other than dominant I-type temperaments) is different. An examination of the means of interpersonal satisfaction of the groups shows that the differences are, indeed, in the hypothesized direction (Table 20). Thus, it can be stated that when salespeople's dominant temperaments are I-types as determined by Graphs 1 and 3 of the PPA, interpersonal satisfaction of purchasing agents is higher than for salespeople with other (not I-type) dominant temperament types.

Table 20. Results of Test of H2: Mean Interpersonal Satisfaction for Dyads with Salespeople with Dominant I-type and not I-type Temperaments.

	Dominant I-type Temperament	Dominant not I-type Temperament	p
Graph I	6.70	6.36	0.002
Graph III	6.66	6.35	0.007

Table 21. Numbers of Salespeople with Primary I-type Temperaments and Secondary D or not-D Temperament Types.

Salespeople Dominant I-type Temperaments	Secondary D-type Temperament	Secondary not D-type Temperament
Graph I	26	54
Graph III	43	53

A similar hypothesis in which salespeople's dominant temperaments are I-types and their secondary temperaments are D-types is stated as follows:

H2a. Salespeople with dominant I-type temperaments and secondary D-type temperaments generate higher levels of interpersonal satisfaction among purchasing agents than do salespeople with dominant I-type temperaments and other secondary-type temperaments.

Table 21 shows the number of salespeople with dominant I-type temperaments further classified into groups containing those with secondary D-type temperaments and those with secondary temperament types that are not D-types (i.e. in which the secondary temperament types are C or S). Utilizing the t-test for the difference between means with a one-tailed test, the null hypothesis was rejected for dyads based on the results of Graph 1, though not for dyads based on Graph 3 ($p = 0.07$ and 0.35, respectively); these results are shown in Table 22. Thus, interpersonal satisfaction of purchasing agents is higher in dyads with salespeople with primary I-type and secondary D-type temperaments than in dyads with salespeople with primary I-type temperaments and secondary temperaments that are not D-types, when temperament types are determined by Graph 1.

The number of dyads with salespeople having primary I-type temperaments and secondary D-type temperaments as determined by Graph 1 was 25. This is somewhat lower than 30, the generally accepted number for application of the Central Limit Theorem. Differences between $n = 25$ and $n = 30$ are relatively

Table 22. Results of Test of H2A: Test of Mean Interpersonal Satisfaction for Dominant I-type Temperaments with Secondary D- and Secondary not D-type Temperaments.

	Dominant I-type Secondary D-type Temps.	Dominant I-type Secondary not D-type Temperament	Significance of Difference p
Graph I	6.84	6.63	0.07
Graph III	6.69	6.63	0.35

minor (Watson et al., 1990). Nevertheless, to test the stability of this result, calculations were made with five hypothetical observations added to sample (to bring the total to 30). Even with the hypothetical observations, the results of the test did not change, regardless of the values of interpersonal satisfaction used for the hypothetical observations.

The third major hypothesis concerns the relationship between dyads in which salespeople and purchasing agents have matched dominant value system types and was stated:

> **H3.** When customer and salesperson dominant value system-types match, interpersonal satisfaction among purchasing agents is higher than when the dominant value system-types do not match.

The number of dyads with matched dominant value system types was 120 while 69 dyads had mismatched dominant value system types. Utilizing the t-test for the difference between means with a one-tailed test, the null hypothesis was rejected ($p = 0.07$); these results are shown in Table 23. Thus, interpersonal satisfaction of purchasing agents is higher in dyads with matched dominant value system types than in dyads with mismatched dominant value system types.

The fourth major hypothesis involved salespeople with one specific dominant value system type, synthesizers, and was written as follows:

> **H4.** Salespeople with dominant synthesizer value systems generate higher levels of interpersonal satisfaction among purchasing agents than do salespeople with other dominant value system-types.

In the entire sample of 189 dyads, only one had a salesperson with a dominant synthesizer value system. Since testing of this hypothesis required dyads containing salespeople with dominant S-type value systems, insufficient data were available to test this hypothesis.

A related hypothesis concerning salespeople with dominant synthesizer value systems was stated as:

Table 23. Results of Test of H3: Test of Mean Interpersonal Satisfaction for Dyads with Matched and Unmatched Dominant Value System-Types.

	n	Mean Interpersonal Satisfaction
Matched dominant value system types	120	6.57
Mismatched dominant value system types	69	6.39
Total	189	

$p = 0.07$.

Table 24. Results of Simple Regression of Intensity of Salesperson S-type Value System on Interpersonal Satisfaction for All Dyads.

Coefficient (m)	−0.026
Standardized coefficient	−0.039
Correlation coefficient (R)	0.039
F	0.42
p	0.52
Degrees of freedom	1,187
Adjusted R^2 (R^{*2})	0.0038

> **H4a.** Among salespeople with dominant S-type value systems, the intensity with which the value system is held is positively correlated with interpersonal satisfaction of purchasing agents.

Since insufficient data existed to test H4, insufficient data also existed to test H4A. Nevertheless, a simple regression was performed on all 189 dyads between interpersonal satisfaction and the intensity with which salespeople held the S-type value system. Utilizing the F-test with 1 and 187 degrees of freedom, the null hypothesis of no relationship could not be rejected ($p = 0.60$). The results are shown in Table 24. Thus, it cannot be determined whether the intensity of the S-component of salespeople's value systems (regardless of the dominant value system type) is correlated with interpersonal satisfaction.

The final hypothesis proposed for testing was stated as:

> **H5.** Among purchasing agents, interpersonal satisfaction is positively related to repurchase intentions.

A simple regression between purchasing agents' interpersonal satisfaction and their intentions to repurchase was performed on the entire sample of 189 dyads. The results of this analysis are shown in Table 25.

Table 25. Results of Test of H5: Results of Simple Regression of Interpersonal Satisfaction on Repurchase Intention.

Coefficient (m)	0.58
Standardized coefficient	0.60
Correlation coefficient (R)	0.60
F	108
p	0.00
Degrees of freedom	1,187
Adjusted R^2 (R^{*2})	0.36

Utilizing the F-test with 1 and 191 degrees of freedom, the null hypothesis of no relationship was rejected ($p = 0.00$). Thus, intention to repurchase is correlated with purchasing agent interpersonal satisfaction and the correlation coefficient, R, of 0.60 indicates a moderately strong correlation of the variables. The adjusted R^2 (R^{*2}) indicates that the regression accounts for 36% of the variance in intention to repurchase.

Power Calculations

A final analysis was performed to determine the effect size (ES) and the power of the statistical inferences for each of the hypotheses. For each hypothesis, based on the size and variance of the sample, calculations of the ES were made. With that information, the power could be determined from appropriate tables (Cohen, 1988). The results are shown in Table 26.

The ES is a unitless measure of the size of the independent variable's effect on the dependent variable. An ES smaller than 0.2 is generally considered small, an ES of 0.5 is considered moderate, and an ES of 0.8 or larger is considered large. For a given effect size, increasing the number of observations increases the power of the test. Given the small ES and the relatively small sample size available for testing of H1, the power was low and, thus, the likelihood of being able to reject the null hypothesis was low. It is significant that the powers of all the tests of the hypotheses that were accepted were 0.54 or greater, in spite of relatively modest effect sizes (as low as 0.21 for H3). Thus, the ability to accept the research hypotheses for H2, H2a, H3, and H5, is especially noteworthy.

Table 26. Power and Effect Sizes for Hypotheses.

	n	ES	Power
H1 (graph I)[a]	72	0.08	0.20
H1 (graph III)[a]	75	0.17	0.38
H2 (graph I)[a,b]	91	0.41	0.93
H2 (graph III)[b]	94	0.37	0.88
H2A (graph I)[a,b]	34	0.47	0.74
H2A (graph III)[a]	46	0.08	0.17
H3[a,b]	88	0.21	0.54
H4	N/A	N/A	N/A
H4A	N/A	N/A	N/A
H5[b]	189	0.60	>0.99

[a] Harmonic mean, n, used.
[b] Hypothesis supported.

Table 27. Summary of Hypotheses Tested and Test Results.

Hypothesis	Test/Results	Result
H1: Matched dominant temperament types correlate with higher I/S	*t*-test $p > 0.10$	Not supported
H2: Salespeople with dominant I-type temperaments correlate with higher I/S	*t*-test $p = 0.002$ (GI) $p = 0.007$ (GIII)	Supported for Graphs I and III
H2A: Salespeople with dominant I-type and secondary D-type temperaments correlate with higher I/S	*t*-test $p = 0.07$ (GI) $p = 0.35$ (GIII)	Supported for Graph I
H3: Matched dominant value system types correlate with higher I/S	*t*-test $p = 0.07$	Supported
H4: Salespeople with dominant S-type value systems correlate with higher I/S	*t*-test	Insufficient data
H4A: Intensity of salespeople's dominant S-type value systems is correlated with I/S	Regression	Insufficient data
H5: Interpersonal satisfaction is positively related to repurchase intention	Regression *F*-test $p = 0.00$	Supported $R^{*2} = 0.36$

For convenience, the results of the tests of hypotheses are summarized and presented in Table 27. The discussion now turns to the implications of the results.

DISCUSSION AND IMPLICATIONS

This section begins with a summary of the purpose of the study and a discussion of the research results. This is followed by a discussion of the contributions of the research, the managerial implications, the limitations of the study, and concludes with suggestions for extensions of the study and future research.

Purpose and Context

The model presented and tested in this study integrates several streams of research. The social exchange paradigm suggests that social behavior both influences and is influenced by the behaviors of each person in face-to-face interactions (Webster, 1968). Applied social exchange recognizes the contribution by Evans (1963) that dyads, rather than simply buyers or sellers, should be examined when investigating buyer behavior. In spite of this theoretical basis for dyadic research, few such studies have actually been undertaken. One of the purposes of this study was to increase the body of knowledge in this area.

A second major stream of research that this study built upon is that of customer satisfaction. The multi-attribute model of customer satisfaction suggests that overall satisfaction is comprised of many components (Crosby & Stephens, 1987; Oliver, 1993; Woodruff, Cadotte & Jenkins, 1983). Interpersonal satisfaction, the satisfaction that the buyer experiences with the relationship with the seller, is thought to be one such component (Oliver & Swan, 1989a). This study examined the relationship of interpersonal satisfaction to repurchase intention, the ultimate goal of satisfaction (Patterson, Johnson & Spreng, 1997).

A third major stream of research built upon by the model investigated in this study is that of personal selling. Unlike satisfaction, selling is not a construct but is an activity. Despite its recognized importance to businesses of all kinds, little research has been undertaken in this area of marketing (Bagozzi, 1976). This study sought to add to our knowledge in this area by examining the effects of temperaments and value systems on interpersonal satisfaction in business-to-business sales relationships.

Personalities can be classified according to different schemes. The MBTI is one instrument for classifying personality traits that has been used in sales research with some success (e.g. Dion et al., 1994; McIntyre, 1991). The instrument used in this study, the PPA, has not previously been used in research in marketing. What makes the PPA potentially valuable in the field of selling is that its classifications are based, in part, on fixed, overt, behavioral traits called temperaments. Different temperaments have different communication styles – both verbal and non-verbal – making it possible to identify the preferred communication style of each buyer. Another purpose of this study was to introduce this typology of temperaments to the marketing literature.

Value systems are the second component of personality (in addition to temperaments) examined in this study. While temperaments represent the overt behavioral characteristics of personality, value systems relate to motives influencing decisions and choices (England, 1967). In the same manner that individual behavioral traits can be classified into distinct patterns, individual values can be grouped into value systems. One of the important characteristics of value systems is their differential influence on communication, especially communication content. Attempts have been made to use individual values in marketing, especially for market segmentation (e.g. Kahle, 1986; Prakash & Munson, 1985). The construct of value systems, however, has not previously been utilized in the field. This study has introduced the value systems construct to the marketing literature.

Communication plays a central and vital role in personal selling (Williams, Spiro & Fine, 1990). Of the four aspects of communication – content, code, rules, and style (Andersen, 1972) – content, the ideas conveyed in the message, has been the most commonly examined in the field of personal selling. However, Steinberg and Miller (1975) suggest that nonverbal behavior is at least as important as the words people are mouthing. As mentioned previously, different non-verbal, as well as verbal, behaviors are one of the distinguishing characteristics of the Marston typology of behavioral types. One of the important reasons for using the Marston typology is that it recognizes the significance of different aspects of communication.

Given this context and these goals of the research, a cross-sectional study was designed to test the effects of temperaments and value systems of both buyers and sellers on customers' levels of interpersonal satisfaction. Invitations to participate in the study were sent to 7240 members of the National Association of Purchasing Management (NAPM) via e-mail. Respondents were asked to select the next salesperson that called upon them and then both were to complete questionnaires contained on a website. The questionnaires were analyzed and the data supported four of the seven hypotheses tested. These results substantiated aspects of the model and demonstrated the value of pursuing further research in this area.

DISCUSSION OF THE RESEARCH FINDINGS

This subsection discusses the findings of the study. First the characteristics of the individual respondents are discussed, then the characteristics of the dyads are examined, and, finally, the tests of each of the hypotheses are addressed. The discussion starts with the purchasing agents since they were responsible for selecting salespeople to participate.

Self-Selection of Purchasing Agent Participants

One factor that may help to explain the high proportion of purchasing agents with dominant T-type value systems is the nature of the work involved in purchasing. Most of the tasks require little in the way of innovation or creativity; the jobs involve judgement and expertise applied in a routine fashion. Individuals with dominant T-type value systems are well suited to this type of work. It may be that a majority of the population has chosen their jobs, at least in part, because of the "fit" of their value systems with the characteristics of the work they are performing. The sample may accurately reflect the value systems of the population. Thus, the entire population may have self-selected by choosing the type of job they perform, by choosing the type of organizations that they work for, and through their membership in the NAPM.

A distinguishing characteristic of people with dominant T-type value systems is that they typically accept power structures, figures of authority, and group norms. Such individuals would fit easily and comfortably into larger organizations which typically have hierarchical power structures – exactly the size of organizations that would be expected to have dedicated purchasing agents. Thus, the population from which the participants were drawn may have wittingly or unwittingly chosen their jobs, at least in part, because of the "fit" of their value systems with the characteristics of the organizational structures in which they operate. Similarly, the population has elected to join the NAPM and accept that organization's structure. The sample may accurately reflect the value systems of the population.

The high proportion of respondents with dominant T-type value systems necessarily meant that other dominant value system types were not as well represented in the sample. This restriction of range in dominant value system types prevented testing the statistical significance of differences in interpersonal satisfaction between groups with different dominant value system types (and thus prevented testing of H4 and H4a). In addition, this had implications for the external validity of the findings.

Sample Dyads

Since the unit of analysis in this study was dyads, it was important to examine the characteristics of the dyads in the sample. This section reviews features of those dyads and factors that may have influenced the selection of participants.

Factors Influencing Salesperson Participation

In planning this study it was considered likely that, absent instructions to the contrary, purchasing agents would select salespeople to participate in the study that those purchasing agents felt especially comfortable asking a favor of. Such a selection of salespeople would have resulted in an under-representation of salespeople with whom purchasing agents were uncomfortable asking favors of. It was speculated that such a selection process would have resulted in few unmatched dyads. Therefore, purchasing agents were instructed to select the first salesperson that called on them after they received the invitation to participate in the study. In this manner, it was expected that a broader sample of relationship types would be obtained.

As Table 15 shows, the numbers of purchasing agents in the sample reporting each of the seven levels of interpersonal satisfaction was highly concentrated at the "satisfied" end of the scale (mean interpersonal satisfaction was 6.5 on a 7-point scale). Only three purchasing agents reported interpersonal satisfaction lower than four (neither satisfied nor dissatisfied) on the scale of seven and those three all reported levels of three. Such skewed satisfaction ratings are consistent with the findings of Peterson and Wilson (1992).

Purchasing agents may be satisfied with virtually all of the salespeople that call on them, but this seems unlikely. Some suppliers, whether because of a buying company's policy (for example, "we buy only IBM computers") or because of unique competitive advantages (such as patent protection for products or manufacturing processes), cannot be dropped. Salespeople for these companies would have to be dealt with regardless of how dissatisfying the interpersonal relationships were to purchasing agents.

Discussions and correspondence with some of the purchasing agents that participated in the survey and in the questionnaire-pretest revealed that a random selection of salespeople was not likely in spite of the instructions. For example, purchasing agents expressed that they were unlikely to ask cold-callers to participate because the purchasing agents did not want to feel a sense of obligation to people that they did not know or already deal with. Rather, purchasing agents expressed the likelihood that they would select salespeople they felt comfortable asking a favor of. (While the instructions stated that a random selection of salespeople was needed, one purchasing agent, in explaining his choice of salespeople, said that a salesperson that he liked was what was sought by the survey.) It is not known how this relates to interpersonal satisfaction but a strong correlation seems likely.

An additional factor influencing the selection of participants is that salespeople may be more likely to call upon purchasing agents with whom the salespeople have satisfactory relationships and less likely to call on customers with whom they have unsatisfactory relationships. The more frequently that salespeople call on purchasing agents with whom they have high interpersonal satisfaction (as opposed to purchasing agents with whom they have low interpersonal satisfaction), the greater the likelihood that a high-satisfaction salesperson would have been the next one in the door. If this is the case, then high-satisfaction salespeople would be more likely to be selected to participate in the study. This seems especially likely if salespeople experience interpersonal satisfaction with purchasing agents in much the same manner as the purchasing agents experience those same relationships. This appears to be the case; the mean level of interpersonal satisfaction felt by salespeople with the purchasing agents was 6.45 and by purchasing agents with the salespeople, 6.50. Using the paired-samples t-test, the difference is not statistically significant ($p = 0.50$).

While non-random selection of participants appears to have occurred, gender does not appear to have influenced the selection. Female purchasing agents selected a higher proportion of saleswomen (26%) to participate than did male purchasing agents (20%) but the difference is not statistically significant ($p = 0.66$). Similarly, the means of interpersonal satisfaction generated by salesmen and saleswomen (6.5 and 6.6) are not significantly different ($p = 0.47$).

The underlying causes of non-random selection could be the very effects being examined in this study. If matched dominant value system types and matched dominant temperament types result in greater comfort levels than result from mismatched types, a high proportion of matched-type dyads would be expected in the sample. The data support this rationale. For example, the number of dyads matched on dominant value systems is 120 out of the total of 189 or 64%. Without knowing the proportions of dominant value system types in the populations of purchasing agents and salespeople, it is not possible to test the likelihood of this percentage being obtained by chance.

Characteristics of Dyads

Over 70% of all dyads in the sample were matched on temperaments, value systems, or both. Because this is the first study involving these typologies, little is known of the proportions of the various dominant types in the populations of purchasing agents and salespeople. Thus, it is not possible to compare the characteristics of this sample with characteristics of other samples. Nevertheless, the number of dyads matched on at least one characteristic seems remarkably high and seems unlikely to have occurred by chance.

The previous section explained the non-random selection of salesperson participants. In addition, the high proportion of purchasing agents with dominant T-type value systems in the sample has been discussed. The combination of these factors provides an explanation for the high proportion of dyads with matched dominant temperament and value system types. The explanation is that many of the purchasing agents that participated selected salespeople because the purchasing agents were comfortable asking a favor (participation in a survey) of those salespeople. As was previously stated, people with similar dominant value system types would intuitively tend to accept each other more than people with different dominant value system-types and they would have greater commitment to each other and to their respective relationships. Similarly, people with similar dominant temperament types would understand each other more readily than people with dissimilar dominant temperament types. Thus, the very effects hypothesized in H1 and H3 seem likely to have been responsible for the high proportion of matched dyads. That there were more dyads matched on value systems (120) than on temperaments (49 on Graph 1) suggests that value systems have the stronger effect. This is borne out by the strong support found for H3 (matched value systems), which was accepted, as opposed to the more modest evidence found for H1 (matched temperaments), which was not accepted.

Since the patterns of dominant value system types and dominant temperament system types are similar among salespeople and purchasing agents, it seems likely that the process used by purchasing agents to select salespeople to participate in the survey is at least partially accountable for this similarity. Attention is now turned to the factors influencing salesperson participation in the study.

Effects of Non-Random Selection of Salespeople

Non-random selections of salesperson participants would have the effect of eliminating from the sample salespeople with whom the purchasing agents felt low interpersonal satisfaction. This is one possible cause of the restriction in range of the interpersonal satisfaction variable. The result is that in testing the mean levels of interpersonal satisfaction between matched and unmatched dyads, the differences in the means will be very small since virtually all purchasing agents in the dyads experienced relatively high levels of interpersonal satisfaction. Thus, inclusion in the sample of only dyads in which purchasing agents experienced high interpersonal satisfaction masks the effects of the relationships being examined in this study.

The finding that buyers' and sellers' dominant value system types (Table 11) and dominant temperament types (Table 13) are concentrated in one or two personality types is consistent with the findings of Dion, Easterling and DiLorenzo-Aiss (1994). Dion et al. studied the same population (i.e. members of the NAPM and salespeople that called upon them) though they utilized the MBTI for classification of personality types. The earlier study accounted for the concentration of personality types by speculating that the task requirements were similar for both groups and so people of similar personalities were attracted to the two types of jobs. Though this cannot be ruled out, it seems more likely that the same factors influencing the self-selection of purchasing agents and the selection of salesperson participants in this study were at work in the prior study as well. The factors described above are believed to explain the concentration of personality types.

In summary, this analysis of the characteristics of the sample dyads indicates that for a number of reasons, a non-random sample was obtained. As discussed, this selection of salesperson participants may mask the effects being examined in this study. And while the non-random selection of salespeople may limit the external validity of the study, it should not invalidate the findings or conclusions of this study. To the contrary, this non-random selection of salesperson participants seems an important finding and helps explain the results of the Dion et al. study. Furthermore, where statistically significant results were found, they are especially noteworthy because the hypothesized effects were demonstrated in spite of conditions which tended to mask those effects.

Hypothesis Tests

The hypotheses tested and the results of those tests were summarized in Table 27. We now discuss those results and some of their implications.

Hypothesis H1

The first hypothesis suggests that when salesperson and purchasing agent dominant temperament types are matched purchasing agent interpersonal satisfaction was higher than when the temperament types were mismatched. A *t*-test of significance of the difference between means was conducted; results were not statistically significant ($p > 0.10$) for dyads determined by Graphs 1 and 3 of the PPA. The lack of support for this hypothesis could be due to the small effect size (0.08).

As mentioned, the mean levels of interpersonal satisfaction of purchasing agents in dyads that were matched and mismatched as determined by Graphs 1 and 3 of the PPA were tested. There were 31 dyads matched on dominant temperament types based on both Graphs 1 and 3. This represents 44% of the 70 dyads that are matched on either or both of the PPA graphs. While it is not known how this proportion compares to the population, it is clear that results based on dominant temperament types using the two graphs are not entirely independent. Thus, it is not surprising that the results are similar for both groups.

Hypothesis H2

The second hypothesis tests whether salespeople with dominant I-type temperaments generated higher levels of interpersonal satisfaction among purchasing agents than did salespeople with other dominant temperament types. As with the first hypothesis, tests were conducted for dyads determined by the outputs of Graphs 1 and 3.

As expected, the results of Graph 1 were statistically significant as were the results of Graph 3. This can be explained by the purposes of the two graphs. Graph 1 depicts the individual's behavior at work and adaptation to the work environment while Graph 3 depicts the individual's preferred or permanent behavioral mode and the way that the individual sees himself or herself. Both salespeople and purchasing agents are in work modes when they interact, thus Graph 1 was expected to be important. That Graph 3 is likewise statistically significant demonstrates that the ways that individuals see themselves are also important determinants of their behaviors. As would be expected with the high commonality between groups of dyads matched on Graphs 1 and 3 (as discussed with respect to H1), the results of the two groups were similar.

The support for H2 is an important finding because it suggests a method for prospectively identifying salespeople that create increased feelings of interpersonal satisfaction of customers towards salespeople. As sales managers hire new salespeople they can screen potential hires for dominant temperament types that increase the likelihood of providing interpersonal satisfaction to customers (i.e. hire candidates possessing dominant I-type temperaments). Such screening would be done utilizing the PPA. Interpersonal satisfaction may be important in and of itself, but it takes on added importance when the results of the test of H5, discussed subsequently, are considered; hypothesis H5 shows that increased interpersonal satisfaction is correlated with increased intention to repurchase.

Hypothesis H2a

It was hypothesized (H2a) that salespeople with dominant I-type temperaments and secondary D-type temperaments would create higher levels of interpersonal satisfaction than salespeople with dominant I-type temperaments and other (S and C) secondary temperament-types. While the number of dyads with salespeople with dominant I-type and secondary D-type temperaments as determined by Graph 1 was 25, somewhat lower than 30 which is the number that would be preferred for a large sample, the number of dyads with salespeople with dominant I-type and secondary D-type temperaments as determined by Graph 3 was 42. The results of the statistical tests for this hypothesis were mixed. While the null hypothesis was rejected for dyads matched on Graph I, it could not be rejected for dyads matched on Graph 3 ($p = 0.04$ and 0.32, respectively). As shown previously, despite $n < 30$ for Graph 1, the result is stable.

This finding is significant because it shows that there are characteristics of temperament types other than those associated with the I-archetype that are beneficial for creating interpersonal satisfaction. It is also significant because this is the only hypothesis for which the results of dyads matched on Graph 3 differ from the results for dyads matched on Graph 1. This demonstrates that results based on the two graphs are not interchangeable and that the graphs are able to discriminate between modes of behavior. It also confirms the prediction that Graph 1 of the PPA is the appropriate graph to utilize for analysis in work settings.

Hypothesis H3

H3 hypothesizes that interpersonal satisfaction in dyads in which salespeople and purchasing agents had matched dominant value system types would be higher than dyads in which the dominant value system types were mismatched. As previously discussed, there is restriction of range on interpersonal satisfaction experienced by purchasing agents; over 92% of purchasing agents reported levels of interpersonal satisfaction of 6 or 7 on the 7-point scale (Table 15). This restriction of range makes differences between groups small and difficult to detect. Additionally, over 75% of all respondents (75.1% of purchasing agents and 76.2% of salespeople) exhibited dominant T-type value systems. As previously discussed, this seems to be due to non-random selection of salesperson participants and is related to the high levels of interpersonal satisfaction. Thus, the finding of support for H3 is especially significant because support for the hypothesis was demonstrated in spite of factors tending to mask the effect. Additionally, since this study introduces the value system construct to the marketing literature, this finding is doubly important.

Hypothesis H4

The fourth major hypothesis proposed that salespeople with dominant S-type value systems generate higher levels of interpersonal satisfaction than do salespeople with other dominant value system types. Only one salesperson with a dominant S-type value system was selected to participate in the study (Table 11). Thus insufficient data were available to test this hypothesis. The lack of sufficient numbers of salespeople with dominant S-type value systems seems likely to be related to the non-random selection of salesperson participants.

Hypothesis H4a

H4a proposes that the intensity of salespeople's dominant S-type value systems was correlated with interpersonal satisfaction of purchasing agents. In the absence of sufficient data to test this hypothesis, a multiple linear regression was performed by regressing interpersonal satisfaction of purchasing agents on the intensity of the four value system types of salespeople. The purpose of this was to determine whether the intensities of each of the different value system types, regardless of the dominant value system type, were predictive of purchasing agent interpersonal satisfaction. The test of significance of the coefficient of multiple determination returned a value of $F = 0.60$ ($p = 0.66$). Thus, these data do not support a conclusion that the intensities of the four value system types are related to interpersonal satisfaction. Stepwise regression determined that results were not statistically significant for any of the four value system types.

Hypothesis H5

The final hypothesis tested in this research posits that purchasing agent interpersonal satisfaction is positively correlated with repurchase intention. In order to test this, a simple linear regression was performed of repurchase intention onto interpersonal satisfaction. As the results in Table 25 show, the null hypothesis can be rejected; thus, the hypothesis is supported and accounts for 36% of the variance in repurchase intention.

The correlation between interpersonal satisfaction and repurchase intention is an important finding because marketers want to know how to increase the likelihood that a customer will continue to purchase goods and services. This finding demonstrates the strong positive correlation of interpersonal satisfaction with salespeople on repurchase intention.

Contributions of the Research

This research makes several contributions to the field of marketing and to knowledge of selling and sales relationships. The primary contributions are:

- the influences of temperaments and value systems of buyers and sellers were modeled in marketing relationships,
- the impact of temperaments on the creation of interpersonal satisfaction was demonstrated,
- the impact of interpersonal satisfaction on repurchase intention was demonstrated,
- and the value system construct has been introduced into the marketing literature.

The theoretical framework developed and tested in this study incorporates personality constructs to explain creation of interpersonal satisfaction, overall customer satisfaction, and repurchase intention, in dyadic marketing relationships. The personality constructs of temperaments and value systems of both buyers and sellers were modeled as determinants of satisfaction with the relationship between the parties. This framework is important because it suggests causal linkages to interpersonal satisfaction and repurchase intention. This view contrasts with such factors as trust, liking, and perceived similarity which have been the focus of previous studies but which are experiential and not characteristics of a salesperson. Thus, while a salesperson can be hired for his or her temperament and value system profiles there is no way to make a hiring selection based on the trust, liking, or perceived similarity that the salesperson may generate.

The second major contribution of this research is the demonstration that aspects of the personalities of both buyers and sellers are determinative factors in the creation of interpersonal satisfaction in buyer/seller relationships. Other researchers, notably Oliver and Swan (1989a), have examined interpersonal satisfaction. Their primary purpose, however, was to determine the degree to which inputs and outcomes affected perceived levels of equity. This research has shown that a different model, personalities (in particular, value systems and temperaments) of *both* participants (i.e. buyers and sellers) are important determinants of interpersonal satisfaction.

Anecdotal evidence supports the importance of personality to the success of salespeople. Even the stereotype of salespeople suggests that they are extroverts, optimists, and smile and talk easily. The results of this research demonstrate that the anecdotal evidence and stereotyping have some basis in fact. Salespeople with dominant I-type temperaments fit this description and were shown to generate higher levels of interpersonal satisfaction than were other dominant temperament types. In addition, this finding provides further support for the multi-attribute model of customer satisfaction.

The third major contribution of this research is the finding that customer interpersonal satisfaction is strongly correlated with repurchase intention. Marketers are concerned with customer satisfaction because satisfied customers have a higher likelihood of repurchasing than do dissatisfied customers (Oliver & Swan, 1989a).

The fourth major contribution of this study is the introduction of the value system construct into the marketing literature. Heretofore, values have been used primarily as a means of segmenting markets (e.g. VALS2). Evidence provided here suggests that this construct can provide powerful insights into marketing relationships. Applications of value systems range from sales relationships to marketing strategies and compensation systems.

STRATEGY IMPLICATIONS

This study has both short- and long-term implications for managers. The findings provide managers with clear and powerful information on the short-term importance of achieving high levels of interpersonal satisfaction: highly satisfied customers are more likely to repurchase than less satisfied customers. And long-term, satisfied customers are more likely not to defect (Schlesinger & Heskett, 1991).

All sales managers use subjective techniques such as interviews and evaluations of role playing situations for selection of salespeople. Anecdotal evidence suggests that salespeople are evaluated on use of proper grammar, tone of voice ("is the candidate pleasant to listen to?"), confidence, creativity, and being "pushy or inappropriately aggressive" (Rasmuson, 1999). This study demonstrates that salespeople with dominant I-type temperaments achieve higher levels of interpersonal satisfaction than do other salespeople. Thus, utilization of this information as an objective selection criterion would be logical.

In addition, matched dominant value system types were shown to be associated with increased interpersonal satisfaction. The data in this study were too limited to establish with clarity whether one dominant value system archetype achieves higher levels of interpersonal satisfaction than the others. Further studies are required before managers use value system information for hiring decisions.

Going beyond the hiring decision, the interpersonal satisfaction achieved by the present sales force may be partially dependent upon the value system profiles of existing customers. Managers need to be mindful of potentially matched and mismatched value system profiles in assignments of major accounts to individual salespeople.

Anecdotal evidence abounds that while most sales managers are vitally interested in better training programs for their salespeople, they have no conceptual framework for teaching salespeople how to build relationships with customers or improve interpersonal satisfaction. For example, at UPS, the large shipping concern, new salespeople go into the field with experienced reps at their sides. The e-commerce sales support manager says, "We let them get a bit of egg on their face. You learn best when you end up falling down" (Marchetti, 2000). Presumably the trainees learn the experienced trainers' good traits and not the bad ones.

Training based on the temperament archetypes would be very beneficial for salespeople. Understanding the temperament types and associated behaviors will provide salespeople with a framework for assessing and responding in an appropriate manner to their customers' communications styles (both verbal and non-verbal). Practice in the techniques of adaptive behavior should be provided so that these techniques become second nature. In this way salespeople will learn to meet the human needs of their customers. Additional training in the differing motivations of customers with the four different dominant value system types would be beneficial.

Another word of caution is in order. Use of personality profiles for hiring salespeople is not a short-term strategy and does not provide a quick fix. It will not achieve bottom line results by the end of the quarter. Furthermore, managers who expect to achieve high and lasting levels of customer satisfaction simply through the selection of the "right" salespeople will be in for a disappointment. Leaders define the strategies guiding employee actions and they establish the environments in which employees function. Unless managers constantly demonstrate strong dedication to customers and their satisfaction, and unless managers provide the resources and latitude necessary to provide satisfaction, the strategy will fail (Davidow and Uttal, 1989). Those who "walk the walk" as opposed to merely "talking the talk" hold value system profiles in which satisfaction is part of their beings.

Limitations of the Study

A limitation of this study is that the sample must be considered a convenience sample. This is because respondents volunteered to participate, only members of the NAPM with e-mail access were sampled, and the incentives may have differentially attracted respondents with different value systems. Use of a convenience sample does present a threat to the external validity of the study. Surveys are intrusive. Questionnaires may not elicit responses to the questions asked, respondents may not provide accurate responses, the questions asked may not accurately tap the intended constructs, and respondents may be unwilling to answer questions or may not be aware of their motives (Malhotra, 1996). All of these may have been present in the study and would threaten its face or construct validity.

The Personal Profile Analysis and the Values Profile System are both psychometric instruments validated by their respective creators. As with all such measures, however, they have critics. Social desirability of responses is the primary threat to validity.

On-line responses were received from 94% of the respondents. While such a computer literate sample may limit the external validity of the results, this sampling bias will be less significant as more people learn to use computers and to navigate the Internet (Coomber, 1997). Furthermore, many NAPM members without e-mail addresses on file with the organization may nevertheless be computer literate.

Selection bias appears to have been present in the sample of respondents. While the sample demonstrated no significant differences from the NAPM membership on demographics, it did exhibit restriction of range on both dominant value system types and on the level of interpersonal satisfaction reported by purchasing agents.

Items were worded carefully and were pretested, yet researcher bias is a potential threat to both internal and external validity. The use of inferential statistics represented another potential threat to validity. Sample-to-sample variations, known as chance sampling fluctuations (Hamburg, 1991), may have resulted in a sample in which no effects were statistically significant even though there was an effect in the population at large. Small effect size (Cohen, 1988) could have had a similar effect.

Another possible threat to validity is that of history. By agreeing to participate in the study, salespeople may have briefly raised purchasing agents' underlying interpersonal satisfaction levels. If this effect was present, had a longer time passed between recruitment of salespeople and completion of the questionnaires, the threat might have diminished.

Implications for Future Research

Several avenues are available for building on the results of this research study. Among them are:

- incorporating additional variables into the model that was tested;
- further testing of the relationships specified in the present model;
- exploration of the application and impacts of the constructs tested in this study on other areas of marketing;
- and, examination of methodological issues raised by electronic data gathering.

Model Enhancement

The study described here was undertaken to demonstrate that temperaments and value systems are related to interpersonal satisfaction. The mechanism through which this relationship is accomplished has not, however, been investigated. Personal characteristics (e.g. Bagozzi, 1979), cognitive style (e.g. McIntyre, 1991), trust (e.g. Swan & Nolan, 1985), perceived similarity (e.g. Mathews, Wilson & Monoky, 1972), similarity of attitudes (e.g. Johnson & Johnson, 1972), communications (e.g. Kale & Barnes, 1992), liking (e.g. Swan et al., 1988), and expertise (e.g. Weitz, 1981b) may be mediating variables. Research into which factors are the most important would be beneficial to practitioners and would allow the model presented here to be made more specific and complete.

Interpersonal satisfaction requires further clarification and testing. There may be component parts of interpersonal satisfaction such as friendliness and liking on a social or personal level as opposed to an appreciation of another person for the assistance that he or she provides in achieving a goal or end on a professional level. While both of these behaviors may fit under the rubric of interpersonal satisfaction, they have very different implications for the types of salespeople that customers would find satisfying. Differing circumstances may also lead the same customer to seek each component (i.e. personal and professional), though at different times.

Additional Model Testing

This research demonstrates that matched dominant temperament and value system types are related to interpersonal satisfaction. The first question stemming from this finding is, what are the effects of a broader spectrum of dominant value system types and levels of interpersonal satisfaction? Specifically, for example, it is important to know what influence salespeople with dominant S-type value systems have on interpersonal satisfaction. The knowledge gained from such additional data holds potential for improving the external validity of the model.

This study suggests that interpersonal satisfaction is a component of the overall customer satisfaction emotion. Customer satisfaction correlates positively with repurchase intentions. The research conducted in this study shows that interpersonal satisfaction is, likewise, correlated positively with repurchase intentions, further suggesting that interpersonal satisfaction is a component of overall customer satisfaction. Additional research is required to determine what proportion of customer satisfaction is represented by interpersonal satisfaction and, if that proportion varies, under what circumstances it does so.

The model can be extended to include specific salesperson characteristics that might be associated with increased interpersonal satisfaction in given situations. For example, most studies of personal characteristics and personality traits assume that more of a given attribute is better. As has been mentioned previously, extreme behavioral patterns may not be the most beneficial levels for creation of interpersonal satisfaction, that is, a non-linear relationship may exist. For example, a dominant I-type temperament was shown to be associated with higher interpersonal satisfaction. Yet a salesperson with an extremely high I-temperament component could not be quiet long enough for a customer to ask a question – hardly a satisfying experience. A careful study of the relationship of temperament and value system component intensities to levels of interpersonal satisfaction is needed. Determination of the optimal intensities of temperament and value system archetypes would be an important contribution. It may well be that optimal levels differ depending on the types of goods or services being offered. For example, higher levels of I-type temperaments may be beneficial when selling standard or commodity-type goods that have little in the way of product characteristics to differentiate them while a lower level of the I-component might be beneficial for more custom-produced items.

Another extension of the model could utilize temperaments as determined by Graph 2 of the PPA. Graph 2 represents the individual's mode of behavior at home or at play. To the extent that golfing, dinners, parties, and other social activities are involved in sales efforts, Graph 2 of the PPA might be an important component in salespeople's personality profiles. It could be useful information to know the extent to which dominant Graph 2 temperament types relate to customer satisfaction in such social settings.

An important question to answer is the degree to which salespeople can adapt their overt behaviors to match those of their customers. And when they do, what impact does it have on interpersonal satisfaction? The answers hold important implications for the types of sales training to provide salespeople.

This study examines the impact of interpersonal satisfaction on repurchase intentions. It will be important to examine the impact of interpersonal satisfaction on initial purchase behavior.

Another area that could benefit from the application of the temperament and value system constructs is sales management. A study examining salesperson performance as the dependent variable instead of interpersonal satisfaction could provide useful insights. If the temperament and value system profiles that create higher interpersonal satisfaction are the same ones that are associated with high performing salespeople this would provide additional support for the model. In addition, it would provide useful and important information to students of management about selection and hiring criteria.

A different aspect of sales management involves the selection of sales managers. It may be that the personality profiles of successful sales managers are quite different from those of successful salespeople because the job requirements and functions are different. An examination of the temperament and value system profiles of successful sales managers could provide useful information for upper management. Promoting the most successful salespeople out of the field and into the office could be counterproductive.

Frequently sales compensation plans are ineffective because they are based on false assumptions about salespeople (Moynahan, 1980). Research into relationships between dominant value system types and preferred compensation systems could be very beneficial for managers. Salespeople with different dominant value system types have different motivations and could be expected to respond differently to various compensation schemes. For example, Challengers are expected to prefer straight commission sales while Synthesizers are expected to prefer straight salary or salary plus bonus. Testing preferences for different compensation plans would allow sales managers to design plans that attract and hold salespeople that have desirable value system profiles.

Adaptation to Other Areas of Marketing

The focus of this study is on the application of the temperament and value system constructs to the area of selling and sales management. The constructs are so powerful that they will find applications in other segments of the marketing field.

In advertising and sales promotion the temperament and value system constructs may provide a useful framework for analyzing and classifying audiences and market segments. Ad campaigns, based on the value system archetypes, may be made to appeal to different audiences. For example, several years ago Oldsmobile had a distinctly middle-aged appeal and clientele – primarily Traditionalists.

Their advertising campaign, "This is not your father's Oldsmobile" offended the Traditionalists (quietly loyal, not used to rapid change) who make up the bulk of the market while trying to appeal to younger, hipper, urban professionals (Challengers). The campaign had no appeal to the new target audience – there was nothing in it for them. Perhaps an understanding of value systems could have prevented this failure.

In the field of services, personal characteristics of service providers are so commingled with the services themselves that it seems that personality types would be as important as in sales positions. Are some profiles more appropriate to some service sectors than others? How important are the profiles of service providers in the process of selecting a service provider? For a service provider, what personality characteristics are important in recovering from a service failure? Quickly establishing empathy with the customer (I-type temperament) and a willingness to break rules and cut red tape (Synthesizer value system) in order to achieve customer satisfaction may be attributes more associated with some personality profiles than with others. Furthermore, in the context of professional services, partners at major accounting firms are essentially salespeople. But how important are the personalities of auditors, tax staff, or consultants that also have client interaction? Would an accountant with a low C-component temperament profile (i.e. not displaying much precision) have much credibility with clients (to say nothing of the match of individual characteristics to the needs of the job being performed)?

The constructs might also have applications in the area of channels of distribution. For example, do value system types apply to organizations? If so, do matched or mismatched organizational value systems help explain the success or failure of channel partners? This could provide a framework for evaluating organizations for channel partners and other strategic alliances.

In the field of consumer behavior there is anecdotal evidence of the impact of the personality of salespeople at the retail level where customer interactions are very short compared to business-to-business relationships. As in selling and sales management, these constructs could provide a framework for the selection of salespeople, especially for goods and services involving longer-term relationships such as stockbrokerage, insurance, and furniture. Research could help determine the potential to use the temperaments and value systems for segmenting markets, for developing product line extensions, and for creating new products.

SUMMARY

This paper explores the effects of temperaments and value systems on buyer-seller relationships in the business-to-business marketing arena. Important associations between dominant temperament types and customer interpersonal satisfaction, dominant value system types and customer interpersonal satisfaction, and interpersonal satisfaction and intention to repurchase were discovered. In addition, there were strong hints that additional relationships exist but results were not statistically significant in the sample obtained.

This study achieves its goals of developing and testing a model of the influences of temperaments and value systems on marketing relationships. Results showed that this conceptualization successfully depicted and predicted some of the processes. Significantly, the study also demonstrated the importance of interpersonal satisfaction in the overall satisfaction construct.

As we enter the 21st century, the marketplace is becoming more competitive than ever before. At the same time, improved manufacturing technology is decreasing marketers' abilities to differentiate their offerings from those of competitors. In this confusing and turbulent environment, sellers will find that personal relationships, connecting one person to another, will drive buyers' selections.

Personal selling is the only medium that allows marketers to adapt their messages to the unique requirements of each individual customer and to influence how those customers interact at a personal level. Relationships with salespeople are a sustainable competitive advantage for achieving customer satisfaction. Used as an offensive tool, customer satisfaction can attract new customers and used as a defensive tool it can create brand loyalty and prevent customer defections. This research provides marketers with important new information for creating, improving, and sustaining high levels of customer satisfaction by showing that dominant temperament and value system types play important roles in the creation of greater interpersonal satisfaction and increasing repurchase intentions.

REFERENCES

Andersen, K. E. (1972). *Introduction of communication theory and practice*. Menlo Park, CA: Cummings Publishing.

Anderson, E. W., Fornell, C., & Lehmann, D. R. (1994). Customer satisfaction, market share, and profitability: Findings from Sweden. *Journal of Marketing*, *58*(3), 53–66.

Baehr, M. E., & Williams, G. B. (1968). Prediction of sales success from factorially determined dimensions of personal background data. *Journal of Applied Psychology*, *52*(2), 98–103.

Bagozzi, R. P. (1976). Toward a general theory for the explanation of the performance of salespeople. Ph.D. Dissertation, Northwestern University.

Bagozzi, R. P. (1978). Sales performance and satisfaction as a function of individual difference, interpersonal, and situational factors. *Journal of Marketing Research*, *15*(November), 517–531.

Bagozzi, R. P. (1979). Toward a formal theory of marketing exchanges. In: O. C. Ferrell, S. W. Brown & C. W. Lamb, Jr. (Eds), *Conceptual and Theoretical Developments in Marketing* (pp. 431–447). Chicago: American Marketing Association.

Basi, R. K. (1999). WWW response rates to socio-demographic items. *Journal of the Market Research Society*, *41*(4), 397–401.

Beatty, S. E., Kahle, L. R., Homer, P., & Misra, S. (1985). Alternative measurement approaches to consumer values: The list of values and the Rokeach value survey. *Psychology and Marketing*, *2*(3), 181–200.

Berscheid, E. (1966). Opinion change and communicator-communicatee similarity and dissimilarity. *Journal of Personality and Social Psychology*, *4*(6), 670–680.

Bitner, M. J., Booms, B. H., & Tetreault, M. S. (1990). The service encounter: Diagnosing favorable and unfavorable incidents. *Journal of Marketing*, *54*(January), 71–84.

Bitner, M. J., & Hubbert, A. R. (1994). Encounter satisfaction vs. overall satisfaction vs. quality. In: R. T. Rust & R. L. Oliver (Eds), *Service Quality: New Directions in Theory and Practice* (pp. 72–94).

Blair, J., & Presser, S. (1993). Survey procedures for conducting cognitive interviews to pretest questionnaires: A review of theory and practice. Proceedings of the Section on Survey Research Methods, Alexandria, VA: American Statistical Association.

Bluen, S. D., Barling, J., & Burns, W. (1990). Predicting sales performance, job satisfaction, and depression by using the achievement strivings and impatience – irritability dimension of type A behavior. *Journal of Applied Psychology*, *75*(2), 212–216.

Bolton, R. N. (1993). Pretesting questionnaires: Content analyses of respondents' concurrent verbal protocols. *Marketing Science*, *12*(3), 280–303.

Bradley, N. (1999). Sampling for internet surveys. An examination of respondent selection for internet research. *Journal of the Market Research Society*, *41*(4), 387–395.

Brock, T. C. (1965). Communicator-recipient similarity and decision change. *Journal of Personality and Social Psychology*, *1*(6), 650–654.

Chapple, E. D., & Donald, G., Jr. (1947). An evaluation of department store salespeople by the interaction chronograph. *Journal of Marketing* (October), 173–185.

Churchill, G. A., Jr. (1991). *Marketing research: Methodological foundations* (5th ed.). Hinsdale, IL: Dryden.

Churchill, G. A., Jr., Collins, R. H., & Strang, W. A. (1975). Should retail salespersons be similar to their customers? *Journal of Retailing*, *51*(3), 29–42.

Churchill, G. A., Jr., Ford, N. M., & Walker, O. C., Jr. (1979). Personal characteristics of salespeople and the attractiveness of alternative rewards. *Journal of Business Research*, *7*(1), 25–50.

Churchill, G. A., Jr., Ford, N. M., Hartley, S. W., & Walker, O. C., Jr. (1985). The determinants of salesperson performance: A meta-analysis. *Journal of Marketing Research*, *22*(2), 103–118.

Churchill, G. A., & Surprenant, C. (1982). An investigation into the determinants of customer satisfaction. *Journal of Marketing Research*, *19*(November), 491–504.

Clabaugh, M. G., Jr., & Forbes, J. L. (1992). *Professional selling: A relationship approach*. New York: West.

Cohen, J. (1988). *Statistical power analysis for the behavioral sciences*. Hillsdale, NJ: Lawrence Erlbaum Associates.

Coomber, R. (1997). Using the internet for survey research. *Sociological Research Online*, *2*(2), http://www.socresonline.org.uk/socresonline/2/2/2.html

Cooper, A. R., Cooper, M. B., & Duhan, D. F. (1989). Measurement instrument development using two competing concepts of customer satisfaction. *Journal of Consumer Satisfaction, Dissatisfaction and Complaining Behavior*, *2*, 28–35.

Cotham, J. C., III (1968). Job attitudes and sales performance of major appliance salesmen. *Journal of Marketing Research*, *V*(November), 370–375.

Coyne, K. (1989). Beyond service fads – meaningful strategies for the real world. *Sloan Management Review*, *30*(Summer), 69–76.

Cronin, J. J. (1994). Analysis of the buyer-seller dyad: The social relations model. *Journal of Personal Selling & Sales Management*, *XIV*(3), 69–77.

Crosby, L. A., Evans, K. R., & Cowles, D. (1990). Relationship quality in services selling: An interpersonal influence perspective. *Journal of Marketing*, *54*(July), 68–81.

Crosby, L. A., & Stephens, N. (1987). Effects of relationship marketing on satisfaction, retention, and prices in the life insurance industry. *Journal of Marketing Research*, *XXIV*(November), 404–411.

Davidow, W. H., & Uttal, B. (1989). *Total customer service*. New York: HarperCollins.

Dion, P. A., Easterling, D., & DiLorenzo-Aiss, J. (1994). Buyer and seller personality similarity: A new look at an old topic. Proceedings of the Southern Marketing Association, New Orleans (pp. 396–402).

Dion, P. A., & Notarantonio, E. M. (1992). Salesperson communication style: The neglected dimension in sales performance. *The Journal of Business Communication*, *29*(1), 63–77.

Divita, S. F. (1993). Understand buyer/seller behavior: Implications for effective selling and customer satisfaction. Unpublished notes, Washington, DC: The George Washington University.

Divita, S. F. (1995). New ways to improve buyer-seller relationships . . . and sales productivity. Unpublished seminar notes, April 12, 1995, Washington, DC: The George Washington University.

Divita, S. F. (n.d.). Interpreting customers' non-verbal language: Key to gaining competitive advantage. Working Paper, The George Washington University.

Drolet, A. L., & Morrison, D. G. (2001). Do we really need multiple-item measures in service research? *Journal of Service Research*, *3*(3), 196–204.

Dunnette, M. D., & Kirchner, W. K. (1960). Psychological test differences between industrial salesmen and retail salesmen. *Journal of Applied Psychology*, *44*(2), 121–125.

Dwyer, F. R., Schurr, P. H., & Oh, S. (1987). Developing buyer-seller relationships. *Journal of Marketing*, *51*(April), 11–27.

Engel, J. F., Blackwell, R. D., & Miniard, P. W. (1995). *Consumer behavior*. Philadelphia: Dryden Press.

England, G. W. (1967). Personal value systems of American managers. *Academy of Management Journal*, *10*(March), 53 68.

Evans, F. B. (1963). Selling as a dyadic relationship – a new approach. *American Behavioral Scientist*, *6*(6), 76–79.

Fornell, C. (1992). A national customer satisfaction barometer: The Swedish experience. *Journal of Marketing*, *56*(January), 6–21.

Frazier, G. L., Spekman, R. E., & O'Neal, C. R. (1988). Just-in-time exchange relationships in industrial markets. *Journal of Marketing*, *52*(October), 52–67.

French, C. L. (1960). Correlates of success in retail selling. *American Journal of Sociology* (September), 128–134.

Ghiselli, E. E. (1973). The validity of aptitude tests in personnel selection. *Personnel Psychology*, *26*(Winter), 461–477.

Grewal, D., & Sharma, A. (1991). The effect of salesforce behavior on customer satisfaction: An interactive framework. *Journal of Personal Selling and Sales Management*, *9*(3), 13–23.

Grikscheit, G. G. (1971). An investigation of the ability of salesmen to monitor feedback. Ph.D. Dissertation, Michigan State University.

Guion, R. M., & Gottier, R. F. (1965). Validity of personality measures in personnel selection. *Personnel Psychology*, *18*, 135–164.

Hamburg, M. (1991). *Statistical analysis for decision making*. Washington, DC: Harcourt Brace Jovanovich.

Harrell, T. W. (1960). The relation of test scores to sales criteria. *Personnel Psychology*, *13*(Spring), 65–69.

Hawes, J. M., Mast, K. E., & Swan, J. E. (1989). Trust earning perceptions of sellers and buyers. *Journal of Personal Selling & Sales Management*, *IX*(Spring), 1–8.

Hempel, D. J. (1977). Consumer satisfaction with the home buying process: Conceptualization and measurement. In: H. K. Hunt (Ed.), *Conceptualization and Measurement of Consumer Satisfaction and Dissatisfaction* (pp. 275–299). Cambridge, MA: Marketing Science Institute.

Howard, J. A., & Sheth, J. N. (1968). A theory of buyer behavior. In: H. H. Kassarjian & T. S. Robertson (Eds), *Perspectives in Consumer Behavior* (pp. 467–487). Glenview, IL: Scott, Foresman.

Howells, G. W. (1968). The successful salesman: A personality analysis. *British Journal of Marketing*, *2*, 13–23.

Iacobucci, D., & Hopkins, N. (1992). Modeling dyadic interactions and networks in marketing. *Journal of Marketing*, *XXIX*(February), 5–17.

Irvine, S. H. (1989). *Personal profile analysis: Technical handbook*. Ormskirk, Lancs., England: Thomas Lyster Ltd.

Johnson, M. D., & Fornell, C. (1991). A framework for comparing customer satisfaction across individuals and product categories. *Journal of Economic Psychology*, *12*, 267–286.

Johnson, D. W., & Johnson, S. (1972). The effects of attitude similarity, expectation of goal facilitation, and actual goal facilitation on interpersonal attraction. *Journal of Experimental Social Psychology*, *8*, 197–206.

Johnston, W. J. (1981). Dyadic communication patterns in industrial buying behavior. In: P. H. Reingen & A. G. Woodside (Eds), *Buyer – Seller Interactions: Empirical Research and Normative Issues* (pp. 11–22). Chicago: American Marketing Association.

Jolson, M. A. (1975). The underestimated potential of the canned sales presentation. *Journal of Marketing*, *39*(January), 75–78.

Kahle, L. R. (1983). *Social values and social change: Adaptation to life in America*. New York: Praeger.

Kahle, L. R. (1986). The nine nations of North America and the value basis of geographic segmentation. *Journal of Marketing*, *50*(April), 37–47.

Kale, S. H., & Barnes, J. W. (1992). Understanding the domain of cross-national buyer-seller interactions. *Journal of International Business Studies* (1st Quarter), 101–132.

Kaplan, S. J. (1983). *The Kaplan report*. Tucson, AZ: TIMS Management Systems.

Kassarjian, H. H. (1971). Personality and consumer behavior: A review. *Journal of Marketing Research*, *VIII*(November), 409–418.

Kassarjian, H. H. (1979). Personality: The longest fad. *Advances in Consumer Research*, *VI*, 122–124.

Kassarjian, H. H., & Sheffet, M. J. (1975). Personality and consumer behavior: One more time. In: E. M. Mazze (Ed.), *Combined Proceedings* (No. 37, pp. 197–201). Chicago: American Marketing Association.

Kassarjian, H. H., & Sheffet, M. J. (1981). Personality and consumer behavior: An update. In: H. H. Kassarjian & T. S. Robertson (Eds), *Perspectives in Consumer Behavior* (pp. 160–180). Glenview, IL: Scott Foresman.

Kirchner, W. K., McElwain, C. S., & Dunnette, M. D. (1960). A note on the relationship between age and sales effectiveness. *Journal of Applied Psychology*, *44*(2), 92–93.

Kluckhohn, F. R., & Strodtbeck, F. L. (1961). *Variations in value orientations*. Westport, CT: Greenwood Press.

Kragness, M., & Rening, L. (1996). A comparison of the *Personal Profile System*® and the *Myers-Briggs Type Indicator*®. Minneapolis, MN: Carlson Learning Company.

Malhotra, N. K. (1996). *Marketing research an applied orientation*. Upper Saddle River, NJ: Prentice Hall.

Manning, G. L., & Reece, B. L. (2001). *Selling today, building quality partnerships*. Upper Saddle River, NJ: Prentice Hall.

Marchetti, M. (2000). Can you build a sales force? *Sales and Marketing Management* (January), 56–61.

Marston, W. M. (1928). *Emotions of normal people*. Harcourt, Brace, New York. Reprint, Ormskirk, Lancs., England: Thomas Lyster, 1989.

Maslow, A. H. (1954). *Motivation and personality*. New York: Harper & Row.

Massey, M. (1979). *The people puzzle: Understanding yourself and others*. Reston, VA: Reston Publishing.

Massey, M., & O'Connor, M. J. (1989). *Values profile system*. Minneapolis, MN: Carlson Learning Company.

Mathews, H. L., Wilson, D. T., & Monoky, J. F., Jr. (1972). Bargaining behavior in a buyer-seller dyad. *Journal of Marketing Research, IX*(February), 103–105.

McIntyre, R. P. (1991). The impact of Jungian cognitive style on marketing interactions. Ph.D. Dissertation, Arizona State University.

Merenda, P. F., & Clarke, W. V. (1959). The predictive efficiency of temperament characteristics and personal history variables in determining success of life insurance agents. *Journal of Applied Psychology*, *43*(6), 360–366.

Miles, M. P., Arnold, D. R., & Nash, H. W. (1990). Adaptive communication: The adaptation of the seller's interpersonal style to the stage of the dyad's relationship and the buyer's communication style. *Journal of Personal Selling & Sales Management*, *X*(February), 21–27.

Mitroff, I. I., & Mitroff, D. D. (1979). Interpersonal communication for knowledge utilization. *Knowledge: Creation, Diffusion, Utilization*, *1*(2), 203–217.

Morgan, R. M., & Hunt, S. D. (1994). The commitment-trust theory of relationship marketing. *Journal of Marketing*, *58*(July), 20–38.

Mosel, J. N. (1952). Prediction of department store sales performance from personal data. *Journal of Applied Psychology*, *36*(February), 8–10.

Moynahan, J. K. (1980). *Designing an effective sales compensation program*. New York: American Management Associations.

NAPM (1998). *1998 NAPM membership demographics*. Tempe, AZ: National Association of Purchasing Management.

O'Connor, M. J. (1986). *The "TICS" model: Trainer and consultant's reference encyclopedia*. Minneapolis, MN: Performax Systems International.

Oliver, R. L. (1980). A cognitive model of the antecedents and consequences of satisfaction decisions. *Journal of Marketing Research*, *17*(November), 460–469.

Oliver, R. L. (1993). Cognitive, affective, and attribute bases of the satisfaction response. *Journal of Consumer Research*, *20*(December), 418–430.

Oliver, R. L. (1997). *Satisfaction: A behavioral perspective on the consumer*. New York: McGraw-Hill.

Oliver, R. L., & Swan, J. E. (1989a). Consumer perceptions of interpersonal equity and satisfaction in transactions: A field survey approach. *Journal of Marketing*, *53*(April), 21–35.

Oliver, R. L., & Swan, J. E. (1989b). Equity and disconfirmation perceptions as influences on merchant and product satisfaction. *Journal of Consumer Research*, *16*(December), 372–383.

Patterson, P. G., Johnson, L. W., & Spreng, R. A. (1997). Modeling the determinants of customer satisfaction for business-to-business professional services. *Journal of the Academy of Marketing Science*, *25*(1), 4–17.

Performax (1985). *Values analysis profile trainer manual*. Minneapolis, MN: Performax Systems International.

Performax (1987). *The "DiSC" model: Trainer and consultant's reference encyclopedia*. Minneapolis, MN: Carlson Learning Systems.

Peterson, R. A., Cannito, M. P., & Brown, S. P. (1995). An exploratory investigation of voice characteristics and selling effectiveness. *Journal of Personal Selling & Sales Management, XV*(1), 1–15.

Peterson, R. A., & Wilson, W. R. (1992). Measuring customer satisfaction: Fact and artifact. *Journal of the Academy of Marketing Science, 20*(1), 61–71.

Prakash, V., & Munson, J. M. (1985). Values, expectations from the marketing system and product expectations. *Psychology and Marketing, 2*(4), 279–296.

Presser, S., & Blair, J. (1994). Survey pretesting: Do different methods produce different results? In: *Sociological Methodology* (Vol. 24, pp. 73–104). Washington, DC: American Sociological Association.

Pruden, H. O., & Peterson, R. A. (1971). Personality and performance-satisfaction of industrial salesmen. *Journal of Marketing Research, VIII*(November), 501–504.

Rasmuson, E. (1999). Can inside sales be a rep's true calling. *Sales and Marketing Management* (April), 82.

Riordan, E. A., Oliver, R. L., & Donnelly, J. H., Jr. (1977). The unsold prospect: Dyadic and attitudinal determinants. *Journal of Marketing Research, XIV*(November), 530–537.

Rogers, H. P., Peyton, R. M., & Berl, R. L. (1992). Measurement and evaluation of satisfaction processes in a dyadic setting. *Journal of Consumer Satisfaction, Dissatisfaction and Complaining Behavior, 5*, 12–23.

Rokeach, M. J. (1971). The measurement of values and value systems. In: G. Abcarian & J. W. Soule (Eds), *Social Psychology and Political Behavior: Problems and Prospects* (pp. 20–39). Columbus, Ohio: Charles E. Merrill.

Rokeach, M. J. (1973). *The nature of human values*. New York: Free Press.

Schlesinger, L. A., & Heskett, J. L. (1991). The service-driven company. *Harvard Business Review* (September–October), 71–81.

Schultz, R. S. (1935). Test selected salesmen are successful. *Personnel Journal, 14*(October), 139–142.

Sheth, J. N. (1976). Buyer seller interactions: A conceptual framework. *Advances in Consumer Research, 3*, 382–386.

Sheth, J. N., & Parvatiyar, S. (1995). Relationship marketing in consumer markets: Antecedents and consequences. *Journal of the Academy of Marketing Science, 23*(4), 255–271.

Smith, A. K., Bolton, R. N., & Wagner, J. (1999). *Journal of Marketing Research, 36*(August), 356–372.

Spiro, R. L., & Weitz, B. A. (1990). Adaptive selling: Conceptualization, measurement, and nomological validity. *Journal of Marketing Research, XXVII*(February), 61–69.

Steinberg, M., & Miller, G. R. (1975). Interpersonal communication: A sharing process. In: G. H. Hanneman & W. J. McEwen (Eds), *Communication and Behavior* (pp. 126–148). Reading, MA: Addison Wesley.

Sudman, S., & Bradburn, N. M. (1991). *Asking questions*. San Francisco: Jossey-Bass.

Sujan, H., Weitz, B. A., & Sujan, M. (1988). Increasing sales productivity by getting salespeople to work smarter. *Journal of Personal Selling and Sales Management, 8*(2), 9–19.

Sukhdial, A. S., Chakraborty, G., & Steger, E. K. (1995). Measuring values can sharpen segmentation in the luxury auto market. *Journal of Advertising Research, 35*(1), 9–22.

Swan, J. E., Trawick, I. F., Jr., Rink, D. R., & Roberts, J. J. (1988). Measuring dimension of purchaser trust of industrial salespeople. *Journal of Personal Selling & Sales Management, VIII*(May), 1–9.

Szymanski, D. M. (1988). Determinants of selling effectiveness: The importance of declarative knowledge to the personal selling concept. *Journal of Marketing*, *52*(1), 64–77.

Tanofsky, R., Shepps, R. R., & O'Neill, P. J. (1969). Pattern analysis of biographical predictors of success as an insurance salesman. *Journal of Applied Psychology*, *53*(2), 136–139.

TIMS Management Systems (1993). *The personal profile analysis*. Tucson, AZ: TIMS Management Systems.

TIMS Management Systems (n.d.). *Personal profile analysis: A technical handbook*. Tucson, AZ: TIMS Management Systems.

Toftoy, C. N., & Shakalova, M. V. (1995). Meeting the transition head on. New ways of thinking. Washington, DC: The George Washington University School of Business and Public Management, Working Paper Series No. 95-12.

Tosi, H. L. (1966). The effects of expectation levels and role consensus on the buyer-seller dyad. *Journal of Business*, *39*(October), 516–529.

Tse, A. C. B. (1998). Comparing the response rate, response speed and response quality of two methods of sending questionnaires: E-mail vs. mail. *Journal of the Market Research Society*, *40*(4), 353–361.

Tse, D. K., & Wilton, P. C. (1988). Models of consumer satisfaction formation: An extension. *Journal of Marketing Research*, *XXV*(May), 204–212.

U.S. Census Bureau (2001). *Statistical abstract of the United States: 2001* (121st ed.). Washington, DC.

Vinson, D. E., Scott, J. E., & Lamont, L. M. (1977). The role of personal values in marketing and consumer behavior. *Journal of Marketing* (April), 44–50.

Warshaw, P. R., & Davis, F. D. (1985). Disentangling behavioral intention and behavioral expectation. *Journal of Experimental Social Psychology*, *21*, 213–228.

Watson, C. J., Billingsley, P., Croft, D. J., & Huntsberger, D. V. (1990). *Statistics for management and economics*. Boston: Allyn and Bacon.

Webster, F. E., Jr. (1965). Modeling the industrial buying process. *Journal of Marketing Research*, *II*(November), 370–376.

Webster, F. E., Jr. (1968). Interpersonal communication and salesman effectiveness. *Journal of Marketing*, *32*(July), 7–13.

Weible, R., & Wallace, J. (1998). Cyber research: The impact of the internet on data collection. *Marketing Research*, *10*(3), 19–24.

Weitz, B. A. (1978). Relationship between salesperson performance and understanding of customer decision making. *Journal of Marketing Research*, *XV*(November), 501–516.

Weitz, B. A. (1979). A critical review of personal selling research: The need for contingency approaches. In: G. Albaum & G. A. Churchill (Eds), *Critical Issues in Sales Management: State-of-the-Art and Future Research Needs* (pp. 76–126). University of Oregon.

Weitz, B. A. (1981a). Adaptive selling behavior for effective interpersonal influence. In: P. H. Reingen & A. G. Woodside (Eds), *Buyer-Seller Interactions: Empirical Research and Normative Issues* (pp. 115–123). Chicago: American Marketing Association.

Weitz, B. A. (1981b). Effectiveness in sales interactions: A contingency framework. *Journal of Marketing*, *45*, 85–103.

Weitz, B. A., Sujan, H., & Sujan, M. (1986). Knowledge, motivation, and adaptive behavior: A framework for improving selling effectiveness. *Journal of Marketing*, *50*(October), 174–191.

Westbrook, R. A., & Reilly, M. D. (1983). Value-percept disparity: An alternative to the disconfirmation of expectations theory of consumer satisfaction. In: R. P. Bagozzi & A. M. Tybout (Eds), *Advances in Consumer Research* (Vol. 3, pp. 256–261). Ann Arbor, MI: Association for Consumer Research.

Williams, K. C., & Spiro, R. L. (1985). Communication style in the salesperson-customer dyad. *Journal of Marketing Research, XXII*(November), 434–442.

Williams, K. C., Spiro, R. L., & Fine, L. M. (1990). The customer-salesperson dyad: An interaction/communication model and review. *Journal of Personal Selling and Sales Management, X*(Summer), 29–43.

Wilson, D. T., & Ghingold, M. (1981). Similarity-dissimilarity: A re-examination. In: P. H. Reingen & A. G. Woodside (Eds), *Buyer-Seller Interactions: Empirical Research and Normative Issues* (pp. 88–99). Chicago: American Marketing Association.

Woodruff, R. B., Cadotte, E. R., & Jenkins, R. L. (1983). Modeling consumer satisfaction processes using experience-based norms. *Journal of Marketing Research, 20*(August), 296–304.

Woodside, A. G., & Davenport, J. W., Jr. (1974). The effect of salesman similarity and expertise on consumer purchasing behavior. *Journal of Marketing Research, XI*(May), 198–202.

Woodside, A. G., Frey, L. L., & Daly, R. T. (1989). Linking service quality, customer satisfaction, and behavioral intention. *Journal of Health Care Marketing, 9*(December), 5–17.

Zeithaml, V. A., Berry, L. L., & Parasuraman, A. (1996). The behavioral consequences of service quality. *Journal of Marketing, 60*, 31–46.

TRUST AND BUSINESS-TO-BUSINESS E-COMMERCE COMMUNICATIONS AND PERFORMANCE

Pauline Ratnasingam

ABSTRACT

The emphasis on inter-organizational systems gave rise to concerns about inter-organizational relationships as trading partners became aware of the socio-political factors and trust that affect their relationships. This paper examines the importance of inter-organizational-trust in business-to-business E-commerce organizations. It examines how inter-organizational relationships impact trading partner trust, perceived benefits, perceived risks, and technology trust mechanisms in E-commerce that can in turn influence outcomes of business-to-business E-commerce. This paper develops a conceptual model and tests the model using a case study research methodology. The aim is to solicit qualitative in depth understanding of inter-organizational-trust in business-to-business E-commerce. Eight organizations from a cross section of industries that formed four bi-directional dyads participated in the third stage of this study. The first two stages include exploratory case studies in three organizations in the automotive industry that applied EDI via Value-Added-Networks in 1997, and a nationwide survey of organizations that examined the extent of E-commerce adoption in Australia and New Zealand in 1998. The findings identify the need for trustworthy business relationships in an E-commerce environment.

Evaluating Marketing Actions and Outcomes
Advances in Business Marketing and Purchasing, Volume 12, 359–434

ISSN: 1069-0964/doi:10.1016/S1069-0964(03)12006-6

INTRODUCTION

The prominence of trust in E-business has been widely touted by practitioners and academicians alike (Hart & Saunders, 1998; Heil, Bennis & Stephens, 2000; Jarvenpaa & Tractinsky, 1999; Keen, 2000). In particular, the motivation to use Internet-based information technology has revolutionized the way information is shared among organizations, resulting in radical transformations of organizational practices for procurement, deliveries, and financial transactions. E-commerce is changing both the technological, as well as relationship landscape of businesses. The current use of the Internet for business-to-business E-commerce participation is expected to grow at spectacular rates. E-commerce participation refers to the extent an organization has adopted, integrated and used E-commerce technologies and applications. Forrester research predicts that business-to-business (B2B) transactions will reach US$2.7 trillion in 2004 (Blackmon, 2000).

Kalakota and Robinson (2000) suggest that virtually every business today is stretched to the limit, while attempting to maintain viability and profitability in the face of unparalleled uncertainty and change. The proliferation of advanced E-commerce technologies and the lack of universal standards and policies to guide trading partners have left most organizations adopting E-commerce lacking the necessary knowledge and expertise (Miller & Shamsie, 1999; Raman, 1996; Storrosten, 1998). Indeed, uncertainties inherent in the current E-commerce environment give rise to a lack of trust in E-commerce relationships, thereby creating barriers to trade. For example, Parkhe (1998), and Ring and Van de Ven (1994) identified two types of uncertainties: (1) uncertainty regarding unknown future events; and (2) uncertainty regarding trading partners' responses to future events. Uncertainties reduce confidence both in the reliability of business-to-business transactions transmitted electronically and, more importantly, in the trading parties themselves.

A perceived lack of trust in E-commerce transactions on the Internet could be a reason for the slow adoption of E-commerce (Norlan & Norlan Institute-KPMG, 1999; Sabo, 1997; Storrosten, 1998). Many businesses perceive that E-commerce transactions on the Internet are insecure and unreliable. Despite the assurances of technological security mechanisms (such as encryption and authorization mechanisms, digital signatures, and certification authorities), trading partners in business-to-business E-commerce do not seem to trust the personnel involved in the transactions (CommerceNet, 1997; Fung & Lee, 1999; Marcella, Stone & Sampias, 1998; Ratnasingam, 1998). This is because the E-business environment is notably characterized by: (a) the impersonal nature of the online environment; (b) the extensive use of communication technology as opposed to face-to-face transactions; (c) the implicit uncertainty of using an open technical infrastructure for transactions; and (d) the newness of the electronic transaction medium. To manage these uncertainties and ensure future opportunities for improving coordination, organizations need to build trust relationships with their trading partners. According to Keen (1999), trust among trading partners is the currency of E-commerce. He notes "We are moving from an IT economy to a trust economy" (Keen, 1999, p. 1). Similarly, O'Hara-Deveraux and Johansen state that "trust is the glue of global work space and technology does not do much to create relationships" (1984, pp. 243–244).

The current age demands trust within organizations (Doney & Cannon, 1997; Dyer & Chu, 2000; Handy, 1995; Kramer & Tyler, 1996; Zaheer, Mc Evily & Perrone, 1998) and between organizations (Barney & Hansen, 1994; Bromiley & Cummings, 1995; Geyskens, Steenkamp & Kumar, 1998; Gulati, 1995; Moorman, Zaltman & Deshpande, 1993). Trust is important in exchange relations when uncertainty and risk are present (Coleman, 1990; Mayer, Davis & Schoorman, 1995). Trust influences trading partner behaviors (Schurr & Ozanne, 1985) and has been shown to be of high significance in uncertain environments such as Internet-based E-commerce environments (Fung & Lee, 1999). Trust, plays an important role in E-commerce participation and it is based on fair dealing, and a sense of reciprocity (Gulati, 1995; Jarvenpaa, Tractinsky & Vitale, 2000; Ring & Van de Ven, 1994). Trust has been related to desirable outcomes such as; firm performance, reduced conflicts, and opportunistic behaviors (Zaheer, Mc Evily & Perrone, 1998); competitive advantages (Barney & Hansen, 1994); satisfaction (Geyskens, Steenkamp & Kumar, 1998); and other favorable economic outcomes (Doney & Cannon, 1997). Researchers have argued that efficiencies within complex systems of coordinated actions are only possible when trading partners maintain effective collaborative relationships (Ganesan, 1994; Granovetter, 1985). Similarly, Morgan and Hunt (1994) suggest that to be an effective competitor (in the global economy), requires one to be a trusted cooperator (in some network). Indeed, Ganesan (1994) suggests that a retailer's trust on their vendors could affect long-term trading relationships in three ways. First, it reduces the perception of risks associated with opportunistic behaviors by the retailer. Second, it increases the confidence of the retailer that short-term inequities will be resolved over a long period; and finally it reduces the transaction costs in an exchange relationship. This line of reasoning is consistent with the findings of Infoweek research which suggests that information sharing and collaboration pays off (Infoweek, 2001a, b). Three hundred and seventy-five IT managers were interviewed and the findings confirm that organizations that share information with customers, suppliers, trading partners have the capacity to create collaborative chains, that contribute to their customer satisfaction, revenue and ultimately their profit margins.

The study of trust in business relationships in other disciplines is not new. Much prior analysis of trust was centered on traditional business exchanges that involved frequent face-to-face interactions (Ganesan, 1994; Kumar, 1996; Morgan & Hunt, 1994; Smith & Barclay, 1997). To date we lack systematic examination and developing a model of inter-organizational-trust for business-to-business E-commerce. Much of the literature on trust focuses on competitive advantages, and neglects factors such as trading partner interactions and trust behavior. Past research is also limited in the process involved in trust building, its impact on organizational and social context within which business-to-business E-commerce operations take place (Ring & Van de Ven, 1994; Rousseau, Sitkin, Burt & Camerer, 1998; Sako, 1998; Smeltzer, 1997). Research in Electronic Data Interchange (EDI) adoption suggests that interdependencies between trading partners can lead to an imbalance of power between the smaller suppliers and their more powerful buyers (Hart & Saunders, 1997; Helper, 1991; Kumar, 1996; Webster, 1995). It is in this environment of dual uncertainty that trust becomes important for business-to-business E-commerce.

Despite the growth of E-commerce for businesses, only limited research exists that explains how relationships and trust evolve between organizations (Sako, 1998; Smeltzer, 1997). This paper lights the importance of inter-organizational-trust in business-to-business E-commerce. Our primary objective is to identify successful trust behavior by examining how trust evolves from one stage to the next stage. The study is the first to formally examine a holistic model that incorporates the strengths and weaknesses of trust in dyadic relationships, within a business-to-business E-commerce context. In this study, inter-organizational-trust is understood as its impact on perceived benefits, perceived risks of E-commerce, thereby influencing outcomes in business-to-business E-commerce participation.

The core research question is: how does inter-organizational-trust impact perceived benefits, and perceived risks of E-commerce thereby influencing E-commerce, participation? In answering this question we first develop a theoretical framework from theories in multi-disciplines.

The rest of the paper is structured as follows. The next section reviews the relevant literature on trust, and justifies the research hypotheses relating to perceived benefits and perceived risks leading to outcomes in E-commerce participation. Drawing upon these discussions, we develop a research model with six research propositions. We then describe the research methodology, including procedures for data collection and analysis. The results of the multiple case studies are then reported. Using the results, we propose a model of inter-organizational-trust within bi-directional dyads in business-to-business E-commerce. The paper concludes by discussing the findings of this study, contributions, theoretical and managerial implications, limitations, and suggestions for future research.

LITERATURE REVIEW

The research model builds from theories in multi-disciplines that include; marketing, management, sociology, information systems, and E-commerce literature. The theoretical perspectives provides a focus on trust in business relationships, perceived benefits, perceived risks, technology trust mechanisms and outcomes in E-commerce participation that formed the constructs of the research model discussed in the next section.

Trust in Business Relationships

Trust is a concept that receives attention in several different social science literatures – psychology, sociology, political sciences, economics, marketing, anthropology, and history (see Anderson & Narus, 1990; Cummings & Bromiley, 1996; Doney & Cannon, 1997; Dwyer, Schurr & Oh, 1987; Gambetta, 1988; Ganesan, 1994; Lewicki & Bunker, 1996; Moorman, Deshpande & Zaltman, 1993; Williamson, 1975). The role of trust in business receives increasing attention from management researchers and practitioners' alike (Hosmer, 1995; Kramer & Tyler, 1996; Mayer, Davis & Schoorman, 1995; Zaheer, McEvily & Perrone, 1998).

Numerous researchers propose that trust is essential for interpersonal and group behavior, managerial effectiveness, economic exchanges, and social or political stability. Yet, according to a majority of these researchers, this concept is never defined precisely, as each discipline offers a unique insight (Doney & Cannon, 1997). A lack of clarity in the definition of trust has led to an overall picture of confusion, ambiguity, conflicting interpretations, and absence of reliable principles (Hosmer, 1995). In fact, no study has attempted to develop a complete and comprehensive theoretical framework of factors that influence trading partner trust in business-to-business E-commerce participation. Table 1 presents the definitions of trust from previous research that led to three types of trading partner trust applied in this study.

Characteristics in the Definition of Trust

The following characteristics of trust are derived from the definitions of trust in Table 1:

- A rational (or objective) view based on an economic perspective emphasizing the confidence in the predictability of one's expectations. The focus is on the credible and confident expectations that arise in relation to trading partners expertise, skills, reliability, and intentions (Anderson & Weitz, 1989; Barney &

Table 1. Definitions of Trust from Previous Research.

Authors Source	Discipline	Definition of Trust
Anderson and Narus (1990)	Marketing	A firm's belief that another company will perform actions that will result in positive outcomes for the firm, as well as not taking unexpected actions that would result in negative outcomes for the firm.
Barney and Hansen (1994)	Management	Mutual confidence that no party in an exchange will exploit one another's vulnerabilities.
Bromiley and Cummings (1992)	Management	Expectation that another individual or group will; (1) have good faith and make efforts to behave in accordance with any commitments, both explicit or implicit; (2) be honest in whatever negotiations preceding those commitments; and (3) not take excessive advantage of others even when the opportunity (to renegotiate) is available.
Deutsch (1958)	Sociology	Actions that increase one's vulnerability to the other.
Doney and Cannon (1997)	Psychology	Perceived credibility and benevolence of a target of trust.
Dyer and Chu (2000)	Management	One party's confidence that the other party in the exchange relationship will not exploit its vulnerabilities.
Fukuyama (1995)	Sociology	Exceptions that arise within a community of regular, honest and cooperative parties, based on commonly shared norms, on the part of other members of that community.
Gabarro (1987)	Management	Consistency of behavior such that judgement about trust in working relationships is based on the accumulation of interactions, specific incidents, problems, and events.
Gambetta (1988, p. 217)	Sociology	Probability that one economic actor will make decisions and take actions that will be beneficial or at least not detrimental to another.
Ganesan (1994)	Marketing	Willingness to rely on an exchange partner in confidence.
Hosmer (1995)	Management (Organizational behavior theory)	Expectation by one person, group, or firm upon voluntarily accepted duty on the part of another person, group, or firm to recognize and protect the rights and interests of all others engaged in a joint endeavor or economic exchange.
Keen (1999)	Information Systems and Management	Confidence in the business relationship. The definition is extended to include risk, and it focuses on the relationships that directly involve computers and telecommunications thus creating a trust bond (security, safety, honesty, consumer-protection laws, contracts, privacy, reputation, brand, mutual self-interest).
Kumar (1996)	Marketing and Management	Trust is stronger than fear. Partners that trust each other generate greater profits, serve customers better, and are more adaptable.

Table 1. *(Continued)*

Authors Source	Discipline	Definition of Trust
Lewicki and Bunker (1996)	Management and Sociology	A state involving confident, positive expectation about another's motives with respect to oneself in situations entailing risk.
Lewis and Weigert (1985)	Sociology	Undertaking of a risky course of action on the confident expectation that all persons involved in the action will act competently and dutifully.
Mayer, Davis and Schoorman (1995)	Management	Willingness of a party to be vulnerable to the actions of another party based on the expectation that the other will perform a particular action important to the trustor, irrespective of the ability to monitor or control that other party.
McAllister (1995)	Management	Cognition based on the concept that we choose whom we will trust, in what respects, and under what circumstances; affective foundations of trust consist of emotional bonds between trading partners.
Mishra (1996)	Management	A party's willingness to be vulnerable to another party based on the belief that the latter party is: (a) competent; (b) open; (c) concerned; and (d) reliable (Mishra, 1996).
Moorman, Deshpande and Zaltman (1993)	Marketing	Willingness to rely on an exchange partner with whom one has confidence. Also trust has been viewed as: (1) a belief, sentiment or expectation; and as (2) a behavioral intention that reflects reliance on trading partners and involves vulnerability and uncertainty on the part of the trustor.
Morgan and Hunt (1994)	Management	Trust exists when one party has confidence in an exchange partner's reliability and integrity.
O'Brien (1995)	Management	An expectation about the positive actions of other people, without being able to influence or monitor the outcome.
Ring and Van de Ven (1994)	Management	Trust as confidence implies: (a) the behavior of another will conform to one's expectation, and (b) the goodwill of another.
Sabel (1993, p. 1133)	Psychology	The mutual confidence that no party to an exchange will exploit the other's vulnerability. Trust is today widely regarded as a precondition for competitive success.
Sako (1998)	Sociology	An expectation held by an agent that its trading partner will behave in a mutually acceptable manner (including an expectation that neither party will exploit the other's vulnerabilities).
Schurr and Ozanne (1985, p. 940)	Marketing	The belief that a party's word or promise is reliable and a party will fulfill its obligations in an exchange relationship.
Zucker (1986, p. 50)	Sociology	A set of logical expectations shared by everyone involved in an economic exchange.

Hansen, 1994; Dwyer, Schurr & Oh, 1987; Ganesan, 1994; Moorman, Zaltman & Deshpande, 1992; Morgan & Hunt, 1994; Van de Ven, 1992). In E-commerce this view refers to organizational credibility (i.e. the extent to which a buyer believes that the supplier has the required expertise) and competence to perform the job effectively and reliably (Doney & Cannon, 1997).
- A relational (or subjective) view based on a social perspective emphasizing confidence in trading partner's goodwill. Goodwill between trading partners focuses on faith and moral integrity (Jones & George, 1998; Lewicki & Bunker, 1996; Mayer, Davis & Schoorman, 1995; McKnight, Cummings & Chervany, 1998). Researchers have also claimed that trust is a behavior reflecting reliance which involves risks, uncertainties, and vulnerabilities on the part of the trustor (Coleman, 1990; Das & Teng, 1996; Lewis & Weigert, 1985; Parkhe, 1998). Parkhe (1998) identified trust with an element of risk, uncertainty, and vulnerability. Parkhe (1998) was consistent with other researchers who claimed that trust involves risks taking (Chiles & McMackin, 1996; Coleman, 1990; Das & Teng, 1998; Deutsch, 1958; Koller, 1988; Lewis & Weigert, 1985; Sitkin & Pablo, 1992).

We adapt Mayer, Davis and Schoorman's (1995) definition of trust because in order for trust to occur there must be an element of risk and a positive outcome. Mayer et al. (1995, p. 712) defined trust as "*the willingness of a party to be vulnerable to actions of another party based on the expectation that the other will perform a particular action important to the trustor, irrespective of the ability to monitor or control that other party*." In the next section, we discuss trust behaviors and characteristics that contributed to three types of trust applied in this study.

Gabarro (1987) suggests that working and social relationships develop over time and can vary in stability, mutuality, and efficiency. His findings suggest that trust in open communications are seen as clear, consistent communication, and when trading partners keep their word (promise). This process is an "interpersonal contract" governed by a set of mutual expectations concerning performance, roles, trust, and influence. Gabarro's (1987) findings contribute to how trading partners communicate over a period of time.

Mayer, Davis and Schoorman (1995) suggest three types of trust behaviors that can help establish trustworthy relationships. They include: (1) ability, skills, competence of trading partners that enable the trustor to have some influence on the trustee; (2) integrity in the trustor's perception that occurs when the trustee adheres to a set of principles making them reliable and predictable (i.e. if they behave consistently); and (3) benevolence or wanting to do good to the trustor, aside from an egocentric profit motive. Benevolence suggests that the trustee has some specific attachment to the trustor.

On the other hand, McAllister (1995) focuses on two principal forms of interpersonal trust: (1) cognition-based trust (or rational trust) grounded in an individuals' belief about peer reliability and dependability; and (2) affect-based trust (or emotional trust) grounded in reciprocated interpersonal care and concerns or feelings of closeness and goodwill. Positive aspects of cognitive-based trust were seen from reliable, competent, and fair consistent behaviors of both trading partners. Over time these evolved into affective or emotional trust. Affective-based trust includes faith, concern, openness, encouragement, and information sharing between trading partners. McAllister's (1995) study contributes to cognitive and affective forms of trust. Mishra (1996) defines trust as one trading partner's willingness to be vulnerable to another trading partner, based on the belief that the latter party is competent, open, caring, and reliable. Mishra's findings contributed to dimensions of trust. Rousseau, Sitkin, Burt and Camerer (1998) referred to calculus trust and institutional trust for early stages of trust development, versus relational trust for later stages of trust development. These dimensions help to reinforce trust and build interest in sustaining the other organization as a trading partner over time.

Sako (1998) identifies three types of trust: (1) contractual trust which hinges on the other trading partner's ability to abide by contractual agreements. Contractual trust rests on shared norms of honesty, promise keeping, a shared understanding of professional conduct, and technical and managerial standards; (2) competence trust relies on the other trading partner's likelihood of following through with her or his promises. It is the ability of trading partners to adhere to the business operations; and (3) goodwill trust relies on the other trading partner's commitment to take initiatives for mutual benefit and refraining from unfair advantages. Goodwill trust can only exist if there is consensus on the principle of fairness. It is seen as highly cooperative (attempting to satisfy another trading partner's needs), and highly assertive (attempting to satisfy one's own needs). Sako's (1998) study contributed to three types of trust behaviors suggesting that trust develop in stages.

Similarly, McKnight and Chervany (2001) propose three constructs of trust; (1) dispositional trust that comes from trait psychology suggesting that actions are molded by certain childhood attributes that become more or less stable over time; (2) institutional trust that comes from sociology literature referring to behaviors that are situationally constructed (i.e. situation specific); and (3) interpersonal trust that relies on the party's trusting beliefs and trusting intentions. Based on the above characteristics of trust behaviors, we derived at three types of trading partner trust.

Three Types of Trading Partner Trust

Competence trust emphasizes trust in trading partners competence (i.e. their skills, technical knowledge, and ability) to operate business-to-business E-commerce applications correctly, and to do what they were supposed to do. Trading partners who demonstrate competence, and have the right skills and expertise in operating E-commerce systems, tend to contribute to high quality goods and services (as in timely delivery of accurate information) to other trading partners (Helper, 1991; Webster, 1995). Competence trust is based on an economic foundation and competitive advantage from savings in costs and time.

Predictability trust is an extension of competence-based trust and is based on a foundation of familiarity. Predictability trust emphasizes belief in trading partners' consistent behaviors that provide sufficient basis for other trading partners to make predictions and judgements based on past experiences. Hence, a series of positive consistent reliable behaviors make trading partners predictable and therefore trustworthy.

Goodwill trust develops from both competence and predictability trust. When reliability and dependability expectations are met, trust moves to affective foundations that include emotional bonds such as, care and concern. It is based on a foundation of empathy, and is characterized by an increased level of cooperation, open communication, information sharing, and commitment, that in turn increases participation in E-commerce.

Table 2 summarizes the trust behaviors and characteristics of the three types of trading partner trust applied in this study.

Technology Trust Mechanisms

Technology trust mechanisms in E-commerce refer to control safeguards and protection services that provide assurances and guarantees. For example, Tan and Thoen (1998) suggest the term "control trust" to refer to embedded protocols, policies, procedures in E-commerce that help to reduce the risk of opportunistic behaviors among consumers and Web retailers. Similarly, Lee and Turban (2001) measure trustworthiness of Internet shopping based on consumer evaluations of technical competence and Internet performance levels (such as speed, reliability and availability). Whereas the traditional notion of trust focused on trading partner relationships, trust in E-business also incorporates the notion of technology trust, which is defined as "the subjective probability by which organizations believe that the underlying technology infrastructure and control mechanisms are capable of facilitating transactions according to their confident expectations" (Ratnasingam & Pavlou, 2002). Within the context of E-commerce, formal governance mechanisms such as contracts, regulations, and policies generate technology mechanisms-based trust (also known as technology trust). Williamson (1993) refers this to governance mechanisms or institutional arrangements as "calculative trust." Thus, contracts and other governance mechanisms seen as

Table 2. Three Types of Trust in Business Relationships.

Trust Perspectives	1st Stage Competence Trading Partner Trust Economic Foundation	2nd Stage Predictability Trading Partner Trust Familiarity Foundation	3rd Stage Goodwill Trading Partner Trust Empathic Foundation
Gabarro (1987)	Character Role competence	Judgement	Motives/Intentions
Mayer, Davis and Schoorman (1995)	Ability	Integrity	Benevolence
McAllister (1995)	Cognitive	Transition from cognitive to affective based trust	Affective
Mishra (1996)	Competence	Reliability	Openness Care Concern
Rousseau et al. (1998)	Calculus institutional	Transition from calculus to relational trust	Relational
Sako (1998)	Contractual and competence trust	Transition from contractual, competence trust to relational trust	Relational
McKnight and Chervany (2001)	Dispositional and institutional based trust	Transition from dispositional, institutional based trust to interpersonal trust	Interpersonal

performance assessment criteria were used to measure the ability and competence of trading partners to conduct E-commerce.

McKnight and Chervany (2001), suggest "beliefs that the Internet has legal or regulatory protection for consumers (institution-based trust) should influence trust in a particular e-vendor (interpersonal trust)," (McKnight & Chervany, 2001, p. 13). Shapiro (1987) also describes institutional-based trust as the belief that a party has about the security of a situation because of guarantees, safety nets, and other structures. The sub-concepts of technology trust include:

Confidentiality mechanisms that reveal data only to authorized parties, who either have a legitimate need to know or have access to the system. Confidentiality in business-to-business E-commerce transactions is achieved by encrypting the messages. Alternatively, disclosure of transaction content may lead to the loss of confidentiality (privacy) of sensitive information whether accidentally or deliberately divulged onto an E-commerce network or an EDI mailbox storage system (Jamieson, 1996; Marcella, Stone & Sampias, 1998).

Integrity mechanisms provide assurance that E-commerce messages and transactions are complete, accurate, and unaltered (Bhimani, 1996; Jamieson,

1996; Parker, 1995). Unauthorized access to E-commerce systems can lead to the modification of messages or records of either trading partner. For example, errors in the processing and communication of E-commerce systems can result in the transmission of incorrect trading information or inaccurate reporting to management. Application and accounting controls are used to ensure accuracy, completeness, and authorization of inbound transactions from receipt to database update, and outbound transactions from generation to transmission. Accounting controls identify, assemble, analyze, classify, record, and report an organization's transactions that maintain accountability for the related assets and liabilities (EDICA, 1990; Marcella, Stone & Sampias, 1998).

Authentication mechanisms establish that trading partners are who they claim they are. Data origin authentication ensures that messages are received from a valid trading partner, and confirms that the trading partner is valid, true, genuine, and worthy of acceptance by reasons of conformity. Authentication requires that first, the sender can be sure that the message reaches the intended recipient, and only the intended recipient, and secondly, the recipient can be sure that the message came from the sender and not an imposter. The organization's security plan needs to include authentication procedures, as a lack of these could lead to valuable to prevent revealing sensitive information to competitors. Encryption mechanisms provide authentication features that provide security and audit reviews. These reviews ensure that E-commerce messages are received only from authorized trading partners (Gentry, 1994; Marcella, Stone & Sampias, 1998; Parker, 1995).

Non-repudiation mechanisms prevent the receiver or the originator of an E-commerce transactions' from denying that the transaction was received or sent. Non-repudiation of origin protects the message receiver against the sender denying the message was sent. Non-repudiation of receipt protects the message sender from the receiver denying that the message was received (Dosdale, 1994; Jamieson, 1996). For example, Ba and Pavlou (2001) suggest that credibility can be generated if appropriate feedback mechanisms on the Internet are implemented. Non-repudiation can also be achieved by using the Secure Functional Acknowledgment Message (FUNACK) protocol.

Availability mechanisms provide legitimate access to E-commerce systems and deliver information only to authorized trading partners, when required, without any interruptions. Service level agreements specify hours of operations, maximum down time, and response time to maintain the availability of E-commerce systems. Disruptions to E-commerce systems can come from both natural and man-made disasters. These could lead to system breakdowns and errors. Inadvertent or deliberate corruption of E-commerce applications could affect transactions, thereby impacting trading partner satisfaction, supplier relations, and perhaps business continuity. Availability issues are addressed by fault tolerance, duplication of communications links, and back-up systems that prevent denial of services to authorized trading partners (Bhimani, 1996).

Access controls provide authorization mechanisms to protect E-commerce messages against weaknesses (such as loss of messages) in the transmission media. They also protect sending trading partners against internal fraud or manipulation (Jamieson, 1996). Access controls are achieved by implementing a secure operating system and segregating crucial E-commerce functions (such as, inquiry, receipt, and payment) from unauthorized employees. In addition, implementing adequate and regular audit reviews and record retention procedures help to establish access controls (Marcella, Stone & Sampias, 1998).

Implementing best business practices help deter, prevent, detect and recover quickly from risks in E-commerce. Best business practices provide E-business security by monitoring suspicious activity and detecting intrusions. Best business practices include the extent and quality of top management commitment, enforcement of written policies, procedures, standards, contingency measures, risk analysis and management strategies. For example, Tallon, Kraemer and Gurbaxani (2000) argue that "management practices" have an important role in the process of IT strategies intent towards a firm's performance thus suggesting that best business practices can increase technology trust and ultimately E-commerce performance.

Perceived Benefits of E-commerce

Perceived benefits of E-commerce discussed in the research literature are remarkably similar (Dyer & Chu, 2000; Kalakota & Whinston, 1996; Nath, Akmanligil, Hjelm, Sakaguch & Schultz, 1998; Premkumar, Ramamurthy & Nilakanta, 1994; Raman, 1996; Senn, 2000; Sydow, 1998; Vijayasarathy & Robey, 1997; Zaheer, McEvily & Perrone, 1998). We identify three types of perceived benefits of E-commerce namely perceived economic, relationship-related and strategic benefits of E-commerce.

Perceived economic benefits arise from improvements and efficiencies in business processes, as a result of speed and automation of E-commerce applications. These benefits occur because transactions are sent electronically from one application to another. They include a reduction of transaction costs and administrative expenses, time-savings from a faster trading cycle, and improved accuracy because the receiving trading partner need not re-key the data (Fearson & Phillip, 1998; Iacovou, Benbasat & Dexter, 1995; Senn, 1999).

For example, Mukhopdyay, Kerke and Kalathur (1995), studied Chrysler assembly centers and identify that EDI use improves the quality of information exchanged and reduced inventory, transportation, and administrative costs. If properly configured, Internet applications can increase and enhance trading partners' productivity, and increase their capability to communicate globally. As more information and services are added to an organization's intranet, business decisions are made more quickly. Intranets contribute to reduced costs of distributing corporate information such as newsletters and memos. Extranets on the other hand offer richer capabilities for information transfers, open new revenue prospects and decrease costs and cycle times by providing real-time tracking and monitoring information. Small to medium-sized businesses can now leverage Internet systems to reduce paperwork and interface with their trading partners, suppliers, and customers in real-time by taking advantage of "leaner and meaner" extranets (Jevans, 1999; Riggins & Rhee, 1998; Senn, 2000). Extranet benefits include using familiar Internet tools and interface, increased communication between trading partners that allow for both internal, external communication, and real-time transaction recording. Transactions are duplicated across both trading partners databases, thus facilitating a high degree of information sharing and enabling decision makers to make more informed decisions.

In addition there is greater flexibility, and rapid customized product development as trading partners monitor E-commerce systems. An extranet also improves communications among customers, suppliers, and collaborators who find it effective in decreasing overheads and increasing revenue because it permits precise and effective management of service and products. Buyer's benefits arise primarily from structural characteristics such as availability of information, provision of search mechanisms, and online product trial, all of which can reduce uncertainty in the purchase decision. Businesses benefit from the potential of the web as a medium for marketing (Hoffman & Novak, 1995; Riggins & Rhee, 1999; Sharp, 1998).

Perceived relationship-related benefits refer to satisfaction that trading partners achieve in their trading partner relationships and from using E-commerce technologies. They include increased operational efficiencies, better customer service, improved inter-organizational relationships, and increased ability to remain in the competitive edge. E-commerce systems such as EDI, intranets, and extranets assist buyer-seller trading partner relationships and improve supplier reliability by improving delivery performance, ensuring an acceptable quality, and correct quantity of goods (Walton, 1997). Technical connections derived from E-commerce applications between organizations are often strong, but trading partners in E-commerce organizations play an even more important role in building social bonds. Continued and repeated transactions lead to stronger relationships between trading partners, thereby tying trading partners together economically, technically, and socially. Relationship-related benefits show the presence of effective coordination, communication and commitment among trading partners.

Coordination reflects the set of tasks one trading partner expects from another trading partner. Successful trading partner relationships are marked by coordinated actions directed at mutual objectives consistent with organizational interests (Mohr & Spekman, 1994). High levels of communication permit disclosure of an organization's needs, priorities, prices, delivery, and terms of agreement and it also enhances an organization's reputation as being fair and equitable (Schurr & Ozanne, 1985). In fact, Mohr and Spekman (1994) identified three aspects of communication that helps to improve coordination. First, communication quality, defined as accuracy, timeliness, adequacy, correctness, and credibility of information exchanged. Second, is the extent of critical and proprietary information shared between trading partners. By sharing information and by being knowledgeable about each other's business, trading partners are able to act independently in maintaining the relationship over time. The systematic availability of information allows people to complete their tasks more effectively and is associated with increased levels of satisfaction, which is an important predictor of a successful trading partner relationship. The third aspect is the extent of participation in planning and goal setting between trading partners. When one trading partner's actions influence the ability of the other to effectively compete, the need for participation in specifying roles, responsibilities and expectations increases (Mohr & Spekman, 1994). Thus, coordination leads to a smooth flow of communication and satisfaction among trading partners that in turn leads to stronger commitment.

Commitment is another relational benefit, and is characterized by the willingness of trading partners to exert effort on behalf of their relationship. A high level of commitment contributes to a situation where both trading parties can achieve individual and joint goals without raising their opportunistic behaviors. Thus, relationship-related benefits are derived from effective coordination, communication and commitment.

Perceived strategic benefits refer to the long-term gains an organization achieves from developing closer ties with their trading partners by using E-commerce to improve its competitive position. Perceived strategic benefits include a compressed business cycle, intensified relationships with trading partners and the development of corporate strategies (Fearson & Phillip, 1998). Yet, perceived strategic benefits are often unseen and difficult to quantify. E-commerce provides decision-making support where strategic use of information becomes available in a computer-usable format. Changes to business processes from E-commerce contribute directly operational improvements (Jamieson, 1996; Kalakota, 1996). For example, E-commerce allows for time-based competitive moves such as quick-response retailing, just-in-time manufacturing and close-to-zero inventories (Kalakota & Robinson, 2000). In addition, E-commerce helps to achieve the broader goals for improving an organization's image, strengthening its reputation, increasing long-term investments, and reaching new markets (Riggins & Rhee, 1998; Senn, 2000).

Perceived Risks of E-commerce

Perceived risks of E-commerce refer to weaknesses in E-commerce technologies, networks, E-commerce environment and mistrust among trading partner relationships that can increase business risks from the trading partners. Risk has been defined as "the possibility of an adverse outcome, and uncertainty over the occurrence, timing or magnitude of that adverse outcome" (Covello & Merkhofer, 1994). Perceived risks have been negatively associated with transaction intentions (Jarvenpaa, Tractinsky & Vitale, 2000), inter-organizational partnerships (Leverick & Cooper, 1998), and joint ventures (Gambrisch, 1993). The fact that the Internet was initially designed for scientific research to primarily share information and not to support business processes implies that no security mechanisms were included (Chellappa, 2001). The widespread use of E-commerce not only changes the way businesses are conducted, but has also introduced new risks that need to be addressed. The Internet emphasizes an open communication that has many inherent security flaws. For example, Internet-based EDI security is still an administrative nightmare with problems from eavesdropping, password sniffing, data modification, spoofing and repudiation (Bhimani, 1996; Drummond, 1995). Other E-commerce risks include snooping, misuse, theft, corruption of information, theft of identity and personal threats (Stewart, 1998).

E-commerce security risks can occur either internally or externally; can be from human or non-human sources that could be accidental or intentional. Risks can be caused by the disclosure, destruction, and modification of E-commerce transactions, or by denial of service attacks that lead to availability problems and the violation of confidential data. Ring and Van de Ven (1994) suggest various terms to describe risks such as technological, commercial, corporate risks. Similarly, Das and Teng (1996) used the term "performance risk" to account for the possibility, that objectives of inter-organizational relationships are not achieved although all partners cooperated. We identify three types of perceived risks of E-commerce namely technology performance-related risks, relational risks and general risks.

Perceived technology performance-related risks associated with access to E-commerce infrastructure and involve the hazards of not achieving the performance objectives of a trading partner relationship (Das & Teng, 1996; Lemos, 2001). Trading partners are subject to security attacks and intrusions by hackers. The security break-ins not only result in revenue losses for businesses, but also projects adverse perceptions of E-commerce security (Chellappa & Pavlou,

2002). Information transmitted may be vulnerable at various points including the trading partner's in-house applications, interface, translation software, network connection, or communication management as well as the carrier's network and mailbox services. This is because E-commerce systems do not operate unilaterally, and networks that connect trading partners are often shared networks. Cross-vulnerabilities that exist between interdependent trading partners in an E-commerce network can put organizations at risk due to the "domino effect" of one trading partner's security failure comprising the integrity of the other trading partner's system (Jamieson, 1996; Marcella, Stone & Sampias, 1998).

Threats to network security come from the Internet and from an organization's internal networks. Messages on internal networks may be intercepted due to improper configuration, overly restrictive access controls, or failure to closely monitor the network traffic. It is estimated that eighty to ninety-five percent of the total number of security incidents are due to insider attacks (Brensen, 1996; Chan & Davis, 2000). Security techniques in existing software and hardware cannot completely assure security. Hence, internal networks connected to the Internet become exposed to outside intruders thus contributing to technology performance-related risks.

Perceived relational risks closely relate to risks derived from mistrust (Williamson, 1993). Relational risks develop from a lack of experience and technical knowledge about security, concerns about the auditability of E-commerce, task uncertainties, environment uncertainties, false impressions of unreliability, and concerns about the enforceability of transaction records in the electronic trade area. The automation of E-commerce systems where transactions are processed at high speed and volumes, has led to reduced opportunities to spot problems using human intuition (ICAEW, 1992). Thus, timely resolution of errors or problems may be hampered by the potential loss of audit trails. This could make reconstruction and reconciliation of records difficult (Bhimani, 1996; EDICA, 1990; Jamieson, 1996). Further, coercive power is often exercised when trading partners lack cooperation (Ratnasingam, 2000). For example, trading partner A is likely to use coercive power when trading partner B, does not cooperate. Power can focus on the control of an organization's critical or strategic activities. Implicit in this issue is the concept of organizational inter-dependence (Saunders, 1990). Hart and Saunders (1997) based on a study of power and trust in EDI adoption, suggest that organizations with greater power can influence their trading partners to adopt EDI. Their findings indicate that use of power was negatively related to the volume of EDI transactions. While electronic networks may facilitate easier exchanges, they may not necessarily lead to increases in the frequency of business-to-business transactions. Thus, power exists on two levels: (1) as a motive; and (2) as a behavior.

Relational risks such as production delays, disrupted cash flows, legal liability and loss of profitability can affect anticipated cost savings and business continuity (Hart & Saunders, 1998; Jamieson, 1996; Marcella, Stone & Sampias, 1998).

Perceived general risks derive primarily from poor business practices. They arise in the following ways.

- Eavesdropping attacks on a network that can result in the theft of information such as account balances and billing information;
- Password *sniffing* attacks to gain access to a system containing proprietary graphic algorithms;
- Data modification attacks used to modify the contents of certain transactions;
- Spoofing when unauthorized personnel masquerade as another party. In one such situation, a criminal can set up a storefront and collect thousands or even millions of account numbers or other information from unsuspecting consumers, and;
- Repudiation which can cause major problems with billing systems and transaction processing agreements.

Even if it is feasible to objectively measure the degree of security inherent in every transaction (CERT, 2000; Miyazaki & Fernandez, 2000), it is unclear whether this measurement would readily correspond into the exact trading partner perception about the security of their business transactions.

Outcomes of E-commerce Adoption

The extent of E-commerce adoption in this study is measured by the extent to which an organization has adopted, integrated, and used E-commerce. It is examined from two perspectives; first, the extent of E-commerce performance and second, the extent of trading partner relationship trust development.

Extent of E-commerce performance refers to the extent to which both trading partners perceive their relationship to be effective in realizing performance objectives. E-commerce performance is measured as a percentage of an organization's business using E-commerce, volume, and dollar value of the transactions. It is the mutually held trading partner perceptions and agreement of their sales performance and satisfaction.

For example, Massetti and Zmud (1995) classified EDI adoption and diffusion into volume (number of document exchanges using EDI) diversity (number of distinct document types a company handles with its trading partners), breadth (number of established connections with external trading partners), and depth (degree of electronic consolidation between two or more trading partners).

Bensaou and Venkatraman (1996) suggest the importance of the "fit" between trust in the trading partners and the socio-political climate, which impacts the performance of inter-organizational-systems. Similarly, electronic partnerships demand a conducive transaction climate and trust among trading partners (Anderson & Narus, 1990; Smith & Barclay, 1997). Hence, perceived E-commerce performance can be modeled as a predictor of satisfaction. Trading partners are likely to be more satisfied with relationships that are effective.

The extent of the development of trading partner trust relationships refers to the extent to which both trading partners in a relationship are satisfied. The development of trust in trading partner relationships is measured by the intensity of communications, cooperation, and commitment. Trading partner satisfaction in E-commerce participation exists when trading partners are engaged in long-term business investments, thereby leading to business continuity and improved reputations.

In E-commerce, trading partner trust relationship development is based on:

- The degree of initial success each trading party had experienced in E-commerce participation;
- Well defined roles for all trading parties;
- Realistic expectations; and
- A well designed trading partner agreement, used as a guideline and not as a source of pressure.

Table 3 defines the constructs and sub-concepts of technology trust mechanisms, perceived benefits of E-commerce, perceived risks of E-commerce, and outcomes in E-commerce participation.

Figure 1 presents the theoretical framework of inter-organizational-trust in E-commerce participation. The arrows in the theoretical framework demonstrate the cause and effect of trading partner relationships. The research question designed for this study is: *How does inter-organizational-trust impacts perceived benefits and perceived risks of E-commerce thereby influencing outcomes in E-commerce participation?* In the next section, we justify the research propositions shown in the theoretical framework.

The Relationship Between Trading Partner Trust and Perceived Benefits of E-commerce

The proposed link between trading partner trust and perceived benefits is popular among scholars who examined trust in business relationships and organizations (Barney & Hansen, 1994; Cummings & Bromiley, 1996; Doney & Cannon, 1997;

Table 3. Definitions of the Constructs and Sub-concepts in the Conceptual Model.

Constructs and Sub-concepts	Definition
Trust in trading partners	Trust behaviors that determine competence, predictability, and goodwill types trading partner trust.
Competence trust	Reliance upon the ability, skills, knowledge, and competence of a trading partner to perform business-to-business E-commerce correctly and completely.
Predictability trust	Reliance upon consistent behaviors of trading partner that allows another trading partner to make predictions and judgments due to prior experiences.
Goodwill trust	Reliance upon the care, concern, honesty, and benevolence shown by trading partners that allow other trading partners to further invest in their trading partner relationships.
Technology trust mechanisms in E-commerce	Technology trust mechanisms in E-commerce are trust assurances, security safeguards and protection services provided by E-commerce technology, organizations, human and third party services.
Confidentiality	Protection of E-commerce transactions and message content against unauthorized reading, copying, or disclosure.
Integrity	Accuracy and assurance that E-commerce transactions have not been altered or deleted.
Authentication	Quality of being authoritative, valid, true, genuine, worthy of acceptance or belief by reason of conformity to the fact that reality is present.
Non-repudiation	Originator of E-commerce transactions cannot deny receiving or sending that transaction.
Access controls	Protection of E-commerce transactions against weaknesses in the transmission media and protection of the sender against internal fraud or manipulation.
Availability	Assurance that passes or conveys E-commerce transactions without interruption by providing authorized users with E-commerce systems.
Best business practices	Policies, procedures and standards that ensure smooth functioning of E-commerce.
Perceived benefits of E-commerce	Perceived benefits of E-commerce are gains received by organizations that have adopted E-commerce.
Perceived economic benefits of E-commerce	Benefits derived from direct savings in costs and time.
Perceived relationship related benefits of E-commerce	Benefits derived from closer trading partner relationship such as: open communications, information sharing, cooperation, and commitment.
Perceived strategic benefits of E-commerce	Benefits derived from long-term business investments and improved reputation of trading partners.

Table 3. *(Continued)*

Constructs and Sub-concepts	Definition
Perceived risks of E-commerce	Perceived risks of E-commerce are the potential weakness, barriers, and losses faced by organizations that have adopted E-commerce.
Perceived technology performance related risks of E-commerce	Risks derived from misuse of E-commerce technologies, viruses, and lack of confidentiality, integrity, unauthorized access, and availability mechanisms.
Perceived relational risks of E-commerce	Risks derived from trading partner's lack of knowledge and training in E-commerce.
Perceived general risks of E-commerce	Risks derived from poor business practices, environmental risks, lack of standards, and audit policies.
Outcomes in E-commerce participation	Extent an E-commerce organization engages in business-to-business E-commerce participation.
Extent of E-commerce performance	Extent of volume, dollar value, and types of business transactions exchanged between trading partners.
Extent of trading partner trust relationship development	Extent of positive affective state derived from all aspects of a trading partner relationship.

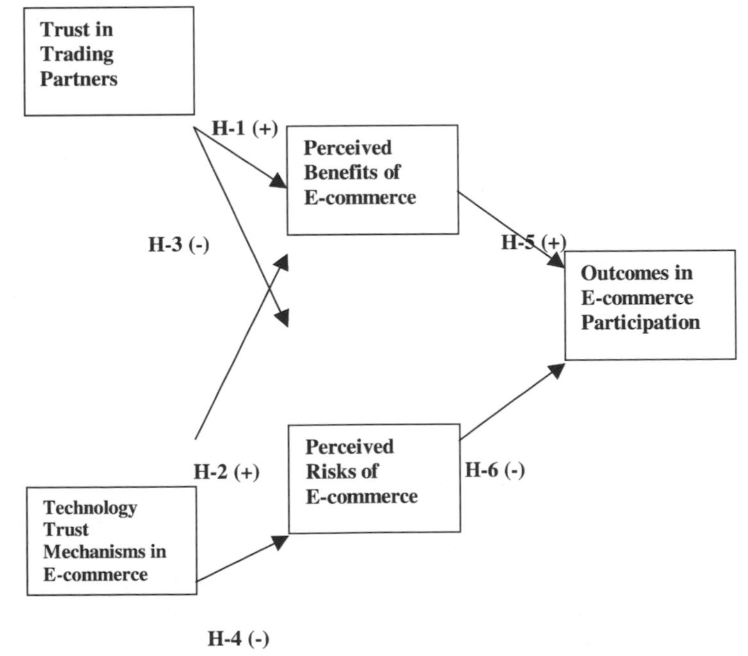

Fig. 1. Theoretical Framework of Inter-Organizational-Trust in E-commerce Participation.

Morgan & Hunt, 1994; Ring & Van de Ven, 1994; Smith & Barclay, 1997; Zaheer, McEvily & Perrone, 1998).

Competence trust demonstrates a trading partner's ability and skills to operate business-to-business E-commerce applications. Consistent positive role competence contributes to the development of credibility among trading partners. When trading partners send and receive E-commerce transactions correctly, and in a timely manner, they contribute to economic benefits derived from time and cost savings and from avoiding re-training or requesting trading partners to re-send the same transaction correctly. Economic benefits lead to direct savings in costs and time, from the efficiencies and automation of E-commerce and applications. They reduce the need for manual processing and contribute to cost savings from less administrative labor costs, time savings from speed and automation, and simplicity from structured standard business processes that reduce and eliminate uncertainties (Riggins & Rhee, 1998; Senn, 2000).

Credibility from a series of consistent positive competence trust promotes reliability and willingness of trading partners to trust each other, and value their relationship, thereby making them predictable and trustworthy. Predictability trust in turn increases trading partners' confidence and encourages them to share information (i.e. timely, accurate, honest, and relevant information). Trading partners who receive timely information are able to make informed decisions, and are more likely to be satisfied in their relationships. Trading partners exhibit collaboration, cooperation, communication openness, and satisfaction. Morgan and Hunt (1994) conducted a study of distribution channels and found that high levels of cooperation among trading partners contributed to satisfaction. Cooperation reduces conflict, increases communication and trading partner satisfaction and capitalizes on relationship-related benefits (Anderson & Narus, 1990; Doney & Cannon, 1997; Ganesan, 1994; Morgan & Hunt, 1994). Satisfaction, in turn, reinforces trading partner trust and reduces opportunistic behaviors such as (cheating, lying or giving inaccurate information). Perceived relationship-related benefits are thus derived from closer trading partner relationships, cooperation and satisfaction (Berry, 1999; Berry & Paramsuraman, 1991).

As predictability trust gradually develops, trading partners experience strategic benefits (seen from improved reputation and image of their organizations). Empirical evidence supports the link between suppliers' reputation and buyers' trust (Doney & Cannon, 1997; Ganesan, 1994; Hill, 1990). For example, in a study of industrial channel dyads, a retailer's favorable perception of their vendor's reputation led to increased credibility and goodwill trust (Ganesan, 1994). Goodwill trust leads to long-term trading partner relationships, because it shifts the focus to future conditions and encourages trading partners to increase investments and contribute to strategic benefits in E-commerce participation

(Anderson & Weitz, 1989; Morgan & Hunt, 1994). Perceived strategic benefits arise from positive images and reputations a trading partner receives from the satisfaction in receiving products and services from other trading partners. Examples of perceived strategic benefits include; closer ties, bigger investments (as manifested in increased volume, dollar value, and types of E-commerce transactions) cooperation, and commitment. Therefore we propose that:

> **RP-1.** Trading partner trust relates positively with perceived benefits of E-commerce.

The Relationship Between Technology Trust Mechanisms in E-commerce and Perceived Benefits of E-commerce

The proposed link between technology trust mechanisms in E-commerce and its impact on perceived benefits is popular among previous researchers who examined security services in E-commerce, and governance mechanisms as a form of establishing initial trust among trading partners (Jamieson, 1996; Marcella, Stone & Sampias, 1998; McKnight & Chervany, 2001; Tan & Thoen, 1998).

Technology trust includes digital signatures, encryption mechanisms (public key infrastructure), authorization mechanisms (User IDs and passwords), and best business practices that enforce regular audit, top management commitment, standards, and contingency procedures (Bhimani, 1996; Cavalli, 1995; Jamieson, 1996; Parker, 1995; Ratnasingam & Pavlou, 2003). Technology trust mechanisms prevent transactions from being intercepted, manipulated, or deleted thus contributing to accurate, complete, correct, and timely business transactions. This, in turn, leads to economic benefits from savings in time and costs (Riggins & Rhee, 1998; Senn, 2000). For example, Mukhopadyay, Kekre and Kalathur (1995), conducted a study of nine Chrysler assembly centers and found that EDI improved the quality of information exchanged, and reduced inventory, transportation, and administrative costs. Their findings indicate that EDI use had helped Chrysler to realize a benefit of over $100 per vehicle, thus amounting to annual savings of $200 million. Therefore, the speed and automation from E-commerce technologies enable trading partners to not only achieve accurate, timely information, but also to increase their productivity and profitability (Nath, Akmanligil, Hjelm, Sakaguch & Schultz, 1998; Premkumar, Ramamurthy & Crum, 1997; Senn, 2000).

Similarly, functional acknowledgments in the form of e-mail feedbacks or other E-commerce protocols provide reliable and timely feedback mechanisms (non-repudiation mechanisms) that increase trading partners confidence, reliability, and dependability thus contributing to relationship-related benefits.

For example, trading partners use reference (sequence) numbers that identify where the transactions came from, thus contributing to accurate, complete and correct information (i.e. integrity mechanisms). E-commerce applications enable product and service differentiation and establish tighter links with suppliers, distributors and customers (trading partners). Consequently, both trading partners and customers achieve relationship-related benefits such as satisfaction and from high quality products and services.

The existence of top management commitment encourage trading partners to abide by best business practices (Tallon, Kraemer & Gurbaxani, 2000). This strengthens trading partners' reputation for exercising high standards, quality, and fairness that in turn increases trading partner satisfaction (Cannon & Perrault, 1999; Geyskens, Steenkamp & Kumar, 1998), and contributes to strategic benefits. Therefore we propose that:

RP-2. Technology trust mechanisms in E-commerce associate positively with perceived benefits of E-commerce.

The Relationship Between Trading Partner Trust and Perceived Risks of E-commerce

The literature on trust suggests that regardless of the analysis level, trading partners remain vulnerable to some extent (Coleman, 1990; Doney & Cannon, 1997; Williamson, 1975). Perceived risks have been negatively associated with transaction intentions (Jarvenpaa & Tractinsky, 1999), interfirm partnerships and joint ventures (Leverick & Cooper, 1998). Williamson's (1975), transaction cost theory rests on the assumption that organizations must act as if trust is absent, as information asymmetry may give rise to opportunistic behaviors and uncertainties. Cummings and Bromiley (1996) suggest that organizations engage in a situation of risk, although organizational actors behave on good faith effort, and are honest in electronic exchanges.

Interdependencies between trading partners create task uncertainties that contribute to technology performance-related risks derived from incompatible systems, and a lack of knowledge, skills, training and ability to transact correctly – i.e. a lack of competence trust (Das & Teng, 1996; Ratnasingam, 2000; Ring & Van de Ven, 1994). Similarly a situation of imbalance of power between trading parties may permit one trading partner to exercise coercive power and exhibit opportunistic behaviors with respect to the other (Hart & Saunders, 1997; Helper, 1991; Hill, 1990). It can create mistrust among trading partners that inhibit cooperation and encourages conflicts among trading partners, thus contributing to a lack of predictability trust. Examples of coercive power used by powerful buyers in the automotive industry include; slow delivery of vehicles to their distributors, slow payment on warranty work, unfair distribution of vehicles, threat of termination and bureaucratic red tape (Langfield-Smith & Greenwood, 1998; Webster, 1995). This practice has been associated in EDI adoption with the catch phrase "*EDI or die*," meaning that trading partners are required to implement EDI systems or customers will not trade with them at all (Helper, 1991; Langfield-Smith & Greenwood, 1998; Webster, 1995). A dissatisfied trading partner will be suspicious of the other trading partner's intentions and motives and will demonstrate reluctance to share and engage in open communications, thus contributing to perceived relational risks (Das & Teng, 1996; Ring & Van de Ven, 1994). Hill (1990) suggests that a lack of trust affects the reputation of trading partners, and has an economic cost. In the long run, this leads to fewer opportunities for smaller trading partners to develop their knowledge and expertise of E-commerce use thus contributing to relational and general risks from business discontinuity (from poor business practices). Trading partners who have a reputation for terminating relationships, and seeking high profits provide a signal to their retailers that they are solely interested in their own profits thus contributing to relational risks. Opportunistic behaviors such as (cheating, lying, and blaming) may yield short-term benefits, but there are long-term costs from a lack of goodwill trust among trading partners, that may inhibit future acquisitions of cost-reducing and/or quality enhancing assets, thus contributing to perceived general risks (Dwyer, Schurr & Oh, 1987; Ganesan, 1994; Hosmer, 1995; Kumar, 1996; Smith & Barclay, 1997). Therefore, we propose that:

> **RP-3.** Trading partner trust relates negatively with perceived risks of E-commerce.

The Relationship Between Technology Trust Mechanisms in E-commerce and Perceived Risks of E-commerce

In online business-to-business marketplaces, the risks Internet creates through identity and product uncertainty, physical separation, and the newness of the medium has been attributed as an important barrier to online transactions (Bakos, 1998). Lack of technology trust mechanisms in E-commerce may introduce vulnerabilities that make E-commerce a risky course of action (Hart & Saunders, 1997; Marcella, Stone & Sampias, 1998). For example, unauthorized access to E-commerce systems provide increased opportunities for malicious parties to modify the records of single organizations or of their trading partners, thus leading to integrity issues (Parker, 1995). Trading partners may act on those messages assuming that they came from a genuine authorized trading partner. For example, a denial of service attack can occur when a malicious party (internal or external) cripples the network server's ability to respond to requests (usually by flooding the server with many requests). Similarly, an organization may encounter software failure when their transactions were deliberately infected by viruses. These threats contribute to technology performance-related risks.

Relational risks can occur from uncertainties in the E-commerce environment and mistrust among trading partners. For example a trading partner may have acted on the information transmitted by a malicious hacker only to find out later that his/her privacy was invaded. This could in turn lead to conflicts, imbalance of power, thus leading to relational risks. Similarly, poor business practices derived from a lack of proper training, inadequate audit and back up retention policies and procedures, or poor contingency planning procedures may lead to general risks. For example, the costs associated due to unavailable servers can severely impact businesses if real-time transactions were involved. Inadvertent or deliberate corruption of E-commerce records could impact trading partner satisfaction and perhaps their ultimate business continuity. Therefore, we propose that:

RP-4. Technology trust mechanisms in E-commerce relate negatively with perceived risks of E-commerce.

The Relationship Between Perceived Benefits and Outcomes in E-commerce Participation

Past research suggest that perceived benefits of E-commerce contribute to increased outcomes in E-commerce. E-commerce technologies provide speed and automation of business processes that reduce transactions and administrative costs thus contributing to economic benefits. Similarly, trading partners who exhibit competence trust are able to provide real-time accurate and complete information to their buyers who can log into the supplier's extranet application, track shipment details, and estimate arrival dates of goods they ordered (Riggins & Rhee, 1998; Senn, 2000). Buyers are able to satisfy their suppliers' needs by delivering the goods on time. This, in turn, contributes to information sharing and open communications that lead to relationship-related benefits from trading partner satisfaction. For example, Vijayasarathy and Robey (1997) examined EDI adoption found that EDI transforms relationships' between organizations, and that economic advantages of EDI are likely to depend on EDI's impact on trading partner relationships and trust.

Increased satisfaction from E-commerce performance in turn contributes to high levels of predictability and goodwill trading partner trust. A series of consistent, high-quality services lead to strategic benefits such as long-term investments (as in increased volume, diversity, and dollar value of E-commerce transactions), and increased reputation of organizations (Anderson & Narus, 1990; Doney & Cannon, 1997; Kumar, 1996; Morgan & Hunt, 1994; Smith & Barclay, 1997). For example, evidence of such trust-based performance was found by comparing supplier relationships in the automotive industry in Japan and the United States. It was found that Toyota's relationship with their suppliers was deeply embedded in long-standing networks of social and economic relations characterized by high levels of goodwill trust thus contributing to perceived economic, relationship-related and strategic benefits (Barney & Hansen, 1994). Therefore we propose that:

RP-5. Perceived benefits of E-commerce relate positively with outcomes in E-commerce participation.

Perceived Risks and Outcomes in E-commerce Participation

Trading partners may be operating with incompatible systems that lacked trusted mechanisms. This may give rise to technology performance-related risks derived from incompatible infrastructure and loss in the transmission media from a lack of competence trading partner trust (Jamieson, 1996; Marcella, Stone & Sampias, 1998; Parker, 1995).

Trading partners (as in suppliers) make sacrifices and show concern for other trading partners (who are their retailers and manufacturers) develop a reputation for fairness within their industry. Consistent behaviors from trading partners develop predictability trading partner trust and provide signals of their future actions. On the other hand, trading partners who have a reputation for terminating relationships and seeking high profits, provide signals to their retailers that they are solely interested in their own profits and contribute to relational risks derived from coercive power and opportunistic behaviors (Helper, 1991; Langfield-Smith & Greenwood, 1998; Webster, 1995).

Similarly, trading partners who found themselves participating in inequitable relationships felt angry and resentful. Such feelings of dissatisfaction may result in suspicious opportunistic behaviors, mistrust and conflicts. Trading partners may view each other as untrustworthy and exploitative. Consequently, trading partners will not trust each other and will not commit to long-term trading partner relationships. This inhibits participation in E-commerce and contributes to relational risks and general risks from poor business practices such as (a lack of top management commitment, audit, quality standards and risk management strategies) (Dwyer, Schurr & Oh, 1987; Ganesan, 1994; Smith & Barclay, 1997). Therefore, we propose that:

RP-6. Perceived risks in E-commerce relate negatively with participation in E-commerce.

The next section describes the research method used to test the theoretical framework.

RESEARCH METHOD

Studying inter-organizational-trust in business-to-business E-commerce participation requires the context of real organizations using E-commerce systems or operating in an E-commerce environment. Yin (1994) suggests that a case study research approach enables an understanding of the nature and complexity of processes occurring, by answering *how* and *why* research questions. Given the newness of the concept of trust in business-to-business E-commerce context, initial exploratory studies were conducted in two stages. The first stage refers to in-depth case studies in three organizations in the automotive industry in 1997. The organizations played the role of buyers and suppliers using EDI via value-added networks. The findings implied the importance of building trustworthy business relationships, and the impact of power among the buyers. Furthermore governance mechanisms in the form of trading partner contracts are important as trading partners rely heavily on written policies and procedures. A second study was conducted using a mail survey sponsored by Norlan and Norlan Institute, KPMG. The aim of this study was to examine the extent of E-commerce adoption in Australian and New Zealand. The findings explicitly indicated that Australia and New Zealand is two to three years behind that of the U.S. and Europe. The main reasons were a lack of trust, perceived risks of the Internet and a lack of technical knowledge and expertise. The third study discussed in this paper examined eight organizations that formed four bi-directional dyads using qualitative multiple case studies.

Justification for a Case Study Research Method

This study uses multiple case studies as it is appropriate for testing the research propositions (see Yin, 1994). Yin suggests that, "multiple case designs have distinct advantages and disadvantages in comparison to single case designs . . . a major insight is to consider multiple cases as one would consider multiple experiments – that is, to follow a 'replication' logic. This is far different from a mistaken analogy in the past, which incorrectly considers multiple cases to be similar to the multiple respondents in a survey (or to the multiple subjects within an experiment) – that is to follow a sampling logic" (Yin, 1989, p. 51).

Similarly, Benbasat, Goldstein and Mead (1987), provided a clear rationale for using multiple case studies. Multiple case designs are desirable when the intent of the research is descriptive, theory building, or theory testing . . . Multiple case designs allow for cross case analysis and extension of theory. Of course, multiple cases yield more general research results (Benbasat, Goldstein & Mead, 1987, p. 373).

Multiple case studies were chosen because they allow an in-depth analysis of the concept in a real life situation, thus enabling trust behaviors and trading partner interactions to be observed. The organizations consisted of both large and small-medium-enterprises, public and private sector organizations. Some used simple applications (i.e. electronic applications without the Web) while others used more Web-based electronic applications and E-commerce technologies. This paved the way for a cross-case analysis of the findings that in turn contributed to meaningful generalizations. Case studies allowed us to examine various aspects of trading partner trust in an E-commerce environment (i.e. within a natural setting of organizations), and enabled extensive data collection from different sources (triangulation).

Data Collection

Data collection proceeded in the following manner. The questionnaire was first pre-tested with a group of academics and E-commerce practitioners using e-mail and telephone interviews. The aim of pre-testing the semi-structured questionnaire was to refine the jargon (commercial language) to suit case sites, and to ensure that we have covered the questions adequately. A semi-structured questionnaire comprising questions for the constructs in the conceptual model was designed to collect rich qualitative data. A copy of the case study questionnaire is enclosed in the appendix. The unit of analysis in this study is an organization (a directional-dyad). Eight organizations that formed four bi-directional dyads (i.e. two organizations) participated in this study.

Multiple sources for collecting data were applied. First, semi-structured interviews lasting between 1 and 3 hours per visit and between 4 and 6 sessions were conducted at each organization. Where respondents were too busy for face-to-face interviews, telephone interviews (lasting between 40 and 60 minutes) were conducted, in order to clarify their responses received earlier through an

Table 4. Characteristics of the Case Study Participants.

Case Study Organizations	Job Title of the Participants	Age Range	Extent of E-commerce Experience
NZ Customs	Intranet Administrators (4)	40–55	10 years
NZ Customs	Sales Consultants (3)	30–50	7–10 years
NZ Customs	Security Analysts (3)	40–55	8–10 years
Internet service provider	Director (1)	40–55	15 years
Internet service provider	Accountant (1)	35–45	9 years
Customs broker	Chief Executive Officer (1)	40–60	10 years
Customs broker	Administrator (1)	25–40	5 years
Importer	Accountant (1)	30–40	7 years
Importer	Inventory Manager (1)	29–45	8 years
Cisco	Accounting Manager (1)	40–50	9 years
Cisco	Sales Consultants (4)	40–50	6 years
Cisco	Internet Business Solutions Group Administrators (4)	35–45	5 years
Cisco	E-commerce Coordinators (1)	35–45	6 years
Compaq	Network Sales Specialist (1)	45–55	9 years
Compaq	Sales Consultants (2)	50–65	9 years
Compaq	Inventory Manager (1)	45–55	9 years
Siemens	Key Accounting Manager (1)	40–50	6 years
Siemens	Marketing Manager (1)	45–55	9 years
Siemens	Network Manager (1)	35–45	6 years
Siemens	Customer Service Manager (1)	20–29	2 years
Telecom	Accounting Manager (1)	40–50	9 years
Telecom	Inventory Manager (1)	35–45	6 years

e-mail questionnaire. Observations took the form of sitting with participants when they were on and off the phones, at meetings and taking notes of their work practices. In addition, document analysis involved, analyzing internal policies, E-commerce implementation reports, trading partner agreements, and security procedures for additional information on trading partner interactions. Table 4 presents the characteristics of the case study participants.

Data Analysis

The criteria for interpreting the findings were carried out via pattern matching, and explanation building (as in narrative descriptions and causal explanations). Multiple sources of data permitted cross checking and multiple perspectives on important issues that help maintain a chain of evidence and increase the reliability of the data collected. Triangulation was used to establish rigor, and reflect an in-depth understanding of inter-organizational-trust in E-commerce participation. The data collected was mapped, cross-referenced and compared with the predictions. Whenever a difference in the participant's opinions occurred, further explanations for their differences were sought and examined. The first step identified recurring patterns and themes from the data collected during interviews. Then the similarities and differences between the participants were identified. This was followed by a cross-case analysis, which paved the way for analytic generalizations and predictions. For example, the findings from competence trust were matched against economic benefits, and technology performance-related risks to see if they correlated. Similarly, the findings from trading partner predictability and goodwill trust were matched against the findings of perceived relationship-related benefits and relational risks. Internal validity was achieved by applying credibility, and seeking participants to describe their responses.

FINDINGS

This section describes the background information of the eight organizations that formed four bi-directional dyads (BDDs). The organizations were selected from a cross-section of industries that included, NZ Customs, their Internet service provider, a customs broker and importer involved in customs clearance. It was then followed by four organizations in the computer, data communications and telecommunications industries.

Background Information of Bi-directional Dyad-A (BDD-A)

Bi-directional dyad (BDD-A) is between NZ Customs and their Internet service provider (ISP). NZ Customs outsource part of their business-to-business E-commerce process to the ISP. NZ Customs implemented CusMod (Customs Modernization), a complex sophisticated alert system which performs intelligence testing via message queue series (a priority-based software). CusMod integrates information, and electronic processes, in order to identify and process goods and passengers. Cusmod's main business functions include, providing clearance service, and information regarding imports and exports (of goods, services and people) coming in and leaving the country (both nationally and internationally). Business transactions include cargo information, shipping documentation, clearance documents, and passenger information (both flight and sea) that are transmitted using Cusmod. All incoming transactions have to go through NZ Customs Internet-service-provider. NZ Customs has more than 200 trading partners including custom brokers (agents), regular importers and exporters.

Internet Service Provider (Electronic Commerce Network Limited (ECN)) is New Zealand's leading trusted electronic business intermediary. ECN's main role is to facilitate technical and operational processes for organizations that want to adopt business-to-business E-commerce. ECN provides services' that enable business transactions between applications across any network and organizations. They provide services that enable business transactions across any network between applications. In addition, the ISP provides other services such as 24 hours × 7 availability of the network and maintenance of network, help desk, maintenance of trading partners details (as in correct information and privacy of trading partner's details), fault reporting, and maintaining direct debit authority schedule.

Background Information of Bi-directional Dyad B (BDD-B)

Bi-directional dyad B is between the customs broker and the importer. The customs broker clears goods for the importer through NZ Customs. Although, the customs broker has been using the "Trade Manager" for the past three years, they have been trading with the importer for the past ten years. The customs broker manages a small company with seven employees. They used "Trade Manager" designed to meet the needs of New Zealand exporters, and importers. Trade manager applies Microsoft Access and Visual Basic applications in order to provide real-time tracking information. The system enables the creation of purchase orders that help to manage shipping procedures. Exporters use it to prepare export documentation, including invoices, shipper's letters of instruction, picking and packing lists, order acknowledgments, certificates of origin, and customs declaration. Importers use Trade Manager to manage their orders, keep a database of all their shipments, and calculate accurate landed costs. All events are recorded against each job, order, on a date/time basis, and memos are created and referenced/filed to each job for future reference that enables importers to achieve lower clearance costs.

The importer also manages a small company with thirteen employees. The importer imports kitchen gadgets, plastic, baby wear, cosmetics, and distributes them to the big five supermarkets in New Zealand (including Woolworth, New Worlds, Big Fresh, Countdown, and Pak & Save).

Background Information of Bi-directional Dyad C (BDD-C)

Bi-directional dyad C is between Cisco NZ and Compaq NZ. Cisco NZ supplies computer and data communication products to Compaq NZ who integrates systems for their end customers. Cisco is a large international company that established a small branch office in Wellington, New Zealand seven years ago. Their reach is international and their product line is data and communication. Cisco had a sales volume of US$480 billion, with 25,000 employees worldwide. Cisco Systems, Inc. is the second largest company in the world after Microsoft.

Cisco NZ joined forces with its head office in San Jose, California (U.S.) to implement its E-commerce application. Cisco's business-to-business E-commerce extranet application, called Cisco Connections Online (CCO), has built-in functions and business transactions such as purchase orders for equipment, delivery, and product information from web-sites. Secondary elements include ordering for equipment, delivery, and ability to check lead track time. Cisco's registered trading partners can download product, equipment, and pricing information from CCO. CCO's interactive, networked services offer immediate and open access to Cisco's information, resources, and systems, anytime and anywhere. Cisco embraces the Global Networked Business model to implement innovative tools and systems to share information with diverse company stakeholders, such as suppliers, distributors, customers, and employees. By using CCO, Cisco connects trading partners to their manufacturing resource planning system, Internetworking Product Center (IPC), and Partner Initiated Customer Access (PICA). Most of Cisco's trading partners are system integrators. They include Logical, Datacom, Compaq, IBM, Telecom, Unisys, Clear, Fujistu, and Computer Link. Cisco's trading partners are contracted to trade between three and five years and Compaq NZ is one of Cisco's main trading partner.

Compaq NZ is a large company with 300 employees and has both a national and global reach. Compaq NZ sells computers, application software, hardware, networks, databases, undertakes systems integration, and develops database application systems. Compaq's main role is to manufacture computer systems integration parts and provide computer services. The Wellington branch is responsible for implementing business-to-business E-commerce via e-mail and files, but mostly uses CCO (Cisco's extranet application). Compaq manages Cisco's orders relating to system integration, applies network equipment, direct products and prices to meet end customer needs. Compaq has five branches in New Zealand; one each in Wellington, Auckland, Christchurch, Hamilton, and Dunedin. Compaq is one of the key channel trading partners, and they purchase computer and data communication parts from Cisco.

Background Information of Bi-directional Dyad D (BDD-D)

Bi-directional dyad D is between Siemens NZ and Telecom NZ. Siemens Information and Communication Networks Group was established in 1997.

Siemens NZ supplies telecommunication products to Telecom NZ. Siemens NZ consists of Siemens Communication Systems, Siemens Nixdorf, and Siemens Building Technologies Ltd. Siemens NZ is one of the world's leading providers of end-to-end solutions for voice, data, and mobile communication networks. With more than sixteen years experience in the New Zealand market, it is firmly established as a leading provider of electrical, electronic, information, and communications technology. Siemens NZ has since learned that success requires relationship marketing, up-to-date designs and attention to costs. The Information and Communication Network Group provides products, systems solutions, service, and support for installation and maintenance of complete corporate and service provider networks. Siemens NZ's products include keyphone systems, PABX systems, Call Center applications, voicemail, Integrated Cordless phones, Computer Telephone Integration (CTI), Interactive Voice Response and Video Conferencing equipment. Siemens Information and Communication Networks Group is a key supplier of broadband network technology to Telecom NZ. Siemens NZ's extranet E-commerce application called "Mainstream Express" is a user-friendly software with an informal structure and a set of menu options. Mainstream Express provides the latest product information, technical updates, news, and sales tools, fun and games. In addition, Mainstream Express has customized information and the presence of a website with ordering and tracking.

Telecom NZ purchases telecommunication parts from Siemens NZ and manufactures telecommunication products for their end customers. Telecom products include mobile phones, Internet access, Telco service, and all forms of telecommunications, mobile, and Internet services. Telecom NZ is a large international organization, with 700 employees situated in the Wellington branch. The Logistics group and network delivery section of Telecom handles the operation of applications and implementation from the corporate supply group. The Telecom staff log onto Siemens extranet application, Mainstream Express in order to obtain product information, place orders, and retrieve real-time tracking information. Table 5 provides a summary of the bi-directional dyads that participated in this study.

In the next section, we discuss the findings of the study as we examine each research proposition.

Trading Partner Trust and Perceived Benefits of Electronic Commerce

RP-1. Trading partner trust relates positively with perceived benefits of E-commerce.

Table 5. Summary of the Bi-directional Dyads That Participated in This Study.

Organizations That Participated in This Study	Name of Organization	Main Role and Size of the Organization	Type of Industry	No. of Respondents	Type of E-commerce Application
Bi-directional dyad A	NZ Customs → Internet service provider	Provides customs clearance service	Public service	10	CusMod using EDI X25 and other means via ISP
Uni-directional dyad 1		Large			
Bi-directional dyad A	Internet service provider → NZ Customs	E-commerce services	Internet service provider	2	Facilitates CusMod
Uni-directional dyad 2		Small-Medium Enterprise (SME)			
Bi-directional dyad B	Customs broker → Importer	Trade facilitator SME	Customs brokerage	2	Trade Manager using Visual Basic – Microsoft
Uni-directional dyad 3					
Bi-directional dyad B	Importer → Customs broker	Retailing SME	Retailing and service	2	Trade Manager using Visual Basic – Microsoft
Uni-directional dyad 4					
Bi-directional dyad C	Cisco NZ → Compaq NZ	Supplier SME	Computer and data communications	10	Extranet Cisco connection online
Uni-directional dyad 5					
Bi-directional dyad C	Compaq NZ → Cisco NZ	Buyer	Computer and data communications	4	Extranet Cisco connection online
Uni-directional dyad 6		Large			
Bi-directional dyad D	Siemens NZ → Telecom NZ	Supplier SME		4	Extranet main stream express
Uni-directional dyad 7					
Bi-directional dyad D	Telecom NZ → Siemens NZ	Buyer	Telecommunications	2	Extranet main stream express
Uni-directional dyad 8		Large			

The impact of competence, predictability and goodwill trust on perceived economic, relationship-related and strategic benefits was strongly supported by most of the bi-directional dyads. Competence trust was rated high because NZ Customs outsourced part of their customs clearance process to their ISP (Electronic Commerce Network Ltd) who was responsible for training NZ Customs staff. The NZ Customs' trading partners (consisting of custom brokers, exporters, and importers) have shown their ability and skills to operate business-to-business E-commerce transactions. NZ Customs' intranet administrator indicated:

> Of course, like any other new system we were trained to use the CusMod. Initially there were errors, as both our staff and trading partners had to learn to use CusMod. Trading partners had to bear the transaction costs if they continued to make errors, as they were required to re-send the same transaction correctly again. In the long run, trading partners realized the additional costs, and made every effort to get it correct. For example, when we first implemented export entries using CusMod, in October 1996, the error rate was 40–50%, but now it has dropped to 10%. This saves us a lot of time on phone calls, and enabled us to focus on other important aspects leading to strategic decisions, thus contributing to economic and strategic benefits. Furthermore, CusMod is 90% automated as the authorization and clearance process is conducted automatically. There is about 10% human intervention, which comes from an alert (i.e. in the case of drug use or fraud). Economic benefits experienced by NZ Customs staff include elimination of duplication and reduced delays in approving cleared goods, thus having a more productive workforce with fewer personnel. In addition, better intelligence systems allowed risks to be more readily and accurately identified. CusMod saved operating costs while improving service, as well as introducing common tariff classifications that enabled shipment of cargoes to and from international markets. This, in turn, benefits importers and exporters who received clearance electronically, as they need not pay for couriers, and no paper was involved thus contributing to importers and exporters economic benefits.

The findings provided evidence that although, NZ Customs did provide initial support to their importers and exporters, it can be a complex area for new trading partners. Furthermore, NZ Customs received support from the government, and had the financial resources, and top management commitment to outsource part of their business-to-business E-commerce processes to their Internet Service Provider. NZ Customs consultant indicated:

> We work as a team and perceive benefits as a win-win situation. Positive feelings towards our trading partners is high, because NZ Customs played an influential role in information sharing because of our strong international networks and reputation for ethical behavior, efficiency, effectiveness, and innovation.

In bi-directional dyad B, both the customs broker and the importer were two small organizations that applied a user-friendly, Microsoft application Trade Manager. The customs broker stated:

> We experienced economic benefits from a major reduction of paper flow, automatic storage of information through computer backup procedures, direct clearance cost savings, reduction of

> clerical work, and reduction by one day in transit time. We cooperate and communicate openly, as we use the same application. We also pay indemnity insurance for the importer's paper work (as a way of sharing risks).

The importer on the other hand indicated that:

> The customs broker has shown a willingness to share information, and provided support when we first implemented Trade Manager. The customs broker's staff came over and gave us some training. They did show care and concern not only in using the technology, but also in making important strategic decisions, such as designing a cost-effective approach for shipping and transport.

The automated clearance process led to fewer errors and transactions were cleared more quickly, thus contributing to economic benefits from savings in time and cost for the importer. In addition, the customs broker provided free software installation and training for the importer who experienced relationship-related benefits from the past experiences they had had with the customs broker. Given that both the customs broker and importer have been trading for almost a decade, although they implemented the E-commerce system for less than two years, goodwill trust established from their trading partner relationships contributed to strategic benefits as they were willing to invest in their relationships.

With Cisco-Compaq NZ (bi-directional dyad C), competence trust was rated medium to high because although, Cisco NZ, being the world's leader of E-commerce, had powerful and sophisticated tools with embedded checking mechanisms that detected errors made by Compaq, initial mistakes were observed. The process involving creating an order for the products and components was complex. Furthermore, Cisco NZ's Internet Business Solutions group managed all queries about technical difficulties and clarifications through their extranet application (Cisco Connection Online). Although, Compaq has been using CCO for the past eighteen months, Compaq had a trading partner relationship with Cisco for the past ten years. Compaq Network Specialists manager stated:

> CCO enables us to check the prices, request a discount if necessary, and electronically receive an estimated time of arrival before even confirming the order. The system automatically gives the number of days as to when the goods will be delivered, thereby enabling Compaq staff to undertake an inquiry on the system in advance. The updated information on Cisco's website is accurate and reliable. Cisco kept their business promises which assisted Compaq to make better strategic decisions, thus contributing to economic benefits from savings in time and costs (from telephone calls).
>
> Competence trust was shown by Cisco staff who had the ability to do their job. Cisco staff are the 'pros' as they know what they are doing. Their IT support people are excellent, very responsive, and timely in providing complete and accurate information. In addition, their online tools have checking mechanisms that enable Compaq to detect errors. Compaq staff said that trading with Cisco is better than trading with most of Cisco's competitors, particularly when

> it comes to problem solving which creates a confidence in handling conflict resolution and clarifying issues. I guess that's why Cisco is number two in the world and it is definitely based on reputation and goodwill trust.

It can be seen that although, the focus may point towards a high level of competence trust, technology and financial resources, relationships (such as honesty in providing reliable information) were found to be important. The time-savings allowed them to concentrate on strategic planning (goodwill trust), that contributed to economic, relationship-related, and strategic benefits.

Competence trust was found to be important by most of the bi-directional dyads except for bi-directional dyad D (Siemens-Telecom NZ). The increased turnover of Telecom staff created difficulties for Siemens staff as they had to re-train the new staff. Siemens staff indicated that:

> Although Telecom staff did have the competence, they continued to send us incomplete purchase orders with errors. We had to provide support and training to Telecom staff all the time. Each time an order came in we checked and verified the order manually. If the order came in incomplete, it will not do the job so we informed Telecom NZ from using e-mail that we cannot accept the order and explaining the reasons. We educate them so that their customers will have confidence in them. We even go with their sales team to their clients and explain the system by giving seminar presentations relating to the functionality that adds value to the products we supply. Further, we write an e-mail news service, which Telecom obtains three times a week, in addition to our newsletters.

Thus, the complexity of the parts and the lack of knowledge and expertise about the products led to uncertainties when placing an order. On the other hand, Telecom staff indicated that:

> Siemens NZ staff are competent and reliable, and they do provide us with adequate operational support in the form of training. Siemens NZ had access to our forecasting information, business development, and we know what price we have been charged for their products, as it was pre-arranged in the contract.

Telecom's end customers were unaware or were not concerned as to where the parts for their mobile phones or telecommunication equipment came from as long as the equipment works.

The findings indicate that trading partner trust did play an important role in contributing to benefits. Furthermore, a history of consistent positive trust behaviors enabled the development of predictability trust and relationship-related benefits. Although, the technology initially contributes to economic benefits, one can argue that trading partners continue to trade for long periods of time (ten to fifteen years) even before the E-commerce technologies were implemented thus contributing to trading partner satisfaction and strategic benefits.

Technology Trust Mechanisms in E-commerce and Perceived Benefits of E-commerce

RP-2. Technology trust mechanisms in E-commerce associate positively with perceived benefits of E-commerce.

The impact of technology trust mechanisms on perceived economic, relationship-related and strategic benefits was strongly supported by most of the bi-directional dyads. Bi-directional dyad A (NZ Customs) experienced perceived economic benefits because their CusMod system had embedded automated checking mechanisms and protocols that enable detection of errors. Furthermore, their Internet service provider supplied the technical expertise, compatible systems, training and support that encouraged relationship-related benefits. NZ Customs also receives support from the government. A NZ Customs consultant indicated:

> The CusMod system operates round the clock 24 hours × 7. When NZ Customs first implemented CusMod, they had to face a forty-eight hour turnaround time for clearance, but now it is just under twenty minutes.
>
> We represent the nation, and trading partners are definitely our driving forces for adopting E-commerce. NZ Custom's staff conducts regular meetings with industry groups, and conduct business surveys on how CusMod is operating. We have introduced a national call center with a toll-fee number in Auckland where information is made available, free of charge for both importers and exporters. We also have a solid framework (that is a business model) which incorporates government legislative bodies, and builds trust, thus enforcing best business practices.

The customs broker and importer in bi-directional dyad B, also experienced benefits from using Trade Manager for reasons of confidentiality, access controls, and best business practices. The importer indicated that:

> Confirmations' and acknowledgments were received using the e-mail embedded in the Trade Manager. In case of urgent orders we use the telephone, e-mail, or fax. We used separate log-on procedures and applied segregation of duties in order to ensure different levels of authorization mechanisms in the importer's organization. The receptionist, administration staff, and drivers do not have access to the Trade Manager. Furthermore, we exercised best business practices in the form of daily backups and followed the Importers Institute standard. We maintained a backup hardcopy version of all purchase order numbers, invoice numbers, reference numbers.

In the case of bi-directional dyad C, Cisco NZ's extranet application is a powerful tool with embedded checking mechanisms used to detect and correct errors, thus contributing to economic benefits (from savings in time and cost). Cisco accounting manager indicated that:

> Initially, the error rate was 80%, has now reduced to less than 15%, thus contributing to economic benefits from savings in time and cost (telephone calls, e-mail, faxes for correcting mistakes).

> CCO provides benefits in the form of real-time order tracking information, and estimated arrival dates of goods. In fact, Cisco participants admitted that 35–40% of the revenue came from Compaq NZ.

Similarly, Compaq too gained from the online tracking of information that was made available from the time an order was placed until the goods were delivered. Compaq participant indicated that:

> Indeed, CCO assisted us in making accurate strategic decisions and building our reputation, as we were able to fulfill our business promises, thus contributing to both economic and relationship-related benefits. We have reference numbers with Cisco NZ that served as unique identifiers thereby providing authentication. We believe that top management commitment is critical especially when it comes to the budget. We do have to meet with the standards set out by Cisco (industry and universal) and policies. We are part of the ISO 9000, but not all of the divisions. We have a trading partner agreement, and undertake regular audit checks in order to manage risks. We have a business risk management group of consultants who design contingency procedures for our E-commerce systems and operations.

Consistent achievement of economic benefits led to relationship-related benefits derived from trading partner satisfaction. In bi-directional dyad D (Siemens-Telecom NZ) experienced economic benefits from savings in time and cost as the product information was made available on their extranet application. Most of the bi-directional dyads experienced benefits from technology trust mechanisms because of their initial investments in implementing E-commerce as they were able to provide adequate training and ensure that their systems were secure and met with their business requirements and high quality standards. The findings implied that technology trust mechanisms encouraged the development of trading partner trust relationships.

Trading Partner Trust and Perceived Risks of E-commerce Participation

RP-3. Trading partner trust relates negatively with perceived risks of E-commerce.

The impact of competence, predictability, and goodwill trading partner trust on perceived technology performance-related, relational, and general risks was not strongly supported by most of the bi-directional dyads. Most E-commerce systems and applications are user-friendly systems with standardized routines using a set of menu options that contributed to low technology-performance related risks. Furthermore, the organizations enforce best business practices (such as high quality standards, written policies, procedures, audit, and risk management strategies). In bi-directional dyad A, NZ Customs rated technology performance-related risks to be low to begin with because they had outsourced part of their business-to-business E-commerce process to their Internet Service Provider (ISP), who was responsible for the compatibility and security of E-commerce applications. The NZ Customs intranet administrator indicated:

> The ISP is responsible for the compatibility of the system between trading partners (importers and exporters) and the CusMod system.

Hence, compatibility of the system was not an issue because the ISP facilitated the movement of transactions between NZ Customs and their trading partners. Although the ISP facilitated the technological needs for customs clearance, a few trading partners experienced relational risks. For example, the NZ Customs intranet administrator indicated:

> The development of CusMod and changes associated with business processes and the introduction of the Internet have left some companies reluctant to change and others hammering on our door demanding change, thus contributing to relational risks. We are aware of the culture shock and are trying to be patient with them. Some of our trading partners are loud, hostile, and even aggressive (i.e. related to functional conflict). For example, the Inland Revenue Department (IRD) wanted a list of all our employees and their income to be submitted through a secure web site. The IRD demonstrated absolute power and authority and we had no choice to exercise either politically or financially. We are trying to build partnerships, and while some trading partners are great at demonstrating change, most of them are not.

Like any change, NZ Customs experienced relational risks in the form of coercive power from senior authorities, reluctance to change and organizational inertia from their trading partners (importers and exporters).

In the case of the customs broker and importer in bi-directional dyad B technology performance-related were rated low because their systems were not connected to the Internet. The importer indicated that:

> We did face initial uncertainties in using the Trade Manager that led to our dependence on our customs broker. These interdependencies gradually led to an imbalance of power. Although our customs broker provided us with free software and initial training, we were left in a difficult position. It is not something that you have to outsource, but we had to change our internal business processes in order to facilitate and simplify the business processes of the customs broker, which takes some time to get it completely right. We do face situations of conflict and handling discrepancies, but it is more a functional conflict relating to business processes rather than a personal conflict.
>
> Our customs broker did appear to exercise opportunistic behaviors by increasing the charges and costs for clearance, which led to functional conflicts derived from a misunderstanding in calculating charges, due to the conversion of currencies, and taxes. In most cases the customs broker had to explain to us on how they derived the figures.

The importer was concerned over relational risks because they were suspicious that their customs broker exercised poor business practices. The importer faxed invoices that revealed their details, prices of goods, types, and quantity of the goods ordered to the customs broker.

Bi-directional dyad C (Cisco-Compaq) experienced low technology performance-related risks because CCO was only implemented by Cisco and Compaq staff only had to log onto the web-site using authorization mechanisms. Cisco's consultants indicated that:

> Conflicts occur always as nobody wants to kill each other. Occasionally, they do get frustrated when they want something to be done. They negotiate and with cohesion depending on how important the situation is.

Compaq staff indicated that:

> Cisco staff did create a situation of imbalance of power, as they would like to have the names of our end customers. But when we chose not to give to them because we would like to have a bit of competition by buying from other suppliers, Cisco staff did not like it.

Similarly, bi-directional dyad D (Siemens-Telecom NZ) rated technology performance-related risks to be low because they had outsourced their extranet application to a private server. However, Siemens had some concerns over relational risks with Telecom staff because of a lack of predictability trust derived from the high turnover of staff.

General risks were rated low and were not applicable by most of the bi-directional dyads, because they enforced best business practices (in the form of regular and adequate audits, backups, high quality, standards, and training).

Technology Trust Mechanisms and Perceived Risks of E-commerce

RP-4. Technology trust mechanisms in E-commerce relate negatively with perceived risks of E-commerce.

The impact of technology trust mechanisms on perceived technology performance-related, relational, and general risks of E-commerce was not strongly supported by most of the bi-directional dyads. Technology trust mechanisms in E-commerce and its impact on perceived risks were found to be insignificant by most of the organizations because of the compatibility of the E-commerce systems. Furthermore, the E-commerce applications came with embedded security mechanisms and protocols that detect errors, thus saving a lot of time transmitting E-commerce transactions. For example, in bi-directional A (NZ Customs-ISP) intelligence testing was implemented to eliminate unauthorized log-on procedures or passwords that could interfere with the maintenance and use of technology.

Most of the risks were not applicable for bi-directional dyad B (Customs Broker-Importer), because the Customs broker was aware of the legal implications involved in the customs clearance process and therefore abides by the standards of the Customs Act. However, the importer was concerned about confidentiality of their shipment information. The importer indicated that:

> Our risks lie in the shipment information being leaked out to other competitors by our customs broker, whom they say sometimes fax documents. The fax revealed the quantity of stock imported for each delivery and can be seen by their employees or any other unauthorized personnel. There are risks in applying poor business practices (lack of audits, back-ups), particularly in a small firm where most of the employees performed multi-tasks.

Cisco NZ rated their technology performance-related risks to be low because of their highly sophisticated extranet application and also indicated that most of the risks did not apply to them. In bi-directional dyad D, Telecom NZ also experienced fewer technology performance-related risks because the compatibility of E-commerce systems was not a concern, as the extranet application was implemented by Siemens NZ.

Perceived Benefits of E-commerce and Outcomes in E-commerce Participation

RP-5. Perceived benefits of E-commerce relate positively with outcomes in E-commerce participation.

The impact of perceived economic, relationship-related, and strategic benefits on the outcomes of E-commerce participation was strongly supported by most of the bi-directional dyads. One possible explanation for this is the automation and speed that E-commerce applications provided increased E-commerce performance, thus contributing to economic benefits. Most of the E-commerce operations involved standardized routine processes. Bi-directional dyad A (NZ Customs-ISP), represents the nation and 98% of their business was conducted electronically.

The NZ Customs intranet administrator indicated that:

> We have moved from a mere gatekeeper for E-commerce transactions to service transactions and business relationships. We try to maintain business continuity and trading partner relationships. Similarly, we perceived an increase in the level of cooperation in our trading partners, as we received support from the government (as in policies to abide). We also perceived an increase in the level of commitment from our trading partners, as the reputation of our organization increased. Without trust there is no effective communication and without effective communication there is no business-to-business E-commerce.

Economic benefits experienced by the customs broker were not found to be significant because they only acted as a trade facilitator and were not directly involved in the importing and distribution of the products or goods. The customs broker indicated that:

> We are now able to compete with the one-stop shop services offered by multi-national freight forwarders and have become a serious competitor for trading partners. Trust is explicit and is built in the trading partner relationship.

The importer found E-commerce important for them because they received real-time information that contributed to strategic benefits as they were able to plan their future business operations. Seventy to eighty percent of Cisco's business involved E-commerce. Cisco consultants indicated that:

> Our main role is to establish customer preference, and increase the number of channel partners by trading relationship trust, which in turn increases E-commerce participation.

Compaq NZ indicated that twenty percent of their E-commerce came from Cisco. Compaq's findings indicated an increase in E-commerce performance, although trading partner trust relationship development was not obvious because Compaq was competing with Cisco NZ. Compaq participants indicated that:

> We cannot promise that Cisco was our driving force for adopting E-commerce, as we adopted E-commerce to stay competitive in the business. We have our own web page, products, catalog, and use e-mail with our customers. We, too, have explicit trading agreements with our trading partners' regarding roles and responsibilities and have maintained long-term trading partner relationships with them. We are Cisco's gold trading partner; that is one of Cisco's top ranked trading partners to trade with.

Siemens found that their strategic benefits were significant because of the uniqueness of their telecommunication products. Telecom NZ did not find their relationship-related and strategic benefits to be significant because their end-customers were unaware as to who actually manufactured the parts of their product. Telecom participants indicated that:

> We are interested in solutions not individual items of sale, and encourage our trading partners to think outside the box with sound knowledge. We are in a competitive global environment and have to abide by our trading contract in order to maintain confidential information and intellectual property which allows us to disclose information, and protect our network from being abused and that they can be trusted to solve our problems. Siemens NZ is not totally the driving force for Telecom's E-commerce. Of course we look for speed, simplicity, cost which meets our purpose for undertaking E-commerce. Among other things, that can be on line purchasing and the provision of invoices. Telecom end customers are unaware as to where the technology came from. All they are interested to see is that the phone line works when it needs to.

The findings implied that perceived benefits from both E-commerce applications and trading partner trust relationships increased outcomes in E-commerce participation.

Perceived Risks of E-commerce and Outcomes in E-commerce Participation

RP-6. Perceived risks of E-commerce relate negatively with outcomes in E-commerce participation.

The impact of perceived technology performance-related, relational, and general risks of E-commerce on the outcomes of E-commerce participation was not strongly supported by most of the directional-dyads. In the case of NZ Customs, the Customs Excise Act was vital to their significant modernization and customs risk management. Most of the risks did not impact bi-directional dyad C (Cisco-Compaq NZ), because they enforced best business practices and their standards were universally acceptable. The impact of perceived risks was found to be insignificant by most of the bi-directional dyads, except for the Siemens-Telecom NZ. One possible explanation for this is that the ordering process was a complex one. Furthermore, the high turnover of staff by Telecom created added difficulties for Siemens NZ, as they had to re-train their new staff all over again. Thus, a lack of consistent competence trust demonstrated from Telecom staff created a situation of dissatisfaction and contributed to relational and general risks. On the whole, the findings provided evidence that perceived risks of E-commerce was not strongly supported.

Cross-Case Analyses of the Bi-directional Dyads

Miles and Huberman (1994) suggest that cross-case analyses enhance generalization and deepen the understanding of the study. This section discusses the findings from a cross-case analysis of four bi-directional dyads.

RP-1. Trading partner trust relate positively with perceived benefits of E-commerce.

The impact of competence, predictability, and goodwill trading partner trust on perceived economic, indirect, relationship-related and strategic benefits was strongly supported by most of the bi-directional dyads. Table 6 provides a summary of the impact of trading partner trust on perceived benefits of E-commerce. Competence trust was rated high by most of the bi-directional dyads except for bi-directional dyad D (Siemens-Telecom NZ). One possible explanation for this could be that in Siemens-Telecom NZ dyad, the impact on economic benefits was rated low to medium. Siemens NZ outsourced a private server for their extranet application. The increased turnover of Telecom staff created difficulties for Siemens staff as they had to re-train the new staff. In addition, the complexity of the parts and the lack of knowledge and expertise about the products led to

Table 6. Research Proposition 1 – Trading Partner Trust and Perceived Benefits.

Three Types of Trading Partner Trust	Bi-directional Dyad A NZ Customs and their ISP Directional Dyads 1–2	Bi-directional Dyad B Customs Broker and Importer Directional Dyads 3–4	Bi-directional Dyad C Cisco NZ and Compaq NZ Directional Dyads 5–6	Bi-directional Dyad D Siemens NZ and Telecom NZ Directional Dyads 7–8
Competence trust				
Economic direct benefits	H	H	M–H	L–M
Predictability trust				
Indirect personal benefits	H	H	H	H
Goodwill trust				
Relationship related symbolic and strategic benefits	M	H	M	M–H

Notes: L = Low (0–3); M = Medium (4–6); H = High (7–10).

uncertainties when placing an order. On the other hand, Telecom's end customers were unaware or were not concerned as to where the parts for their mobile phones or telecommunication equipment came from as long as the equipment works.

In the case of NZ Customs and their ISP (bi-directional dyad A), the ability of their trading partners to use E-commerce applications correctly after an initial period of making mistakes made them realize economic benefits (i.e. savings from time and costs in re-sending the same transaction twice). The findings show that although NZ Customs did provide initial support to their importers and exporters, it can be a complex area for new trading partners. Furthermore, NZ Customs received support from the government, and had the financial resources, and top management commitment to outsource part of their business-to-business E-commerce processes to their Internet Service Provider. In bi-directional dyad B, both the customs broker and the importer were two small organizations that applied a user-friendly, Microsoft application Trade Manager. The automated clearance process led to fewer errors and transactions were cleared more quickly, thus contributing to economic benefits from savings in time and cost for the importer. In addition, the customs broker provided free software installation and training for the importer who experienced relationship-related benefits from the past experiences they had had with the customs broker. With Cisco-Compaq NZ (bi-directional dyad C), competence trust was rated medium to high because although, Cisco NZ, being the world's leader of E-commerce, had powerful and sophisticated tools with embedded checking mechanisms that detected errors made by Compaq, initial mistakes and training was observed. The process involved in creating an order for the products and components was a complex one. Furthermore, Cisco NZ's Internet Business Solutions group managed all queries about technical difficulties and clarifications through their extranet application (Cisco Connection Online). Although, Compaq has been using CCO for the past eighteen months, Compaq had a trading partner relationship with Cisco for the past ten years. It can be seen that the focus may point towards a high level of competence trust, technology and financial resources, relationships (such as honesty in providing reliable information) were found to be important. They were able to accomplish online what used to be done over the telephone and e-mail. The time saving allowed them to concentrate on strategic planning (goodwill trust), which contributed to economic, relationship-related, and strategic benefits.

Although, the technology initially contributed to economic benefits, one can argue that trading partners continued to trade for long periods of time (ten to fifteen years) even before the E-commerce technologies were implemented that contributed to trading partner satisfaction. Thus, trading partner trust was found to be important in E-commerce participation. Table 6 presents the relationship between trading partner trust and perceived benefits of E-commerce in the bi-directional dyads.

Trading Partner Trust and Perceived Risks of E-commerce Participation

RP-2. Trading partner trust associate negatively with perceived risks of E-commerce.

The impact of competence, predictability, and goodwill trading partner trust on perceived technology performance-related, relational, and general risks was not strongly supported by most of the bi-directional dyads. Table 7 presents the impact of trading partner trust on perceived risks of E-commerce. Most E-commerce systems and applications were user-friendly and came with standardized routines using a set of menu options that contributed to low technology-performance related risks. The risks derived from a lack of competence were rated low by most of the bi-directional dyads. Bi-directional dyad A (NZ Customs) rated technology performance-related risks to be low because they had outsourced part of their business-to-business E-commerce process to their Internet Service Provider, who could exhibit opportunistic behaviors derived from expert knowledge that contributed to relational risks. In bi-directional dyad C (Cisco-Compaq) rated technology performance-related risks to be low because CCO was only implemented by Cisco and Compaq staff only had to log onto the web-site using authorization mechanisms. Similarly, Siemens-Telecom NZ (bi-directional

Table 7. Research Proposition 2 – Trading Partner Trust and Perceived Risks.

Three Types of Trading Partner Trust	Bi-directional Dyad A NZ Customs and Their ISP Directional Dyads 1–2	Bi-directional Dyad B Customs Broker and Importer Directional Dyads 3–4	Bi-directional Dyad C Cisco NZ and Compaq NZ Directional Dyads 5–6	Bi-directional Dyad D Siemens NZ and Telecom NZ Directional Dyads 7–8
Competence trust				
Technology performance related risks	L–M	L–M	L	L
Predictability trust				
Relational risks	M–H	L–M	L	M
Goodwill trust				
General risks	N/A	N/A	L–N/A	L–N/A

Notes: L = Low (0–3); M = Medium (4–6); H = High (7–10).

dyad D) rated technology performance-related risks to be low because they, too, had outsourced their extranet application to a private server. However, they rated relational risks to be medium due to a lack of predictability trust arising from a high turnover of staff. Telecom staff on the other hand also rated technology performance-related risks to be low because they had to log onto a private server.

Relational risks were rated low to medium by the importer (bi-directional dyad B) because they were suspicious that their customs broker exercised poor business practices. The importer faxed invoices that revealed their details, prices of goods, types, and quantity of the goods ordered to the customs broker.

General risks were rated low and were not applicable by most of the bi-directional dyads, because of the implementation of best business practices (including audits, backups, high quality, standards, and training). The findings provided evidence that trading partner trust does impact risk. Trading partner relationship trust contributed to low risks in most of the directional dyads. Table 7 presents the relationship between trading partner trust and perceived risks of E-commerce in the bi-directional dyads.

Technology Trust Mechanisms in E-commerce and Perceived Benefits of E-commerce

RP-3. Technology trust mechanisms in E-commerce associate positively with perceived benefits of E-commerce.

Table 8. Research Proposition 3 – Technology Trust Mechanisms and Perceived Benefits.

Technology Trust Mechanisms in E-commerce	Bi-directional Dyad A NZ Customs and Their ISP Directional Dyads 1–2	Bi-directional Dyad B Customs Broker and Importer Directional Dyads 3–4	Bi-directional Dyad C Cisco NZ and Compaq NZ Directional Dyads 5–6	Bi-directional Dyad D Siemens NZ and Telecom NZ Directional Dyads 7–8
Confidentiality				
Direct economic benefits	H	M	H	H
Integrity				
Direct economic benefits	H	H	H	H
Authentication				
Indirect personal benefits	H	H	H	H
Non-repudiation				
Indirect personal and relationship related benefits	H	H	H	H
Availability				
Direct economic benefits	H	M	H	H
Access controls				
Direct economic benefits	H	M	H	H
Best business practices				
Relationship related and symbolic strategic benefits	H	M	H	H

Notes: L = Low (0–3); M = Medium (4–6); H = High (7–10).

The impact of technology trust mechanisms on perceived economic, indirect, relationship-related and strategic benefits of E-commerce was strongly supported by most of the bi-directional dyads. Table 8 presents the impact of trust and security-based mechanisms on perceived benefits. Most of the bi-directional dyads rated perceived benefits to be high because the main goal of E-commerce systems and applications was to provide efficiency, thus contributing to economic benefits. Bi-directional dyad A (NZ Customs) rated perceived economic benefits to be high because the CusMod system had embedded automated checking mechanisms and protocols that enable detection of errors. Furthermore, their ISP provided the technical expertise, compatible systems, training and support that help build relational benefits. On the other hand, the customs broker and importer rated perceived benefits to be medium to high for reasons of confidentiality, access controls, and best business practices. One possible explanation for this is that their E-commerce application Trade Manager was not attached to their network. Most of the bi-directional dyads experienced benefits from trust and security-based mechanisms because of their initial investments in implementing E-commerce as they were able to provide adequate training and ensure that their systems were secure and met their business requirements and high quality standards. In the case of bi-directional dyad C, Cisco NZ's extranet application was a powerful tool with embedded checking mechanisms that detected and corrected errors, contributing to economic benefits from savings in time and cost. Similarly, Compaq too gained from the online tracking of information that was made available from the time an order was placed until the goods were delivered. Consistent achievement of economic benefits led to relationship-related benefits from trading partner satisfaction. In the case of bi-directional dyad D (Siemens-Telecom NZ), they, too, experienced economic benefits from savings in time and cost as product information was made available on their extranet application. Siemens extranet web site application only permitted information access (i.e. read only) that could not be modified or written. The findings implied that technology trust mechanisms enabled trading partner trust to be developed. Table 8 presents the relationship between technology trust mechanisms and perceived benefits of E-commerce in the bi-directional dyads.

Technology Trust Mechanisms in E-commerce and Perceived Risks of E-commerce

RP-4. Technology trust mechanisms in E-commerce relates negatively with perceived risks of E-commerce.

The impact of technology trust mechanisms on perceived technology performance-related, relational, and general risks of E-commerce were not strongly supported by most of the bi-directional dyads. Table 9 presents the impact of technology trust mechanisms on perceived risks of E-commerce. Technology trust mechanisms in E-commerce and its impact on perceived risks was rated low by all the organizations because compatibility of the systems was not an issue between trading partners in a dyad. Furthermore, there were efficiencies from E-commerce applications that came with embedded security mechanisms. These mechanisms

Table 9. Technology Trust Mechanisms and Perceived Risks of E-commerce.

Technology Trust Mechanisms in E-commerce	Bi-directional Dyad A NZ Customs and Their ISP Directional Dyads 1–2	Bi-directional Dyad B Customs Broker and Importer Directional Dyads 3–4	Bi-directional Dyad C Cisco NZ and Compaq NZ Directional Dyads 5–6	Bi-directional Dyad D Siemens NZ and Telecom NZ Directional Dyads 7–8
Confidentiality				
Technology performance related risks	L	M	L	L
Integrity				
Technology performance related risks	L–M	L	L	L
Authentication				
Relational risks	N/A	L	L	L
Non-repudiation				
Relational risks	N/A	L	L	L
Availability				
Technology performance and Relational risks	N/A	L	L	L
Access controls				
Technology performance related risks	L	N/A	L	L
Best business practices				
General risks	N/A	L	N/A	L

Notes: L = Low (0–3); M = Medium (4–6); H = High (7–10).

enabled errors to be quickly detected and corrected, saving a lot of time transmitting E-commerce transactions. For example, in bi-directional A (NZ Customs-ISP) intelligent testing was implemented to eliminate unauthorized log-on procedures or passwords that could interfere with the maintenance and use of technology.

Most of the technology trust mechanisms were not applicable for bi-directional dyad B (Customs Broker-Importer), because the Customs broker is aware of the legal implications involved in the customs clearance process and therefore abides by the standards of the Customs Act. However, the importer was concerned about confidentiality of their shipment information being leaked out to other competitors by the customs broker because of poor business practices. Cisco NZ found their technology performance-related risks to be low because of their highly sophisticated extranet application and also indicated that most of the risks did not apply to them. Cisco NZ found their technology performance-related risks to be low because of their highly sophisticated extranet application and both Cisco NZ and Compaq NZ indicated that most of the risks did not apply to them. Furthermore, high quality standards were practiced. In bi-directional dyad D, Telecom NZ rated technology performance-related risks to be low because the compatibility of E-commerce systems was not a concern, as the extranet application was implemented by the trading partner (Siemens NZ). The findings provided evidence of three factors. First, efficiencies of E-commerce technologies and applications reduced compatibility problems and technology performance-related risks. Second, due to past experience, relational risks were rated low as the organizations knew who they were dealing with. Finally, most of the bi-directional dyads enforced best business practices and had to abide by industry standards that provided high quality services that contributed to low general risks. Table 9 presents the relationship between technology trust mechanisms and perceived risks of E-commerce in the bi-directional dyads.

Perceived Benefits of E-commerce and the Extent of E-commerce Participation

RP-5. Perceived benefits of E-commerce relate positively with E-commerce participation.

The impact of perceived economic, indirect, relationship-related, and strategic benefits on the extent of E-commerce participation was strongly supported by most of the bi-directional dyads. Table 10 presents the impact of perceived benefits of E-commerce on the extent of E-commerce participation. One possible explanation for this is the automation and speed that E-commerce applications provided increased E-commerce performance, thus contributing to economic benefits. Most of the E-commerce operations involved standardized routine processes. Most of the trading partners were trading manually or used other means of communicating before trading with E-commerce applications. Furthermore, trading partners have met each other before. Bi-directional dyad A (NZ Customs-ISP), represents the nation and 98% of their business was conducted electronically. Economic benefits in bi-directional dyad B were rated medium because the customs broker only acted as a trade facilitator and was not directly involved in the importing and

Table 10. Research Proposition 5 – Perceived Benefits of E-commerce and the Extent of Participation in E-commerce.

Perceived Benefits of E-commerce	Bi-directional Dyad A NZ Customs and Their ISP Directional Dyads 1–2	Bi-directional Dyad B Customs Broker and Importer Directional Dyads 3–4	Bi-directional Dyad C Cisco NZ and Compaq NZ Directional Dyads 5–6	Bi-directional Dyad D Siemens NZ and Telecom NZ Directional dyads 7–8
Perceived direct benefits				
Extent of E-commerce performance	H	M–H	H	H
Perceived indirect benefits				
Extent of E-commerce performance and trading partner trust relationship development	H	H	H	H
Perceived relationship related benefits of E-commerce				
Extent of trading partner trust relationship development	H	H	H	M
Perceived strategic benefits				
Extent of trading partner trust relationship development	H	M	H	M–H

Notes: L = Low (0–3); M = Medium (4–6); H = High (7–10).

distribution of the products or goods. The importer found E-commerce important for them because they received real-time information. Similarly, strategic benefits and their impact on E-commerce participation were rated medium by the importer. In bi-directional dyad C, seventy to eighty percent of Cisco's business involved E-commerce. Compaq NZ indicated that twenty percent of their E-commerce comes from Cisco. Compaq's findings indicated an increase in E-commerce performance, although trading partner trust relationship development was not obvious because Compaq was competing with Cisco NZ. Siemens rated strategic benefits to be high because of the uniqueness of the telecommunication products. Telecom NZ bi-directional dyad D, rated relationship-related and strategic benefits to be medium because their end-customers were unaware as to who actually manufactured the parts of their product. The findings implied that perceived benefits from both E-commerce applications and trading partner trust relationships increased participation in E-commerce. Table 10 presents the relationship between perceived benefits and outcomes in E-commerce participation in the bi-directional dyads.

Perceived Risks of E-commerce and the Extent of E-commerce Participation

RP-6. Perceived risks of E-commerce associate negatively with E-commerce participation.

The impact of perceived technology performance-related, relational, and general risks of E-commerce and its impact on the extent of E-commerce participation was not strongly supported by most of the directional-dyads. Table 11 presents the impact of perceived risks on E-commerce participation. The impact of perceived risks on the extent of E-commerce participation was rated low by most of the bi-directional dyads, except for the Siemens-Telecom NZ, which rated the impact of perceived risks on E-commerce participation to be medium. One possible explanation for this is that the ordering process was a complex one. Furthermore, the high turnover of staff by Telecom created added difficulties for Siemens NZ, as they had to re-train their new staff all over again. Thus, a lack of consistent competence trust demonstrated from Telecom staff created a situation of dissatisfaction and contributed to relational and general risks Most of the risks did not impact bi-directional dyad C (Cisco-Compaq NZ), because they enforced best business practices and their standards were universally acceptable. In the case of NZ Customs the Customs Excise Act was vital to their significant modernization and customs risk management. On the whole the findings provided evidence that perceived risks to E-commerce were not strongly supported by other bi-directional dyads. The next section discusses the similarities and differences in the bi-directional dyads. Table 11 presents the relationship between perceived risks and outcomes in E-commerce participation in the bi-directional dyads.

Similarities and Differences of the Bi-directional Dyads

The similarities and differences in the findings came from the type of industry and E-commerce applications used. Organizations that implemented extranet applications (Cisco-Compaq NZ and Siemens-Telecom NZ) had to deal with

Table 11. Research Proposition 6 – Perceived Risks of E-commerce and the Extent of Participation in E-commerce.

Perceived Risks of E-commerce	Bi-directional Dyad A NZ Customs and Their ISP Directional Dyads 1–2	Bi-directional Dyad B Customs Broker and Importer Directional Dyads 3–4	Bi-directional Dyad C Cisco NZ and Compaq NZ Directional Dyads 5–6	Bi-directional Dyad D Siemens NZ and Telecom NZ Directional Dyads 7–8
Perceived technology performance related risks				
Extent of E-commerce performance	L	L	L–N/A	M
Perceived relational risks				
Extent of trading partner trust relationship development	L	L	L	M
Perceived general risks				
Extent of E-commerce participation and trading partner trust relationship development	L	M	L–N/A	M

Notes: L = Low (0–3); M = Medium (4–6); H = High (7–10).

complex parts and a difficult ordering process. Furthermore, the implementation of the E-commerce extranet application was carried out by only one trading party (i.e. the supplier – Cisco NZ and Siemens NZ), who is responsible for updating it with real-time tracking information of the products and orders placed by their buyers. These initial one-sided implementation cost, motivated suppliers, at times to apply coercive power and exhibit opportunistic behaviors (in the form of high prices) to their buyers, thus contributing to relational risks of E-commerce.

The similarities between NZ Customs, and the customs broker and importer were long-term trading partner relationships. The impact of E-commerce participation by smaller organizations such as the customs broker and importer was successful. One possible explanation for this was the simplicity of their E-commerce application, and for security reasons their E-commerce application was not connected to the networks. The differences between NZ Customs and their Internet Service Provider include outsourcing part of their E-commerce processes to their Internet Service Provider.

DISCUSSION

The findings contributed to a model of inter-organizational-trust in business-to-business E-commerce. This research thus has important implications for research and practice. Specifically, the theoretical framework and findings suggest that in order for organizations to sustain and manage collaborative trading partner relationships businesses must establish trustworthy relationships with their trading partners.

Model of Inter-Organizational-Trust within Bi-directional Dyads in E-commerce Participation

The model of inter-organizational-trust within bi-directional dyads in E-commerce participation was developed from the findings. Figure 2 depicts the model of inter-organizational-trust within bi-directional dyads in E-commerce participation. The

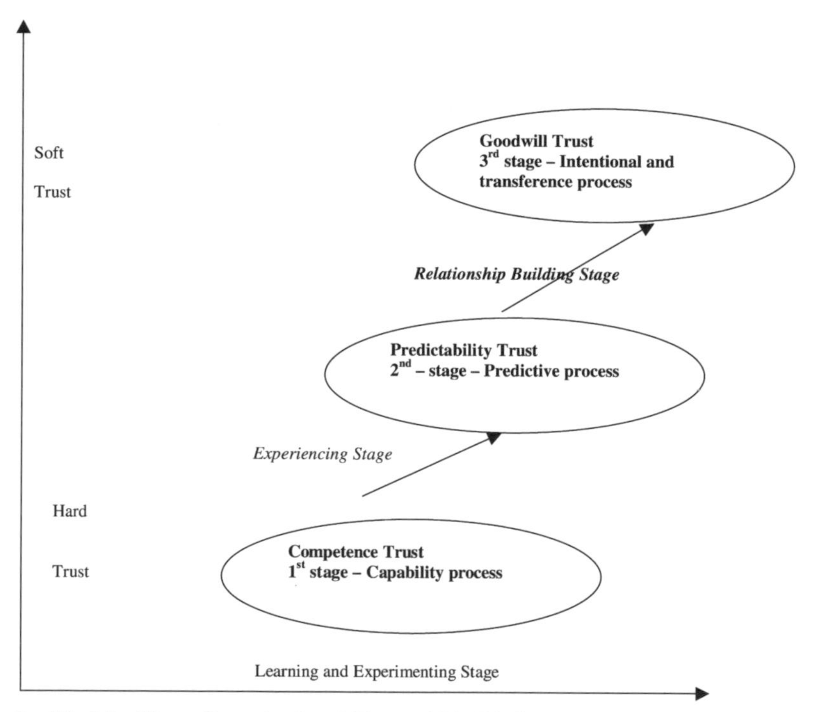

Fig. 2. Model of Inter-Organizational-Trust within Bi-directional Dyads in E-commerce Participation.

model of inter-organizational-trust within bi-directional dyads in E-commerce participation identifies a gradual development of inter-organizational-trust in three stages.

Stage 1: Competence Trading Partner Trust – Capability Process
In the first stage, new E-commerce adopters typically concentrate on training their trading partners to use E-commerce applications correctly, thus taking a bottom up approach in analyzing trust behaviors (such as trading partner skills, knowledge, ability, and product expertise). The emphasis is on trading partners' competence, skills, and ability to operate business-to-business E-commerce applications. The focus is on a capability process, which examines competence of trading partners in sending and receiving business-to-business E-commerce transactions. This contributes to transactional (objective) trust and economic benefits from competence trust, and also from technology trust mechanisms (in the form of governance mechanisms). Economic benefits are derived from savings in time and costs, as competent trading partners were able to send and receive transactions correctly. For example, Ring and Van de Ven (1994) propose that a decision to continue inter-organizational relationships were based on an assessment of economic efficiency and fairness of past transactions. Alternatively, trading partners who operate using incompatible E-commerce applications or apply poor business practices (as in inadequate audit, quality standards, training and skills) are constantly making errors during transacting and are contributing to technology performance-related risks.

Stage 2: Predictability Trading Partner Trust – Predictive Process
During the second stage, consistent positive behaviors from reliable trading partners contribute to trustworthy behaviors and predictability trust. This, in turn, increases trading partner satisfaction and contributes to relationship-related benefits such as a willingness to cooperate, commit, and share information. The predictive process applies a middle-out approach for analyzing trust behaviors of trading partners because it was based on the knowledge of past experiences. The focus now gradually evolves from transactional (objective) trust to relational (subjective) trust, which emphasizes trading partner behaviors (such as reliability, honesty and predictability). Alternatively, trading partners who do not exhibit consistent, reliable, positive trust contribute to relational risks derived from uncertainties, lack of knowledge, dependencies, situations of imbalance of power, and opportunistic behaviors leading to conflicts.

Stage 3: Goodwill Trading Partner Trust – Intentional and Transference Process
Finally, in the third stage, goodwill trust focuses on institutional (organizational) reputation and brand names, accomplished by enforcing best business practices.

Positive consistent behaviors encourage trading partners to invest in their trading partner relationships (i.e. renew the trading partners contract), increase E-commerce performance (as in volume, dollar value of goods), and reduce opportunistic behaviors (such as power, control, and the imposition of strict deadlines). Thus, positive repeated behaviors increase trading partner satisfaction and enables goodwill trust to develop. Satisfaction with delivered products and service has been suggested and empirically documented as affecting the buyer's decision to continue a relationship (Anderson & Narus, 1990) and conversely reduce the likelihood of exit for the relationship and negative word of mouth. The focus now is on the organization as it takes a top-down approach in analyzing their trading partner trust behaviors that include top management commitment, high-quality standards, open communications, information sharing, and long-term trading partner relationships, that contribute to strategic benefits. Alternatively, poor business practices (such as a lack of audit reviews, lack of poor risk management strategies or lack of top management commitment) may lead to general risks. Research on trading partner trust has shown that trust is transferable. Thus, trust in this stage is based on an intentional and transference process (Doney & Cannon, 1997; Lewicki & Bunker, 1995; McKnight & Chervany, 2001; Rousseau, Sitkin, Burt & Camerer, 1998).

Figure 2 depicts the model of inter-organizational-trust within bi-directional dyads in E-commerce participation.

Although business-to-business E-commerce systems and applications facilitate the development of initial competence trust, trust needs time to develop, as it evolves gradually from one stage to the next stage. The model enables trading partners to identify which stage of trust they and their trading partners belong to. Table 12 outlines the characteristics of the three stages of inter-organizational-trust and highlight the key benefits and risks experienced by the four bi-directional dyads.

The model of inter-organizational-trust within bi-directional dyads in E-commerce participation identified characteristics that determined positive and negative consequences of inter-organizational-trust. Most previous research examined behavioral characteristics in business relationships. It can be concluded that inter-organizational-trust is important for E-commerce participation and this model contributed not only to trust behavioral characteristics and processes involved at each stage of trust building, but also to benefits and risks that organizations face in business-to-business E-commerce participation (adoption, integration and use). E-commerce practitioners will benefit from this information, as they will be able to assess the likely effects and effectiveness of trust if applied correctly in their organization. This study contributed to the determinants of successful E-commerce participation.

Table 12. Model of Inter-Organizational-Trust within Bi-directional Dyads in E-commerce Participation.

Three Stages of Inter-Organizational-Trust	Bi-directional Dyads A–D Focus on the Supplier or Provider of a Service	Bi-directional Dyads A–D Focus on the Buyer or Receiver of a Service
Competence trading partner trust characteristics (economic foundation)	• Enforces trading partner agreement • Provides initial training • Exhibits tolerance for mistakes • Applies persuasive power • Focuses on operational goals • Emphasizes on trust and security based mechanisms from E-commerce applications	• Experiments with new E-commerce applications • Makes costly mistakes • Experiences an increase in communication and trading partner interactions (via telephone calls, e-mail and fax) • Relies on IT technical support
Benefits of competence trading partner trust	• Less time spent on training buyers • More time was spent on strategic planning and updating information on the extranet application	• Experiences economic benefits from savings in time and cost from telephone calls • Buyers were able to obtain online product information and real-time tracking information
Risks of competence trading partner trust	• Risks were not significant from the suppliers side	• Experiences technology performance related risks from mistakes • Buyers made initial mistakes that did cost them time and money
Predictability trading partner trust characteristics (familiarity foundation)	• Consistent behaviors in providing high quality service • Focuses on mid-range tactical goals	• Experiences the ability to operate correctly from previous mistakes • Exhibits consistent behavior
Benefits of predictability trading partner trust	• Experiences relationship-related benefits from trading partner satisfaction • Relationship-related benefits such as buyers willingness to communicate and share information was found	• Experiences relationship-related benefits from accurate real-time information and end-customers satisfaction
Risks of predictability trading partner trust	• Experiences relational risks such as business conflicts did occur at times due to uncertainties from the suppliers	• Experiences relational risks from suppliers who was perceived to exercise coercive power and opportunistic behaviors

Table 12. (*Continued*)

Three Stages of Inter-Organizational-Trust	Bi-directional Dyads A–D Focus on the Supplier or Provider of a Service	Bi-directional Dyads A–D Focus on the Buyer or Receiver of a Service
Goodwill trading partner trust characteristics (Empathy Foundation)	• Experienced increased reputation of their organization • Brand names • Aims to sign long-term contracts • Plans strategic long-term goals • Experiences increase in E-commerce performance (that is increased volume, dollar value of goods) • Exhibits willingness to share information, open communication	• Exhibits cooperation • Exhibits commitment • Experiences an increase in trading partner satisfaction • Exhibits willingness to put every effort to invest in the relationship • Engages in long term investments
Benefits of goodwill trading partner trust	• Experiences increase in trading partner trust • Willingness to engage buyers in long-term investment	• Experiences strategic benefits and increased reputation as buyers were willing to coordinate and commit to long-term relationships with suppliers
Risks of goodwill trading partner trust	• Exhibits less opportunistic behavior – Suppliers felt that buyers could leak out sensitive information	• Experienced a demand from suppliers to enforce best business practices in the form of high quality standards

CONCLUSIONS

In this study we examine trust from both trading partners' perspectives (that is both buyer-supplier and/or trustor-trustee) within a bi-directional dyad. Based on the cross-case analyses, it can be concluded that a positive impact of inter-organizational-trust in E-commerce performance is necessary for long-term trading partner trust relationship development.

While this study emphasized on trading partner trust in business-to-business E-commerce, it is not the purpose of this paper to undermine the importance of technology trust mechanisms of E-commerce (i.e. safeguards and assurances provided by E-commerce technologies). By including technology trust mechanisms, perceived benefits of E-commerce, and perceived risks of E-commerce constructs in the theoretical framework, we extend the current literature to focus on the consequences of trust. We have demonstrated in this paper that building trust alone is not only necessary, but also a sufficient condition for successful business-to-business E-commerce. In addition, reaping benefits, mitigating and controlling risks in E-commerce is also important by way of best business practices enforced by technology trust mechanisms.

Theoretical Implications

A number of theoretical and practical implications arise from this study. Drawing on the results from eight case studies, the paper generates an understanding of trading partner trust behaviors. The study contributes to theory as rather than inferring characteristics of E-commerce adoption from a technical and economic background, this study examined behavioral characteristics of trading partners in business-to-business E-commerce from theories in multi-disciplines. They include economic as well as the socio-political, behavioral, and organizational perspectives of inter-organizational-trust in E-commerce participation.

The primary emphasis of prior research was on transaction economics, its competitive advantages and/or external pressure (socio-political). This study focused on the importance of inter-organizational-trust in E-commerce adoption. The findings of the study led to the development of a model of inter-organizational-trust within bi-directional dyads in E-commerce adoption. The model of inter-organizational-trust within bi-directional dyads in E-commerce participation identified three stages of trust will enable trading partners to identify which stage of trust, they and their trading partners belong to. The model thus acts as a checking mechanism for both growth and development of inter-organizational-trust in business-to-business E-commerce participation.

Practical Implications

This study also provides implications for practice and managers. Trading partners can now clearly assess the effectiveness of their trust mechanisms not only of themselves but that of their trading partners. The main implications of this study are the insights it provides about the relationship between trust and levels of participation in E-commerce. We believe that the findings of this study would enable current and potential E-commerce adopters to design more effective strategies, trading partner agreements, and partnering/relationship charters, in order to increase their level of trading partner interactions, and coordination, thus eventually increasing trust and E-commerce participation.

Limitations and Suggestions for Future Research

The multiple case studies represent a first effort in IS research to formally examine behavioral characteristics of trust in business-to-business E-commerce. Interpret the results of this study in view of these limitations. First, the findings and implications of the research may be constrained by the research context of bi-directional dyads. Though our findings support the general theoretical framework, undertaking similar studies in the form of longitudinal research would be useful in order to enhance or refute our empirical findings. In addition, the dynamic and constantly changing context of E-commerce environment affect inter-organizational-relationships and trust in the future. This research examines only a sub-set of the many possible relationships between trust and its antecedents, consequences and moderating variables. Future research should take a more extensive approach to cover all possible positive and negative antecedents of trust in E-commerce.

REFERENCES

Anderson, J. C., & Narus, J. A. (1990). A model of distributor firm and manufacturer firm working partnerships. *Journal of Marketing*, *54*(January), 42–58.

Anderson, E., & Weitz, B. A. (1989). Determinants of continuity in conventional channel dyads. *Marketing Science*, *8*(4), 310–323.

Ba, S., & Pavlou, P. A. (2001). Evidence of the effect of trust building technology in electronic markets, price premiums, and buyer behavior. *MIS Quarterly*, *26*(3), 243–268.

Bakos, Y. (1998). The emerging role of electronic marketplaces on the internet. *Communications of the ACM*, *41*(8), 35–48.

Barney, J. B., & Hansen, M. H. (1994). Trustworthiness as a source of competitive advantage. *Strategic Management Journal*, *15*, 175–216.

Benbasat, I., Goldstein, D. K., & Mead, M. (1987). The case research strategy in studies of information systems. *MIS Quarterly* (September), 368–383.

Bensaou, M., & Venkatraman, N. (1996). Inter-organizational relationships and information technology: A conceptual synthesis and a research framework. *European Journal of Information Systems*, *5*, 84–91.

Berry, L. L. (1999). *Discovering the soul of service: The nine drivers of sustainable business successes*. New York: Free Press.

Berry, L. L., & Paramsuraman, A. (1991). *Marketing services competing through quality*. New York: Free Press.

Bhimani, A. (1996). Securing the commercial internet. *Communications of the ACM*, *39*(6), 29–35.

Blackmon, D. A. (2000). Where the money is? *Wall Street Journal* (pp. 30–32). New York.

Brensen, M. (1996). An organizational perspective on changing buyer-supplier relations: A critical review of the evidence. *Organization Articles*, *3*(1), 121–146.

Bromiley, P., & Cummings, L. L. (1995). Transaction costs in organizations with trust. Working Paper 28, Strategic Management Research Center, University of Minnesota, Minneapolis.

Cannon, J. P., & Perrault, W. D. (1999). Buyer-seller relationships in business markets. *Journal of Marketing Research, 36*, 439–460.

Cavalli, A. (1995). Electronic commerce over the internet and the increasing need for security. *Trade-Wave*, December 8.

CERT, C.E.R.T.C.C. (2000). *Infosec Outlook*. p. 1.

Chan, S., & Davis, T. (2000). Partnering on extranets for strategic advantage. *Information Systems Management, 17*(1), 58–64.

Chellappa, R. K. (2001). Working Paper, *ebizlab*, Los Angeles.

Chellappa, R., & Pavlou, P. A. (2002). Perceived information security, financial liability, and consumer trust in electronic commerce transactions. *Journal of Logistics Information Management, Special Issue on 'Information Security', 15*(5–6), 358–368.

Chiles, T. H., & McMackin, J. F. (1996). Integrating variable risk preferences, trust, and transaction costs economics. *Academy of Management Review, 21*(1), 73–99.

Coleman, J. S. (1990). *Foundations of social theory*. Cambridge, MA: Belknap Press.

CommerceNet (1997). Barriers and inhibitors to the widespread adoption of internet commerce. CommerceNet Research Report No. 97-05, April.

Covello, V. T., & Merkhofer, M. W. (1994). *Risk assessment methods*. New York: Plenum Press.

Cummings, L. L., & Bromiley, P. (1996). The organizational trust inventory (OTI): Development and validation. In: R. M. Kramer & T. R. Tyler (Eds), *Trust in Organizations: Frontiers of Theory and Research* (pp. 302–320). Thousand Oaks, CA: Sage.

Das, T. K., & Teng, B.-S. (1996). Risk types and inter-firm alliance structures. *Journal of Management Studies, 33*(4), 827–843.

Das, T. K., & Teng, B.-S. (1998). Between trust and control: Developing confidence in partner cooperation in alliances. *Academy of Management Review, 23*(3), 491–512.

Deutsch, M. (1958). The effect of motivational orientation upon trust and suspicion. *Human Relations* (13), 123–139.

Doney, P. M., & Cannon, J. P. (1997). An examination of the nature of trust in buyer-seller relationships. *Journal of Marketing* (April), 35–51.

Dosdale, T. (1994, July). Security in EDIFACT systems. *Computer Communications, 17*(7), 532–537.

Drummond, R. (1995, December). Safe and secure electronic commerce. *Network Computing, 7*(19), 116–121.

Dwyer, R. F., Schurr, P. H., & Oh, S. (1987). Developing buyer-seller relationships. *Journal of Marketing* (51), 11–27.

Dyer, J. H., & Chu, W. C. (2000). The determinants of trust in supplier auto-maker relationships in the U.S., Japan and Korea. *Journal of International Business Studies, 31*(2), 259–285.

EDICA (1990). EDI Control Guide – Make your business more competitive. *EDI Council of Australia and EDP Auditors Association.*

Fearson, C., & Phillip, G. (1998). Self-assessment as a means of measuring strategic and operational benefits from EDI: The development of a conceptual framework. *European Journal of Information Systems* (7), 5–16.

Fung, R., & Lee, M. (1999). Trust in electronic commerce: Exploring antecedents factors. Proceedings of the 5th Americas Conference on Information Systems (pp. 517–519).

Gabarro, J. (1987). *The dynamics of taking charge*. Boston: Harvard Business School Press.

Gambetta, D. (1988). *Trust: Making and breaking cooperative relations*. New York: Basil Blackwell.

Gambrisch, H. (1993). Difficulties in establishing joint ventures in Eastern Europe. *Eastern European Economics, 31*(4), 19–51.

Ganesan, S. (1994). Determinants of long-term orientation in buyer-seller relationships. *Journal of Marketing*, *58*(April), 1–19.

Gentry, D. J. (1994). Selecting financial EDI software. *TMA Journal*, 40–45.

Geyskens, I., Steenkamp, J. B., & Kumar, N. (1998). Generalizations about trust in marketing channel relationships using meta-analysis. *International Journal in Marketing*, *15*, 223–248.

Granovetter, M. (1985). Economic action and social structure: The problem of embeddedness. *American Journal of Sociology*, *91*(3), 185–205.

Gulati, R. (1995). Does familiarity breed trust? The implications of repeated ties for contractual choice in alliances. *Academy of Management Journal*, *38*(1), 85–112.

Handy, C. (1995). Trust and the virtual organization. *Harvard Business Review*, *73*(3), 40–50.

Hart, P., & Saunders, C. (1997). Power and trust: Critical factors in the adoption and use of electronic data interchange. *Organization Science*, *8*(1), 23–42.

Hart, P., & Saunders, C. (1998). Emerging electronic partnerships: Antecedents and dimensions of EDI use from the supplier's perspective. *Journal of Management Information Systems*, *14*(4), 87–111.

Heil, G., Stephens, D., & Bennis, W. (2000). *Douglas McGregor, revisited: Managing the human side of the enterprise*. Somerset, NJ: John Wiley and Sons.

Helper, S. (1991). How much has really changed between U.S. automakers and their suppliers? *Sloan Management Review*, *32*(4), 15–28.

Hill, C. W. L. (1990). Cooperation, opportunism, and the invisible hand: Implication for transaction cost theory. *Academy of Management Review*, *15*, 500–513.

Hoffman, D. L., & Novak, T. P. (1995). Marketing in Hypermedia Computer-Mediated Environments: Conceptual Foundation, Revised July 11 (http://www2000.ogsm.vanderbilt.edu/cmepapaer.revision.july11.1995/cmepaper.html).

Hosmer, L. T. (1995). Trust: The connecting link between organizational theory and philosophical ethics. *Academic Management Review*, *20*(2), 379–403.

Iacovou, C. L., Benbasat, I., & Dexter, A. S. (1995). Electronic data interchange and small organizations: Adoption and impact of technology. *MIS Quarterly*, *19*(4), 465–485.

ICAEW (1992). Institute of Chartered Accountants in England and Wales – EDI Working Party. *Harnessing EDI, Controlling the Business Risks*. Institute of Chartered Accountants in England and Wales, U.K., pp. 1–43.

Infoweek (2001a). Information sharing and collaboration: A matter of trust, May.

Infoweek (2001b). Mantra of collaboration: Trust but verify, July 30.

Jamieson, R. (1996). Auditing and Electronic Commerce. *EDI Forum*. Perth, Western Australia.

Jarvenpaa, S. L., & Tractinsky, N. (1999). Consumer trust in an internet store: A cross-cultural validation. *Journal of Computer Mediated Communication*, *5*(2).

Jarvenpaa, S., Tractinsky, N., & Vitale, J. (2000). Consumer trust in an internet store. *Information Technology and Management Journal*, *1*(2), 45–71.

Jevans, D. (1999). Extranets Rev Up EDI. *ECOM World*, http://www.ecomworld.com/html/entrpriz/020199-3.htm

Jones, G., & George, J. (1998). The experience and evolution of trust: Implication for cooperation and teamwork. *Academy of Management Review*, *23*(3), 531–546.

Kalakota, R. (1996, Spring). Manager's guide to electronic commerce. Addison-Wesley (forthcoming) previewed in: E. Burns (Ed.), *Defining Electronic Commerce: EDICAST*. Issue 28, February/March, p. 5.

Kalakota, R., & Robinson, M. (2000, December). *E-business 2.0: Roadmap for success*. Pearson Education.

Kalakota, R., & Whinston, A. B. (1996). *Frontiers of electronic commerce*. Addison-Wesley.

Keen, P. G. W. (1999). Electronic commerce: How fast, how soon? http://strategis.ic.gc.ca/SSG/mi06348e.html

Keen, P. G. W. (2000). Ensuring E-trust. *Computerworld*, March 13.

Koller, M. (1988). Risk as a determinant of trust. *Basic and Applied Social Psychology*, *9*(4), 265–276.

Kramer, R. M., & Tyler, T. R. (1996). *Trust in organizations: Frontiers of theory and research*. Thousand Oaks, CA: Sage.

Kumar, N. (1996). The power of trust in manufacturer-retailer relationships. *Harvard Business Review* (November–December), 92–106.

Langfield-Smith, K., & Greenwood, M. R. (1998). Developing co-operative buyer-supplier relationships: A case study of Toyota. *Journal of Management Studies*, *35*(3), 331–353.

Lee, M. K. O., & Turban, E. (2001). A trust model for consumer internet shopping. *International Journal of Electronic Commerce*, *6*(1), 75–92.

Lemos, R. (2001). Egghead hack costs millions: Companies paid big bucks to reissue credit cards, January 9.

Leverick, F., & Cooper, R. (1998). Partnerships in motor industry: Opportunities and risks for suppliers. *Long Range Planning*, *31*(1), 72–81.

Lewicki, R. J., & Bunker, B. B. (1995). Trust in relationships: A model of trust development and decline. In: B. B. Bunker & J. Z. Rubin (Eds), *Conflict, Cooperation and Justice* (pp. 133–173). San Francisco, CA: Jossey-Bass.

Lewicki, R. J., & Bunker, B. B. (1996). Developing and maintaining trust in work relationships. In: R. M. Kramer & T. R. Tyler (Eds), *Trust in Organizations: Frontiers of Theory and Research* (pp. 114–139). Thousand Oaks, CA: Sage.

Lewis, J. D., & Weigert, A. (1985). Trust as a social reality. *Social Forces*, *63*(4), 967–985.

Marcella, A. J., Stone, L., & Sampias, W. J. (1998). *Electronic commerce: Control issues for securing virtual enterprises*. The Institute of Internal Auditors.

Massetti, B., & Zmud, R. (1995). Measuring the extent of EDI usage in complex organizations: Strategies and illustrative examples. *MIS Quarterly*, *20*(3), 331–345.

Mayer, R. C., Davis, J. H., & Schoorman, F. D. (1995). An integrative model of organizational trust. *Academy of Management Review*, *20*(3), 709–734.

McAllister, D. J. (1995). Affect- and cognition-based trust as foundations for interpersonal cooperation in organizations. *Academy of Management Journal*, *38*(1), 24–59.

McKnight, D. H., & Chervany, N. L. (2001). What trust means in E-commerce customer relationships: An interdisciplinary conceptual typology. *International Journal of Electronic Commerce*, *6*(2).

McKnight, D. H., Cummings, L. L., & Chervany, N. L. (1998). Initial trust formation in new organizational relationships. *Academy of Management Review*, *23*(3), 473–490.

Miles, M. B., & Huberman, A. M. (1994). Qualitative data analysis, *an expanded sourcebook* (2nd ed.). Sage.

Miller, D., & Shamsie, J. (1999). Strategic responses to three kinds of uncertainty: Product line simplicity at the Hollywood film studios. *Journal of Management*, *25*(1), 97–116.

Mishra, A. K. (1996). Organizational responses to crisis – the centrality of trust. In: R. M. Kramer & Tyler (Eds), *Trust in Organizations: Frontiers of Theory and Research* (pp. 261–287). Thousand Oaks, CA: Sage.

Miyazaki, A. D., & Fernandez, A. (2000). Internet privacy and security: An examination of online retailer disclosures. *Journal of Public Policy and Markets*, *19*(Spring), 54–61.

Mohr, J., & Spekman, R. (1994). Characteristics of partnership success: Partnership attributes, communication behavior, and conflict resolution techniques. *Strategic Management Journal* (15), 135–152.

Moorman, R., Zaltman, G., & Deshpande, R. (1993). Factors affecting trust in market research relationships. *Journal of Marketing, 57*(January), 81–101.

Morgan, R. M., & Hunt, S. D. (1994). The commitment-trust theory of relationship marketing. *Journal of Marketing, 58*, 20–38.

Mukhopadyay, T., Kekre, S., & Kalathur, S. (1995). Business value of information technology: A study of electronic data interchange. *MIS Quarterly, 19*(2), 137–156.

Nath, R., Akmanligil, M., Hjelm, K., Sakaguch, T., & Schultz, M. (1998). Electronic commerce and the internet: Issues, problems and perspectives. *International Journal of Information Management, 18*(2), 91–101.

Norlan and Norton Institute, KPMG (1999). *Electronic commerce – the future is here.*

O'Hara-Devereaux, M., & Johansen, B. (1984). *Global work: Bridging distance, culture, and time.* San Francisco: Jossey-Bass.

Parker, D. B. (1995). A new framework for information security to avoid information anarchy. *IFIP*, 155–164.

Parkhe, A. (1998). Understanding trust in international alliances. *Journal of World Business, 33*(3), 219–240.

Premkumar, G., Ramamurthy, K., & Crum, M. (1997). Determinants of EDI adoption in the transportation industry. *European Journal of Information Systems, 6*, 107–121.

Raman, D. (1996). *Cyber-assisted business – EDI as the backbone of electronic commerce.* EDI-TIE.

Ratnasingam, P. (1998). Internet based EDI trust and security. *Information Management and Computers Security, 6*(1), 33–39.

Ratnasingam, P. (2000). The influence of power among trading partners in business to business electronic commerce. *Internet Research, 1*, 56–62.

Ratnasingam, P., & Pavlou, P. (2002). Technology trust: The next value creator in business to business E-commerce. International Resources Management Association Conference. Seattle, Washington, May 19–22.

Ratnasingam, P., & Pavlou, P. (2003). Technology trust in internet-based interorganizational electronic commerce. *Journal of Electronic Commerce in Organizations* (Inaugral Issue), *1*(1), 17–41.

Riggins, F. J., & Rhee, H. S. (1998). Toward a unified view of electronic commerce. *Communications of the ACM, 41*(10), 88–95.

Riggins, F. J., & Rhee, H. S. (1999). Developing the learning network using extranets. *International Journal of Electronic Commerce, 14*(1), 65–84.

Ring, P. S., & Van de Ven, A. H. (1994). Developing processes of cooperative inter-organizational relationships. *Academy of Management Review* (19), 90–118.

Rousseau, D. M., Sitkin, S. B., Burt, R. S., & Camerer, C. (1998). Not so different after all: A cross-discipline view of trust. *The Academy of Management Review, 23*(3), 393–404.

Sabo, D. (1997). *Industry pulse: Electronic commerce barriers survey results.* ITAA.

Sako, M. (1998). Does trust improve business performance? In: C. M. Lane & Bachmann (Eds), *Trust Within and Between Organizations – Conceptual Issues and Empirical Applications.* Oxford University Press.

Saunders, C. (1990). The strategic contingencies theory of power: Multiple perspectives. *Journal of Management Studies, 27*(1), 1–18.

Schurr, P. H., & Ozanne, J. L. (1985, March). Influence on exchange processes: Buyer's perception of a seller's trustworthiness and bargaining toughness. *Journal of Consumer Research* (11), 939–953.

Senn, J. A. (1999). The evolution of business to business commerce models – The influence of new information technology models. *WECWIS*, 153–158.

Senn, J. A. (2000). Business to business E-commerce. *Information Systems Management* (Spring), 23–32.

Shapiro, S. P. (1987). The social control of impersonal trust. *American Journal of Sociology*, *93*(3), 623–658.

Sharp, D. E. (1998). Extranets: Borderless internet/intranet networking, strategic directions. *Information Systems Management* (Summer), 31–35.

Sitkin, S. B., & Pablo, A. L. (1992). Re-conceptualizing the determinants of risk behavior. *Academy of Management Review*, *17*(1), 9–38.

Smeltzer, L. (1997). The meaning and origin of trust in buyer-seller relationships. *International Journal of Purchasing and Materials Management*, *33*(1), 40–48.

Smith, J. B., & Barclay, D. W. (1997). The effects of organizational differences and trust on the effectiveness of selling partner relationships. *Journal of Marketing* (51), 3–21.

Stewart, T. R. (1998). Selected E-business issues – Perspectives on business in cyberspace. *Deloitte Touche Tohmatsu* (September), 1–26.

Storrosten, M. (1998). Barriers to electronic commerce. European Multimedia, Microprocessor Systems and Electronic Commerce Conference and Exhibition. Bordeaux, France.

Sydow, J. (1998). Understanding the constitution of inter-organizational-trust. In: C. Lane & R. Bachmann (Eds), *Trust within and between Organizations, Conceptual Issues and Empirical Applications*.

Tallon, P. P., Kraemer, K. L., & Gurbaxani, V. (2000). Executives perceptions of the business value of information technology: A process-oriented approach. *Journal of Management Information Systems* (16), 145–173.

Tan, Y.-H., & Thoen, W. (1998). Towards a generic model of trust for electronic commerce. *International Journal of Electronic Commerce* (3), 65–81.

Van de Ven, A. (1992). Suggestions for studying strategy process. *Strategic Management Journal*, *13*(Summer Special Issue), 169–188.

Vijayasarathy, L. R., & Robey, D. (1997). The effect of EDI on market channel relationship in retailing. *Information and Management* (33), 73–86.

Walton, S. V. (1997). The relationship between EDI and supplier reliability. *International Journal of Purchasing and Materials Management*, *33*(3), 30–35.

Webster, J. (1995). Networks of collaboration or conflict? Electronic data interchange and power in the supply chain. *Journal of Strategic Information Systems*, *4*(1), 31–42.

Williamson, O. E. (1975). *Markets and hierarchies*. New York: Free Press.

Williamson, O. E. (1993). Opportunism and its critics. *Managerial and Decision Economics* (14), 97–107.

Yin, R. K. (1989). *Case study research: Design and methods*. Sage.

Yin, R. K. (1994). *Case study research: Design and methods* (2nd ed.). Thousand Oaks, CA: Sage.

Zaheer, A., McEvily, B., & Perrone, V. (1998). Does trust matter? Exploring the effects of inter-organizational and interpersonal trust on performance. *Organization Science*, *9*(2), 141–159.

APPENDIX

Case Study Questionnaire

Date:

To whom it may concern

Dear Sirs

Re: A Study of Inter-Organizational-Trust in Business-to-Business E-commerce Participation

The purpose of this study is to examine the impact of inter-organizational-trust in E-commerce participation. E-commerce participation refers to the extent an organization adopts, integrates or implements business-to-business E-commerce. Inter-organizational-trust is actually interpersonal trust, hence it also refers to trading partner trust. We aim to examine how and why inter-organizational-trust (or trading partner trust) influences the perception of benefits and risks in E-commerce, thus leading to the extent of E-commerce participation. We believe that you will find it interesting and useful to participate in this study. It is our hope that this knowledge will help increase business-to-business E-commerce participation.

The questionnaire consists of four sections:

Section 1

This section seeks to obtain background (demographic) information about your organization. We seek information relating to factors that motivated your organization to adopt E-commerce and determine antecedent trust behaviors relating to trading partner relationships, technology trust mechanisms in E-commerce participation.

Section 2

This section seeks information about your organization's perception of benefits in E-commerce participation (that is perceived benefits derived from both trading partner relationships, trust and security-based mechanisms in E-commerce).

Section 3

This section seeks information about your organization's perception of risks in E-commerce participation (that is perceived risks derived from both trading partner relationships, and technology trust mechanisms in E-commerce).

Section 4

This section seeks information about your organization's extent of E-commerce participation.

Confidentiality

All responses will be kept in its strictest confidence. No individuals or organizations will be named in any outputs, nor will demographic information be revealed, such that the individual organizations can be identified. When the results of this study are published, it will be impossible to identify specific individuals or organizations, unless prior permission was received.

Summary Results

We will send a summary of the results to all organizations that participate in this study. The summary will provide conclusions related to the extent of inter-organizational-trust impacts perceived benefits and risks, thus leading to business-to-business E-commerce participation. We truly appreciate the time and effort you have put into this study. Your response will be of considerable help to this study.

Thank You

Ms Pauline Ratnasingam

Research Topic: The Importance of Inter-Organizational-Trust in E-commerce Participation

Research Question: How and why does inter-organizational-trust (trading partner trust) influence the perception of benefits and risks in business-to-business E-commerce, thus leading to the extent of its participation (adoption and integration)?

Section 1: (A) Demographic Section

(1) Name of your organization?
(2) Your job title?
(3) What is your organization's reach? (Local, regional, national, or global)?
(4) What is the size of your organization – (large or Small-Medium-Enterprise)?
(5) Type of industry and sector your organization is involved in (public or private)?

(6) What is your organization's product line?
(7) What is the main role of your organization? (Buyer, seller, manufacturer, or supplier)?
(8) Who is responsible for implementing E-commerce and is involved in E-commerce operations in your organization?
(9) What types of business transactions are actively supported by E-commerce in your organization?
(10) What types of E-commerce technologies/applications did your organization implement or will be implementing?
(11) How many trading partners does your organization have?
(12) How did your organization choose its trading partners?
(13) How long has your organization been trading with these trading partners?
(14) How do you maintain your trading partners (renewal of contracts)?
(15) What other measures are used?

(B) Trading Partner Trust

Trading Partner Trust – refers to the expectation that one trading partner will abide to the trading contract, is honest, and will act in a way not to take advantage of other trading partners. Trading partner trust is categorized as low, moderate, and high.

Please indicate your organization's reflections on behavioral characteristics relating to trading partner trust relationships. If yes, how, why, and in what situations do you relate these behaviors to trading partner trust (please provide examples and evidence)? Are there other trust behaviors that your organization faces in E-commerce participation?

Competence Trading Partner Trust

(1) Trading partner's ability, skills, and level of competence in business-to-business E-commerce operations.
(2) Trading partner depends on your organization.

Predictability Trading Partner Trust

(3) Trading partner's consistent behavior in business interactions.
(4) Trading partner's reliability in keeping business promises.
(5) Trading partner's adherence to policies, terms of contract, and trading partner agreements.
(6) Predictability of your trading partner.

Goodwill Trading Partner Trust

(7) Trading partner's willingness to share information and provide support relating to E-commerce adoption.
(8) Trading partner demonstrates care and concern in important decisions.
(9) Trading partner is committed to business arrangements and exhibits cooperation.
(10) Positive feelings towards your trading partner.
(11) Long-term trading relationships with your trading partner.
(12) Your organization is willing to put in more effort and invest in your trading partner relationships.
(13) Trading partner is honest in providing information and shows accuracy in meeting deadlines.
(14) Trading partner behavior in a situation of conflict and handling discrepancies exhibits?

Does your organization feel anger, frustration, resentment, or hostility towards your trading partner?

(15) Trading partner in a situation of pressure or imbalance of power.
(16) Trading partner considers security concerns.
(17) Trading partner is the driving force for adopting E-commerce.
(18) There are explicit agreements with the trading partners regarding roles and responsibilities.

How do you maintain your trading partner relationships? Is it short-term or long-term? What do you look for? How, why and in what situations?

On a scale of 10 what would you rate the level of trading partner trust? (Low = 0–3, Medium = 4–6, High = 7–10)

What did you rate the level of trading partner trust to be?

How, why, and in what situations?

Are there other antecedent trust behaviors your organization perceived in your trading partners?

(C) Technology Trust Mechanisms in E-commerce

Technology trust mechanisms in E-commerce refer to trust assurances as in confidence in the security protection services provided by E-commerce technologies.

Please indicate if your organization has adopted the following technology trust mechanisms in E-commerce. If yes, how, why, and in what situations were they implemented (please provide examples, evidence)? Are there any other trust and security-based mechanisms that your organization has implemented?

Confidentiality

(1) Firewalls
(2) Encryption mechanisms

Integrity

(3) System integrity tests and audits
(4) Sequence numbers in messages
(5) Application controls
(6) Accounting controls
(7) Web seal assurances

Authentication

(8) Formal logon procedures (user-IDs and passwords)

Non-repudiation

(9) Message receipt confirmations and acknowledgments
(10) Digital signatures

Access Control

(11) Network access controls
(12) Authorization mechanisms

Availability

(13) Segregation of duties

Best Business Practices

(14) Top management commitment
(15) Standards (industry and universal) and policies
(16) Trading Partner Agreement
(17) Audit check
(18) Training and education of staff
(19) Risk analysis and audit involvement
(20) Contingency procedures

On a scale of 1–10, how would you rate the level of technology trust mechanisms in E-commerce? (Low = 0–3, Medium = 4–6, High = 7–10)

What did you rate the level of technology trust mechanisms in E-commerce?

How, why, and in what situations?

Are there any other technology trust mechanisms in E-commerce your organization has implemented?

How does trading partner trust influence the perception of technology trust mechanisms in E-commerce?

Section 2: Perceived Benefits in E-commerce Participation

Perceived Benefits refer to gains that your organization may receive from adopting E-commerce. The perceived benefits are derived from both your trading partner relationships and from the E-commerce technology. Perceived benefits are categorized as direct (economic), indirect, relationship-related, and strategic benefits.

Please indicate if your organization faces the following benefits? If yes, how, why, and in what situations do you relate to the perceived benefits (please provide examples, evidence). Are there any other perceived benefits your organization faces? How does trading partner trust influence the perception of benefits in your organization?

Perceived Economic Benefits

(1) Reduced operation, transaction, and administrative costs
(2) Reduced error rates and improved accuracy of information exchanged
(3) Faster response to orders and creating reduced lead time
(4) Reduced inventory levels and optimized supply chain

Perceived Relationship-related Benefits

(5) Improved customer service and product quality
(6) Improved productivity, improved profitability, and increased sales
(7) Gaining competitive advantage
(8) Sharing of risks with your trading partner
(9) Improved communication and cooperation with your trading partners
(10) Sharing of information that is accurate, timely, speedy, complete, and relevant
(11) Increased level of commitment with your trading partners

Perceived Strategic Benefits

(12) Improved organizational image and reputation
(13) Increased long-term investments and continued trading partner relationships

On a scale of 10, how would you rate the level of perceived benefits in E-commerce? (Low = 0–3, Medium = 4–6, High = 7–10)

What is the impact of these perceived benefits as a result of trading partner trust and trust and security-based mechanisms in your organization?

What did you rate the level of perceived benefits to be?

How, why, and in what situations?

Are there any other perceived benefits in E-commerce your organization faces?

Section 3: Perceived Risks in E-commerce Participation

Perceived risks refer to barriers and obstacles your organization faces as a result of adopting E-commerce. Perceived risks are derived from both trading partner relationships and from the E-commerce technology. Perceived risks are categorized as technology performance-related risks, relational risks and general risks.

Please indicate if your organization faces the following perceived risks? If yes, how, why, and in what situations do you relate to these perceived risks (please provide examples, evidence)? Are there any other perceived risks that your organization faces? How does trading partner trust influence the perception of risks in your organization?

Perceived Technology Performance-related Risks

(1) Compatibility problems with hardware and software
(2) Infrastructure and initial implementation costs
(3) Confidentiality concerns due to viruses
(4) Lack of adequate accounting controls
(5) Internal security error (lack in integrity as in delayed and inaccurate messages)
(6) Complexity in operating business transactions
(7) Uncertainties (task and environment)

Perceived Relational Risks

(8) Trading partner reluctance to change
(9) Lack of training, knowledge, and awareness
(10) Poor reputation of trading partner
(11) Trading partner demonstrating a conflicting attitude
(12) Lack of trust in your trading partner
(13) Trading partner demonstrating opportunistic behaviors
(14) Partnership uncertainty

Perceived General Risks

(15) Lack of security in your trading partner's system
(16) Difficulty in identifying or quantifying costs and benefits
(17) Repudiation
(18) Authenticity of your trading partner
(19) Availability of technology
(20) Lack of a standard infrastructure (for data and payments)
(21) Lack of government policies
(22) Poor business practices

On a scale of 10 how would you rate the level of perceived risks in E-commerce to be? (Low = 0–3, Medium = 4–6, High = 7–10)

What is the impact of these perceived risks as a result of trading partner trust and trust and security-based mechanisms in your organization?

What did you rate the level of perceived risks in E-commerce to be?

How, why, and in what situations?

Are there any other perceived risks your organization faces?

Section 4: Participation in E-commerce

Participation in E-commerce refers to the extent your organization has adopted E-commerce. Participation in E-commerce is categorized as performance in E-commerce and the extent of mutual satisfaction your organization has with its trading partners.

Please reflect on your organization's extent of E-commerce participation. If yes, how, why, and in what situations do you relate to E-commerce participation (please provide examples, evidence)? Are there any other factors that your organization faces which contribute to E-commerce participation?

Extent of E-commerce Performance

(1) How important is E-commerce for your organization?
(2) What percentage of your business involves the use of E-commerce?
(3) What is the annual monetary value of E-commerce transactions in NZ$?
(4) What is the annual number of E-commerce transactions?

On a scale of 10 how would rate the extent of E-commerce performance in your organization? (Low = 0–3, Medium = 4–6, High = 7–10)

What did you rank the extent of E-commerce performance in your organization?

How, why, and in what situations?

Are there any other performance factors relating to E-commerce your organization has achieved?

How does trading partner trust influence the perception of E-commerce performance?

How do perceived benefits impact E-commerce performance?

How do perceived risks impact E-commerce performance?

Extent of Trading Partner Trust Relationships Development

(5) The trading partner will continue to be a major source of revenue for us.
(6) Has the number of trading partners increased?
(7) Do you perceive your organization to engage in long-term business investments with your trading partner?
(8) Do you perceive an increase in the level of open communications in your trading partner?
(9) Do you perceive an increase in the level of cooperation in your trading partner?
(10) Do you perceive an increase in the level of commitment in your trading partner?
(11) Has the reputation of your organization increased as a result of your trading partner?

On a scale of 10 how would you rate the extent of satisfaction in your trading partner relationships? (Low = 0–3, Medium = 4–6, High = 7–10)

What did you rate the level of satisfaction in your trading partner relationships to be?

How, why, and in what situations?

Are there other factors relating to trading partner satisfaction that your organization experienced?

How do trading partners trust, influence the perception of satisfaction in your trading partner relationships?

How do perceived benefits influence the perception of satisfaction in your trading partner relationships?

How do perceived risks influence the perception of satisfaction in your trading partner relationships?

EXAMINING INTERVENTIONALIZATION OF THE PROFESSIONAL SERVICES FIRM

Maria Anne Skaates

1. INTRODUCTION TO THE STUDY AND ITS METHODOLOGY AND RESEARCH QUESTIONS

At the beginning of the nineties, the Danish construction market was in the midst of a severe slump (Eurostat, 1995). At the same time, the German market was beginning to boom, due to the process of unifying the two German states (European Construction Research, 1995). Because of the poor home market circumstances, many Danish construction industry actors, including individual architects and architectural firms, attempted to find work in Germany (Halskov, 1995). However, the aspirations of most of these actors were dashed. By 1996, many of the largest Danish civil engineering and contracting firms had lost billions of Danish *kroner*, and a great number of small firms, typically architectural firms or subcontractors in the construction process, had also experienced severe losses, some of which had jeopardized the very existence of these firms (*ibid.*). This turn of events surprised both insiders in the Danish construction industry and the general Danish population as both groups believed that Denmark has high construction standards and that the most of the firms that had attempted operations in Germany were technically competent and had sound domestic business policies.

On the basis of the above scenario, this paper focuses on the Germany-related internationalization activities of three Danish architectural firms that were exceptions, in that they achieved a degree of success on the German market

Evaluating Marketing Actions and Outcomes
Advances in Business Marketing and Purchasing, Volume 12, 435–513

ISSN: 1069-0964/doi:10.1016/S1069-0964(03)12007-8

in the 1990s. The internationalization that is focused upon is *market-seeking* internationalization (Erramilli & Rao, 1990; Majkgård & Sharma, 1998), that is, firms' independent internationalization into foreign exchange networks. In contrast, *client-following* internationalization, that is, firms following exchange partners into the exchange partners' international networks (see Erramilli & Rao, 1990; Majkgård & Sharma, 1998), is only treated *en passant*, when it appears as an empirical phenomenon.

Furthermore the focus is upon the *internationalization* of architectural firms and not merely on exports. This is because, for the case of architectural services, it is problematic to speak of exports defined as, e.g. "the sale of goods and services in another country than the country in which they were produced" (Luostarinen & Welch, 1990, p. 20) because at least some of the production of architectural services most often happens in the same place as the sale (see e.g. Sharma, 1991; Winch, 1997).

Qualitative methods are used in the data collection process. Qualitative methods were chosen because the research focuses on firms from one small country (Denmark) of only 5.2 million inhabitants, meaning that the firm sample size would have been too small to merit a quantitative study design. Further, including firms from several different countries in a quantitative sample was not a feasible option, as substantial evidence (Button & Fleming, 1992; Dræbye, 1998; Oliver-Taylor, 1993; Stevens, 1998; Winch, 1997) suggests that architectural firms are not strictly comparable across countries. Concerning this issue and from an Australian perspective, Stevens (1998, p. 29) has stated the following:

> In France, architects rarely prepare construction drawings, and may never set foot on site. In Australia and other Commonwealth nations, the measurement and costing of large buildings is conducted by quantity surveyors, an occupation totally independent of architects. Norwegian architects also invariably handle town planning. Not only does the division of labor vary, but also the sort of client handled by the architect. In Italy, almost all small-scale construction is handled by the *geometria*, who we would consider surveyors, and the division of labor between architects and civil engineers is very indistinct. Spanish architects deal with highly technical buildings, such as industrial plants, that English-speaking architects tend to leave to civil engineers. Similarly, architects in the Benelux countries produce technical drawings that in the U.K. or USA would be handled by engineers.

Thus, it was necessary to specify the exact meaning of the terms "architectural services" and "internationalization" in the German-Danish context of this paper. Unfortunately no suitably comprehensive definition of "architectural services" for the Danish and German market could be found in pre-existing literature. I therefore constructed my own definition for this study. According to my definition, the individual services offered by architectural service firms in Denmark as well as Germany may encompass the following areas:

(1) Designing new buildings;
(2) Designing additions or improvements to existing buildings, e.g. renovations;
(3) Managing or assisting in the construction of new buildings;
(4) Managing or assisting in the construction of additions or improvements to existing buildings;
(5) Urban planning related services;
(6) Designing lawns, gardens, playgrounds or other outdoor areas (these areas belong to the sub-field of landscape architecture);
(7) Building inspection-related services, e.g. in connection with e.g. property sales/rentals or *forces majeures* damage;
(8) Management of facilities;
(9) Performing preliminary site studies for clients considering construction projects.

As concerns the aforementioned distinction between internationalization and exports, service types 3, 4, 7–9 are *inseparable* (see, e.g. Shostack, 1997), i.e. they must be performed on site and thus cannot be exported. On the other hand, design and urban planning-related services (types 1, 2, 5, and 6) may to some extent be performed at a distance from the site and thus also exported, as long as the architect(s) performing the service has(have) received the necessary information about the site from the customer.

Furthermore, service types 1–6 and 9 are project-related, meaning that they are related to transaction(s) "concerning a functioning whole which is delivered to the buyer" (Holstius, 1987, p. 8), i.e. the construction or renovation of a building or a landscape. However, in the project marketing and project management literature, project operations can also be subdivided into *partial projects, turnkey projects* and *turnkey plus projects* (Holstius, 1987; Luostarinen & Welch, 1990). Partial projects include partial system deliveries such as the delivery of the design of a building or the delivery of the waste disposal system for a factory in construction. In a turnkey project, a complete unit is delivered to the buyer. Turnkey plus projects are complete unit deliveries plus supplementary services such as, e.g. personnel training or operation services in the first years after delivery.

As regards consulting, architectural and engineering design services, Holstius (1987, pp. 49 and 56) has stated the following in relation to the three different types of projects listed above:

> The provision of consulting services differs from the delivery of construction, machinery and turnkey projects in that those consulting services can be directed specifically towards one or more of the stages in the project cycle. [...] the services rendered by a consulting firm can range from a more limited assignment at one of the project stages to a full turkey projects.

On the basis of Holstius' definition, it can be seen that architectural services most often are partial projects in the context of a larger, turnkey construction project. Thus, the project marketing literature may prove highly relevant to understanding the marketing and internationalization practice of architectural firms; this issue will therefore be examined further in Section 3 of this summary.

In the empirical study, three case firms are used as representatives for the three categories of Danish architectural firms that managed to achieve success in Germany during the nineties. The first firm belongs to the *internationally renowned* category, as it has been active on international markets for years and has won prizes in many countries. The second firm is *nationally renowned* in that it is one of Denmark's largest and most well established firms. It has won many Danish prizes in previous decades and has built some of the largest, most prestigious projects in its home country. Finally, the third firm is smaller and is *an innovative new firm* that has only existed on the Danish market since 1986.

The sources of data collected for the study include:

- Danish, German, and Pan-European or EU industrial statistics and studies of construction industry exports and internationalization;
- Articles in Danish, German, and Pan-European professional publications, i.e. articles written in the following languages: Danish, English, French, German and Swedish;
- Firm- and organization-specific documentary data such as annual reports, brochures, minutes of meetings, strategy plans;
- Studies undertaken by Danish and German architects' organizations;
- A total of 50 semi-structured qualitative interviews, which are listed in this summary's appendix. The interviews were conducted in Danish and German following the methodological suggestions of Kvale (1996).

To achieve greater internal validity, *data triangulation* (Yin, 1994) was used, the data were collected on more specific subjects from at least three different sources using established methods in relation to each source (*ibid.*, Silverman, 1993). Furthermore, multiple *key informants* (see Heide & John, 1992) were interviewed concerning *critical incidents* (see Hedaa & Törnroos, 1997).

However, an inherent weakness of the critical incident method is its tendency to represent the past as a series of discrete events, i.e. specific stimuli and responses, with intervening periods of no action. This entails the danger of excess simplification, reduction and even misrepresentation of the past, because less critical incidents or more "everyday" events, which are not mentioned by and/or taken for granted by the interviewees (see, e.g. Bourdieu, 1990), may have also played a role in the developments described by the respondents. Moreover, a second weakness of the critical incident method is its inability to overcome the problems of selective memory bias.

To overcome these weaknesses to the greatest extent possible, the interviews were structured such that the interviewer's questions included some of the key themes from the existing body of project marketing and architectural service literature under scrutiny. However, the questioning at the same time aimed to be open and not leading, to allow for the discovery of paradoxes, surprises, and the contra-theoretical insights of respondents, which could then be incorporated in subsequent questioning and data collection (see Maaløe, 1996, pp. 183–187). An extensive researchers' journal of the entire research process was kept, as suggested by, e.g. Alasuutari (1995) and Maaløe (1996), in order to keep track of the changes in the explanations examined and to formalize and document the parallel processes of collecting empirical material, scrutinizing theory and generating explanations to the greatest extent possible. Thus, the chosen interview method is not phenomenological, as suggested by McCracken (1988), but instead interpretive (see Kvale, 1996).

The common reason for choosing multiple case firms is "analytical generalization" (see, e.g. Alasuutari, 1995, pp. 120–132; Flick et al., 1995, pp. 446–450; Maaløe, 1996, pp. 71–75; Yin, 1994, p. 31), which in the words of Yin (*ibid.*) means the following:

> Multiple cases, in this sense, should be considered like multiple experiments (or multiple surveys). Under these circumstances, the method of generalization is "analytic generalization," in which a previously developed theory is used as a template with which to compare the empirical results of the case study. If two or more cases are shown to support the same theory, replication may be claimed. The empirical results may be considered yet more potent if two or more cases support the same theory but do not support an equally plausible, *rival* theory. (The *italics* are Yin's.)

Due to major firm-internal differences and the explorative-integrative nature of the study, which will be described below, the multiple cases in this study should not be considered 100% analogous to multiple experiments as described above. From a critical realist ontology (see, e.g. Easton, 1998; Yin, 1994), this raises some questions of external validity. However, my three case study firms are prime case study firms in that they have been chosen from a total population of four Danish architectural firms who opened a subsidiary on the German market in the 1990s. Because the population of Danish architectural firms that have attempted internationalization on the German market and obtained a reasonable degree of success is quite small yet very diverse, I am convinced that aiming for full analytical generalization would not have been feasible. Furthermore, I deal with the issue of external validity in more depth in the concluding section (Section 4) of this summary, in my remarks on the scientific contribution of the study (Subsection 4.1).

With regard to the treatment of existing theories and conceptualizations, the study is explorative-integrative (Maaløe, 1996) and abductive (Denzin, 1978, p. 109; Dubois & Gadde, 1999) and not deductive, as suggested by Yin (1994, see above). This means that in the study, inductive and deductive methodologies have been combined through the concurrent examination of empirical observations concerning the three case firms and of existing theories (i.e. contributions from industrial marketing, project marketing and Bourdieu's (1979, 1986a, b) sociology of art and cultural and social capital framework). The theories scrutinized include many theories, models and conceptualizations generated or tested in qualitative studies. In these instances, the qualitatively developed work is also scrutinized in the light of further empirical evidence, to generate more knowledge concerning external validity of the pre-existing findings.

However, the research questions themselves are descriptive. They have been chosen on the basis of a pilot study composed of a literature study and the first 14 of the 50 interviews listed in this summary's Appendix. (The pilot study is described in full-length in Appendix. It focused on the problems that the Danish architectural firms were less successful in attempting internationalization that they had on the German market during the nineties. The pilot study shows that the less successful firms generally had had difficulties detecting needs and establishing relationships to potential clients and cooperation partners in Germany as well as problems coping with the laws and conventions of the German construction industry.)

As for the study's descriptive research questions, the first two are of a general nature:

(1) How did the German market for architectural services develop during the nineties in terms of, e.g. institutions, total market size and growth rates at the national and federal state (in German: *Länder*) levels?
(2) Who were the major actors involved either directly or indirectly in the Danish internationalization of architectural services to Germany in the 1990s and to what extent and how did they cooperate with one another?

The purpose of these two questions is to provide an overview of the situation in the German market at the macro-economic, macro-structural and key German-Danish actor levels. Also the purpose is to lay a broad contextual foundation upon which the experiences of the three case firms can be scrutinized in relation to relevant international business and business-to-business marketing literature. The remaining three research questions concern the three case firms; they are as follows:

(3) What sort of specific knowledge about the German market and specific projects on the German market did the case firms use to obtain architectural projects in Germany?

(4) How were concrete architectural project jobs obtained by the case firms?
(5) What role did previous project work play when the case firms obtained specific projects in the German market?

In order to better explain how these three case study questions have been analyzed, relevant basic characteristics of architectural services and architectural projects are briefly presented in the next section of this summary. This second summary section also contains an introduction to the work of the Industrial Marketing and Purchasing (IMP) Group and its affiliated subgroup, the International Network for Project Marketing and Systems Selling (INPM), as the work of these research communities is heavily scrutinized. Finally, the second section concludes with a more thorough introduction to the three case firms. Thereafter the abductive links between previous contributions and the analysis of the empirical material related to the study's research questions are presented in Section 3 of this summary. Finally, Section 4 contains the concluding remarks of this summary, including its implications for further scientific study and for firms' and managers' practice. Finally, Section 5 is the list of references.

2. INTRODUCTION TO KEY DEFINITIONS AND THE THEORETICAL FRAMEWORK AND CASE FIRMS OF THE STUDY

The presentation of theories, definitions, and models resembles the extensive literature review genealogies more common to French- and German-language social science research than to English-language business administration literature (for an example of a genealogy, see Gemünden, 1990). Using genealogies was chosen due to the eclectic and descriptive nature of the literature on marketing and internationalization on business-to-business and project markets (see e.g. Backhaus & Büschken, 1997; Günter & Bonaccorsi, 1996).

Architectural services are defined as is presented in the introductory first section of this summary. Furthermore, relevant internationalization theories and contributions concerning knowledge in organizations are presented; these are introduced and discussed in this summary's Subsections 3.2 and 3.3. Finally, drawing on literature concerning the professional services[1] (e.g. Alvesson, 1995; Løwendahl, 2000; Sharma, 1991), architectural services (e.g. Albertsen, 1996; Baus, 1997; Harrigan & Neel, 1996; Marquart, 1997; Sommer, 1996; Stevens, 1998; von Gerkan, 1995; Winch & Schneider, 1993), construction projects (e.g. Day, 1994; Hellgren & Stjernberg, 1995; Oxley & Poskitt, 1996), and project marketing (e.g. Backhaus, 1995; Cova, 1990; Holstius, 1989; Mattsson, 1973), a general, two-phase model of the key characteristics of construction project-related architectural services is created as the fundament for the rest of the research; it is depicted in Fig. 1.

The model of Fig. 1 takes into account the following distinguishing characteristics of *architectural services* as *professional, technical services*, which are often organized as partial projects (see, e.g. Holstius, 1987; Luostarinen & Welch, 1990) of larger, turnkey construction projects:

General Characteristics related to neither Specific Construction Projects nor Specific Construction Project Processes and Phases:

- Project offerings concern large orders in terms of monetary value.
- Project offerings concern long-term investments.
- Project offerings may include financial packages or other "financial engineering" measures.
- Project offerings often involve bidder/supplier coalitions.
- Project offerings are international in the sense that offerers sometimes come from foreign countries.
- Project offerings are subject to deadlines and time limits.
- Orders for projects are discontinuous.

General Characteristics related to Specific Construction Project Phases:
Project Phase: Key Characteristics:

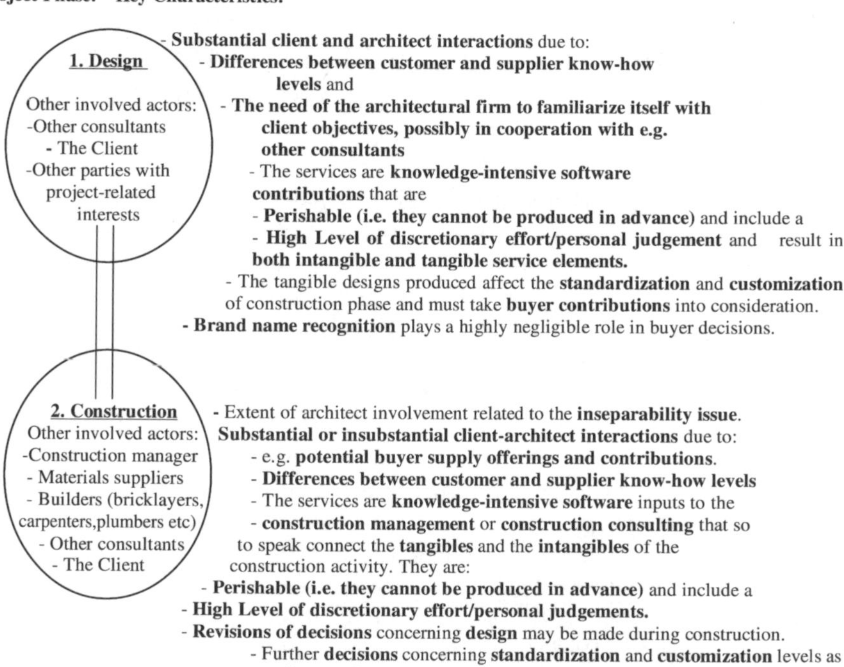

Fig. 1. General Characteristics of Architectural Projects.

(1) Architectural services are highly knowledge intensive (Alvesson, 1995; Løwendahl, 2000; Sharma, 1991; Winch & Schneider, 1993).
(2) Architectural services involve a high degree of customization and thus are non-standardized (Alvesson, 1995; Løwendahl, 2000; Sharma, 1991; Winch & Schneider, 1993).
(3) Architectural services involve a high degree of personal judgement on the part of the architect(s) (Alvesson, 1995; Løwendahl, 2000; Sharma, 1991; see also Starbuck, 1997).
(4) Architectural services involve substantial interaction with clients or their representatives (Alvesson, 1995; Løwendahl, 2000; Sharma, 1991; Winch & Schneider, 1993).
(5) Architectural services involve both intangible and tangible elements; however, most inputs in the production of the tangible elements are intangible (Sharma, 1991).
(6) Architects perceive a problem of inseparability in relation to architectural services, whereas other actors do not necessarily perceive this problem (compare Sharma, 1991 to *Arkitekt- og byggebladet*, November 1992, p. 31; Eurostat, 1995, pp. 24–39).
(7) Brand name is extremely difficult to establish in the field of architectural services (compare Sharma, 1991, to von Gerkan, 1995).
(8) Architectural services are perishable in the sense that advance production is not possible (Alvesson, 1995; Løwendahl, 2000; Sharma, 1991; Winch & Schneider, 1993).
(9) Architectural services are highly dependent upon the efforts of individual architects (Alvesson, 1995).
(10) Architectural services are less dependent on material assets than they are on elements in the minds of the architect(s), their networks, and the routines related to architectural services (Alvesson, 1995; Winch & Schneider, 1993).
(11) The production of architectural services requires loyalty from the architects involved (Alvesson, 1995).
(12) Architectural services are influenced by both the governance structure of the construction market and by rules that are common to artistic fields (compare Albertsen, 1996; Baus, 1997; Marquart, 1997; Stevens, 1998; von Gerkan, 1995; Winch & Schneider, 1993).

Furthermore, the model encompasses the fact that most construction management handbooks and practitioners (see, e.g. Day, 1994; Oxley & Poskitt, 1996) separate the construction project process into two major phases: (1) the design phase; and (2) the construction phase, even though in real life less significant design tasks may also be undertaken during the construction phase (Hellgren & Stjernberg, 1995).

Moreover, as regards the project marketing contributions, the model is broad enough to allow for the following general characteristics from the project marketing literature,[2] which are applicable to architectural partial projects and projects:

(1) Project offerings include hardware and software components (Mattsson, 1973).
(2) Some of the components are standardized (Mattsson, 1973).
(3) Customization is a part of project offerings, yet the extent of customization varies from project to project (compare Backhaus, 1995; Sharma, 1991).
(4) The scope of supply offerings may be varied between project offerings in accordance with the capabilities and means of the buyer (compare Backhaus, 1995; Løwendahl, 2000).
(5) Project offerings concern large orders in terms of monetary value (Backhaus, 1995; Holstius, 1989).
(6) Project offerings concern long-term investments (Backhaus, 1995; Holstius, 1989).
(7) Project offerings may include financial packages or other "financial engineering" measures (Backhaus, 1995).
(8) Project offerings often involve bidder/supplier coalitions (Backhaus, 1995).
(9) Project offerings are international in the sense that offerers sometimes come from foreign countries (Backhaus, 1995).
(10) Project offerings are subject to deadlines and time limits (Holstius, 1989).
(11) Orders for projects are discontinuous (Ahmed, 1993; Backhaus, 1995; Cova & Ghauri, 1996; Mandják & Zoltan, 1998).

Finally, in contrast, there are several characteristics from the project marketing literature that do not apply to all construction projects and are therefore not accounted for in Fig. 1. These include the following:

- There are differences between customer's and supplier's level of know-how with respect to project offerings (from Backhaus, 1995). This is most often not the case if the client is a large company or a municipality with in-house construction procurement experts.
- The objective of the sale is the creation of long-term capital assets (from Holstius, 1989). This characteristic is not always relevant because in some construction projects the main objective is to create or renovate a building that is an example of a particular artistic style for the sake of art or historical authenticity. However, it probably applies to most construction projects.

- Project offerings include an ever-increasing proportion of service elements (from Backhaus, 1995). This is not the case because in the European construction sector, the trend seems to be going toward increased rationalization of service inputs (*Arkitekt- og byggebladet*, February 1993, p. 30; Eurostat, 1995, pp. 24–39).

The very eclectic and spread body of existing project marketing contributions (see Günter & Bonaccorsi, 1996) is assessed in a literature review, to enable the abductive analysis of their potential applicability in the realm of architecture. The literature review is based on a meta-analytical (see Easton, 1995) assessment of studies in English, German, and the Scandinavian languages as well as a few selected French-language studies. In the review, the following types of studies have been accounted for:

(1) Quantitative empirical studies.
(2) Case studies and other qualitative empirical studies.
(3) Conceptual studies and literature reviews.
(4) Practice-oriented contributions.

The included studies were selected on the basis of an assessment of two factors:

- The relevance of the study in question to the research project at hand;
- The standing of the study in question in the academic world (studies judged to be especially "renowned" and/or "seminal" have been included).

Moreover, as concerns the practice-oriented contributions, their inclusion in meta-analytical literature reviews is controversial, as some practice-oriented contributions are scientifically less well founded. Nevertheless I chose to critically examine them in connection with the literature review due to the following reasons:

(1) Practice-oriented contributions influence the cognitive perceptions of researchers and inspired qualitative and quantitative empirical studies as well as conceptual and meta-theoretical work.
(2) The scientific community does not agree about the precise definitions and boundaries between "practice-oriented contributions" and "scientific studies," due to more fundamental disagreement about ontology and epistemology.

The literature review furthermore categorizes the included studies as being either related or non-related to the work of the Industrial Marketing and Purchasing (or IMP) Group (see, e.g. Håkansson, 1982; or Ford, 1997) and its affiliated subgroup, the International Network for Project Marketing and Systems Selling (INPM, see Günter & Bonaccorsi, 1996; or http://research.eap.net/pmri/pmri_people.html).

A total of 17 relevant non-IMP/INPM contributions and 13 relevant IMP/INPM contributions are reviewed. Of these, many will be described briefly in Subsections 3.3–3.5 of this summary, in connection with the presentation of the analysis relating to Research Questions 3–5.

The coverage of the IMP/INPM work on project marketing in the literature review is, however, preceded by a description of the IMP Group, to lay the foundation for the justification of the inclusion of Bourdieu's (1979, 1986a, b) social and cultural capital theory in the theoretical framework of the study. The main points of this presentation are listed in the following paragraphs.

The IMP Group emphasizes that relationships in dyads and networks play a decisive role with regard to marketing practice on business-to-business markets. Furthermore, they emphasize that the characteristics of what is bought and sold on industrial markets are very often determined in relational interaction between the buying and the selling organizations (Håkansson, 1982). According to Easton (1997, pp. 106–110), relationships are seen by IMP researchers as being comprised of four elements: mutual orientation, mutual dependence, bonds, and relationship investments. Further, in the IMP approach, an industry is viewed as "a network of interconnected exchange relationships" (Easton, 1997, p. 104) encompassing "actor bonds," "activity links," and "resource ties" (Håkansson & Johanson, 1993).

However, IMP Group members admit that networks do not always play a major role in industrial markets. In order to depict when networks prevail, Håkansson and Johanson (1993, p. 45) have developed a model of four possible ideal types of industrial governance. These are depicted below in Fig. 2.

In Fig. 2, the ideal types of industrial governance are determined by: (a) two alternative types of actor-internal forces (own interests versus general norms); and (b) two alternative types of actor-external exchange relations (general relations, which are a consequence of the dominance of general interplay between all industry actors, versus specific relations, which imply the dominance of specific long-term relationship-related interactions between individual actors).

		Internal Force is based on	
		Interests	Norms
External Force is Based on	Specific relations	*Network*	*Hierarchy*
	General relations	*Market*	*Culture, profession*

Fig. 2. Håkansson and Johanson's Classification of Governance Structures. *Source:* Håkansson and Johanson (1993, p. 45).

In the *Network*, which is the first ideal type governance structure, activities are governed by actors' different individual interests which are channeled to each other via relationships between specific actors. In the *Hierarchy*, interests have been replaced by norms which individuals follow and which are enforced through specific relations to other actors. In the *Market*, actors also follow their own individual interests, but, in contrast to the *Network*, do not predominantly interact with specific other actors. This means that the actors are, on one hand, freer in relation to one another, yet on the other hand, they cannot take advantage of specific productivity gains that occur through specific joint activities with other actors in the *Network* governance structure. Finally, in the *Culture* or *Profession*, actions are once again governed by norms, yet the external forces that ensure that the norms are followed are based on general relationships that involve all members of the governance structure.

Turning now to project marketing-related INPM contributions, there are several issues specific to project marketing which mandate the selective use as well as modification of the IMP framework, as described above. These issues are encompassed by the so-called D-U-C-framework (see Cova & Ghauri, 1996; Mandják & Zoltan, 1998; Tikkanen, 1998; compare also to Ahmed, 1993, pp. 55–56). Tikkanen (1998, p. 264) lists the D-U-C features as follows:

D. "The *discontinuity* of demand for projects";
U. "The *uniqueness* of each project in technical, financial and socio-political terms";
C. "The *complexity* of each individual project in terms of the number of actors involved throughout the supply process" (The *italics* in the above are Tikkanen's).

The consequence of these characteristics is that the world of project business cannot be a priori placed in the network governance structure, as discontinuity and uniqueness often do not allow for long-term relationships marked by mutual dependence, bonds, and relationship investments (see Easton, 1997) or continuous activity and resource links beyond the completion of an individual project (see Håkansson & Johanson, 1993). Thus, Cova and Ghauri (1996) suggest that project marketing governance structures are usually something between markets and networks.

In Chapter 4, this problem is dealt with by building on and modifying Håkansson and Johanson's (1993, p. 45) model of possible structures of industrial market governance for the specific case of professional, technical service projects such as architectural projects and partial projects. In this type of project, the distinction between "interests" and "norms" is not clear-cut, as the projects are highly *complex* (see the D-U-C model above), yet also subject to many normative requirements and

External force is based on	Specific relations	*The Network (with or without hierarchical elements)*
	General relations	*The Socially Constructed Market*

Fig. 3. Classification of Governance Structures in Professional Services Industries.

often marked by *potential difference in know-how* between the buyer and the seller (see Backhaus, 1995; Løwendahl, 2000; Sharma, 1991). However, this does not imply that the relationship between, e.g. architects and their customers is purely dependent on existing norms, as achieving project sales depends on the *selling of a credible promise*, and, in the longer term, being able to *deliver the credible promise to the satisfaction of the buyer* (Løwendahl, 2000), which also means taking the perceived interests of the buyer into account. Thus, one can conclude that both norms and interests govern the world of professional, technical services projects and that this world thus only has two possible governance structures, as depicted in Fig. 3.

As indicated in Fig. 3, the *Network* is the governance structure that one finds in situations where specific relationships marked by a complex interplay between norms and interests exist across individual projects. On the other hand, the *Socially Constructed Market* functions in situations where relationships between cooperation partners are mainly limited to the concrete project in question due to, e.g. the discontinuous nature of the project demand and the uniqueness of each individual project.

Concepts used to describe the environment of project marketing are defined broadly enough to be used both on network and socially constructed market governance structures. One such concept is Cova, Mazet and Salle's (1996) project marketing *milieu*, which is a "socio-spatial configuration that can be characterized by four elements" (*ibid.*, p. 654):

- a territory;
- a network of heterogeneous actors related to each other within this territory;
- a representation constructed and shared by these actors;
- a set of rules and norms ("the law of the milieu") regulating the interactions between these actors.

However, Tikkanen (1998) has challenged the inclusion of territoriality in the definition of the milieu; he argues that many project marketing milieus could be global. This challenge is dealt with empirically for the specific case of architectural firms in Chapters 6–10 of this study; it will therefore be taken up again in the concluding Section 4 of this summary.

Another concept that meets the above criterion is the *field*; the French sociologist Pierre Bourdieu (1979, 1986a, b) and some institutional theorists (DiMaggio & Powell, 1983; Melin, 1989; Scott, 1994) use it. On the basis of meta-theoretical and conceptual analysis the field concept is almost identical to the milieu concept. Thus, bridges can be built between contributions that use either of the two concepts.

The concepts of *cultural* and *social capital* (Bourdieu, 1986b) are then introduced in Chapter 4 because, in Bourdieu's theory (see *ibid.*), cultural and social capital accumulation takes place and determines an actor's position in a field. These concepts are also useful because they relate to the norms, rules and representations of a specific environment (see *ibid.*) and because they can be applied in both the network and socially constructed market governance structures. Furthermore, in the study, they enable the abductive analysis of the Danish case firms' establishment of credibility and position in their market-seeking activities in the German construction industry milieu.

However, in order to be able to optimally present the theory-related analytical results of the study, a more in-depth presentation of the three case firms is first in order. Case firm 1 belongs to the small group of "internationally renowned and established architectural firms" (see Section 1) in Denmark. The firm, which is one of Denmark's largest architectural firms, was founded in 1971 after the unexpected and premature death of its founder, a world-famous Danish architect (and furniture/industrial designer). It was founded by two of his closest colleagues and assistants, who were Danes. One of these was, however, born in the southern Danish town of Haderslev, which was actually German from 1865 to 1920 (the town's German name is Hadersleben). Thus, this Dane had substantial exposure to the German language and culture during his childhood and teenage years, which he also used in his work for the world-famous Danish architect. Additionally, this architect had won international competitions since the late 1950s; therefore the newly-founded firm and its employees had some years of international experience that could be used – also after the world-famous architect's death. Finally, this architect's international reputation had attracted ambitious foreign-born employees to his firm; some of these continued their work in the firm founded after his death by his successors (see, e.g. Børsen, 1996).

Thus, today, Case Firm 1 has a relatively international body of employees that includes Germans, North Americans, and Asians. Three of its ten currently active partners were born in Germany, two of whom came to Denmark in the late sixties to work for the world-famous architect. (The third German-born partner came the year after his death.) In the following years, the newly established successor firm concentrated on and succeeded in establishing an acceptable balance between carrying on the design tradition of the world-famous architect and developing a name in its own right (Børsen, 1996). Today the firm is active in a broad range of architectural projects – from, e.g. office buildings to museums to theaters to factories to hospitals to housing to university buildings. It has also designed buildings for a Danish multinational corporation (MNC) both at home and abroad for many years, thus it has also been involved in *client-following* international activities (see, e.g. Erramilli & Rao, 1990; Majkgård & Sharma, 1998; and Section 1 of this study) before and during the nineties.

Case firm 1 has also had both German as well as many other foreign clients for several decades, and both German and Danish architects have been responsible for the specific projects in Germany, depending upon the skills required. Additionally, the firm has had one steady northern German customer for many years. The firm has in previous decades had a couple of temporary subsidiaries in Germany, and in 1997 a subsidiary was once again started in Wiesbaden in connection with a specific project in the German federal state of Hesse. This subsidiary has subsequently been moved from Wiesbaden to the nearby German city of Frankfurt am Main in 1999 and is run by one of the firm's German-born partners, who is referred to as Partner AA in the appendix of this study.

In the 1990s, Firm 1 was involved in German projects or received runner-up prizes/honorable mentions in Germany for, e.g. projects concerning university buildings, a museum, a theater, housing, a hospital, urban renewal projects, offices, and administrative buildings. In connection with this research project, I interviewed the firm's three German-born partners (referred to as AA, CC and DD in the appendix) as well as the Danish-born partner responsible for the firm's finances and economic matters (referred to as BB in the appendix). I also received various written documents from the firm.

Case firm 2 was founded in 1924 by a Danish architect who is very well known in Denmark for having designed the campus of a major Danish university. Today, the firm is one of Denmark's large architectural firms with a strong design profile in Denmark as well as broad expertise from a wide range of construction project types – e.g. museums, sports stadiums, universities, research and development centers, factories, hospitals, housing. The firm can be classified as a "nationally renowned and established architectural firm" (see Section 1), as it is very well known in Denmark, but had not made a name for itself abroad, except in the culturally very similar near markets in Norway and Sweden, at the beginning of the nineties. However, similar to the internationally renowned Danish Case Firm 1, Case Firm 2 has also worked to some extent as a client-follower on international markets by designing factories for a (different) Danish MNC that have been built in a host of different countries. Today, Case Firm 2 has offices in several Danish cities as well as in Berlin, Germany and Oslo, Norway.

Case Firm 2 chose to concentrate on the German market in the beginning of the 1990s and subsequently to locate its German office in Berlin in the spring of 1993. The German office in Berlin is managed by a Danish-educated Danish architect (YY in the appendix), who became a partner in the firm on January 1, 1997. The other persons currently employed in the Berlin subsidiary are Germans educated in Germany. The firm has been involved in German projects or received runner-up prizes/honorable mentions in Germany for, e.g. housing projects, a research and development center, a sports stadium, a hospital, and a university building. In the course of my studies I interviewed both the partner who is responsible for the firm's finances and budgets (XX in the appendix) and the partner who heads the German office (YY). I also received various written material from the firm as well as three books about the firm as gifts.

Case Firm 3 is this study's "Innovative Younger Firm," in accordance with the classifications mentioned in Section 1 of this summary. It was founded in 1985 by an architect who also has the Danish vocational and apprenticeship carpentry education; this architect is referred to as MM in the appendix. Case Firm 3 employed approximately 30 persons throughout the nineties, of whom most are either educated as architects or certified building constructors, or have received some other construction industry-relevant education. Case Firm 3's original strength in 1985 was housing. However, the firm has subsequently sought and succeeded in broadening its areas of expertise in Denmark, especially after the so-called "Ølgaard report," which was made public by the Danish government in 1989 (see Hansen, Lund, Nygaard & Lyhne-Knudsen, 1994). The main conclusion of the "Ølgaard report" was that there would be a decreased need for publicly-funded housing in Denmark in the 1990s (see *ibid.*). In the late nineties the firm received orders for many types of construction and urban renewal projects on its domestic market.

Case Firm 3 began its activities on the German market in 1990. Its initial form of activity was a cooperation agreement between itself and a German architectural firm situated in Hamburg. However, Case Firm 3 terminated this agreement after the first year because it did not provide the expected and perceived "fair" share of income from the joint projects.

In the beginning of 1992, a strategy concerning the establishment of a subsidiary in Germany was discussed in Case Firm 3's professional advisory board. After some modifications to the strategy, a subsidiary was established in Berlin. From 1993 to 1999, the subsidiary was led by two successive Danish certified building technicians, who had previous experience on the German market, in cooperation with a German architect educated in Germany. During the nineties, Case Firm 3 was involved in German projects or received prizes in Germany concerning, among other things, gasoline stations, housing, urban renewal, and shopping centers.

For the purposes of this study, Case Firm 3 provided me with a written documents related to the firm as well as a series of interviews with: (a) founding partner (called MM in the appendix to this study); (b) the employee in the firm's Danish office who was responsible for the firm's budgets and finances during the nineties (called NN in the appendix); and (c) the Danish certified building technician who led the German office from 1995 to 1999 (called OO in the appendix).

3. PRESENTATION OF THE EMPIRICAL STUDY

3.1. Analysis Pertaining to Research Question 1 (Chapter 6)

As previously mentioned in Section 1, Research Question 1 was formulated as follows:

(1) How did the German market for architectural services develop during the nineties in terms of, e.g. institutions, total market size and growth rates at the national and federal state (in German: *Länder*) levels?

With regard to this research question, which concerns general developments on the German market, as previously mentioned in Section 1, the German construction market boomed at the beginning of the nineties. The boom stopped in 1995, yet the total level of German construction industry volume was still much higher at the end of the nineties than it had been in the previous decade and the number of architects practising on the German market continued to increase throughout the decade (Deutsches Institut für Wirtschaftforschung). In comparison, the Danish market was severely depressed from 1990 to 1992; it started to pick up after 1993, yet by no means reached a level of record highs in subsequent years (Danmarks Statistik, 1992–1999). These trends are captured in the EU statistical indices of Tables 1 and 2.

The largest European construction industry firms, which generally were large contracting, project management and civil engineering firms, experienced a wave of mergers and acquisitions as well as strategic alliance constellations during the nineties (Lubanski, 1999); however, few architectural firms were directly involved in this wave. Instead, architectural firms and smaller engineering firms on the German market have traditionally been responsible for a greater scope of the construction process than their Danish counterparts, including especially construction management tasks (see Button & Fleming, 1992; Oliver-Taylor, 1993); this is illustrated in Table 3's depiction of the normal allocation of legal responsibility in German and Danish construction projects.

Table 1. European Construction Industry Indices of Production.

	Europe (EUR15)	DK	D
1989	97.1	105.8	94.4
1990	100.0	100.0	100.0
1991	99.0	90.1	104.1
1992	98.6	86.7	115.4
1993	94.5	78.0	119.0
1994	96.0	81.8	131.2
1995	96.0	87.2	129.6

Source: Eurostat: *Industrial Trends. Monthly Statistics*, 12/96, p. 95.

Table 2. European Indices of Construction Permits for Housing.

	Europe (EUR15)	DK	D
1989	96.3	138.1	70.5
1990	100.0	100.0	100.0
1991	95.9	78.5	102.4
1992	96.8	79.9	122.4
1993	99.4	68.1	152.9
1994	110.1	68.5	179.7
1995	103.4	60.4	161.1

Source: Eurostat: *Industrial Trends. Monthly Statistics*, 12/96, p. 96.

However, on the German market, architects experienced increasing competitive pressure from the General Contractor (i.e. in German "*Generalunternehmer/GU*," i.e. general contractor, and "*Generalübernehmer/GÜ*")[3] ways of coordinating construction processes during the nineties, in that private and public German clients

Table 3. Usual Allocation of Legal Responsibility in the Construction of Buildings in the Early 1990s.

Area of Responsibility	Denmark	Germany
Ground	Contractor	Architect + Engineer
Stability calculation	Contractor	Engineer
Bill of quantities	Contractor	Architect
Design	Contractor	Architect
Inspection of materials	Contractor	Architect + Contractor
Direction of the works	Contractor	Architect
Final acceptance	Contractor + Client	Architect

Source: Oliver-Taylor (1993, pp. 25 and 35).

increasingly gave "direction of works" construction management responsibilities to other construction industry actors (Baus, 1997; Rudolph-Cleff & Uhlig, 1997; see also Huovinen & Kiiras, 1998). Yet despite the fact that architects operating in Germany lost some of their traditional domain of work during the nineties, German architects at the end of the nineties still more often managed construction work than their Danish colleagues (see Dræbye, 1998).

With regard to pressure on fees during this time, the German *Honorarordnung für Architekten und Ingenieure*, a fixed set of fees, which enjoys legal status, provided German architects with some protection against falling levels of domestic income (see *ibid.*). In Denmark, this type of legal protection did not exist during the nineties (Albertsen, 1997), and immediately after the Danish implementation of the EU Public Services Directive in 1993, average fees charged by architects fell substantially – perhaps by as much as 50% (*Arkitekt- og byggebladet*, January 1994, pp. 12–14). Additionally, to guard themselves against income drops caused by fluctuations on the much smaller Danish market, Danish architectural firms are to a larger extent more generalists than German architectural firms (see Albertsen, 1997).

As regards exports, Danish architectural exports peaked in 1996. However, throughout the nineties, Germany was the recipient of Danish architectural exports (Hartung, 1997; PAR, 1999), as indicated in Table 4 .

Table 4. Export Turnover of Danish Architectural Firms that are Members or Associated Members of the Danish Council of Practicing Architects (*Praktiserende Arkitekters Råd*/PAR) in Millions of DKK.

	Total Export	Germany, Nordic Council Region	Rest of Europe	Rest of the World
1990[a]	72.0	23.9	14.7	6.8
1991	74.4	NA	NA	NA
1992[b]	54.2	45.5	6.9	1.8
1993	78.0	60.1	12.1	5.8
1994	87.3	66.0	11.9	9.5
1995	97.0	76.1	12.7	8.1
1996	103.1	71.4	13.4	18.2
1997	69.1	53.4	8.6	7.1
1998	73.4	51.7	14.6	7.1

Source: Data provided by Paul Jeppesen, PAR's Secretary for International Affairs in August 1999.

[a] For the year 1990, only DKK 45.5 mill. of the total export turnover of DKK 74.4 mill. has been attributed to one the three different regions. The destination of the final DKK 28.9 mill. of export turnover is not known.

[b] The geographic distribution of the figures for 1992 is based upon a survey questionnaire, which was answered by PAR member firms who stood for 73% of the total export turnover of that year. From 1993, the figures are based on the export turnover figures used to calculate insurance premiums.

The implementation of the EU Public Service Directive (92/50/EEC), which establishes procedural rules for the publication of calls for tender and the awarding of public procurement contracts, was the largest single change in the institutional environment of European architectural firms during the nineties. After implementation, public sector architectural design partial projects beyond a certain threshold value had to be tendered according to the rules of the directive. (In contrast, design-and-build turnkey contracts are tendered using the EU Public Works Directive.)

The scope of services covered by the public services directive is very broad, including, e.g. catering, cleaning, data processing, facility management, garbage collection, psychological counseling, sanitation and even certain financial services. In the world of construction, the directive mainly applies to the architectural and civil engineering services that are used in the planning and conceptualization of buildings and infrastructure projects as well as to the construction and installation tasks of construction projects procured using a "separate trades" construction organization model. The deadline set by the EU Council of Ministers for the implementation of the Directive in EU Member States was July 1, 1993 (*ibid.*).

In Denmark, the public service directive was implemented by the above-mentioned official deadline (Danish Ministry of Commerce and Industry, June 22, 1993). The date of official implementation in Germany was November 1, 1997 (Jochem, 1998); however, many German public authorities were following their interpretation of the rules of the directive before its official German implementation, to avoid EU court cases. For the case of architectural services, the directive mentions four modes of tendering: design contests (Article 13), open or restricted tendering procedures (Article 36) and, in certain cases (Article 11), tendering procedures involving negotiations with selected pre-qualified bidders (*ibid.*):

> *Open procedure* means that all interested firms can submit a tender. [...] In accordance with the *restricted procedure* the contracting authority, by means of a contract notice, invites interested service providers to apply for an invitation to submit tenders (prequalification). [...] The *design contest* is a procedure which can be chosen if the objective is to receive ideas and proposals. Design contests must follow clear and non-discriminatory selection criteria. [...] The *negotiated procedure* can be applied in special cases, where contract specification cannot be established with sufficient precision to permit the award of the contract by selecting one of the other procedures (Danish Association of Consulting Engineers, March 1996, p. 3; the *italics* are from the original text).

Additionally, in Title V, the directive mentions two time schedules, the normal and the accelerated procedure schedules, and states the minimum amount of time that must be allotted for bidding firms for each of these schedules (Council of the European Communities, October 13, 1997). According to Article 20, the accelerated procedure may be used "in cases where urgency renders impracticable the time limits laid down in Article 19." It shortens the time limits that the potential service provider receives to work out its bid and can be applied in relation to the restricted procedure and the negotiated procedure (Council of the European Communities, October 13, 1997; Danish Association of Consulting Engineers, March 1996, p. 2).

Finally, the directive specifies that the selection of the winner of a public procurement procedure (i.e. the open, restricted, and negotiated procedures) may be made on the basis of two criteria (see Article 36 of Council of the European Communities, October 13, 1997), the economically most advantageous tender or the lowest price only. With regard to the economically most advantageous tender, the directive's Article 36 contains a non-exhaustive list of possible criteria: "quality, technical merit, aesthetic and functional characteristics, technical assistance and after-sales service, delivery date, delivery period or period of completion, price" and furthermore specifies the following:

> Where the contract is to be awarded to the economically most advantageous tender, the contracting authority shall state in the contract documents or in the tender notice the award criteria which it intends to apply, where possible in descending order of importance (*ibid.*).

Germany implemented the directive through two ministerial orders, the "*Verordnung über die Vergabebestimmungen für öffentliche Aufträge*" (VgV) and the "*Verdingungsordnung für freiberufliche Leistungen*" (VOF), which is the main ministerial order for the area of architectural and engineering consulting services. The main procedure specified by the VOF is the negotiated procedure because it is assumed that the public authority in question needs to discuss the concrete project with the professional, technical service firms that could potentially do the work, in order to ensure the best "match" between the public authority and the firm and thus the best project (Jochem, 1998; Prinz, 1998).

However, this interpretation was markedly different from the Danish government's interpretation, in that that Danish Ministry of Commerce and Industry had issued a single ministerial decree (Danish Ministry of Commerce and Industry, 1993), which applies for all service types. In direct opposition to the German interpretation, Section 3 of the Danish Ministerial Decree states that the open or restricted procedures of tendering are to be used by Danish authorities. Moreover, the Danish decree's Section 3's Subsection 3 specifies that the negotiated procedure may not be used, excepted in cases where the preconditions stated in the Directive's article 11, Sections 2 and 3 are fulfilled (see Council of the European Communities, 1997). This major difference in interpretation has resulted in major differences in national public tendering practices. Whereas German public sector architectural projects were tendered during the nineties using the negotiated procedure in approximately 75% of the cases; the preferred method of public sector architectural project tendering in Denmark was the restricted procedure, which during that time was used approximately 65% of the time (see Danish Association of Consulting Engineers, 1996–1999). Furthermore, the German authorities used the accelerated procedure in about 40% of all public tendering processes, whereas the Danish authorities only used this procedure in about 5% of the cases (see *ibid.*).

With regard to design contests in Germany, a pre-existing German legal text, *Grundsätze und Richtlinien für Wettbewerbe auf den Gebieten der Raumplanung, des Städtebaus und des Bauwesens* (GRW), was changed in 1995 to accommodate the requirements of the EU Public Service Directive (Neusüss, 1995, p. 174). The major change in the GRW was the abolition of regional criteria for the participation for contests exceeding the EU Public Service Directive's threshold value (*ibid.*, p. 178). Previously, German public authorities had been able to specify that only architects situated in a certain German or European region had the right to participate in a given design contest.

The immediate consequence of this change, combined with the downturn on the German market after 1995, was a huge increase in the number of participants in German design contests; some competitions for relatively small project types such as, e.g. elementary schools resulted in many hundreds of participants (interviews with eight Danish and German respondents, various articles in German newspapers). The more far-reaching consequence of this development was that many German public authorities subsequently reduced the number of design contests, due to the extremely high transaction costs of processing proposals from so many participants. However, the number of design contests in Denmark has, in contrast to Germany, increased, not decreased, since the implementation of the Public Services Directive and the increase in participants has been minor in comparison to the German situation. Thus, in sum, it can be said that despite attempts to harmonize European public sector tendering, Danish architectural firms tendering for projects financed by the public sector in both Germany and Denmark must be able adapt to two very different interpretations of the pubic service directive and two very different resultant competitive situations.

3.2. Analysis Pertaining to Research Question 2 (Chapter 7)

Research Question 2 was formulated in this summary's introductory Section 1 as follows:

(2) Who were the major actors involved either directly or indirectly in the Danish internationalization of architectural services to Germany in the 1990s and to what extent and how did they cooperate with one another?

Thus, this research question deals with the actors involved in the Danish architectural firms' internationalization on the German market and their cooperation. From an IMP Group relationship and network perspective, Johanson and Mattsson (1997, p. 200) have suggested that there are three possible ways in which a firm can internationalize:

(1) "through establishment of positions in relation to counterparts in national nets that are new to the firm, i.e. *international extension*";
(2) "by developing the positions and increasing resource commitments in those nets abroad in which the firm already has positions, i.e. *penetration*";
(3) "by increasing coordination between positions in different national nets, i.e. *international integration*" (The *italics* are Johanson and Mattsson's).

Here it should be noted that Johanson and Mattsson's (*ibid.*) definition of nets is not as rigorous and restrictive as the network definition of other IMP contributions, in which, e.g. the relationships of the networks are seen as being comprised of four very long-term elements: mutual orientation, mutual dependence, bonds, and relationship investments (see Easton, 1997, pp. 106–110, as described in Section 2 of this summary). Thus, Johanson and Mattsson's three possible modes of internationalization may also be used in project marketing situations, such as architectural partial project marketing situations, where the governance structure is probably "something between markets and networks (see Cova & Ghauri, 1996), e.g. the "socially-constructed market" (see Fig. 3).

The empirical data indicates that, during the nineties, Danish architectural firms active in Germany had a relatively large number of social and informational contacts to other Danish construction industry actors who were also active on the German market. These actors included Danish engineering, contracting, and developing firms as well as two of the construction industry attachés of the Danish Ministry of Housing, who were situated in Berlin and Düsseldorf, Germany during the nineties, and the construction industry trade commissioners of the Danish Ministry of Commerce and Industry, who were situated in Hamburg and Rostock, Germany during the nineties. However, these contacts were often ad hoc and did not usually relate to joint project acquisition or cooperation with regard to multiple projects over time.

The official construction industry representatives of the two Danish ministries mainly collected information for Danish private sector firms and the Danish government as well as arranged official visits of the Danish Ministry of Housing to the German federal and various German state-level ministries of housing. These official visits often resulted in the signing of official inter-ministerial "cooperation agreements" in, e.g. the area of housing production or environmentally friendly construction techniques. However, the former Danish construction industry attaché in Berlin (Mr.) Jørgen Skovgaard Vind, regarded these meetings and the signing of cooperation agreements as initial "door openers" for the Danish and German actors in the construction sector (interview with Jørgen Skovgaard Vind, February 24, 1999). Whether private sector actors showed interest for a given cooperation agreement and attempted to establish contracts and form of cooperation within the framework of the given cooperation, agreement was completely up to the private sectors themselves; the Danish officials did not promote or help establish long-term relationships between specific private sector actors or between specific private sector actors and specific public sector actors (*ibid.*).

Thus, the networks of the Danish construction industry firms active on the German market had more of a mere social and informational nature than the characteristic long-term mutual orientation, mutual dependence, and bonds, and the high relationship investments common to the assumptions of the IMP group (see, e.g. Easton, 1997, pp. 106–110). There were also few if any long-standing resource ties and activity links (compare to Håkansson & Johanson, 1993, as described in Section 2 of this summary) among the Danish construction industry firms.

Furthermore, Danish architectural firms, engineering firms, and contractors predominantly went about the task of establishing themselves on the German market alone. As a rule they accumulated their own separate knowledge bases and had their own dealings with German actors concerning both possible subsidiary establishment and the acquisition of projects on the German market. However, from time to time, they did work together, if all the involved parties believed that such cooperation was advantageous. Yet, with specific regard to cooperation with Danish contractors, during the mid-nineties, Danish architectural firms and engineering firms became increasingly skeptical about the advantages of such cooperation in the German market. This is because contractors from all over Europe had entered the German market at about the same time in the beginning of the nineties (Halskov, 1995; Kragballe, 1996). This soon resulted in greatly reduced prices for the contractors as well as many unpleasant surprises due to the foreign contractors' lack of knowledge of or disregard for the German construction industry's laws and norms (see, e.g. Entreprenørforeningen, 1996). As concerns Danish firms, almost all Danish contracting firms, including, most notably, some of the largest and most well-known, e.g. Højgaard & Schultz A/S and Rasmussen and Schiøtz A/S, left the market again in the mid-1990s (Frisch-Jensen, 1999; Halskov, 1995; Kragballe, 1996).

Thus, on the basis of the developments explained in the preceding paragraphs, it can be said the governance structure between the Danish construction industry actors engaged in internationalization on the German market was the *socially constructed market* (see Section 2 of this summary and Fig. 3). Moreover, in the terminology of Johanson and Mattsson (1997), most of the Danish private sector actors were mainly involved in *international extension*, i.e. "the establishment of positions in relation to counterparts in national nets that are new to the firm." However, the first case study firm was an exception to this rule. As it belongs to the *internationally renowned* category of architectural firms, it had been active on the Germans markets for several decades previous to the 1990s and was therefore primarily engaged in *penetration* (see Johanson & Mattsson, 1997) during the nineties.

The above governance structure-related finding may have relevance beyond this particular study or the realm of architecture. It is namely probable that the socially constructed market is fairly common in project marketing situations, due to the aforementioned *discontinuity* and *uniqueness* (see, e.g. Ahmed, 1993; Cova & Ghauri, 1996; Tikkanen, 1998). Furthermore this governance structure may be key in the professional services, due to the previously mentioned role that "socially constructed" factors such as judgment and credibility play in these services (see, e.g. Løwendahl, 2000). Lastly, the socially constructed market governance structure may furthermore be the dominating governance structure for "smaller" types of projects, i.e. projects often offered by architectural and engineering firms as well as by software solution developing firms. Here, in addition to the discontinuity of offerings, there are many potential offerers, i.e. many different architectural and engineering firms, thus making socially constructed market conditions all the more probable. This leads to the following theoretical proposition, which should be tested further in coming scholarly contributions:

- The socially constructed market governance structure prevails in project marketing situations where there are many potential offerers of projects.

3.3. Analysis Pertaining to Research Question 3 (Chapter 8)

The first case study research question, Research Question 3, deals with knowledge of the German market and specific projects on that market. As previously stated in Section 1 of this summary, it is formulated as follows:

(3) What sort of specific knowledge about the German market and specific projects on the German market did the case firms use to obtain architectural projects in Germany?

In connection with this question, the so-called "Uppsala Internationalization Model," which was a result of a number of studies of Swedish manufacturing firms conducted by researchers situated at the Swedish Uppsala University (see e.g. Johanson & Vahlne, 1977), is drawn upon, as this model describes the internationalization process "as a gradual step-by-step commitment to sell and to manufacture internationally as a part of a growth and experiential learning process" (Johanson & Mattsson, 1997, p. 209).

Here it must, however, be noted that the Uppsala Internationalization Model may only be used in a modified form when dealing with service firms such as architectural firms, due to the problems discussed in Section 1 of this study concerning speaking of exports, i.e. "the sale of goods and services in another country than the country in which they were produced" (Luostarinen & Welch, 1990, p. 20), which are caused by the often inseparable (see, e.g. Stostack, 1977), i.e. on-site, nature of the production of architectural services. However, the Uppsala Internationalization Model's suggestions that: (a) a process of experimentation and learning determines the internationalization path of the firm; and (b) this process takes place through interactions with actors on the foreign market (see Petersen & Pedersen, 1996) may also apply to service firms such as architectural firms.

Moreover, as regards knowledge related to internationalization, Axelsson and Johanson (1992, pp. 221, 231–233)[4] mention several factors often ignored by the so-called "textbook view" of firms' internationalization (e.g. Root, 1994):

- Knowledge of the specific actors in the network(s) of the foreign country in question;
- Knowledge of the relative positions of the actors in the foreign country's network(s);
- Knowledge of direct and indirect firm relationships to actors in the foreign country's network(s);
- Knowledge of how the support of these actors could be mobilized in relation to the planned export activities;
- The ability of the export firm's actors to orient themselves, i.e. "obtain an understanding of where different actors including the actor itself stand in relation to each other" (*ibid.*, p. 231);
- The ability of the export firm's actors to position their firm in the network of other firms;
- The ability of the firm to seize export market network opportunities that turn up at irregular or totally unexpected intervals.

The importance of the above factors is analyzed on the basis of the empirical report. In this study: (a) types of knowledge; and (b) potential locations of knowledge are also discussed on the basis of a previous presentation of Nonaka and Takeuchi's (1995) knowledge conversion and sharing processes, which is depicted in Fig. 4, Boisot, Griffiths and Moles' (1997) social learning cycle, which is reproduced in Fig. 5, and Alajoutsijärvi and Tikkanen's (2000) possible

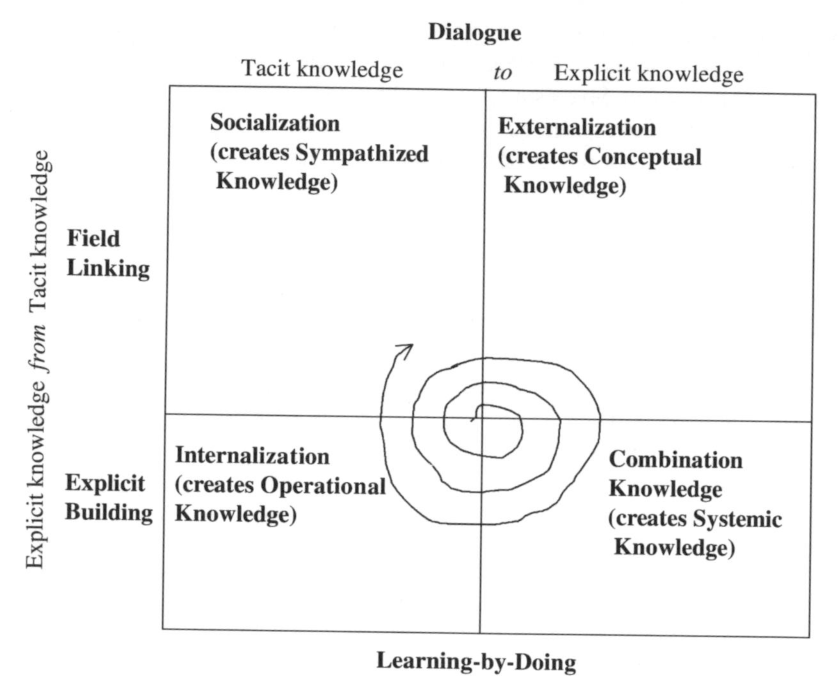

Fig. 4. *Source:* Nonaka and Takeuchi (1995, pp. 64, 71, and 72).

placement of competencies in the context of relationships, which is presented in Fig. 6.

In relation to the above figures, the distinction between tacit (i.e. "subjective," uncodified analogous knowledge of experience and practice, see Polanyi, 1962) and explicit (i.e. "objective," codified, rational, and digital knowledge) is key, as it is assumed that tacit knowledge is more difficult to transfer, making it a more probable source of competitive advantage than explicit, codified knowledge (see, e.g. Sanchez & Heene, 1997). Thus, when comparing the models, Polanyi's (1962) tacit knowledge, which is included in the model of Nonaka and Takeuchi (1995, Fig. 4), is, in the terms of Boisot et al. (1997, Fig. 5), uncoded and either "personal knowledge" or a group's "common sense." "Absorption" in Boisot et al.'s social learning cycle (Fig. 5) is approximately the same as "learning-by-doing" and "internalization" in Nonaka and Takeuchi's model. However, Boisot et al.'s subconscious process of "scanning" to create new "personal knowledge" at the level of the individual or very small groups is not included in Nonaka and Takeuchi's model. On the other hand, Nonaka and Takeuchi's process of "socialization" (spreading tacit knowledge) is not included in Boisot et al.'s model, due to Boisot et al.'s

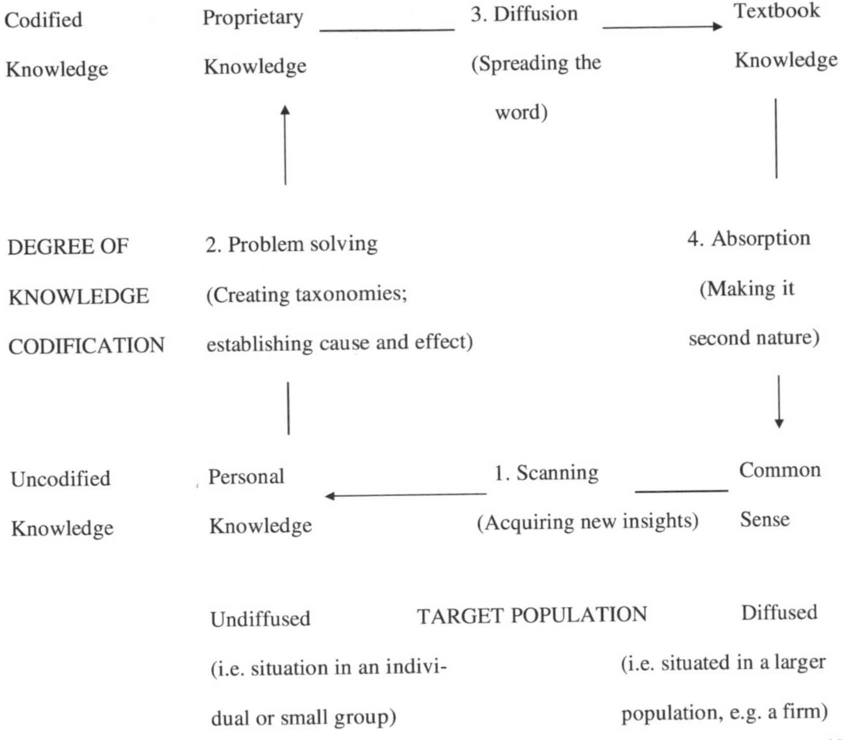

Fig. 5. Boisot et al.'s Social Learning Cycle. *Source:* Boisot et al. (1997, p. 69), Hall (1997, p. 48).

assumption that tacit knowledge is difficult to diffuse beyond very small groups (*ibid.*, p. 69).

Boisot et al.'s "problem solving" and "diffusion" are encompassed by Nonaka and Takeuchi's "dialog" and "linking explicit knowledge." However, diffusion may also take place through one-way communication of codified or explicit knowledge. Thus, Boisot et al.'s "proprietary knowledge" and "textbook knowledge" both contain elements of conceptual and systemic explicit knowledge.

As regards Fig. 6, Alajoutsijärvi and Tikkanen (2000) have built upon Boisot et al.'s "Social Learning Model" by explicitly incorporating the inter-organizational relationships that are the focus of the IMP Group perspective (see, e.g. Section 2 of this summary) in the model and by adding an intermediate level of knowledge as depicted in Fig. 6. Moreover, the arrows of Fig. 6 refer to the four directions of knowledge transfer specified by Boisot et al. in Fig. 5. Additionally, on the horizontal axis, Alajoutsijärvi and Tikkanen have incorporated four specific

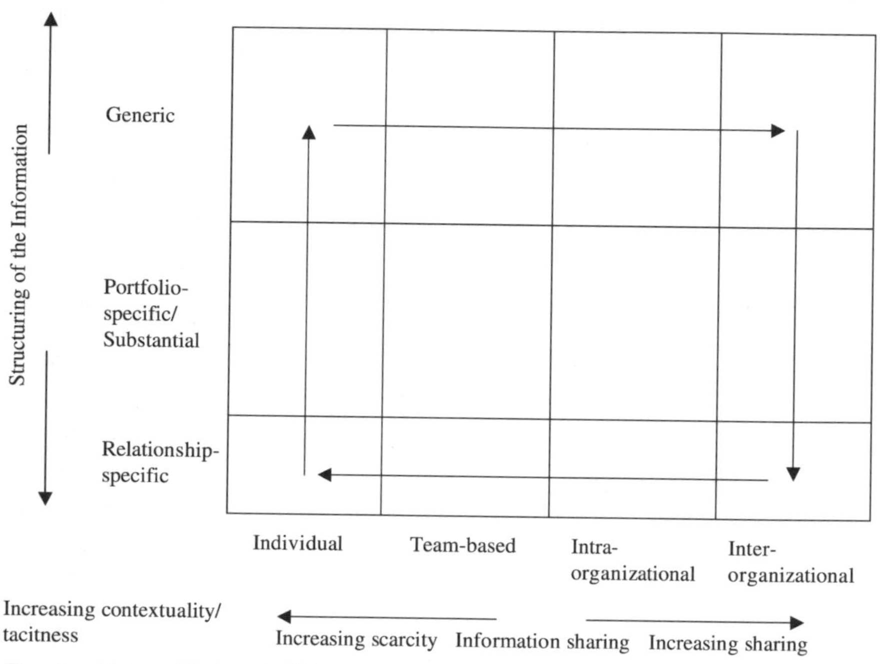

Fig. 6. Alajoutsijärvi and Tikkanen's Competencies in the Context of Relationships. *Source:* Alajoutsijärvi and Tikkanen (2000).

levels at which contextual knowledge can be found. Finally, they have added an intermediate category of knowledge contextuality, portfolio specific or substantial knowledge that they define as follows (*ibid.*):

> Portfolio-specific or substantial competence refers to the knowledge and skills that are transferable and applicable across individuals and individual relationships of the organization. Adapting new technology, for example, may represent a competence within several customer relationships.

As regards my empirical data, the interviews I conducted with the three case firms were the primary source of information about the knowledge these firms used to acquire architectural project orders on the German market. However, as interviews can only deal with knowledge that lies at the level of discursive consciousness (see, e.g. Kvale, 1996), my case study research was not able to capture the knowledge that has remained completely tacit throughout the study. However, originally tacit knowledge may have been articulated through my questioning in interviews, thus perhaps also triggering some of the social learning effects depicted in the above Figs 4–6.

My empirical material indicates that Firm 1, the *internationally renowned firm*, had much more initial Germany-related knowledge than Firms 2 and 3, the *nationally renowned* and *innovative firms*, due to its years of experience on the German market and its many German-born employees. This finding is in accordance with the Uppsala Internationalization model's key pretenses of "a gradual step-by-step commitment"... "as a part of a growth and experiential learning process" (Johanson & Mattsson, 1997, p. 209, as previously discussed).

Key areas in which Firms 2 and 3 had initial deficits are related to the broader knowledge of the German market (e.g. knowledge of the German Deutsches Institut für Normung/DIN building norms, construction law, typical German negotiation procedures including the procedures used in public sector tendering) as well as knowledge of German client's possible interpretations of traditions, values, and tastes. Actors from Firms 2 and 3 were initially dependent upon the goodwill of their German clients due to this lack of knowledge; however, they worked actively to make up for deficits in Germany-specific knowledge, often putting their newly acquired knowledge of Germany to use almost immediately after it had been obtained in efforts to fulfill the norms and practices of the German construction industry. The general knowledge of design traditions, which all three case firms possessed, was, on the other hand, not usually put into direct use in connection with obtaining specific architectural projects, as most German clients also had some knowledge of the design tradition of the architectural firm in question.

Firm 3's acquisitive strategy was based on the personal, tacit knowledge of its founding partner concerning creating and nurturing relationships; however, sometimes Firm 3 did not sufficiently understand the needs of its current or potential customers or how to mobilize them. Firm 2, as a nationally renowned firm, in turn, had to initially realize that its position on the German market was markedly different from its position on the Danish market due to the fact that its Danish references meant little to potential German customers (see also Subsection 3.5 of this summary). After realizing this, Firm 2 had to then acquire further knowledge, to enable it to develop practices and social capital (see Subsection 3.4 of this summary) for acquiring projects without a wealth of references on the German market.

Thus, as concerns Axelsson and Johanson's (1992) previously mentioned discussion of factors often ignored by the so-called "textbook view" of internationalization, on the basis of the information in the previous two paragraphs, it can be said that the following factors were initially relevant for Firms 2 and 3's internationalization, yet were lacking:

- Knowledge of the relative positions of the actors in the foreign country's network(s);
- Knowledge of how the support of these actors could be mobilized in relation to the planned export activities;
- The ability of the export firm's actors to orient themselves, i.e. "obtain an understanding of where different actors including the actor itself stand in relation to each other" (*ibid.*, p. 231);
- The ability of the export firm's actors to position their firm in the network of other firms.

In contrast, because the governance structure of the German and Danish markets are that of the aforementioned "socially-constructed market" (see Fig. 3); the more network-related knowledge areas mentioned by Axelsson and Johanson (1992) played less of a role. However – and once again with regard to Uppsala Internationalization Model, context-specific knowledge of the German construction industry milieu was *a significant prerequisite* for the German project acquisition activities of the Danish case architectural firms. This conclusion goes beyond the theoretical insight of the Uppsala model (see, e.g. Johanson & Vahlne, 1977) concerning the importance of knowledge in the internationalization process of the firm. The Uppsala model merely relates the internationalization path of the firm to an experiential learning process (see *ibid.*; Petersen & Pedersen, 1996); it does not specify that the nature of knowledge *as a prerequisite* for acquisitive activities.

The fact that context-specific knowledge is a prerequisite for project acquisition activities leads to a "Catch-22" situation, in which the knowledge that is a necessary prerequisite for obtaining projects can only be obtained through projects. However, the unique boom situation on the German market at the beginning of the nineties as well as the resultant lack-of-capacity problems helped the two Danish case firms (2 and 3) without previous German market-seeking project experience overcome the "Catch-22" dilemma. During this period, German clients were more open to working with foreign firms that did not have the normal level of contextual-knowledge about the German field. The theoretical implication of this is the following proposition:

- Entering foreign markets in boom situations enables a firm to better overcome potential "Catch-22" situations than entering foreign markets in other circumstances.

Contrary to the theoretical models of Nonaka and Takeuchi (1995), Boisot et al. (1997), and Alajoutsijärvi and Tikkanen (2000), the case firms' knowledge acquisition process did not follow a spiral or cycle of codification/de-codification and inter-firm diffusion. Knowledge was not usually codified through, e.g. reports nor formally taught to other organization members; instead it either slowly diffused through use to the team who was working on the specific project at hand or remained mainly in the minds of the project acquirer. (The structure of the project acquisition work in the three case study firms is depicted in Fig. 7 below, and the main location of knowledge is shown in Fig. 8, which is titled "Tacit Knowledge in the Context of Relationships.") Thus, the knowledge remained in the "from tacit knowledge to tacit knowledge" quadrant of the "Knowledge Conversion and Sharing Matrix" of Nonaka and Takeuchi (1995, see Fig. 4).

Furthermore, concerning Fig. 8, in contrast to the suggestion of Alajoutsijärvi and Tikkanen (2000), codified, generically applicable knowledge did not always have greater utility than relationship-specific tacit knowledge. Thus, this study identified three questions to be addressed in further research:

(1) Under what circumstances are the models of Figs 4–6 valid? (This is a question of model testing in a variety of circumstances.)
(2) What triggers and hinders the social learning cycle proposed by Boisot et al. (1997)?
(3) What is the relationship between the degree of codification of knowledge, the level of applicability of knowledge (i.e. generic vs. relationship-specific), and the utility of the knowledge?

Moreover, with regard to Fig. 8, the figure was constructed due to two reasons:

(1) The responses of respondents showed that they had difficulties in determining the boundary between the generic and the portfolio levels of Alajoutsijärvi and Tikkanen's (2000) levels of information structuring (see Fig. 6). However, knowing the exact boundary was also shown not to be managerially relevant for the process of obtaining concrete architectural projects, as the case firms in question never have to deal with a large and generic market.
(2) The empirical material indicated that the implicit assumption of Alajoutsijärvi and Tikkanen's (2000) (Fig. 6) that information sharing at dyadic inter-organizational level presupposes broad sharing within both of the firms that compose the dyad was found not to be true of two of the three case firms. Therefore the model of Fig. 8 includes a more precise classification of how knowledge may be shared at the inter-organizational level (i.e. between two firms), in that it more explicitly depicts that knowledge shared at the inter-organizational level may only be present in the individual firms at the level of the individual and/or the team.

In relation to point 2, it must also be mentioned that the information sharing at the dyadic inter-organizational level was mainly found in the context of a few

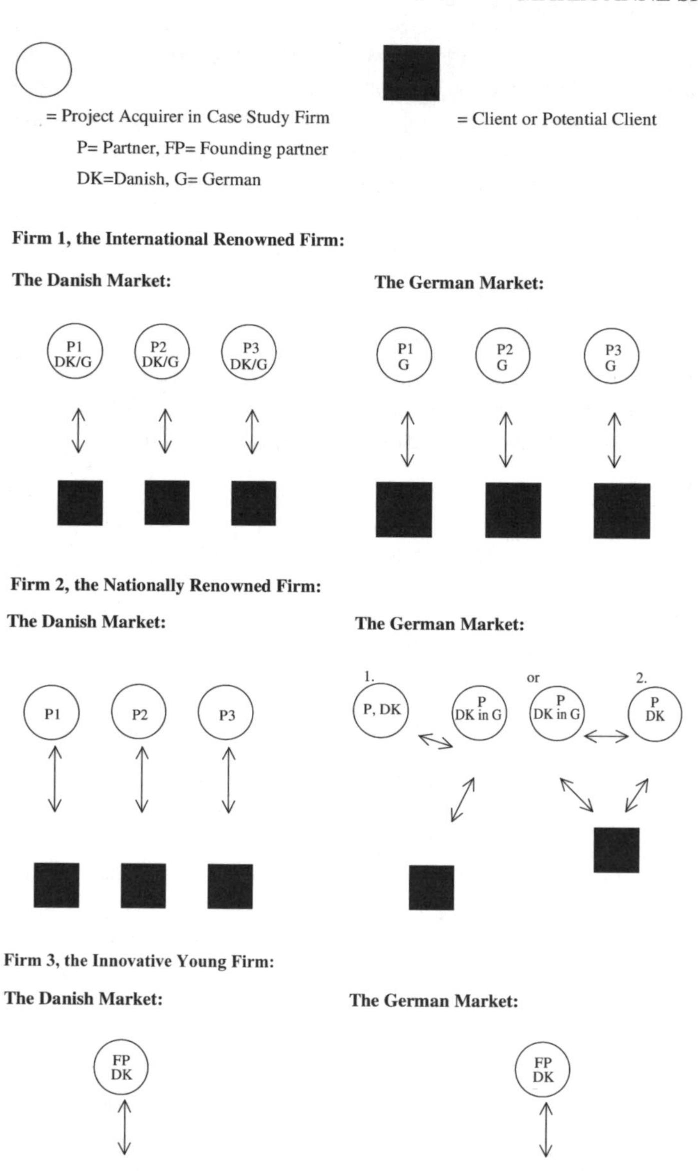

Fig. 7. The Structure of the Case Study Firms' Project Acquisition Work.

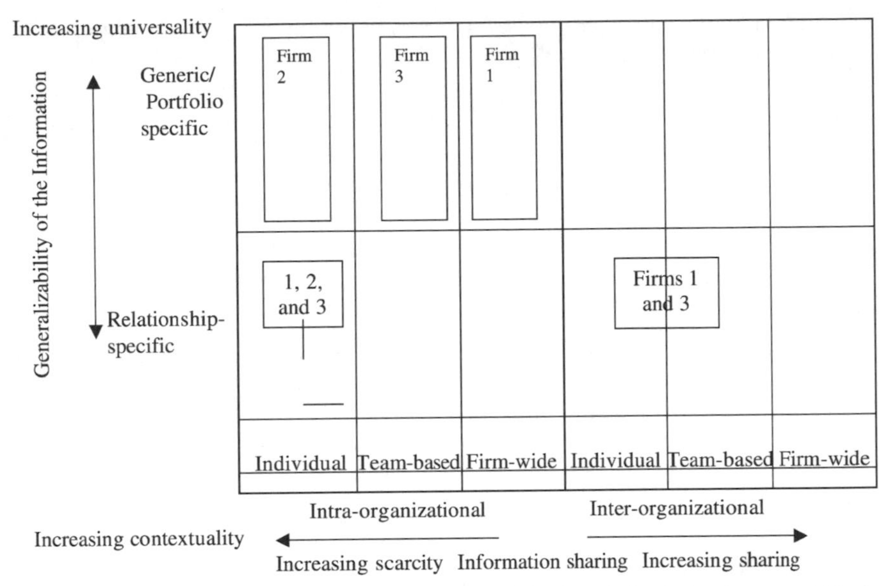

Fig. 8. The Case Study Firm's Germany-Related Tacit Knowledge in the Context of Their Danish-Germen Relationships – Revised Figure.

exceptional client relationships that lasted over a series of multiple projects. Firm 1 has such a long-standing relationship with a specific German client and Firm 3 had a multi-project relationship on the German market for a period of several years during the nineties. These exceptional relationship situations will also be discussed in the next subsection.

Finally, in relation to tacit knowledge, it was found that increases in the tacit and experiential portfolio-specific knowledge lessened the amount of relationship-specific knowledge the two new firms on the German market (2 and 3) needed to acquire in specific situations. This leads to a second proposition:

- For firms in the initial stages of internationalization, increases in the amount of portfolio-specific knowledge held will result in decreases in level of relationship-specific knowledge necessary to obtain.

As the knowledge-sharing processes suggested by the models of Nonaka and Takeuchi (1995, Fig. 4), Boisot et al. (1997, Fig. 5), and Alajoutsijärvi and Tikkanen (2000, Fig. 6) were not found, I also considered the alternative explanations of individuals' learning and information sharing behavior from (mainly American) sociological "boundary spanner" literature (e.g. Adams, 1983; Aldrich & Herker, 1977) and the German IMP Group-related "Relationship Promoter" (in German: "*Beziehungspromotor*") literature (e.g. Gemünden, 1998, 1990; Gemünden & Walter, 1994, 1995a, b).

The "Boundary Spanner" literature (see Adams, 1983; Aldrich & Herker, 1977) focuses on three different functions of organizational actors who interact with actors from other organizations; the second and third of these functions are also in focus in the German "Relationship Promoter" literature:

1. The Information Processing Function.
2. The Agent-of-Influence-over-the-External-Environment Function.
3. The External Representation Function.

In relation to the connection between the information processing function of boundary spanners and the learning processes of the firm, Aldrich and Herker (1977, pp. 218–219) state the following:

> Boundary role incumbents, by virtue of their position, are exposed to large amounts of potentially relevant information. The situation would be overburdening if all information originating in the environment required immediate attention. Boundary roles are a main line of organizational defense against information overload. [...] Expertise in selecting information is consequential, since not all information from the environment is of equal importance. [...]
>
> The process by which information filters through boundary positions into the organization must be examined. Boundary roles serve a dual function in information transmittal, acting as both filters and facilitators. [...] boundary role personnel selectively act on relevant information, filtering information prior to communicating it. [...]
>
> The expertise of boundary role occupants in summarizing and interpreting information may be as important to organizational success as expertise in determining who gets what information, depending upon the uncertainty in the information processed. Information to be communicated often does not consist of simple verifiable "facts". If the conditions beyond the boundary are complexly interrelated and cannot be easily quantified, the boundary role incumbents may engage in 'uncertainty absorption' – drawing inferences from perceived facts and passing on only the inferences."

Additionally, Adam (1976) makes a number of remarks about firm-internal conflicts regarding boundary role persons, as cited in below:

> The BRP who bargains with an external agency on behalf of his organization must not only attempt to reach an agreement with outsiders, but must also obtain agreement from his own group as to what constitutes an acceptable agreement with the external organizations. The BRP is at the crunode of a dynamic, dual conflict in which the outcomes of conflict resolutions attempts (however tentative) in one conflict become inputs to the second conflict, the outcomes of which then become new inputs to the first conflict, and so on (Adams, 1983, p. 1178).
>
> [O]verseas personnel frequently express concern about their distance from their organization, even though they may be enjoying unusual perquisites. The feelings are often reinforced by visits and audits that are clearly in the nature of 'checks' (*ibid.*, pp. 1176–1177).

> Under some conditions, accurate representation of the external world is of paramount importance, as when the function of the information is to permit the organization to adapt to external events. Under other conditions, the function of representation is secondary and designed to subserve another function, such as influence of organization members. This, in effect, constitutes impression management of the BRP's own constituents (*ibid.*, p. 1177).

> When representation by the BRP is manipulatively designed to influence selectively either insiders or outsiders, it may be primarily coping or defensive behavior on his part. It is coping behavior if it is functionally related to the achievement of organization outcomes. [...] Distorted representation is defensive if the primary object of the behavior is to project the BRP as a person and if such service potentially conflicts with the achievement of organization outcomes (*ibid.*, pp. 1177–1178).

> BRPs must *display* [Adams' *italics*] their loyalty and norm-adherence to a greater extent than do other organization members, although, in fact, their loyalty and norm-adherence may under some conditions be greater and more rigid than that of other members. To the degree organizational membership is attractive to BRPs, their display of fealty may be correlated with the extent to which they perceive their fidelity is suspected. The more they feel their behavior and beliefs are suspect, the more they will display their loyalty, the more they will *be* [Adams' *italics*] loyal to the organization and its norms, and the more narrowly, rigidly, and exclusively they will interpret organizational norms and demands. Paradoxically, a corollary consequence of being suspect and of rigid norm interpretation is to *apply* [Adams' *italics*] norms and demands inflexibly in bargaining transactions and, therefore, possibly to reduce bargaining effectiveness, at least over the long run. That is, there may result an intransigent demand for maximum outcomes rather than a question for optimal outcomes (*ibid.*, p. 1179).

> [Another] consequence deriving from the suspicion attached to boundary role positions is conflict for the incumbent. The organizational need for optimal outcomes, whether explicit or not, and the need for BRPs to display their loyalty and norm-adherence are often incompatible. For example, allowing a vendor a given margin of profit in order to achieve an optimal outcome in obtaining organizational inputs may give the appearance that the negotiator is disloyal (*ibid.*).

Some evidence of the phenomena described in the two citations above was found in the material concerning Case Firm 3. However, these phenomena were much more pronounced concerning Case Firm 2, due to two reasons:

(1) As indicated in Fig. 7, for Firm 2, the structure of the project acquisition in the German market was both new and different from the structure of the project acquisition that the firm practiced in the Danish market.
(2) As is discussed in the Subsection 3.5 of this study, the level of prestige of projects that Firm 2 acquired on the German market was markedly lower than the level of prestige connected to many of the firm's acquired projects on the Danish market. This was difficult for some of the partners situated in Denmark to accept.

Furthermore, the interview data indicated that there is a learning loop between the information processing and external representation roles of boundary spanners (see, e.g. Aldrich & Herker, 1977), in that the knowledge actors obtain from information processing is used in external representation and vice versa. Thus, boundary spanners also learn in external representation situations – a factor not sufficiently accounted for by the existing boundary spanner literature. Furthermore, generally speaking, less psychological and organizational distance was found between the boundary spanners active on the German market and the other case firm organizational actors than the reviewed contributions (Adams, 1983; Aldrich & Herker, 1977) suggest. It was hypothesized that the reason for this could be the relatively small size of the studied firms in relation to firms in other industries. This leads us to the study's third theoretical proposition:

- Organizational and psychological distance between boundary spanners and other firm actors will vary positively with firm size.

Similar propositions concerning size have been put forth in general organizational theory (see e.g. Barney & Ouchi, 1986; Milgrom & Roberts, 1992; Williamson, 1979), yet not with specific regard to the boundary spanner role.

Thus, in sum, the contribution of the boundary spanner literature can perhaps contribute to an explanation of why the knowledge-sharing processes suggested by the models of Nonaka and Takeuchi (1995, Fig. 4), Boisot et al. (1997, Fig. 5), and Alajoutsijärvi and Tikkanen (2000, Fig. 6) were not found. However, as previously indicated in this subsection, this is an area in need of further research.

3.4. Analysis pertaining to Research Question 4 (Chapter 9)

Research Question 4 concerns the process of architectural project acquisition and is worded as follows:

(4) How were concrete architectural project jobs obtained by the case firms?

It was formulated based on a pre-existing phase model of project marketing, the "General Marketing Configuration for Project-to-Order Supplier Firms" (Cova, Mazet & Salle, 1994, see Fig. 9).

The Cova et al. (*ibid.*) model is especially powerful due to its ability to depict varying project situations, including the situations where there are no immediate projects at hand and situations where there is concurrent work on several projects. Furthermore, in comparison to previous project marketing models (e.g. the six phase project marketing cycle of Holstius, 1987), the model provides a more detailed depiction of the pre-project phases, as Mandják and Zoltan (1998, p. 484) indicate:

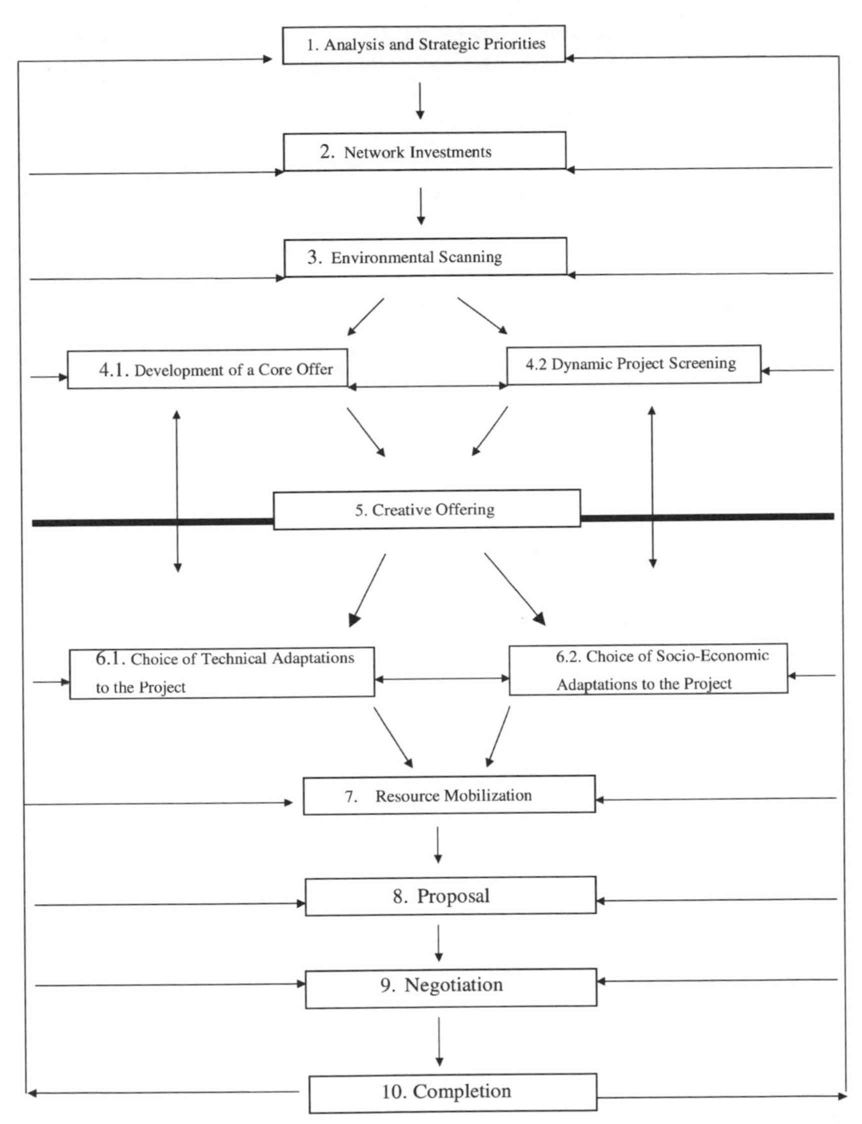

Fig. 9. Cova, Mazet, and Salle's General Marketing Configuration for Project-to-Order Suppliers Firms. *Notes:* Steps 1–5 above and on the thick horizontal line concern marketing-related steps taken independent of a given project (the phase of the anticipation of a project). Steps 5–10 on and below the thick horizontal line concern marketing and coordination efforts related to a specific project that has been awarded or is in the processes of being awarded to the project-selling firm in question (the adaptive phase). *Source:* Cova et al. (1994, p. 40).

> The merit of the model is that it calls attention to the two fundamentally separate phases of the preparatory stage of the project. There is no concrete project in the anticipative stage; the company watches the market in order to gain information necessary for anticipating future projects. This, however, is not a passive process because the company wants to affect and initiate the future projects through the network (Cova et al., 1996; Cova & Hoskins, 1997). Once a future project starts to take shape either in the form of a tender or contract, it is followed by the second stage of preparation, the adaptive stage.

The purpose of Research Question 4 is not to validate or refute the model in its entirety, but rather to use it as the point of departure for a rich description of the process of obtaining project orders on the German market. Therefore, three similar models which depict the architectural project marketing process in the German and Danish private sector, the project marketing process in the German and Danish public sector, and the project marketing process in a German, Danish or German-Danish relationship situation (Figs 10–12) are developed and described in Chapter 9. The developed models are the result of the scrutiny of empirical data, the original Cova et al. (1994) model (see Fig. 9), and the previously mentioned architectural services project phase model (see Fig. 1).

The dotted lines in the final phases of the model of Fig. 10 indicate that the architectural firm may or may not be involved in the construction of the building it has designed, as also previously indicated by, e.g. Fig. 1 and by the discussion of architectural projects and partial or turnkey projects in Section 1 of this summary. Furthermore, there is an implicit theoretical criticism of the original Cova et al. (1994) in the model of Fig. 10, which is also reflected in the models of Figs 11 and 12: Cova et al.'s (*ibid.*) original model (see Fig. 9) depicts strategic analysis and priority setting as activities mainly completed before investments in (social and informational) networks and scanning in the environment are undertaken; this is not in accordance with the case firms' practice. Moreover, as explained in the previous subsection, the answer to Research Question 3 has shown that some scanning and knowledge-acquisition must necessarily precede strategic analysis and determination of strategic priorities. Furthermore, the strategies of architectural firms have a strong emergent (Mintzberg, Otis, Shamisie & Waters, 1988) element as well as an element of tacit socialization (see, e.g. Nonaka & Takeuchi, 1995). Thus, in relation to further refinement of the Cova et al. (1994) framework, it would be relevant to test the following proposition:

- Strategic analysis and priority setting in project marketing firms develops mainly organically and emergently (e.g. Mintzberg et al., 1988).

Paradoxically, initial empirical support for this proposition already exists – in a previous work (Cova, Mazet & Salle, 1993) by the very same authors of the "General Marketing Configuration for Project-to-Order Supplier Firms"! Their

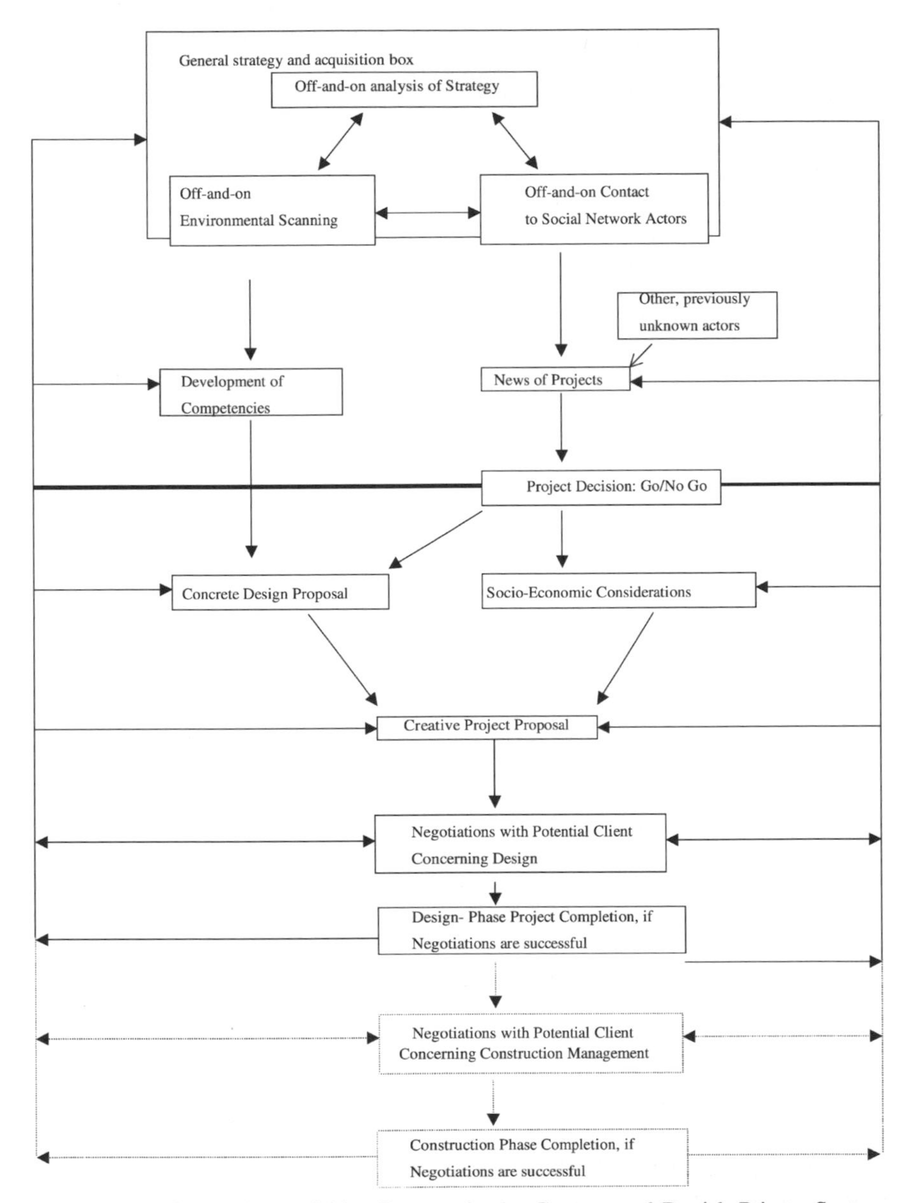

Fig. 10. Architectural Acquisition Process in the German and Danish Private Sectors. *Source:* Author's own conceptualization on the basis of Cova et al. (1994) and case study interviews.

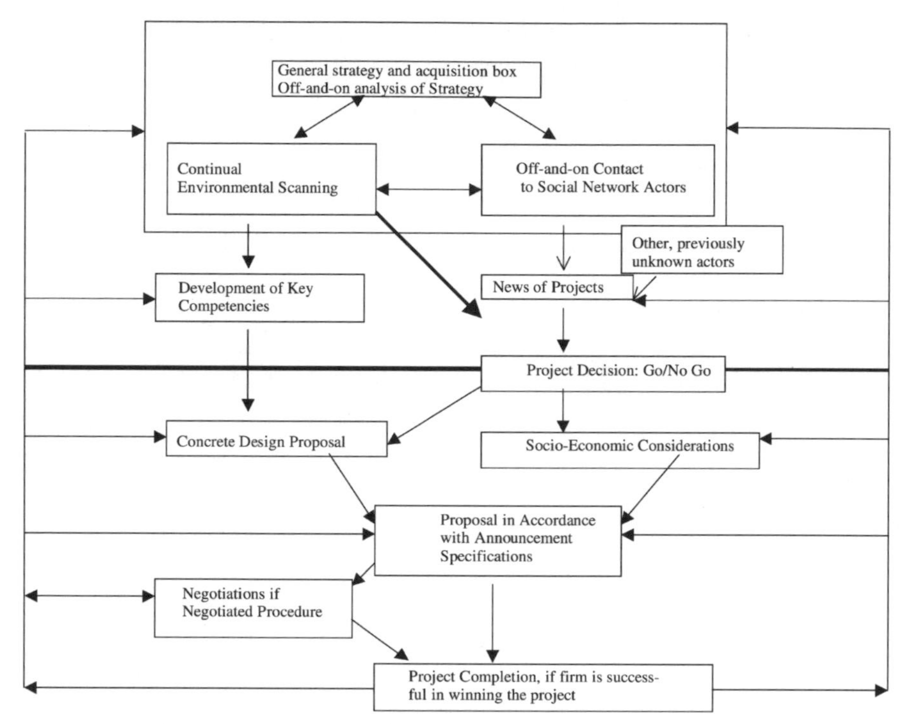

Fig. 11. Architectural Project Acquisition Process in the German and Danish Public Sector. *Source:* Author's own conceptualization on the basis of Cova et al. (1994) and case study interviews.

work contains an assessment of the project business strategic management literature (e.g. Ahmad, 1990; Boughton, 1976; Slatter, 1990), which sought to develop, e.g. Management Information Systems (MIS) "to monitor, gather, analyze, store, and evaluate a wide range of market information [. . .] far beyond the collection of past competitive bid histories. Information about the company, the customer, the competition, and the environment must be obtained and processed to produce knowledge that can be used directly in decision making" (Boughton, 1976).

Cova et al. (1993, 1994) examined: (a) to what extent project marketing firms had attempted to practice the normative suggestions of the project business strategic management literature; and (b) the results of these attempts. Their findings were as follows (*ibid.* 1994, p. 32):

(1) "concerning the screening of projects, companies found difficulties defining stable priorities for a given period of time due to the importance of human

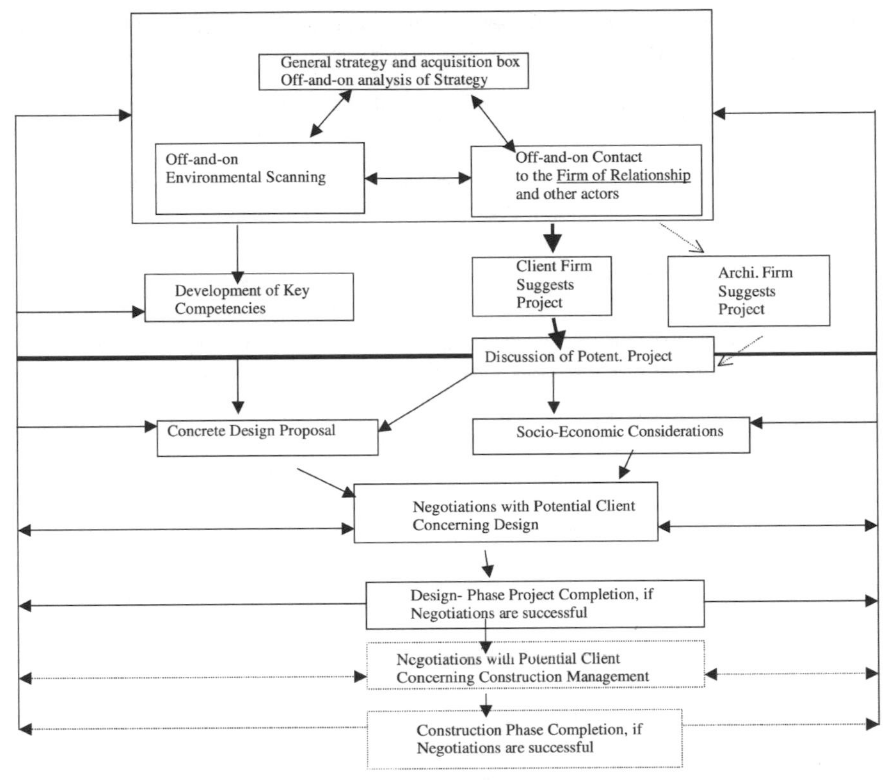

Fig. 12. Architectural Project Acquisition Process in a Danish, German or German-Danish Client Relationship Situation. *Source:* Author's own conceptualization on the basis of Cova et al. (1994) and case study interviews.

factors (intuition, motivation, personal involvement) in the selection of projects [...] In addition, most firms claim their screening strategy is of the 'go/no go' type, but flexible enough to vary and balance the different efforts put into each project. More than sophisticated choices concerning markets and technologies, it is often the financial constraints of the yearly budget allocated to tender preparation that determine the final selection of projects.

(2) Concerning the scanning of project opportunities, companies have progressively acknowledged the importance of interpersonal relationships between individuals [...]
(3) Concerning the analysis and the definition of strategic priorities, difficulties are raised due to the heterogeneity and the rapid evolution of the projects and of their environment."

Based on these results, Cova et al. (*ibid.*) made the following statement concerning the strategic marketing literature they reviewed:

> it does not grasp the complexity of the situation faced by project-to-order supplier firms and may lead to criticisms and contradictions when strictly implemented.

The study's empirical material provides further support for the above statement. Furthermore it provides support for Cova et al.'s (1993) assertion of a fundamental anticipation-flexibility dilemma in project business, which arises due to fundamental uncertainties about the nature of future projects. In this connection, "anticipation" refers to the project supplier firm's strategic need to use resources: (a) to keep itself informed about very specific technological, political, or financial developments in the environment that might only be relevant in a few type of projects out of the total pool of projects that the firm could potentially acquire; and (b) to develop specific policies and programs in relation to these specific developments. "Flexibility," on the other hand, refers to the need to use resources to maintain broad social networks to ensure that that all of the firm's project offerings that are considered to be top quality by important parties in the broader milieu, and to ensure that all individual project offerings can be adapted to e.g. more widely varying cost and customization demands (Bansard, Cova & Salle, 1993, pp. 130–131).

Additionally, in relation to Cova et al.'s (1994) configuration model of Fig. 9 (see also Alajoutsijärvi & Tikkanen, 2000), the results from Research Question 3, as depicted in Figs 10–12, indicate that architectural firms do not develop "core offers" to be used in multiple architectural projects; instead they develop competencies to put to use in multiple architectural projects. (The previous work of Alvesson, 1995; Løwendahl, 2000; Sharma, 1991; Winch & Schneider, 1993 also suggest this.) On the basis of this finding, a further proposition can be put forth for use in future project marketing model refinement:

- Firms that offer customized services that are marketed as projects do not develop "core offers," but rather "core competencies."

Figure 11 concerns the architectural project acquisition process in the public sector, an aspect that was previously discussed in Subsection 3.1. In relation to Germany and Danish public sector architectural projects, the main way that the case firms learn of potential projects is by reading announcements in the relevant public tendering bulletins of the European Union; these announcements are read by the case firms on a continual basis, as is indicated by the boldface filled-in arrow from Fig. 11's box "Continual Environmental Scanning" to the "News of Projects" box. However, it occasionally also happens that relevant public sector projects are mentioned in the case firm's social network or by previously unknown actors who would like to submit a proposal as a team with the case firm in question.

In relation to the previously mentioned result of Research Question 2 (see Subsection 3.2 of this summary) that the German and Danish networks in which the Danish actors participated were more of a social nature than of the nature of networks as typically defined by IMP researchers, the following knowledge-related phenomena were also noted: Only in a very few instances were there long-term relationships between a client and an architectural firm. Two such instances were Firms 1 and 3's more long-term client relationships to customers on the German market, which were initially mentioned in the final paragraphs of Section 2 of this summary and also mentioned in Subsection 3.3. In addition to these two long-term Danish-German relationships, Firms 1 and 2 also have a total of two long-term relationships to two domestic, that is, Danish clients (see the end of Section 2). All of these more long-term relationships are exceptions, in that they are characterized by relationship investments and some technical and economic bonds related to the specific knowledge necessary to build optimally for the particular client in question (see, e.g. the relationship definitions of Easton, 1997 from Section 2). In these exceptional relationships, the process of reaching agreement concerning an architectural project is somewhat different, in that is usually takes less time, due to the parties' knowledge of each other. The process of acquiring architectural projects in relationship situation is depicted in Fig. 12.

Figure 12 shows that the architectural firm in question has relatively continual contact with the client firm with whom it has the relationship. Therefore this client firm is given special mention and even underlined in the general strategy and acquisition box of Fig. 12. Furthermore, due to the relationship, the client firm often starts by discussing a potential project with its "own" architectural firm, instead of approaching a number of alternative architectural firms; this is indicated by the boldface arrows in Fig. 12, yet the architectural firm could in theory also suggest projects to the client firm, as is indicated by the dotted and unfilled arrows. Finally, after the potential project has been discussed, fewer steps are needed to reach the final negotiation stage than is the case in the situation of Figs 10 and 11, as both parties know and understand each other better.

As concerns the capital framework of Bourdieu (1979, 1986a, b), which was initially mentioned in Sections 1 and 2 of this summary, the analysis of Research Question 4 resulted in the following statement: In the socially constructed market of the German construction industry, actors in the case firms generated social capital (*ibid.*), which was used to establish credibility during the process of project acquisitions.

However, in order for the reader to understand the nature and implications of this statement it is necessary for me to explain Bourdieu's (*ibid.*) capital framework and my adaptation of it in this study. Bourdieu (1986b, pp. 248–249) defines social capital as follows:

> social capital is the aggregate of the actual or potential resources which are linked to possession of a durable network of more or less institutionalized relationships of mutual acquaintance and recognition – or in other words, to membership in a group – which provides each of its members with the backing of the collectively-owned capital, a 'credential' which entitles them to credit, in the various senses of the word.

With regard to the operationalization of social capital, which is undertaken in Chapter 4 of this dissertation, the group was defined in my dissertation as being comprised of actors who have similar viewpoints concerning architecture and the construction industry. Moreover, it was stipulated that the groups are small enough to allow the members to have heard of or know each other and each other's viewpoints on architecture and the construction industry. The groups are furthermore a part of the *milieu* (Cova et al., 1996) of the construction industry, a concept that was already treated at the end of this summary's Section 2. The milieu, in turn, encompasses architects and other construction industry actors as well as persons working at schools of architecture or as critics of architecture and members of society at large who show interest in architecture.

Concerning the uses of social capital, it is regarded as a "credential" in this dissertation, in the sense that it entitles its bearers to be regarded as credible to other members of the given group of the field of architecture (see Løwendahl, 2000; Majkgård & Sharma, 1998). Credibility is, in turn, "the actor's perceived ability [i.e. perceived by the other members of the group] to perform something he *claims* he can do on request" (Blomqvist, 1997, the italics are Blomqvist's). Credibility thus includes the client's judgment of the architect in question's knowledge of and ability to judge complex technical and economical factors.

In relation to the process of architectural project acquisitions, social capital was initially generated during contact episodes with actors from the case firms' social networks; in situations in which the case firm actors demonstrated: (a) their knowledge of the rules of the German construction industry milieu; and/or (b) their ability to converse credibly (see also Løwendahl, 2000). Further social capital was generated: (a) during the project specific activities with clients; and (b) through word-of-mouth referrals from previous clients. With regard to project acquisition in the public sector, the EU Public Service Directive (92/50/EEC) specifies the ideal that judgment as to the winner of a tendering is to be made on the basis of the concrete proposals from the tendering firms. However, in reality, the Danish and German implementation of the EU Public Services Directive did not eliminate all of the potential social capital effects. Especially in the German use of the negotiated procedure (see Subsection 3.1 of this summary), social capital could still play a role.

The German "relationship promoter" literature (e.g. Gemünden, 1998, 1990; Gemünden & Walter, 1994, 1995a, b), which was initially mentioned in Subsection 3.3 of this summary, also deals with issues that are tangent to the Bourdivan definition of social capital. With regard to the relational social capital uses in "the Agent-of-Influence-over-the-External-Environment Function" and "the External Representation Function," Gemünden and Walter (1994, pp. 5–6) have created an extensive list of the possible power sources of the relationship promoter:

(a) Persons are attractive partners in social systems because of certain *personal characteristics.*

- Persons who dispose of a certain amount of *expert knowledge* are able to lead professional conversations with potential problem solvers as well as assess their need of problem solving within a sufficient reliability. Persons who are competent in their fields are likelier to be asked for advice and to be accepted by experts as undemanded counsellors as well as mediators. Expert knowledge is helpful in order to influence external partners.
- Relationship promoters acquire or dispose already of sufficient knowledge about the (potential) *co operation partners*. This relates among others to their willingness and ability to co operate with each other as well as to the risks that could endanger the cooperation.
- Persons, between which an *affinity* exists, e.g. relating to the language, value notions, and aims, are more expected to be able and/or willing to maintain exchange relationships.
- Relationship promoters possess the *social competence* to awaken and keep up the willingness to interact with partners, once they are found.
- Relationship promoters develop or dispose of an *identification power* (*referent power*) with respect to their partners which is particularly useful when it comes to exert [*sic!*] influence beyond organizational frontiers.
- Relationship promoters dispose of the necessary *experience* of how to detect appropriate partners and win them over for a cooperation. They are aware of the typical relationship conflicts and pay attention to a foresighted conflict management, where conflicts are spoken out openly in good time and binding agreements are made.

(b) Persons are attractive partners because of a certain *position* in a social system.

- Persons with high, *hierarchically legitimated power* are attractive partners because of their decision competence and their pervasion potential as well as the resources they dispose of, as, e.g. promoters by authority on the side of the user organization with respect to the innovation supplier.

- Position bearers of *lower ranking*, like, e.g. project managers, are favored in the taking over of a relationship promoter role, since possibilities (e.g. a time budget) are granted to them in order to enter into and maintain contacts and relationships to external partners. Furthermore, it can be expected that external partners hold project managers for competent dialogue partners. There are a number of other proposals coming from the business practice. To these belong the "product manager" and the "key account manager." The proposals are embedded in certain management concepts.
- Relationship promoters dispose of a *strong network*. They already know appropriate internal and external co-operation partners or persons who could provide contacts to potential partners. Connect to a *high network centrality* is the access to information and the possibility to control it, which presents a power source for relationship promoters.

To briefly relate the above framework of Gemünden and Walter (1994) to Bourdieu's (1986b) social capital, it can be said that the personal characteristics identified by Gemünden and Walter (1994) help in the initial and subsequent processes of the accumulation of social capital. Their position, in turn, both reflects past processes of the accumulation of social capital and future possibilities for accumulating further social capital.

However, to relate this study's empirical material to the theoretical framework of Gemünden and Walter (*ibid.*), support was found for all of the mentioned personal characteristics as well as the strong network position characteristic, which will be treated in more depth in the next subsection. However, the empirical data did not support Gemünden and Walter's (ibid.) assertion of hierarchically legitimated power as a power source that could be used by certain employees in the Danish architectural firms in question.

3.5. Analysis pertaining to Research Question 5 (Chapter 10)

Research Question 5 deals with the issue of how previous project work influences the acquisition of new projects. It is formulated as follows:

(5) What role did previous project work play when case study firms obtained specific projects on the German market?

In order to describe the analysis of this issue, it is necessary to summarize the rest of the Bourdivan capital framework. Bourdieu's theory also encompasses cultural capital, which refers to value of the practices and physical artifacts that are the result of individuals' socialization within one of many groups within society and this group's efforts to distinguish itself from other groups (Bourdieu, 1986b). Cultural capital manifests itself in three forms (*ibid*., p. 243):

> in the *embodied* state, i.e. in the form of long-lasting dispositions of the mind and body; in the *objectified* state, in the form of cultural goods (pictures, books, dictionaries, instruments, machines, etc.), which are the trace or realization of theories or critiques of these theories, problematics, etc.; and in the *institutionalized* state, [. . . e.g. in the form of] educational qualifications (Bourdieu's *italics*).

In the dissertation's operationalization of cultural capital and in contrast to the theory of Bourdieu (*ibid*.), the *embodied* state of cultural capital (e.g. the dispositions of the mind and body) is excluded to separate and distinguish the terms "cultural capital" and "social capital" as much as possible, as these embodied characteristics have already been included in the operationalization of social capital. The *objectified* and *institutionalized* states are, on the other hand, included: The objectified cultural goods that the architect produces are buildings, parks, etc. Furthermore, with regard to the *institutionalized* state, as a professional, the architect has most often received: (a) formal training and titles which provide him/her with the opportunity to claim the right to practice the profession and thus provide cultural capital, i.e. institutionalized legitimacy. He or she may also have; (b) membership in architects' organizations as well as (c) reference lists which include: (i) projects either in progress or completed as well as; (ii) prizes or honorable mentions awarded in design contests or public tendering procedures. All of these factors provide the architect in question and his or her firm *institutionalized* cultural capital.

The generation of credibility in field of architecture is, however, complex; the architect's (and, consequently, his or her firm's) ability to acquire architectural projects depends upon: (a) his/her credibility in relation to other architects as well as architectural critics; (b) other construction industry actors' judgment of his/her credibility; and (c) complex interaction effects that occur when these two groups voice their opinions about his/her credibility (Albertsen, 1996, p. 2). On the basis of the above, Albertsen (*ibid*.) has identified three possible sub-fields of the field or milieu[5] of architecture:

- The *artistic sub-field*: Members of this sub-field orient themselves predominantly towards architectural and art critics and show little outward concern about market mechanisms, marketing, or profit.
- The *professional sub-field*: Members of this sub-field show interest in design-related ideas and aim to produce architecture that is considered "good design," yet they are also concerned with, e.g. productivity, marketing, technological, and environmental issues. The three case study firms belong to this field.

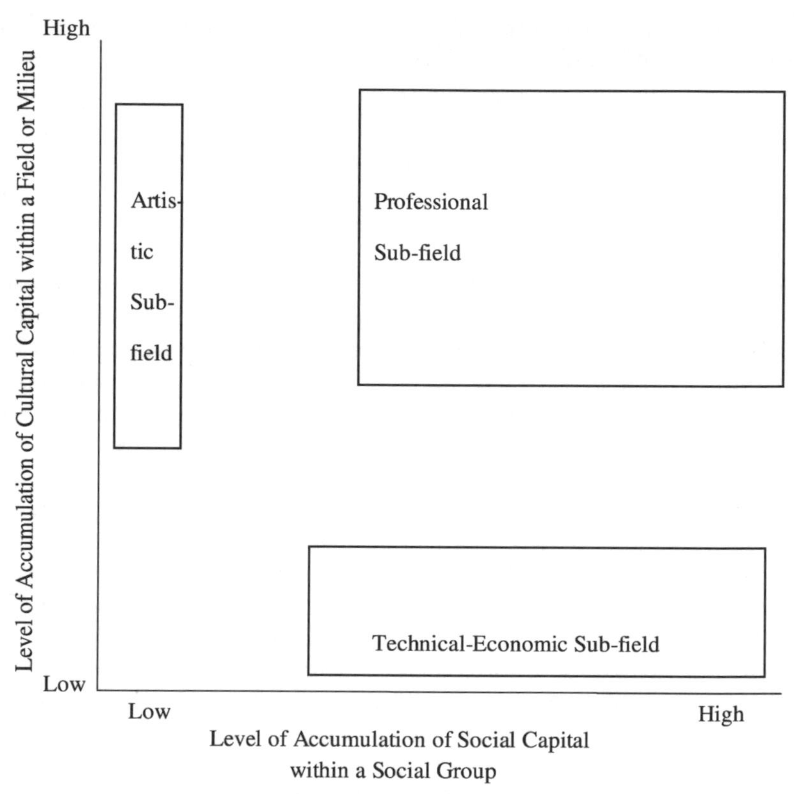

Fig. 13. The Positioning of Albertsen's Three Architectural Sub-fields with Regard to Levels of Accumulated Cultural and Social Capital. *Source:* Author's conception based on Bourdieu (1979, 1986a, b), Albertsen (1996). Further support for the model is also found in Stevens (1998) and von Gerkan (1995).

- The *technical-economic sub-field*: In this sub-field efficiency and productivity are valued most highly. The architectural firms of this sub-field attempt to obtain customers previously served by the professional sub-field; however, architects from the other two sub-fields scorn the work done by the architects of this sub-field due to their "neglect" of the artistic aspect of architecture.

Figure 13 depicts the levels of social and cultural capital that architects in each of these three sub-fields typically possess.

Concerning capital levels, the architects of the *artistic sub-field* (and their firms) have generally low levels of total social and economic capital, yet may have high levels of cultural capital due to the fact that they may have received acclaim from other architects and art critics. The architects (and firms) of the *professional sub-field* have high levels of both social and cultural capitals due to their success in achieving critical acclaim as well as their contacts to both other architects, art critics, and other parties in the construction industry. The high levels of both types of capital generally also translate into a high level of economic capital (Albertsen, 1996). Finally, the architects of the *technical-economic sub-field* have a relatively high-level of social capital in their dealings with other actors in this sub-field, in comparison with their level of cultural capital. They are generally not successful at accumulating cultural capital, as they do not enter and win design contests. The relation between the level of the sub-field and the levels of the groups and networks is as follows: Groups are always a part of a specific sub-field. Social and informational networks are also most often mainly part of one sub-field.

In my dissertation research, architects from the artistic sub-field were, however, excluded. This is because I presumed that they would show little interest in contributing to business administration research due to their disdain for the influence of economic power on architecture. However, in relation to the international capital dynamics of the professional and technical-economic sub-fields, existing contributions suggest that there is a hierarchy of firms as is depicted in Fig. 14.

In order to reach the most elite positions, which result in the most prestigious types of projects, an architectural firm must have a certain degree of artistic renown, which automatically puts it in the professional sub-field. The aspect of artistic renown is also very favorable to a firm's international renown, as almost all reputed journals on architecture, be they in a major world language such as English or Spanish or a minor national language such as Danish, publish articles on foreign buildings that are considered to be "works of art." The same degree of internationalization is not found to the same extent as concerns other issue of concern to construction sector actors, e.g. excellence in construction economics or energy efficiency. However, firms from the techno-economic sub-field may also be quite successful and indeed receive some degree of national renown, as indicated in Fig. 14.

The dissertation's qualitative empirical data provides initial support for the use of the framework of Figs 13 and 14 with regard to the marketing of architectural projects. The initial positions of the three case firms on the Danish and German markets are thus illustrated in Figs 15 and 16.

As previously mentioned, the case firms all belonged to the professional sub-field on the Danish market and also aimed at establishment in (Firms 2 and 3) or were already part of (Firm 1) this sub-field on the German market. During the nineties, they thus aimed at generating both social and cultural capital on the German market (Firms 2 and 3) or at maintaining their levels of capital on this market (Firm 1). This is illustrated in Fig. 17.

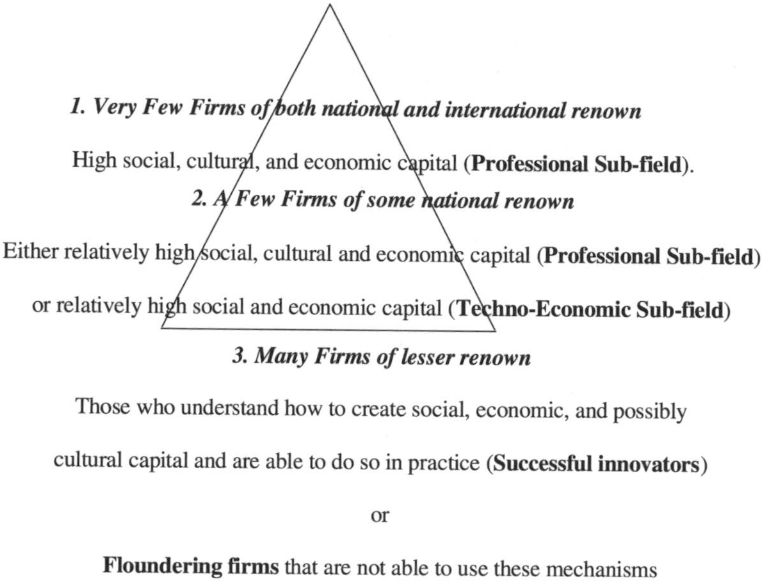

1. Very Few Firms of both national and international renown
High social, cultural, and economic capital (**Professional Sub-field**).
2. A Few Firms of some national renown
Either relatively high social, cultural and economic capital (**Professional Sub-field**)
or relatively high social and economic capital (**Techno-Economic Sub-field**)
3. Many Firms of lesser renown
Those who understand how to create social, economic, and possibly
cultural capital and are able to do so in practice (**Successful innovators**)
or
Floundering firms that are not able to use these mechanisms

Fig. 14. The Hierarchy of Architectural Firms on a National Market. *Source of support for the figure:* Albertsen (1996), Stevens (1998), von Gerkan (1995), various articles in Arkitekt- og byggebladet, Der Architekt and Deutsche Bauzeitung as well as the dissertation study's own empirical data.

1. Very Few Firms of both national and international renown (Case Study Firm 1)
2. A Few Firms of some national renown
(Case Study Firm 2)
3. Many Firms of lesser renown
e.g. Firms climbing the social ladder, (Case Study Firm 3)
or
Local or regional firms without high ambitions,
or
Floundering firms

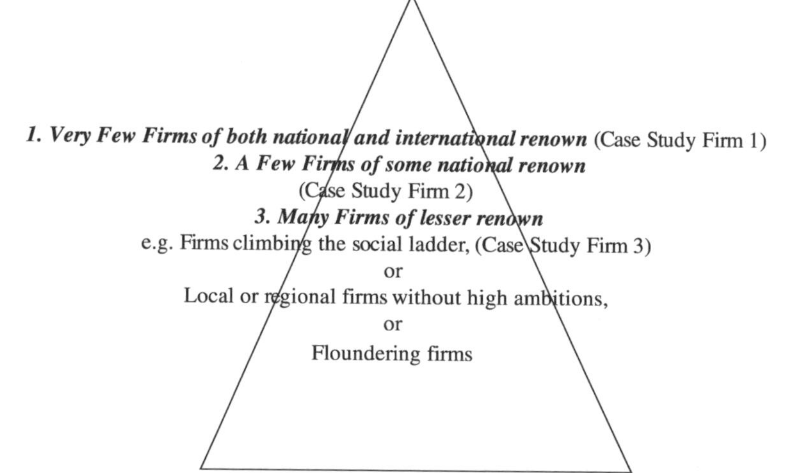

Fig. 15. The Initial Position of the Case Study Firms on the Danish Market.

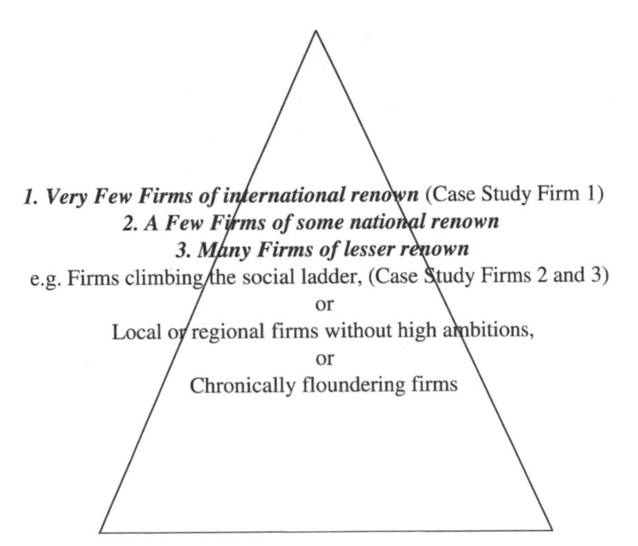

Fig. 16. The Initial Position of the Case Study Firms on the German Market.

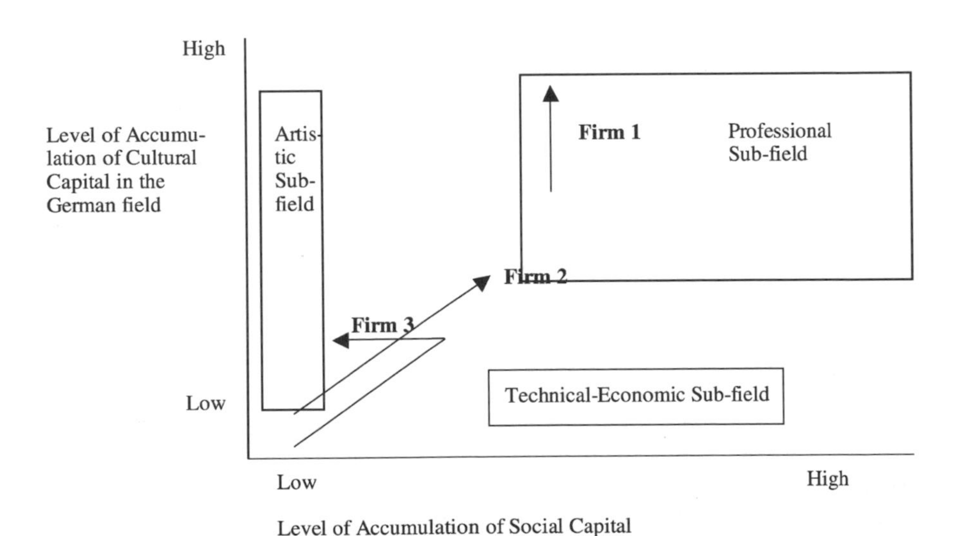

Fig. 17. The Dynamics of the Three Case Study Firms on the German Market During the 1990s in Relation to Albertsen's Three Architectural Sub-fields.

Firm 1 mainly generated additional cultural capital during the nineties by winning German architectural projects; its social capital generating activities were much less emphasized in the 1990s. Firm 1, however, started with a much higher level of social capital due to decades of experience on the German market and the substantial long-standing contacts to their market of its many German-born employees (see Section 2). It thus did not need market presence to maintain its social contacts, and the founding of a subsidiary in Wiesbaden, Germany in connection with the completion of a major renovation project in Frankfurt am Main, Germany did not influence its social capital to any great extent, as the subsidiary's main functions were related to this concrete architectural project (see *ibid.*).

In contrast, Case Study Firms 2 and 3 mainly acquired cultural and social capital after the founding of their offices on the German market. The partners and employees of these offices were able to establish contact with German actors, thus creating social capital, as well as winning some competitions and public tendering procedures, thus establishing cultural capital. Firm 3 was, however, also forced to stop the activities of its subsidiary on the German market to minimize its losses after a German project management court case; this meant that its level of social capital fell again, as presence on the market was necessary for it to maintain many of its social contacts to the actors on the German market. Firm 2, in turn, had to accept that it initially received less prestigious project types on the German market than it was used to on the Danish market, due to its lesser stock of German social and cultural capital.

As regards the use of specific forms of objectified and institutionalized cultural capital, all firms put their cultural capital to use in their reference lists and other promotional material as well as took contacts to the press to inform them of projects they had won as an indirect way of promoting themselves on the basis of previous or current project work. On the other hand, direct promotional efforts related to previous project work, such as asking former clients to promote the architectural firm in question in brochures or advertising, were not used. Similarly clients were not directly asked by the case firms to use social capital-related mechanisms such as word-of-mouth; this happened at client initiative only.

Additionally, it should be emphasized that the accumulation of both cultural and social capital was to a large extent specific to the country of origin, i.e. Germany or Denmark, except in the aforementioned case of internationally recognized artistic achievements, as German clients usually had no contact with construction industry actors in Denmark and did not hold most types of Danish references in high regard. Thus, to some extent, the case firms all had to accumulate social and cultural capital separately for each market, something that, as previously mentioned in Subsection 3.3 of this summary, came as a surprise for the nationally renowned Firm 2. In relation to the concept of the milieu, which was defined in Cova, Mazet and Salle

(1996, p. 654, see also Section 2) as a "socio-spatial configuration" characterized by, among other things, "a territory" and Tikkanen's (1998) challenge of the inclusion of territoriality in the definition of the milieu, the following can be said: The empirical material of this dissertation study supports the territorial delimitation of the milieu, in that the social and information nets of heterogeneous actors mainly were specific to one of the countries in question as were the representations, rules and norms constructed, shared and used by the actors in question.

With regard to the public sector, the final selection process in rounds of public tendering is ideally supposed to be immune from cultural capital effects, as the EU Public Service Directive (92/50/EEC) specifies that judgment is to be made on the basis of the concrete proposal alone. However, similar to the case of social capital (see the end of Subsection 3.4), the empirical material indicates that the effect of cultural capital was also not eliminated completely in the final selection round, as the submissions to public sector restricted and negotiated tendering procedures were not usually anonymized.

On the basis of the above paragraphs of this subsection, it can, in sum, be said that previous project work could provide the three case firms with social or cultural capital or both. Moreover, as depicted in Fig. 17, throughout the nineties the three case firms accumulated cultural capital on the German market.

However, the reference theoretical framework of Salminen (1997) was also scrutinized and used as a supplement to the Bourdivan capital framework in Chapter 10, in cases where the empirical material supported this use. Salminen's (*ibid.*) doctoral dissertation is an in-depth conceptual study of the concept "reference," based on an extensive literature review and a qualitative case study of a Finnish company that delivers and installs turnkey facilities abroad and thus offers turnkey projects (see, e.g. Holstius, 1987; or Luostarinen & Welch, 1990, as discussed in Section 1 of this summary, for definitions of the turnkey projects). As Salminen's (*ibid.*) study is based on qualitative data, the external validity of his framework might be problematic; therefore it was highly relevant to further scrutinize his framework based on further Danish-German data related to architectural partial projects (see Section 1, Holstius, 1987 or Luostarinen & Welch, 1990 for the definition of partial projects).

On the basis of, e.g. Holstius (1987), Salminen (1997, p. 50) defines a reference as follows:

> a deal containing one or several deliveries of products/services, or a part of that deal that either already has been, or will be delivered to a subsequently recognized customer.

According to Salminen (*ibid.*, p. 160), a reference can be used in many ways in relation to either keeping present customers or developing exchange relationships with new customers, as indicated in Fig. 18.

PURPOSE	TASKS	PRACTICES
Keeping present customers	UNIVERSAL TASKS: Internal: – serve as strategic criterion in bidding decisions – prove the functionality of technology to supplier – improve sales force performance External: – break competing supplier relationships – reestablish credibility among old customers – signal service quality – prove the functionality of technology to supplier – develop supplier's image – aid in the access to new market segments SPECIFIC TASKS – create opportunities for further customer contact abroad – make launching customer promote new product/ technology actively – legitimate a new technology paradigm	– articles – press releases – reference lists – promotional material – seminars and conferences – requests to reference customers to promote
Development of new exchange relationships (new customers or sources of supply)	Purchasing/tendering process: – prequalification/shortlisting – winning the final bid	– visits to reference sites – detailed descriptions of similar contracts

Fig. 18. Salminen's Potential Modes of Reference Use in Industrial Marketing. *Source:* Salminen (1997, p. 166).

Furthermore, Salminen (1997, p. 57) lists a number of factors that influence the value of a specific reference, categorized by the degree of specificity in relation to the delivery at hand. Below, as in Saliminen (*ibid.*), the most general are listed first and the most project-specific factors are listed last:

(1) Environmental factors. These relate to, e.g. the effect of the home country of the reference customer firm or the degree of competition in the given industry.
(2) Party-specific factors. These relate to, e.g. salespersons of the potential selling firm, representatives of the purchasing firm and the reference customer firm and their ability to communicate with each other.
(3) The equipment or buildings that the reference customer has purchased (i.e. both the equipment and buildings supplied by the potential selling firm and the equipment and buildings supplied by other firms).
(4) The three usability factors of the specific reference in question: openness, experience, and satisfaction.

The three usability factors are depicted below in Fig. 19.

In relation to Fig. 19, Salminen (*ibid.*, pp. 50–51) has specified the following: "The *first* dimension (experience) means that the utilizability of a reference depends on the age experience of the use of a reference. A deal (reference) can be utilized in reference lists immediately after the deal with a customer has been closed, although its "reference value" is not very high at that time (it is not yet very convincing). After the installation of the equipment has been completed, the value of the reference increases, and after the equipment has been used for one year, the reference can really be utilized. [. . .]

The *second* dimension (openness) means that a reference may not be so secret that the name of the customer cannot be mentioned. The effectiveness of a reference decreases essentially if the name of the customer is kept secret for some reason.

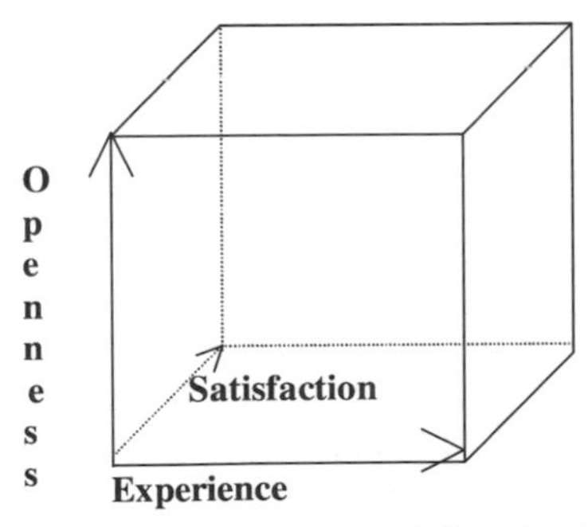

Fig. 19. Salminen's Factors that Determine the Usability of a Specific Reference. *Source:* Salminen (1997, p. 51).

The *third* dimension (satisfaction) means that someone in the buying company has to express the satisfaction of that buying company. This means that a reference must be "good" enough (satisfied customer) and someone has to express that satisfaction" (Salminen's *italic* and boldface type.)

In terms of Bourdieu (1986b), Salminen's (1997) reference value corresponds to the accumulation of project specific *institutionalized* (i.e. prizes) and *objectified* (i.e. buildings) *cultural capital* as well as, with specific regard to dimensions 1 and 3, experience and satisfaction, *social capital* (through e.g. positive word-of-mouth). This study's Danish-German partial project-related data supports Salminen's reference theory with a few exceptions:

(1) Contrary to Salminen's suggestion of the importance of the length of the client's experience to reference value, evidence was found that some architectural design projects not yet completed already have a reference value and that the reference value does not necessarily further increase as the potentially referring client gains experience with the facilities provided by the project.
(2) In relation to Figs 18 and 19, verbal communication between the potential project purchaser and the potentially referring client is the exception, not the rule, as conversations with former clients often do not provide further relevant information for potential project purchasers that could not be acquired through other means, e.g. by personally examining the building in question.
(3) Partially due to point 2 (above) and to the reticence of German and Danish architects concerning advertising their services (see, e.g. Albertsen, 1997; Sommer, 1996), the case architectural firms did not ask their previous customers to promote their architectural business, as suggested in Fig. 18.
(4) References were not used internally to "improve sales force performance," as suggested by Fig. 18. It was usually partners in the firm who were responsible for acquiring contacts; these partners were neither benchmarked by references nor required by their fellow partners to use the firm's references in a specific manner in negotiations with potential clients.
(5) In relation to public sector tendering according to the EU public service directive, references did not play any role in relation to winning the final bid in the two-tier closed procedure, as the references were only examined in the round of pre-qualification.

Finally, in connection with Research Question 5, two further concepts from the literature review of project marketing concepts (in the full-length dissertation's Chapter 3) were scrutinized: the "sleeping relationship" (Hadjikhani, 1996) and the "social construction of demand" (Cova & Hoskins, 1997). In his 1996 article, Hadjikhani examines the nature of project-related buyer-seller relationships in the period of discontinuity, i.e. "the period after project completion in which project buyer and seller are not concerned with resource exchange or negotiation and the buyer's mobility is high." Hadjikhani's (*ibid.*, pp. 332–333) "sleeping relationship" refers to cases of continued buyer-seller dependence after completion of the project itself (due to i.e. the possible future need for improvements or replacement parts with regard to the project):

> During sleeping relationships, contacts based on, e.g. technology-based, financial, and social relationships were used on an off-and-on basis. These contacts were based on trust, and trust was a significant factor in the sense that it affected influenced buyer behavior with regard to ordering new projects. In the sample of case study firms studies, in cases where the level of buyer trust of the sleeping relationship was high, there were incidents where buyer chose the seller to produce new projects.

Relating the above definition to the empirical material, it was found that it is extremely difficult for architectural firms as relatively small firms with many potential competitors to create dependence in relations, which would otherwise result in relationships or even the previously mentioned network governance structure (see Figs 2 and 3), as architectural firms neither can offer unique replacement parts nor achieve a monopoly on future improvements. Thus trust, which is directly related to social and cultural capital, plays an extremely important role in the management of often very discontinuous "sleeping relationships," yet the possibilities for overcoming the discontinuity and achieving long-term relationships are relatively small, as previously indicated in Subsection 3.2.

With regard to the "social construction of demand" issue, Cova and Hoskins (1997, pp. 546–547) state that project marketing firms can follow two courses of action in relation to the rules and norms of the milieu:

(1) The *deterministic approach*, i.e. anticipating and conforming to the competitive arena and the rules of the game as well as conforming to the specification requirements of customers.
(2) The *constructivist approach*, i.e. becoming actively involved in shaping the competitive arena and the rules of the game or in shaping customers' preferences and requirements.

The empirical data indicates that the case firms have mainly followed the determinist approach. Furthermore, the empirical material could be interpreted such that the case firms did not choose the constructivist approach, except in rare instances where they tried to influence a single customer's preferences and requirements, because they did not have the size and power to single-handedly actively shape the competitive arena and the rules of the milieu. These results lead to the final two theoretical propositions to be presented in this summary, which should be tested further in other project marketing studies:

- In relation to the "sleeping relationship" concept, small project-selling firms rely on trust creation more than on creating dependency.
- With regard to the "social construction of demand" issue, small project selling firms mainly follow the deterministic approach.

4. CONCLUSIONS CONCERNING THE CONTRIBUTION OF THIS STUDY

4.1. Conclusions on the Scientific Contribution of the Study

The presentation of the theoretical research results in the previous five sections' was necessarily spread and eclectic. This was in part due to the similar nature of the previous project marketing literature (see Section 2) and in part due to the qualitative, abductive and explorative-integrative study design (see Section 1). However, there are still four ways in which such a descriptive, explorative-integrative study can contribute to scientific progress (see, e.g. Andersen, 1988; Flick et al., 1995):

(1) By providing further empirical evidence supporting previous scientific results and the corresponding theories;
(2) By providing evidence that existing theories and conceptual frameworks may need to be modified;
(3) By providing the basis for theoretical and conceptual syntheses;
(4) By contributing with totally new theories, models, and conceptual definitions.

With regard to points 1 and 2, providing further or contradicting evidence is always positive in relation to both quantitative and qualitative research. However, due to the inherently larger problems of external validity in qualitative research, these points are especially relevant when applied to previous contributions based on qualitative analysis Therefore, this study scrutinizes a number of qualitatively developed concepts and models with the help of additional empirical data. These qualitative contributions include Alajoutsijärvi and Tikkanen's (2000) "Competencies in the Context of Relationships" model (Fig. 6); Boisot et al.'s (1997) "Social Learning Cycle" (Fig. 5), which is also used in Hall (1997); Cova et al.'s (1996) *milieu*, their "General Configuration for Project-to-Order Firms" (Fig. 9, which is from Cova et al., 1994) and their "Anticipation-Flexibility" framework (from Cova et al., 1993, see also Bansard et al., 1993); Nonaka and Takeuchi's (1995) "Knowledge Conversion and Sharing Processes" (Fig. 4); and concepts from Salminen's (1997) dissertation on references, the "sleeping relationship" of Hadjikhani (1996), and the "social construction of demand" issue (see, e.g. Cova & Hoskins, 1997).

For example, with regard to the aforementioned territoriality issue related to the milieu concept (see Tikkanen, 1998, as described in Section 2 of this summary), the empirical treatment of all five research questions provides support for the original geographical delimitation of the milieu concept (see Cova et al., 1996) with regard to European architectural and other construction industry projects. This result should therefore be included in any future research about when it is acceptable to define the boundaries of the milieu (or the similar concept of the field, see Section 2) in geographical terms.

Concerning point 3, theoretical and conceptual synthesis, the model depicted in Fig. 1 is a synthesis that depicts key characteristics of architectural project offerings. As previously described in Section 2, it was developed on the basis of the professional services literature, project marketing literature, and literature concerning construction projects in particular. Furthermore, the commensurability of the project marketing *milieu*-concept (Cova et al., 1996) and Bourdieu's (1986a, b, 1990) and the institutional theoretical (see, e.g. DiMaggio & Powell, 1983; Melin, 1989; Scott, 1994) concept of the field was demonstrated in this dissertation (see Section 2 of this summary), to enable further syntheses. This, in combination with the identification of two possible governance structures for project-based services industries (see Section 2 and Fig. 3), enables this study to draw on and combine insights from both IMP/INPM-contributions and Bourdivan social and cultural capital theory in a project marketing capital accumulation framework (see Subsection 3.5). This application of Bourdieu's capital framework to the contexts of the internationalization of architecture firms and of international project marketing is a novelty. This is because in relation to architectural and other professional services, Bourdivan theory has previously only been used in a quantitative historical study that looked at the previous effects of social relations to renowned architects on the architect in question's success (Stevens, 1998) and in a description of the "pecking order" among Danish architectural firms (Albertsen, 1996). As regards marketing theory, it is also new that Bourdieu's theory is applied to business-to-business project marketing, in that most previous marketing-related contributions (e.g. Hold, 1996; Torres, 1999; Tzokas & Saren, 2001; van Heel, 1998) use Bourdivan theory and the capital framework to examine phenomena on consumer markets. In contrast, this study does not develop any truly new models (point 4, above), but it does present a number of hypothetical prepositions to be tested and/or further scrutinized in future research endeavors.

In connection with these hypothetical prepositions, one final critical scientific issue must be elaborated upon, namely the issue of generalizability. As a qualitative research contribution, this study as a whole is somewhat problematic with regard to external validity (see, e.g. Flick et al., 1995; Kvale, 1996). This is especially true due to major case firm-internal differences and the explorative-integrative nature of the study, which also result in the situation that the multiple cases of this study cannot be considered analogous to multiple experiments, as suggested by Yin (1994, see Section 1). Furthermore, as mentioned above, this study draws on a large number of other qualitative contributions, which are also marked by a lack of knowledge of the boundaries of their external validity.

On the basis of the information in the previous paragraph, the ambition of this study as a whole has neither been to present universally valid results nor to develop grand theory (see Chapter 1). Instead the aim has been to develop a more local conceptual framework based on synthesis and to provide further empirical evidence concerning a multitude of existing frameworks. However, the propositions developed in the study, which were presented in the previous five subsections, may fruitfully be tested in a variety of business service contexts, also by using more statistically generalizable quantitative methods.

Furthermore, it is plausible that the new models depicted in Fig. 1 ("General Characteristics of Architectural Projects," see Section 2), 3 ("Classification of Governance Structures in Professional Service Industries," see Section 2), 13 ("The Position of Albertsen's Three Architectural Sub-fields with Regard to Levels of Accumulated Social and Cultural Capital," see Subsection 3.5), and 14 ("The Hierarchy of Architectural Firms on a National Market") have more general validity, as they were developed based on very extensive literature reviews and/or deductive analysis of a number of conceptual and/or empirical research contributions from a variety of countries, including some quantitative studies.

4.2. Managerial Implications for Architectural and Other Construction Industry Firms

With regard to managerial implications, which are covered in the full-length dissertation's Chapter 12, this study's proposed application of Bourdieu's (1986a) social and cultural capital Salminen's (1997) reference theory to international architectural project marketing is meant as "food for thought" in architectural business managers' evaluation and discussion of their acquisitive practices abroad. This "food for thought" is deemed highly relevant, as managers of architectural firms most often have not received much formal management education (see e.g. Harrigan & Neel, 1996; Løwendahl, 2000; Östnäs, 1984; Östnäs & Svensson, 1986; Sommer, 1996; Stevens, 1998) and there are very few books and articles about marketing architectural services internationally. For this audience, the presentation of such a "food for thought" description is managerially relevant in that it may provoke discussion at a higher level than otherwise: (a) by providing a new vocabulary for variables relevant to the marketing process; and (b) by motivating these leading employees to study their own firms' experiences in the accumulation and use of social and cultural capital and references in order to create managerial heuristics.

In contrast, the framework presented is not a ready-made model of "how-to-market" architectural services, as the undertaken research has been *explorative-integrative* (see Maaløe, 1996), in relation to a lot of previous descriptive contributions. However, the framework may still be fruitfully presented to students of architecture as well as professional architects at, e.g. seminars, to make up for the aforementioned real-life management situation deficits in many architectural and engineering educational programs.

It is also fitting to make several general suggestions to architects and other construction industry actors as to what could fruitfully receive greater focus in their international acquisitive practice on the basis of the study's many interviews. First, with regard to internationalization, the empirical data provides unequivocal support for the statement that there are substantial national variations in the way the construction industry is organized in Western Europe. Therefore acquiring as much advance knowledge of these variations as possible as well as envisioning ways in which one's firm can continue to learn about the variations after beginning operations on a foreign market is highly recommendable for all types of firms internationalizing on European construction markets. Those firms that did not take sufficient measures to this effect did not usually receive a sufficient level of orders and were in some cases also involved in extremely expensive court cases, which jeopardized their very existence.

However, at the same time, it must be remembered that a large portion of this knowledge – e.g. the knowledge concerning negotiations and the unwritten rules of dealing with persons on the foreign market in question – is tacit and must be acquired through experience, as construction industry actors, including recognized international experts, do not usually communicate verbally about these aspects. Furthermore, in relation to this tacit knowledge, there is a fundamental "Catch-22" dilemma, in that this knowledge on one hand normally is a precondition for acquiring projects on the market in question, yet on the other hand can only be acquired through project activities on the market in question. This "Catch-22" dilemma may be especially difficult for architectural firms and other construction industry-related professional service firms (e.g. civil engineering and project management companies) to overcome, as they have little proof of their abilities to fulfill their foreign clients' requirements if they do not have references from the foreign country in question.

In relation to this dilemma, firms should consider whether entering the market in question during a boom period could alleviate this dilemma, because during a boom period, the market in question may be marked by too little capacity, making the potential clients more flexible and open to contraction to foreign firms without previous experience. However, at the same time, if these firms choose to enter the foreign market in a boom period, they must also be prepared at least in the short term to pay extremely high wages to employ some of the native experts on the market in question to help them overcome their tacit knowledge deficits. Wages will be high because these persons will be in especially high demand during the boom.

Moreover, in relation to knowledge management, it is wise for firm actors to ponder beforehand: (1) how to insure a continued increase in the knowledge of the foreign market after operations abroad have been started; and (2) who in the firm will need to possess this knowledge. Considering these issues is important because the diffusion of knowledge to relevant firm actors is not an automatic process, due to many of the dead ends and role conflicts experienced by boundary spanners (see Subsection 3.3 of this summary). Moreover, the following specific knowledge-related developments may need to be addressed before and during the internationalization process, as they inherently encompass the possibility of intra-firm conflict:

(1) The profile of the firm may change and perhaps even become muddy as it internationalizes, as the types of projects it acquires on the new market in its establishment phase may be very different from the types of projects it is used to working with on its home market (see the information about Case Firm 2 in Subsections 3.3 and 3.5).
(2) Not all employees in the firm will have the same knowledge of the foreign market as the relationship promoter/boundary spanner employees who work abroad (see Subsection 3.3). This may cause tension within the firm concerning e.g. goals and means (see *ibid.*).

In relation to point 2, Fig. 8, which depicts the location of knowledge in the context of client relationships, is a further helpful descriptive tool for managers in determining the location and spread of knowledge at given points in time and conversing about it to achieve consensus on future aims among leading employees and/or firm owners.

Furthermore, in relation to the decision of who should head the firm's operations abroad, considerations should be made concerning the personal characteristics necessary for this position (see the characteristics listed by Gemünden & Walter, 1994 in Subsection 3.4). A knowledgeable professional, e.g. a competent and experienced architect or engineer, is not necessarily a good relationship promoter abroad. He or she may have the necessarily skills for solving problems for existing clients, but not be good at creating the social and informational relationships that are a prerequisite for obtaining project orders from previously unknown foreign clients, due to, e.g. a lack of personal presentation skills or the ability to adapt to a foreign environment.

Internationalizing architectural firms may, however, find conversations about this point difficult due to the fact that promotional activities are not so often discussed in daily business conversation (see, e.g. Albertsen, 1996; Harrigan & Neel, 1996; Sommer, 1996). Thus, to facilitate communication about these issues, leading employees or partners in a firm contemplating internationalization should consider working with this point at a higher level of abstraction, by introducing the vocabulary of the "boundary spanner" and "relationship promoter" literature (see Subsection 3.3) in discussions about the firm's internationalization.

Yet another reason for difficulties in discussing promotional activities relates to architectural firms' opinions and prejudices concerning marketing. In this study, many of the respondents from the case study firms as well as the other firms interviewed were skeptical of the role of marketing. Opinions such as "marketing is not relevant for the architectural profession" could be discerned in many conversations. These opinions seemed to be based on an understanding of marketing as encompassing "rigid planning," "product parameter strategies for standardized products," "advertising," and/or "showy selling." These practices are not *comme il faut* for the professional self-understanding of most Danish architects (see, e.g. Albertsen's, 1996 remarks concerning the professional sub-field in Section 1), nor are they necessarily relevant for the marketing of project-related services (see, e.g. Cova & Salle, 1996; Cova et al., 1996; Subsection 3.4).

Marketing can, however, be understood alternatively as: (a) acting constructively on one's understanding the rules of the market in which one operates; and (b) acting constructively on one's understanding the needs and wishes of clients and potential clients (see, e.g. Ford, 1997). This understanding of the term is much more applicable for the sale of professional services in the construction industry; thus it is recommended that architects as well as others in business-to-business markets take this perspective.

Concerning the role of planning, markets for construction industry products and services are relatively volatile, due to the huge influence of the fluctuating cost of borrowing money on the demand for construction investments as well as the continual appearance of new environmental regulations and new building components. This volatility makes it very risky to produce long-term demand forecasts or plans for the future use of capacity. Thus, the often mechanistic prescriptions of the strategic planning school (e.g. Ahmad, 1990; Boughton, 1976; Slatter, 1990, as described in Subsection 3.4) – where one gathers information first, then plans, and then implements the plans – are not especially applicable to the construction industry (see, e.g. Cova et al., 1993 as described in Subsection 3.4). Instead information must be gathered at the same time that one is making, implementing, and constantly revising one's plans in a more *emergent* strategy process (see, e.g. Figs 10–12; Mintzberg et al., 1988).

In connection with the above and drawing on Bourdieu (1990) and Cova and Hoskins (1997), marketing is best understood as being about understanding the rules of the game in a way that enables one to act spontaneously, yet also in a suitable way, in a given situation to ensure a project order. This is because, after, e.g. a firm has decided whether to seek a given project on the basis of its understanding of its goals, its marketing activities are not yet finished. Marketing also encompasses the spur-of-the-moment, spontaneous actions that are a part of the discussions of potential projects with potential clients. During these discussions, one either discusses and plans projects in a *reactive* way, by following the existing recognized rules of the game or the client's specifications without question, or in a *proactive, constructivist* way, i.e. when one becomes active in shaping client preferences and specifications to one's own and/or to clients' potential advantage (see, Cova & Hoskins, 1997; as discussed in Subsection 3.5).

The knowledge that is necessary to market architectural services professionally pertains first, with specific regard to foreign markets, to: (1) the previously mentioned "rules" and "practices" of the market in question at the levels of the client portfolio and individual client (see, e.g. Section 2); and to (2) limited possibilities for manipulating with client preferences and practices (see Subsection 3.5). In relation to (1), several factors commonly ignored by textbooks on firms' internationalization (see, Axelsson & Johanson, 1992) are especially relevant for architectural firms to remember:

- Knowledge of the relative positions of the actors in the foreign country's network(s).
- Knowledge of how the support of these actors could be mobilized in relation to the planned export activities.
- The ability of the export firm's actors to orient themselves, i.e. "obtain an understanding of where different actors including the actor itself stand in relation to each other" (*ibid.*, p. 231).
- The ability of the export firm's actors to position their firm in the network of other firms.

Furthermore, with regard to all markets, domestic or foreign, it is paramount to understand the different sub-fields of the world of architecture (see Section 1 or Albertsen, 1996) and the modes of credibility generations in the eyes of clients (see Subsections 3.3–3.5). In this dissertation, these modes have been described by building the framework based on the cultural and social capital of the French sociologist Pierre Bourdieu (1979, 1986a, b; see Subsections 3.4 and 3.5) and the reference theory of Salminen (1997, see Subsection 3.5), i.e. the frameworks which provide the aforementioned useful "food for thought" to practising architects.

Furthermore, in both foreign and domestic client relations, similar to Harrigan and Neel (1996) and Sommer (1996), the empirical material provides support for the claim that architects (and to some extent also engineering and contracting firms) do not always focus sufficiently on understanding their clients' perceptions of their own needs, wishes, and priorities. Thus, I suggest that architectural and other firms in the construction industry should not shun marketing and economic information gathering per se either domestically or internationally. Although there are a number of very well-known models that are less suitable for the marketing of architectural services (i.e. the marketing-mix models, see Cova & Salle, 1996), models can be found that, when used, aid in the understanding of the marketing and economic side of the business as well as planning and assessing one's business operations both at home and abroad.

If many of the Danish architectural firms of my pilot and case studies had prioritized this type of information more highly and sought methods of acquiring this information, it would have been easier for them to make a credible promise (see Løwendahl, 2000) to clients and thus to acquire more projects. This is especially true with regard to German private sector clients; many Danish architectural firms vastly under-prioritized the collection of this type of information from these clients, perhaps due to a false assumption that they as educated architects were automatically familiar with and understood "the client's best interests" (see Harrigan & Neel, 1996 for further discussion of this issue in an American context).

In connection with the above point, I would argue that architectural firms need to ensure that they are familiar with clients' perceptions to an even greater extent when operating on a large, anonymous foreign market such as the German market than is the case when operating on a small domestic market such as the Danish market. On a small national market, the domestic architectural firms may actually know quite a bit about the viewpoints and preferences of most potential domestic clients because: (a) the market is relatively small and there are thus fewer cliques and subcultures; and (b) they have substantial knowledge of the discourses of the domestic construction industry from years of operating on the market.

Familiarity with clients' perceptions also entails monitoring debates about improvements in the construction national industry in question, which in turn implies that Axelsson and Johanson's (1992, see Subsection 3.3) "knowledge of how the support of these actors could be mobilized in relation to the planned [...] activities" entails much more than simply knowing that one receives projects through contacts or one's network. Via, e.g. knowledge of debates about construction quality and economics (or the lack thereof), one also learns something one's potential clients would like, but have not been offered yet – knowledge which can subsequently be used in a constructivist strategy (see Cova & Hoskins, 1997, as explained in Subsection 3.5).

In connection with this point, it may important for actors from firms in the professional sub-field to also be aware of what is happening in the techno-economic sub-field, in spite of the fact that they as members of the professional sub-field often shun their techno-economic sub-field colleagues (see, Albertsen, 1996; Subsection 3.5). This is because, in the long term, there may be spillover effects from the techno-economic sub-field to the professional sub-field – even though this is something that the members of the professional sub-field are often most loath to accept or admit.

For example, on the German market, the relationship to general contractors may need to be proactively reconsidered by professional sub-field architects, due to the fact that German clients are increasingly procuring construction projects through the German general contractors modes of construction project organization (in German: *Generalunternehmer* and *Generalübernehmer*, see Subsection 3.2) and thus more often than previously building without independent architectural council (see, *ibid.*). This development may be dealt with proactively if architects: (a) define their field broadly, as a field of knowledge or expertise concerning construction (see, e.g. Ahlqvist, 1992), instead of "defending their territory" by refusing to accept new constellations of cooperation; and (b) seek answers to marketing-related questions such as "Which types of solutions are potential clients purchasing instead of the solutions I would propose?" "What advantages and disadvantages do these solutions offer?", and "Would it be advantageous to our firm to offer different types of services based on these new trends, possibly in cooperation with other firms?"

Furthermore, in relation to the aforementioned proactive, constructivist offering (see, Cova & Hoskins, 1997, as described in Subsection 3.5, in a sector now marked by over-capacity (see, e.g. Lubanski, 1999, as described in Subsection 3.2) and the aforementioned diminishing traditional role of the architect as an independent consultant servicing his or her client, developing superior proactive skills may become a pre-condition for survival, as many architectural firms are struggling to keep their heads about water. However, at the same time, the room for proactive experimentation is small; as an individual architectural firm does not usually have the size or power to change the rules of the industry as a whole (see the discussion of this issue in Subsection 3.5). Furthermore the firm may lose its good reputation or its financial stability, if one of its proactive ideas fails badly, as happened in Firm 3's first German construction management project (see *ibid.*).

Finally, construction industry offerings are not only influenced by customer's perceptions, needs, and desires and the more immediate marketing processes focused upon in this dissertation research. The economic rules and conventions of the various viable "worlds of production" of construction industry-related goods and services (see, e.g. Storper & Salais, 1997) also constrain current and future offerings, whether proactive or reactive. It is therefore imperative that architects and other construction industry actors interested in developing a proactive strategy for future project acquisition also understand the fundamental economic rules that govern the world of construction industry production. This will enable them to better assess the possible economic and structural consequences of, e.g. the continued increase in standardization of European construction offerings (see, e.g. Eurostat, 1995, as discussed in Section 2). This understanding, in turn, will enable them to better assess their "market-space" as well as to find ways in which they could organize their production to avoid excessive price pressure and/or pressure to produce solution types that they would prefer not to produce. This type of proactive economic analysis is also preferable to merely reacting to "unfavorable" trends and tendencies when these trends actually appear, a practice that all too many construction industry firms seem to follow today.

NOTES

1. As I use a number of German works in this study, it must be noted that there is no good direct German translation of the term "profession," as there is in Danish and French. In German the term "*freie Berufe*" refers to self-employed practitioners with specialist knowledge whereas the "*akademische Berufe*" are "the old university educated professions of the clergy, divinity, law, and high school teaching" (Stevens, 1998, p. 27).

2. Unfortunately there are also translation problems with regard to the English language term "project marketing." The German "*industrielles Anlagengeschäft*" is not a perfect translation of project marketing, as "*industrielles Anlagengeschäft*" does not include certain types of non-industrial construction activities for the public or private sector such as the building of museums or residential dwellings. However, on the basis of Günter and Bonaccorsi (1996), I mainly make reference to the German language "*industrielles Anlagengeschäft*" literature as the literature most comparable to English-language project marketing and systems selling literature.

A translation problem also exists in relation to French language literature. According to Cova (1990, pp. 9–10), "*systemes*" refers to combinations of goods and services, "*travaux*" is the correct term for the installation work in connection with systems and project sales, and "*projets*" and "*affaires*" are the French words for systems and project sales that involve a combination of installed goods and projects.

3. In the German "*Generalübernehmer*" mode of coordination, a project management firm typically assumes legal responsibility for the entire construction project, yet delegates either: (a) all construction-related tasks of; (b) all specific design- and construction-related tasks to a large number of subcontractors.

4. Further support for this deficit in textbooks is found in Chetty and Eriksson (1999); they actually define foreign market knowledge as "the knowledge of business relationships in the local network."

5. See Section 2 of this summary for a discussion of the commensurability of these two terms.

REFERENCES

Adams, J. S. (1983). The structure and dynamics of behavior in organizational boundary roles. In: M. D. Dunnette (Ed.), *Handbook of Industrial and Organizational Psychology* (pp. 1175–1199). New York: Wiley.

Ahlqvist, B. (1992). Architecture – field of activity or knowledge? In: G. Kazemian (Ed.), *Conference Proceedings. International Conference on Theories and Methods of Design* (pp. 81–86). Chalmers. Sweden: Royal Institute of Technology/Chalmers University of Technology (May 13–15).

Ahmad, I. (1990). Decision-support system for modeling bid/no-bid decision problem. *Journal of Construction Engineering and Management, 116*(4), 595–608.

Ahmed, M. M. (1993). International marketing and purchasing of projects: Interactions and paradoxes. A study of Finnish project exports to the Arab countries. Helsinki: Publications of the Swedish School of Economics and Business Administration. Ph.D. Dissertation.

Alajoutsijärvi, K., & Tikkanen, H. (2000). Competence-based business processes within industrial networks. A theoretical and empirical analysis. In: A. Woodside (Ed.), *Advances in Business Marketing and Purchasing* (Vol. 9, pp. 1–49). Stamford CN USA: JAI Press.

Alasuutari, P. (1995). *Researching culture. Qualitative method and cultural studies*. London: Sage.

Albertsen, N. (1996). Architectural practice. Habitus, field and cultural capital. Nottingham, UK: Paper presented at the conference "Occupations and Professions: Changing Patterns, Definitions, Classifications." University of Nottingham (September 11–13).

Albertsen, N. (1997). Transversal answers for the case of Denmark. Århus, Denmark. Unpublished Working Paper, Aarhus School of Architecture (June).

Aldrich, H., & Herker, D. (1977). Boundary spanning roles and organization structure. *Academy of Management Review, 2*(2), 217–230.

Alvesson, M. (1995). *Management of knowledge-intensive companies*. Berlin: de Gruyter.

Andersen, H. (1988). Videnskabsteori and metodelære for erhvervsøkonomer (2nd ed.). Copenhagen: Samfundslitteratur.

Arkitekt- og byggebladet (1993). Periodical published by Dansk Praktiserende Arkitekter until February 1993, thereafter by Associated Danish Architects (Associerede Danske Arkitekter/ADA).

Axelsson, B., & Johanson, J. (1992). Foreign market entry – the textbook vs. the network view. In: B. Axelsson & G. Easton (Eds), *Industrial Networks – A New View of Reality* (pp. 218–236). London: Routledge.

Backhaus, K. (1995). *Investitionsgütermarketing* (4th ed.). Munich: Verlag Vahlen.

Backhaus, K., & Büschken, J. (1997). What do we know about business-to-business interactions? A synopsis of empirical research on buyer-seller interactions. In: H. G. Gemünden, T. Ritter & A. Walter (Eds), *Relationships and Networks in International Markets* (pp. 13–36). Oxford: Pergamon.

Bansard, D., Bernard, C., & Robert, S. (1993). Project marketing. Beyond competitive bidding strategies. *International Business Review, 2*(2), 125–141.

Barney, J. B., & Ouchi, W. G. (Eds) (1986). *Organizational economics: Towards a new paradigm for understanding and studying organizations*. San Francisco: Jossey-Bass.

Baus, U. (Ed.) (1997). Architekten: Apocalypse now? Die veränderung eines berufsbildes. Stuttgart: Deutsche Verlags-Anstalt.

Blomqvist, K. (1997). The many faces of trust. *Scandinavian Journal of Management, 13*(3), 271–286.

Boisot, M., Griffiths, D., & Moles, V. (1997). The dilemma of competence: Differentiation vs. integration in the pursuit of learning. In: R. Sanchez & A. Heene (Eds), *Strategic Learning and Knowledge Management* (pp. 65–83). New York: Wiley.

Børsen (1995–1999). A national Danish business newspaper (published on weekdays).

Boughton, P. D. (1976). The competitive bidding process: Beyond probability models. *Industrial Marketing Management, 16*, 87–94.

Bourdieu, P. (1979). *La distinction. Critique sociale du judgment*. Paris: Les Editions de Minuit.

Bourdieu, P. (1986a). *Distinction – a social critique of the judgment of taste*. London: Routledge, 1986 (English Translation of Bourdieu 1979).

Bourdieu, P. (1986b). The forms of capital. In: J. G. Richardson (Ed.), *Handbook of Theory and Research for the Sociology of Education* (pp. 241–258). New York: Greenwood Press.

Bourdieu, P. (1990). *The logic of practice*. Stanford, CA: Stanford University Press.

Button, K., & Fleming, M. (1992). The professions in the single European market: A case study of architects in the U.K. *Journal of Common Market Studies, 30*(4), 403–418.

Chetty, S., & Eriksson, K. (1999). Market experience and how it influences learning in international expansion. In: D. McLoughlin & C. Horan (Eds), *Proceedings of the 15th Annual IMP Conference*. Dublin: University College Dublin CD-Rom.

Council of the European Communities (1997). Council Directive 92/50/EEC of 18 June 1992 in relation to the coordination of procedures for the award of public service contracts as amended by European Parliament and Council Directive 97/52/EC of 13 October 1997.

Cova, B. (1990). Marketing international de projects: un panorama des concepts et des techniques. *Revue Française du Marketing, 127/128*, 9–37.

Cova, B., & Ghauri, P. N. (1996). Project marketing. Between mass marketing and networks. Working Paper, the European Seminar on Project Marketing and System Selling.

Cova, B., & Hoskins, S. (1997). A twin-track networking approach to project marketing. *European Management Journal, 15*(5), 546–556.

Cova, B., Mazet, F., & Salle, R. (1993). Towards flexible anticipation: The challenge of project marketing. In: M. J. Baker (Ed.), *Perspectives on Marketing Management* (Vol. 3, pp. 375–399). Chichester, UK: Wiley.

Cova, B., Mazet, F., & Salle, R. (1994). From competitive tendering to strategic marketing: An inductive approach to theory-building. *Journal of Strategic Marketing, 2*, 1–19.

Cova, B., Mazet, F., & Salle, R. (1996). Milieu as the pertinent unit of analysis in project marketing. *International Business Review, 5*(6), 647–664.

Cova, B., & Salle, R. (1996). The marketing of complex industrial services: A pluralist approach. La Londe-Les Maures, France: Conference paper, Actes du 4e Séminaire International de Recherche en Management des Activités de Services.

Danish Association of Consulting Engineers (Foreningen af Rådgivende Ingeniører/F.R.I.) (1996–1999). F.R.I.-Survey of architectural and consulting engineering services 1995–1998: Statistical analysis related to the EU Services Directive (March 1996–1999, published annually).

Danish Ministry of Commerce and Industry (Erhvervsministeriet) (1993). Bekendtgørelse om samordning af fremgangsmåderne ved indgåelse af kontrakter om offentlige indkøb af tjenesteydelser i De Europæiske Fællesskaber (June 22).

Danmarks Statistik (1992–1999). Generel Erhvervsstatistik og Handel. Statistiske Efterretninger. Copenhagen: Danmarks Statistik.

Day, D. W. J. (1994). *Project management and control*. London: MacMillan.

Denzin, N. K. (1978). *The research act*. Englewood Cliffs, NJ: Prentice-Hall.

Der Architekt (1991–1999). Periodical published by the German Federation of Architects (Bund deutscher Architekten/BDA).

Deutsche Bauzeitung. German periodical published by Bund deutscher Baumeister, Architekten und Ingenieure e.V/BDB).

Deutsches Institut für Wirtschaftsforschung (DIW). The German Institute of Economic Research. Statistics found at the following Internet address: http://statfinder.diw-berlin.de/widab_html/bam_diw_bauvolumen.html

DiMaggio, P. J., & Powell, W. W. (1983). The iron cage revisited: Institutional isomorphism and collective rationality in organizational fields. *American Sociological Review, 48*, 147–160.

Dræbye, T. (1998). Forundersøgelse: Byggevirksomheders succes og fiasko på det tyske marked. Copenhagen: Danish Ministry of Commerce (Erhversministeriet)/the Danish Agency for Trade and Industry (Erhvervsfremmestyrelsen).

Dubois, A., & Gadde, L.-E. (1999). Case studies in business market research: An abductive approach. In: D. McLoughlin & C. Horan (Eds), *Proceedings of the 15th Annual IMP Conference*. Dublin: University College CD-Rom.

Easton, G. (1995). Methodology and industrial networks. In: K. E. K. Möller & D. Wilson (Eds), *Business Marketing: An Interaction and Network Perspective* (pp. 411–492). Boston: Kluwer.

Easton, G. (1997). Industrial networks: A review. In: D. Ford (Ed.), *Understanding Business Markets* (2nd ed., pp. 102–126). London: Dryden.

Easton, G. (1998). Case research as a methodology for industrial networks: A realist apologia. In: P. Naudé & P. W. Turnbull (Eds), *Network Dynamics in International Marketing* (pp. 73–87). Oxford: Elsevier.

Entreprenørforeningen (Danish Association of Contractors) (1996). Danske Entreprenører i Tyskland. Copenhagen: Entreprenørforeningen.

Erramilli, M. K., & Rao, C. P. (1990). Choice of foreign market entry modes by service firms: Role of market knowledge. *International Management Review, 30*(2), 135–150.

European Construction Research (1995). Bygge- og ejendomsmarkedet i Vesttyskland. Glostrup, Denmark: European Construction Research.

Eurostat (1995). Architects. In: *Eurostat. Panorama of EU Industry 1995/1996* (pp. 24–40). Luxembourg: Office for Official Publications of the European Communities.

Flick, U., von Kardorff, E., Keupp, H., von Rosenstiel, L., & Wolff, S. (Eds) (1995). *Handbuch Qualitative Sozialforschung: Grundlagen, Konzepte, Methoden und Anwendungen* (2nd ed.). Weinheim, Germany: Beltz, Psychologie-Verl.-Union.

Ford, D. (Ed.) (1997). *Understanding business markets* (2nd ed.). London: Dryden.

Frisch-Jensen, M. (1999). Formandens udkast til konklusioner og anbefalinger. Copenhagen: Material for a seminar on international construction projects held on March 3; organizer: Projekt Renovering, Danish Ministry of Housing.

Gemünden, H. G. (1990). Innovationen in Gechäftsbeziehungen und Netzwerken. Karlsruhe, Germany: IABU, Institut für Angewandte Betriebswirtschaftslehre und Unternehmensführung.

Gemünden, H. G. (1998). 'Promotors' – Key persons for the development and marketing of innovative industrial products. In: K. Backhaus & D. T. Wilson (Eds), *Industrial Marketing: A German-American Perspective* (pp. 134–166). Berlin, New York and Tokyo: Springer Verlag. (This article is also reprinted in: K. Grønhaug & G. Kaufmann (Eds), *Innovation: A Cross-Disciplinary Perspective* (pp. 347–374). Oslo: Norwegian University.)

Gemünden, H. G., & Walter, W. (1994). The relationship promoter – Key person for inter-organizational innovation processes. In: J. N. Sheth & A. Parvatiyar (Eds), *Relationship Marketing: Theory, Methods and Applications. Proceedings of the Second Research Conference on Relationship*

Marketing (pp. 1–15). Atlanta, GA: Center for Relationship Marketing, Goizueta Business School.

Gemünden, H. G., & Walter, A. (1995a). Der Beziehungspromotor. Schlüsselperson für inter-organisationale Innovationsprozesse (Innovationsfähigkeit). *Zeitschrift-für-Betriebswirtschaft, 65*(9), 971–986.

Gemünden, H. G., & Walter, A. (1995b). Der Beziehungspromotor. Schlüsselperson für inter-organisationale Innovationsprozesse. Karlsruhe, Germany: IABU, Institut für Angewandte Betriebswirtschaftslehre und Unternehmensführung.

Günter, B., & Bonaccorsi, A. (1996). Project marketing and systems selling: In search of frameworks and insights. *International Business Review, 5*(6), 531–537.

Hadjikhani, A. (1996). Project marketing and the management of discontinuity. *International Business Review, 5*(3), 319–336.

Håkansson, H. (Ed.) (1982). *International marketing and purchasing of industrial goods – an interaction approach*. Chichester, UK: Wiley.

Håkansson, H., & Johanson, J. (1993). The network as a governance structure: Interfirm cooperation beyond markets and hierarchies. In: G. Grabher (Ed.), *The Embedded Firm* (pp. 35–49). London: Routledge.

Hall, R. (1997). Complex systems, complex learning, and competence building. In: R. Sanchez & A. Heene (Eds), *Strategic Learning and Knowledge Management* (pp. 39–64). New York: Wiley.

Halskov, L. (1995). Tysk eksporteventyr med store skrammer. In the Danish national daily newspaper Berlingske Tidende, Wednesday (October 18).

Hansen, J. W., Knudsen, G., Lund, N.-O., & Lyhne-Knudsen, M. (1994). *Dansk arkitektur. Vilkår, muligheder og udfordringer*. Århus, Denmark: Forlaget Klim.

Harrigan, J. E., & Neel, P. R. (1996). *The executive architect*. New York: Wiley.

Hartung, A. (1997). Vi nyder godt af det tyske husbehov. *Ingeniøren, 15*(11/4).

Hedaa, L., & Törnroos, J.-Å. (1997). Understanding event-based business networks. Copenhagen: Copenhagen Business School. Working Paper, No 1997-10.

Heide, J. B., & John, G. (1992). Do norms matter in marketing relationships? *Journal of Marketing, 54*(2), 32–44.

Hellgren, B., & Stjernberg, T. (1995). Design and implementation in major investments – a project network approach. *Scandinavian Journal of Management, 11*(4), 377–394.

Hold, D. (1996). Social class revisited: Does Bourdieu's theory of taste apply to American consumption? In: *Marketing Today for the 21st century: Proceedings of the Annual Conference* (pp. 1689–1692). Helsinki: European Marketing Academy.

Holstius, K. (1987). Project export. Lappeenranta, Finland: Lappeenranta University of Technology, Research Report 1.

Holstius, K. (1989). Project business as a strategic choice. A theoretical and empirical study of project marketing. Lappeenranta, Finland: Lappeenranta University of Technology, Research Report 12/1989.

Huovinen, P., & Kiiras, J. (1998). Procurement methods as determinant for interactions and networking within the EU building industry. In: A. Halinen-Kaila & N. Nummela (Eds), *14th Annual IMP Conference Proceedings* (Vol. 2, pp. 557–579). Work-in-progress papers, Turku, Finland: Turku School of Economics and Business Administration.

Jochem, R. (1998). Verdingungsordnung für freiberufliche Leistungen – VOF. *Deutsches Architektenblatt, 30*(January), 49–54.

Johanson, J., & Mattsson, L.-G. (1997). Internationalisation in industrial systems – a network approach. In: D. Ford (Ed.), *Understanding Business Markets* (2nd ed., pp. 205–214). London: Dryden.

Johanson, J., & Vahlne, J.-E. (1977). The internationalisation process of the firm – a model of knowledge development and increasing market commitment. *Journal of International Business Studies*, *8*(2), 23–32.

Kragballe, S. (1996). Entreprenørerne trygnet i Tyskland. In the Danish national business newspaper *Børsen* (April 5).

Kvale, S. (1996). *Interviews. An introduction to qualitative research interviewing*. London: Sage.

Lubanski, N. (1999). Europæisering af Arbejdsmarkedet. Bygge- og Anlægssektoren i Tyskland, Danmark og Sverige. Copenhagen: Jurist- og Økonomforbundets Forlag.

Luostarinen, R., & Welch, L. (1990). *International business operations*. Helsinki: Helsinki School of Economics Press.

Løwendahl, B. R. (2000). *Strategic management of professional service firms* (2nd ed.). Copenhagen: Copenhagen Business School Press.

Maaløe, E. (1996). *Case-studier af og om mennesker i organisationer*. Copenhagen: Akademisk Forlag.

Majkgård, A., & Sharma, D. D. (1998). Client-following and market-seeking strategies in the internationalization of service firms. *Journal of Business-to-Business Marketing*, *4*(3), 1–41.

Mandják, T., & Zoltan, V. (1998). The D-U-C Model and the Stages of project marketing process. In: A. Halinen-Kaila & N. Nummela (Eds), *14th IMP Annual Conference Proceedings, Interaction, Relationships and Networks: Visions for the Future* (Vol. 1: Competitive Papers, pp. 471–490). Turku, Finland: Turku School of Economics and Business Administration.

Marquart, C. (1997). *Marketing und Öffentlichkeitsarbeit für Architekten und Planer*. Stuttgart: av-Edition.

Mattsson, L.-G. (1973). System selling as a strategy on industrial markets. *Industrial Marketing Management*, *3*, 107–119.

McCracken, G. (1988). *The long interview*. London: Sage.

Melin, L. (1989). The field-of-force metaphor. In: S. T. Cavusgil, L. Hallén & J. Johanson (Eds), *Advances in International Marketing* (Vol. 3, pp. 161–179). Greenwich, CN: JAI Press.

Milgrom, P., & Roberts, J. (1992). *Economics, organization, and management*. Englewood Cliffs, NJ: Prentice-Hall.

Mintzberg, H., Otis, S., Shamsie, J., & Waters, J. A. (1988). Strategy of design: A study of 'architects in co-partnership.' In: J. H. Grant (Ed.), *Strategic Management Frontiers* (pp. 311–359). Greenwich, CN: JAI Press.

Neusüss, W. G. R. W. (1995). Was ändert sich im Wettbewerbswesen? – Teil 1. Bundesbaublatt. *Heft*, *3*(96), 174–178.

Nonaka, I., & Takeuchi, H. (1995). *The knowledge-creating company. How Japanese companies create the dynamics of innovation*. New York: Oxford University Press.

Oliver-Taylor, E. S. (1993). *The construction industry of the European community*. Beckenham, Kent, UK: STEM SYSTEMS.

Östnäs, A. (1984). *Arkitekterna och deres yrkesutveckling i Sverige*. Gothenburg, Sweden: Gothenburg College of Technology.

Östnäs, A., & Svensson, L. (1986). *Arkitektarbete*. Gothenburg, Sweden: Department of Architecture, Gothenburg College of Technology.

Oxley, R., & Poskitt, J. (1996). *Management techniques applied to the construction industry* (5th ed.). Oxford: Blackwood Science.

PAR (1999). Unpublished export statistics provided to the dissertation's author by Paul K. Jeppesen, the international secretary of Praktiserende Arkitekters Råd/PAR (the Danish Council of Practicing Architects).

Petersen, B., & Pedersen, T. (1996). Twenty years after – support and critique of the Uppsala internationalisation model. In: I. Björkman & M. Forsgren (Eds), *The Nature of the International Firm* (pp. 117–134). Copenhagen: Copenhagen Business School Press.

Polanyi, M. (1962). *Personal knowledge*. Chicago: University of Chicago Press.

Prinz, T. (1998). Gerechtfertigte Richtlinien. *Der Architekt*, *11*, 600–605.

Root, F. R. (1994). *Entry strategies for international markets* (Rev. and Ex. ed.). San Francisco, USA: Lexington Books.

Rudolph-Cleff, A., & Uhlig, G. Neue Anforderungen an den Architekten in Deutschland. Karlsruhe, Germany: University of Karlsruhe, Department of Architecture, German-language version of a manuscript for a French-language lecture in Paris 1997 at Séminaire de recherche prospective "Les conditions et les modalités des commandes architecturales et urbaines: Comparaisons européennes."

Salminen, R. T. (1997). Role of references in international industrial marketing – a theory-building case study about supplier's processes of utilizing references. Lappeenranta, Finland: Lappeenranta University of Technology, Doctoral Dissertation.

Sanchez, R., & Heene, A. (Eds) (1997). *Strategic learning and knowledge management*. New York: Wiley.

Scott, W. R. (1994). Conceptualizing organizational fields: Linking organizations and societal systems. In: H.-U. Derlien, U. Gerhardt & F. W. Scharpf (Eds), *Systemrationalität und Partialinteresse* (pp. 203–221). Baden-Baden, Germany: Nomos Verlagsgesellschaft.

Sharma, D. D. (1991). *International operations of professional firms*. Lund, Sweden: Studentlitteratur.

Shostack, G. L. (1997). Breaking free from product marketing. *Journal of Marketing* (April), 73–80.

Silverman, D. (1993). *Interpreting qualitative data. Methods for analyzing talk, text and interaction*. London: Sage.

Slatter, S. St. P. (1990). Strategic marketing variables under conditions of competitive bidding. *Strategic Management Journal*, *11*(4), 309–317.

Sommer, A.-W. (1996). Auftragsbeschaffung für architekten und ingenieure: neue ideen, bewährte methoden und konzepte, anschauliche beispiele. Cologne: Verlagsgesellschaft Rudolf Müller.

Starbuck, W. H. (1997). Learning by knowledge-intensive firms. In: L. Prusak (Ed.), *Knowledge in Organizations* (pp. 147–175). Newton, MA: Butterworth-Heinemann.

Stevens, G. (1998). *The favored circle: The social foundations of architectural distinction*. Cambridge, MA: Massachusetts Institute of Technology.

Storper, M., & Salais, R. (1997). *Worlds of production*. Cambridge, MA: Harvard University Press.

Tikkanen, H. (1998). Research on international project marketing. A review and implications. In: H. Tikkanen (Ed.), *Marketing and International Business. Essays in Honour of Professor Karin Holstius on her 65th Birthday* (pp. 261–285). Turku, Finland: Turku School of Economics and Business Administration.

Torres, R. (1999). *Image commodification and the evolution of taste: The case of food in the global cultural economy*. Ithaca, NY: Paper published on the Internet at http://www.people.cornell.edu/pages/rjt5/image/image.html. Cornell University, Field of Development Sociology, Department of Rural Sociology.

Tzokas, N., & Saren, M. (2001). *Value transformation in relationship marketing*. Academic paper published online at http://www.crm2day.com/library/ap/ap0005.shtml, Strathclyde, UK: University of Strathclyde, Department of Marketing.

Van Heel, B. (1998). *Marketing in small-scale theatre companies*. London, UK: Paper written to receive the Postgraduate Diploma in Arts Administration of the City University of London.

von Gerkan, M. (1995). *Architektur im dialog: Texte zur architekturpraxis*. Berlin: Ernst & Sohn.

Williamson, O. E. (1979). Transaction cost economics: The governance of contractual relations. *Journal of Law and Economics*, *22*, 233–261.

Winch, G. (1997). The organisation of exports by architectural practices in the European context. Paris: Paper presented at Les Pratiques de l'Architecture: Comparaisons Europeennes et Grands Enjeux, Paris: Plan Construction et Architecture (November).

Winch, G., & Schneider, E. (1993). Managing the knowledge-based organization: The case of architectural practice. *Journal of Management Studies*, *30*(6), 923–937.

Yin, R. K. (1994). *Case study research. Designs and methods* (2nd ed.). Thousand Oaks, CA: Sage.

APPENDIX

The first fourteen interviews were part of the pilot study mentioned in Section 1. Moreover, some names of individuals and firms have been anonymized, in accordance with agreements.

(1) March 6, 1998. Interview with Associate Professor Niels Albertsen, Aarhus School of Architecture, Århus, Denmark.
(2) May 6, 1998. Interview with Annette Blegvad, International Secretary of Danske Arkitekters Landsforbund/DAL (Federation of Danish Architects), Copenhagen, Denmark.
(3) May 7, 1998. Interview with two unemployed Danish architects, one with some non-German international experience, Greater Copenhagen area, Denmark.
(4) May 11, 1998. Interview with Architect Peter Theibel, Special Advisor, the International Office of the Danish Ministry of Housing, Copenhagen, Denmark.
(5) May 11, 1998. Interview with a Danish architect with experience on the German market on the basis of work for a Danish public housing association, Greater Copenhagen area, Denmark.
(6) May 19, 1998. Interview with Paul K. Jeppesen, International Secretary of Praktiserende Arkitekters Råd/PAR (Danish Council of Practising Architects), Copenhagen, Denmark.
(7) May 25, 1998. Interview with Associate Professor Nils Lykke Sørensen, Aarhus School of Architecture, Århus, Denmark.
(8) May 25, 1998. Interview with Ernest Müller, Teacher at the Horsens School of Building Construction's International Program, Horsens, Denmark.
(9) June 3, 1998. Interview with two partners from a nationally renowned, professional sub-field Danish architectural firm. Research Professor Per Jenster, who was at the time employed by the Department of Intercul-

tural Management and Communication of Copenhagen Business School also participated in this interview, Greater Copenhagen Area, Denmark.

(10) June 6, 1998. Interview with a Danish architect with four years of work experience in Berlin, Copenhagen, Denmark.

(11) June 23, 1998. Interview with a Danish architect who is a partner in a firm as well as a Professor at a Danish School of Architecture, Greater Copenhagen Area, Denmark.

(12) July 3, 1998. Interview with a Danish architect who works for a firm that does not export to Germany, Copenhagen, Denmark.

(13) July 27, 1998. Interview with a Danish architect who previously had his own firm and now works as a decorating consulting for a Danish firm that is opening shops in Germany, Greater Copenhagen Area, Denmark.

(14) August 10, 1998. Interview with a Danish architect who has previously worked for a large architectural firm with international projects outside of Germany and who is currently building up his own firm, Greater Copenhagen Area, Denmark.

(15) November 1998. Telephone Interview with Flemming Hallen, Associerede Danske Arkitekter/ADA(Associated Danish Architects)'s Vice President and Spokesperson regarding the EU Public Service Directive.

(16) November 1998. Telephone Interview with Ministerial Principal Bjarne Strand, Spokesperson for the Danish Ministry of Housing regarding the EU Public Service Directive.

(17) November 1998. Telephone Interview with Jacob Scharff, Spokesperson for the Danish Organization of Municipalities concerning the EU Public Service Directive.

(18) November 1998. Telephone Interview with Keld Møller, Managing Director of Praktiserende Arkitekters Råd/PAR (Danish Council of Practising Architects) and PAR's Spokesperson concerning the EU Public Service Directive.

(19) November 1998. Telephone Interview with Jesper Kock, Architectural Competition Secretary of Danske Arkitekters Landsforbund/DAL (Federation of Danish Architects), concerning the EU Public Service Directive.

(20) November 17, 1998. Interview with Partner XX, who is responsible for the finances and book-keeping of Case Firm 2, Copenhagen, Denmark.

(21) November 19, 1998. Interview with Founding Partner MM of Case Firm 3, Århus, Denmark.

(22) November 25, 1998. Interview with Partner XX, Case Firm 2, Copenhagen, Denmark.

(23) November 29, 1998. Interview with German-born Partner AA of Case Firm 1, who ran Case Firm 1's German subsidiary, Greater Copenhagen Area, Denmark.

(24) December 10, 1998. Interview with Partner YY, who was responsible for Case Firm 2's German office during the nineties, Berlin, Germany.
(25) December 29, 1999. Interview with Partner AA, Case Firm 1, Copenhagen, Denmark.
(26) February 18, 1999. Interview with NN, In-House Accounting and Business Administration Expert at Case Firm 3, Århus, Denmark.
(27) February 19, 1999. Interview with Partner XX of Case Firm 2, Århus, Denmark.
(28) February 23, 1998. Interview with OO, Manager of Case Firm 3's German subsidiary, Berlin, Germany.
(29) February 24, 1998. Interview with Construction Industry Attaché Jørgen Skovgaard Vind at the Danish Consulate General in Berlin, Germany.
(30) February 24, 1998. Interview with Elke König, the International Secretary of Bund Deutscher Architekten/BDA (German Federation of Architects), Berlin, Germany.
(31) March 11, 1999. Interview with Partner YY of Case Firm 2, Copenhagen, Denmark.
(32) March 18, 1999. Interview with Partner BB, who is responsible for Case Firm 1's finances, Copenhagen, Denmark.
(33) March 29, 1999. Interview with the Export Manager of a Danish Producer of Building Components, Århus, Denmark.
(34) March 29, 1999. Interview with Partner XX of Case Firm 2, Århus, Denmark.
(35) March 29, 1999. Interview with Partner MM of Case Firm 3, Århus, Denmark.
(36) April 9, 1999. Interview with Partner YY of Case Firm 2, Berlin, Germany.
(37) April 9, 1999. Interview with the Danish Managing Director of the German subsidiary of a large Danish Engineering Firm, Berlin, Germany.
(38) April 12, 1999. Interview with a Danish partner in a Danish-German architectural firm in Berlin, Germany.
(39) April 12, 1999. Joint Interview with Dr. Tillman Prinz, Legal Advisor of Bund Deutscher Architekten/BDA (German Federation of Architects) and Elke König, the International Secretary of BDA, Berlin, Germany.
(40) April 26, 1999. Interview with German-born Partner CC of Case Firm 1, Copenhagen, Denmark.
(41) April 26, 1999. Interview with German-born Partner DD of Case Firm 1, Copenhagen, Denmark.
(42) July 19, 1999. Interview with Architect Sven Silcher, the Representative of Bund Deutscher Architekten/BDA (German Federation of Architects) in the Architects' Council of Europe (ACE), Hamburg, Germany.

(43) July 19, 1999. Interview with Engineer Herbert Michaelis, Vice-President of the Hamburg Chapter of Bund deutscher Baumeister, Architekten und Ingenieure e.V/BDB, Hamburg, Germany.
(44) August 18, 1999. Interview with the German Export Market Development Contact Person of a large Danish Engineering Firm, Greater Copenhagen area.
(45) August 20, 1999. Interview with the Danish-born Managing Director of the Berlin subsidiary of a large Danish Engineering Firm, Greater Copenhagen area.
(46) October 14, 1999. Interview with Architect Mahmood Sairally, the Representative of Bund deutscher Baumeister, Architekten und Ingenieure e.V/BDB in the Architects' Council of Europe, Hamburg, Germany.
(47) October 14, 1999. Interview with an internationally renowned German Architect/Professor of Architecture who is also a partner in a large, internationally renowned German architectural firm, Hamburg, Germany.
(48) October 18, 1999. Joint interview with the German Construction Department Director of the subsidiary of a Danish engineering firm and a Dane who has been employed by a Danish contracting firm on the German market in the 1990s, Berlin, Germany.
(49) October 22, 1999. Interview with the managing director of the German subsidiary of a Danish real estate investment firm that has invested in real estate in Northern Germany in the 1990s, Copenhagen, Denmark.
(50) November 1, 1999. Interview in Copenhagen with Architect Elsebeth Terkelsen, the former Danish Construction Industry Attaché in Düsseldorf, Germany. The interview took place in Copenhagen, Denmark.

RESEARCH ON BUSINESS-TO-BUSINESS CUSTOMER VALUE AND SATISFACTION

Robert B. Woodruff and Daniel J. Flint

ABSTRACT

In today's markets, many organizations feel pressure to become more responsive to their customers. Managing your business to deliver superior value to targeted customers may provide a strong avenue to improved performance. The route from value-based strategies to share holder value can be complicated, however. These strategies have the most direct impact on performance with your customers in the form of customer satisfaction, word of mouth and loyalty. Successful customer performance should translate into higher market performance, as evidenced by a supplier's higher customer retention rates and sales. Finally, market performance provides the engine for increasing company performance or shareholder value. Attaining shareholder value through customer value strategies requires committing major management attention to how best to create, deliver and communicate superior value to targeted customers.

INTRODUCTION

In today's markets, many organizations feel pressure to become more responsive to their customers. Much of the reason stems from suppliers finding that their

Evaluating Marketing Actions and Outcomes
Advances in Business Marketing and Purchasing, Volume 12, 515–547

ISSN: 1069-0964/doi:10.1016/S1069-0964(03)12008-X

customers are more demanding than ever before. Increasing global and domestic competition creates more choices for customers, which shifts negotiating advantage in their direction. At the same time, supplier consolidations combined with customers' efforts to reduce their supplier bases means that larger shares of customers' businesses are awarded to fewer large suppliers. This trend creates greater risk for both suppliers and customers. In addition, business-to-business customers have learned a lot about their needs through the quality initiatives of the last two decades. That has translated into higher expectations placed on suppliers to provide quality products and services while remaining price competitive. Yet, delivering quality no longer provides much of a competitive edge. In today's competitive markets, suppliers must look for other ways to add value for their customers.

More recently, we see competition shifting toward higher customer retention as a way of creating new advantage. Many discovered that they spend less to retain a customer than acquire a new one, and that savings often flow directly to the bottom line (Birch, 1990; Butz & Goodstein, 1996). Furthermore, customer retention creates an asset for suppliers that may significantly affect long term performance (Srivastava, Shervani & Fahey, 1998). For instance, strong customer relationships can help overcome short-term problems with customer service or the onslaught of new competitors (Garver, Gardial & Woodruff, 1999).

In light of these trends, a growing number of writers make the case that managing a business to deliver superior value to targeted customers can lead to improved performance (Gale, 1994; Kaplan & Norton, 1996; Naumann, 1995; Slywotzky & Morrison, 1997; Wayland & Cole, 1997; Woodruff, 1997). However, a complicated path lies between customer value strategies and company performance (Fig. 1). The most immediate effect of successful customer value strategies shows up in customer responses, such as satisfaction, word of mouth, and loyalty. These responses, in turn, impact key company market performances, as measured by customer retention rates, sales, and market share. Finally, market performances provide the engine for increasing a company's financial performance, including cash flow, profitability, and shareholder value. The circle of performances closes when that financial performance creates incentive for new decisions on customer value strategies.

Business-to-business suppliers face formidable challenges when trying to make this customer value strategy/performance linkage work successfully in practice. Even seemingly straightforward questions, such as "Who are the key customers to which we deliver value?" "What kinds of value do they want?" and "How do we develop superior advantage with customer-value strategies?" often have no simple answers. Part of the reason lies in the complexity of business-to-business market dynamics that suppliers face.

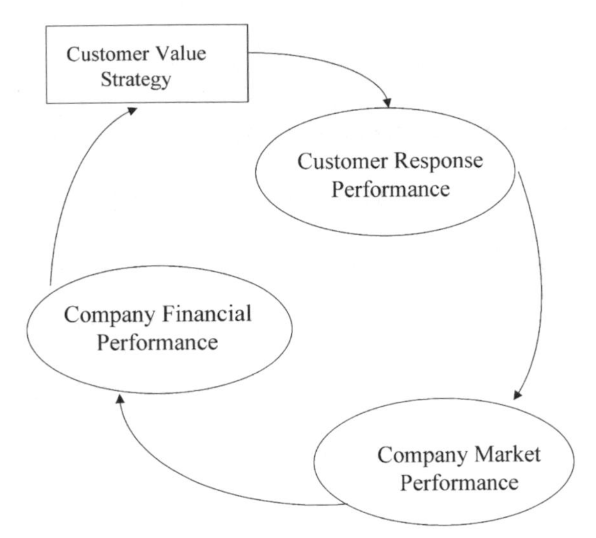

Fig. 1. Customer Value Strategies and Company Performance.

Business-to-Business Market Dynamics

Because business-to-business suppliers are located in a larger chain of organizations reaching downstream to end-users, customers that are one or more steps removed may have to be considered in strategy decisions. A supplier should expect that value demanded by immediate customers would likely be influenced by the needs of downstream customers. This characteristic of demand channels raises difficult questions. For example, how far downstream should a supplier look for value needs? And, how can learning about downstream customers be incorporated into customer value strategies aimed at immediate customers?

Even when a supplier knows which customer organizations to target, there remains the challenge of identifying the influential persons inside customer organizations. Decisions on which suppliers to use may be spread among a buying group, including those from purchasing, design engineering, manufacturing operations, quality assurance, and financial control. Buying group members may also play different roles in the supplier decisions, such as decision maker, product/service user, influential, and decision facilitator. Sorting out who plays what role in each customer organization can be difficult, but essential to do since each role may elicit different kinds of value desired from the supplier.

As we discuss later in the chapter, suppliers may face a complex array of value desired by their customers. Often, customers want much more than product and service quality or competitive pricing, though those value dimensions are critical. They may also seek less tangible aspects of value, such as trust, commitment, and cooperation (e.g. Wilson, 1995). On the social side of value, customers' may only want to deal with supplier's customer contact persons with whom they feel comfortable and relaxed (Garver, Gardial & Woodruff, 1999). Furthermore, customers change the value they seek from suppliers over time, making it difficult to keep up with customer dynamics (Flint, Woodruff & Gardial, 1997).

Finally, a supplier faces competition from its customers as well as from other suppliers. A business-to-business customer may decide, for a variety of reasons, to bring a product or service in-house as a result of a "make-or-buy" decision, with corresponding lost business for the supplier. For example, a third party logistics provider of distribution services may be dropped if a key client customer decides to take over the service internally.

Purpose of the Chapter

All of these complexities create an enormous challenge for business-to-business suppliers who want to become more customer-oriented. What help does current research on customer value and satisfaction provide? This research has a distinctly consumer focus. Most of the studies that examine the nature of customer value use consumer samples and consumer products and services. Similarly, consumer studies dominate customer satisfaction research (Patterson, Johnson & Spreng, 1997), even though more and more research is being reported on customer satisfaction in business-to-business contexts (Mohr-Jackson, 1998; Qualls & Rosa, 1995; Tanner, 1996), as is research on what it means to be close to business customers (Homburg, 1998). We believe that all of this work has much to offer to business-to-business suppliers. However, these suppliers must look for opportunities to apply customer-based strategies to their world. The purpose of this chapter is to facilitate this effort. In the next section, we discuss the conceptual foundations for customer value and satisfaction and their applicability to the business-to-business context. Then, we turn to three activities central to this application: (1) determining industrial customers; (2) understanding industrial customers' perceived value and satisfaction; and (3) managing customer value through customer value strategies. The chapter concludes by suggesting both managerial applications and research implications of business-to-business customer value and satisfaction issues.

CUSTOMER VALUE AND SATISFACTION RESEARCH

As indicated by the large and growing literature on customer value and satisfaction, these topics have taken on the status of mainstream research for marketing and business practice. No doubt, the popularity of this research is linked to companies searching for new ways to develop and sustain superior competitive advantage. More importantly, this research lays conceptual foundations for customer value and satisfaction applications that are essential for business-to-business suppliers who want to become more customer-oriented with strategy. We begin by examining these foundations, first for customer value and then for customer satisfaction.

Customer Value in a Business-to-Business Context

The tremendous current interest in customer value and customer value strategies reflects a push for suppliers to look more outwardly to markets and customers for guidance on strategy initiatives. The most important part of this guidance lies in suppliers' responsibility to understand how targeted customers perceive value when choosing or interacting with suppliers.

Concept of Customer Value

What exactly is customer value, particularly as customers perceive it? Customer value research reveals highly practical insights into this question. Early research focused on the nature of value reflected in consumers' purchase decisions (e.g. Gale, 1994). Theoretically based on multi-attribute attitude theory, customers were thought to associate value with the various attributes of a supplier's product offer (e.g. Hayes, 1992; Oliver, 1997; Zeithaml, Parasuraman & Berry, 1990). These *valued attributes* are customer-preferred characteristics of the seller's product and service offer. In the industrial customers' world, attributes might be things like a supplier's product quality, product performance, competitive prices, on-time delivery, technical service quality, sales person knowledge, and the like.

Customer value research helps us understand how customers combine attributes into an overall assessment of net value. Customers see many supplier attributes as being benefits in the sense of having positive worth to them. Product quality is like that. Other attributes are viewed as costs, such as the supplier's price. So, net customer value would be the customer's perception of attribute benefits less attribute costs (e.g. Anderson & Narus, 1998; Gale, 1994).

An expanded theoretical concept, which we call the *customer value hierarchy* (Gutman, 1982; Woodruff, 1997; Woodruff & Gardial, 1996; Zeithaml, 1988), offers a more complete way of looking at customer value (see Fig. 2). Briefly,

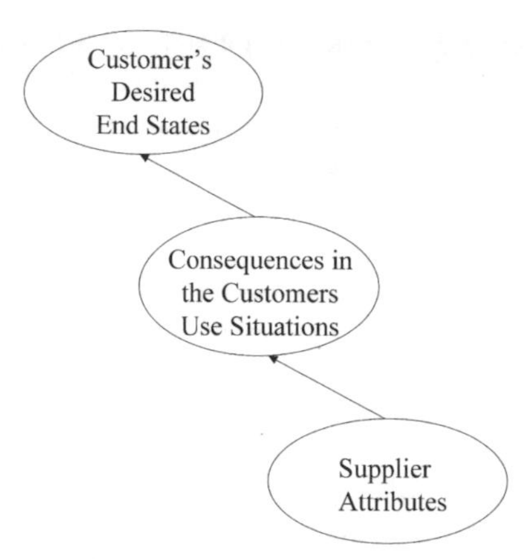

Fig. 2. A Customer Value Hierarchy.

attributes are desired by customers only when they lead to consequences that the customer wants to have happen in its own use situation. For instance, a supplier's customer may need on-time delivery, defined as receipt of an order at the customer's dock within a narrow twenty-four hour delivery window two days after the order is placed. This window allows the customer to feel assured that a scheduled production run occurs as planned and that excess inventory will not take up valuable facility space because orders arrive too early (consequences in the customer's production use situation).

All consequences are outcomes of product/service use. However, some may be positive for the customer, while others are seen as negative. For instance, a customer's production run that occurs on schedule because a supplier delivered its product on time is a positive consequence, but a breakdown of a key piece of the customer's machinery due to defects in a supplier's component product is a negative consequence. This distinction opens strategic options for suppliers to offer value to customers by facilitating positive consequences and/or preventing negative consequences from happening.

Which consequences are most valued by a customer? The customer value hierarchy suggests that customers learn which consequences facilitate achieving specific desired end states. Desired end states are the purposes or goals that a customer wants to accomplish. For instance, a business-to-business customer may want to make money, look good with their customers, be an innovator, and the like. Customers value most those supplier-facilitated use consequences that lead to each important desired end state. As shown in the illustrative customer value hierarchy in Fig. 3, this customer wants strong technical support from the supplier because that support enables the customer to avoid production stoppages, develop

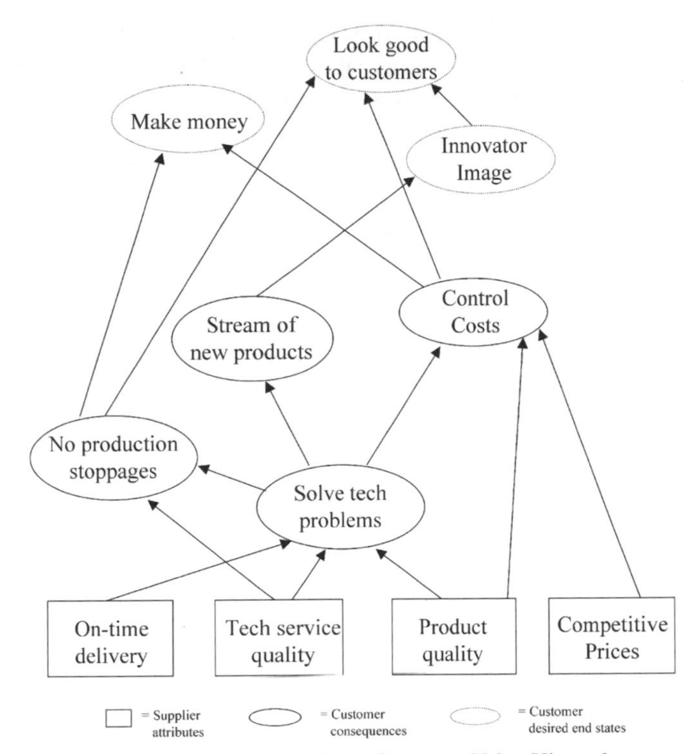

Fig. 3. Business-to-Business Customer Value Hierarchy.

a stream of new products, and control costs. In turn, avoiding product stoppages helps this customer make money and look good to its customers.

The customer value hierarchy concept moves customer value from the world of the supplier (attributes) into the world of the customer (consequences in their use situations and desired end states). It demonstrates that customers see much more to value than just the attributes that a supplier offers. Most importantly, the customer value hierarchy explains why certain attributes of a supplier's offer are desired and why others are seen as costs.

Business-to-Business Customer Value

Applying the customer value hierarchy to a business-to-business context, emerging research reveals two different kinds of insights into customer value. First, business-to-business customers perceive value in at least two different categories – functional value and relationship value. *Functional value* lies in customer value hierarchies centering on product availability and quality, delivery service quality (e.g. EDI, Internet access, and product delivery), and pricing by suppliers (Garver, Gardial & Woodruff, 1999). Figure 3 illustrates a functional value hierarchy. Today's business-to-business suppliers are most familiar with this kind of value. In addition, customers perceive *relationship value* stemming from the quality of the interactions going on between customer and supplier, such as between purchaser and supplier salespersons (e.g. Dwyer, Schurr & Oh, 1987; Garver, Gardial & Woodruff, 1999; Weitz & Jap, 1995; Wilson, 1995). For example, customers want to trust that a supplier's salespersons will follow through on promises made. While we are in an early stage of research to empirically document the various kinds of relationship value, we already know that there is both a business side to it, such as a customer's perceptions of trust and loyalty, and an interpersonal side, such as the customer feeling comfortable with a salesperson, feeling taken care of, and enjoying the relationship (Garver, Gardial & Woodruff, 1999).

The second emerging insight into business-to-business customer value concerns its dynamic nature. Most everyone agrees that, over time, customers change what they value from suppliers. For example, a customer may place increasing emphasis on relationship value as competitive suppliers reach parity on delivering functional value. The challenge for new research will be to better understand how and why such changes occur. Only then will improved techniques be developed to predict customer value change in advance (Flint, Woodruff & Gardial, 1997).

What do we know about value change? Flint, Woodruff and Gardial (2002) discovered that tension within a business-to-business customer organization, as reflected by a heightened sense of urgency, anxiety, and even panic, creates motivation to change what they value from a supplier. This tension derives from externally driven event pressures, such as in the customer's markets (e.g. new quality initiatives by the customer's customers), or internally within a customer organization (e.g. a new top management directive to achieve a higher shareholder value return). These external events can be grouped into five categories which include: (1) changes in customers' customer demands; (2) changes internal to customer organizations; (3) moves made by customers' competitors; (4) changes in suppliers' performance and/or demands; and (5) changes in the macro-environment. Events in these locations produce tension because customers feel that they are inadequately prepared to deal with them. This self-assessment covers customers' evaluations of their and their company's performance, knowledge, and control. As customers attempt to reduce this tension by increasing their knowledge, improving their performance, and attempting to control more aspects of their environment, they begin to recognize their dependence on suppliers for achieving their objectives, resulting in changes in what they value from those suppliers (Flint, Woodruff & Gardial, 2002).

Recent research suggests how customers' perceptions of tension drive value change (Flint, Woodruff & Gardial, 2002). Based on the customer value hierarchy, tension motivates customers to look for new consequences that will lead to desired end states, as a way of relieving that tension. For example, a customer's operations managers may want to hold less parts inventory (a new consequence) in order to lower cost (a related new consequence); so that top management can achieve its goal of making more money (an existing desired end state). These value changes, in turn, are because the customer realizes that it is dependent on parts suppliers to get inventory levels down. At that point, the customer places new demands on its suppliers such as carrying inventory or improving manufacturing processes so that a just-in-time delivery system can exist between the companies. These new demands may be accompanied by customer-initiated strategies to get suppliers to comply with the new demands. For example, customers build partnering relationships with suppliers partially because those relationships help customers motivate suppliers to respond to changes in what they value.

Many customer value changes taking place are value components related to logistics needs, such as EDI, more precise delivery systems, order tracking, facility locations, improvements in forecasting accuracy, and the coordination and delivery of complete systems of components as opposed to merely one's own products (Flint & Mentzer, 2000). These findings depict a significant opportunity for integrating marketing and logistics expertise in order to respond to changes in customers' desired value and create competitive advantages. Such a strategy demands that logisticians "market" the value they bring to the table to internal customers as well as to external customers.

Understanding what customers want (desired value) is only half of the puzzle. The other half comes when a supplier learns how well customers perceive that they are receiving the value that they desire. Customer satisfaction measurement focuses on this kind of perception. Next, we examine the theoretical foundations for customer satisfaction.

Customer Satisfaction in a Business-to-Business Context

Customer satisfaction research has a rich history of contributions to business practice (e.g. Dutka, 1994; Hayes, 1992). As quality initiatives and tools have taken a central place in the arsenal of management techniques, customer satisfaction emerged as the primary tool for ensuring that supplier quality conforms to customer perceived quality (Woodruff, 1997). For that reason, many organizations, in both consumer and business-to-business contexts, have allocated significant resources to measuring customer satisfaction. This research utilizes a well-accepted theoretical model (see Fig. 4). We briefly review this model.

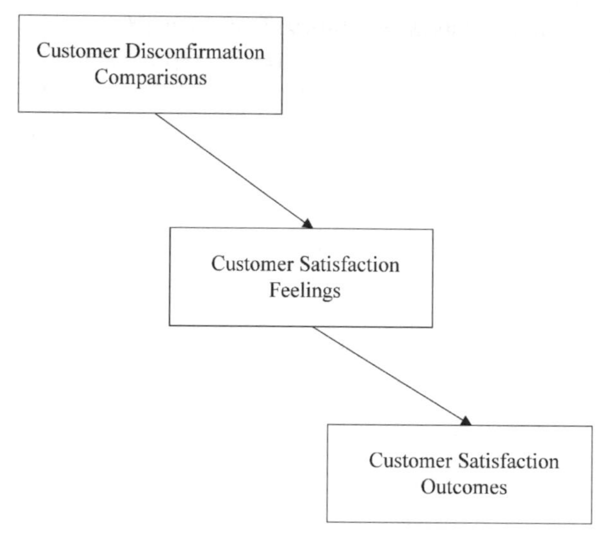

Fig. 4. Customer Satisfaction Model.

Satisfaction Feelings

Starting in the middle of the model, the state of customer satisfaction refers to a customer's overall feeling about the value received from a supplier (Woodruff, 1997). Often, customers use emotion words, like excited, happy, angry, and disappointed, to express this feeling (Dabholkar, 1995; Woodruff & Gardial, 1996). As these words suggest, satisfaction may be positive or negative and may vary from mild to extreme.

Satisfaction Outcomes

Satisfaction feelings motivate customers to behave in the marketplace. Those who are highly satisfied with a supplier are the ones most likely to buy again from that supplier, spread positive word-of-mouth information to other customers, and develop loyalty (Jones & Sasser, 1995; Oliver, 1997). On the other hand, dissatisfaction can lead to negative behavior, such as switching to another supplier, reducing the percent of purchases allocated to that supplier, engaging in negative word of mouth, and even initiating retaliation to get revenge (Garver, Gardial & Woodruff, 1999). Because of this motivational role, successful customer retention strategies place major emphasis on gaining customer satisfaction.

Disconfirmation Comparisons

Research has been helpful in revealing how customers form satisfaction feelings. Customers perceive the extent to which a supplier delivers the value dimensions (i.e. attributes and/or consequences) that they desire. For example, quality assurance personnel at a customer's plant may monitor the percentage of a supplier's parts deliveries that fall outside of tolerance limits on requirements. That percentage is the customer's perception of the supplier's value performance received on that dimension. Whether that performance is evaluated as good or bad depends on how that customer believes that it compares to some standard. Satisfaction research has discovered a variety of different kinds of standards that customers may use (Woodruff et al., 1991). A standard may be the value desired (Spreng et al., 1996), as illustrated by a customer's desired on-time delivery. Or, the standard could be the value delivered by a competitor (Cadotte, Woodruff & Jenkins, 1987; Gale, 1994), the value delivery promised by the supplier, or any of several other beliefs. Overall satisfaction feelings, then, vary directly with the results of this comparison. The more favorably the supplier's performance compares to the standard, the higher the positive satisfaction feeling.

Customer Satisfaction and the Customer Value Hierarchy

As mentioned earlier, the model in Fig. 4 underlies satisfaction measurement practice in industry today. In some cases, a supplier may only want to know overall satisfaction feelings, providing a satisfaction index, in order to assess its performance. However, more typically the supplier will want to know why that overall index is high or low. For this reason, satisfaction measurement in practice provides data on how well customers believe that the supplier performed on each specific value dimension desired (Dutka, 1994; Hayes, 1992). For example, the supplier may ask customers to rate it on on-time delivery performance, product quality, ease of ordering, trust, and so forth.

Because a supplier wants to know how well it is doing on specific value dimensions, customer value data must guide satisfaction measurement (Woodruff, 1997; Woodruff & Gardial, 1996). Most importantly, how a supplier views customer value will determine what kinds of performance are measured. If a supplier considers customer value as key buying attributes important to customer's supplier selection decisions, customer satisfaction measurement will yield data on perceived attribute performance. This measurement corresponds to customer attribute level satisfaction in Fig. 5. Most of satisfaction measurement practice today is limited to this lowest level in a customer value hierarchy.

Suppliers who rely on attribute-only satisfaction measurement may not fully understand how well customers perceive that they experience valued consequences. Success in achieving any particular consequence may well depend on more than one supplier attribute. For example, an illustrative attribute measure might ask customers to rate the supplier's technical support service level on a scale from "poor" to "excellent." However, that rating may not reveal how well customers

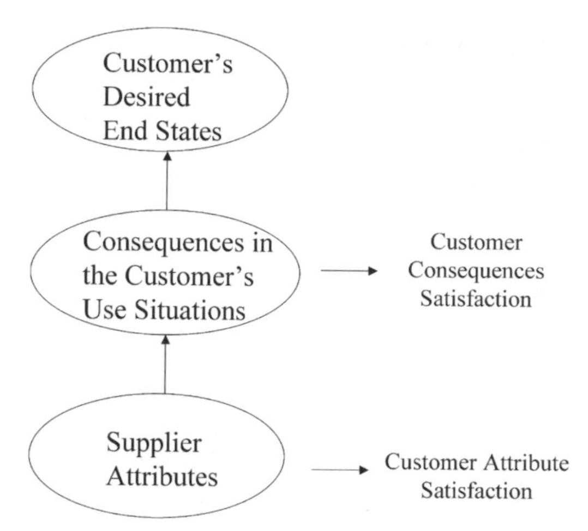

Fig. 5. Customer Value and Satisfaction. *Source:* Adapted from Woodruff and Gardial (1996), *Know Your Customer*.

perceive the supplier as solving their own technical problems (a consequence linked to tech support service in Fig. 3). As shown in Fig. 5 (see customer consequence satisfaction), this latter customer perception only comes from measuring perceived performance at the consequence level in a customer value hierarchy. As the customer value hierarchy concept of value becomes more mainstream, we expect to see satisfaction measurement move up the hierarchy to the consequence level to help managers truly understand why customers feel about the supplier the way they do.

BUSINESS-TO-BUSINESS CUSTOMER VALUE AND SATISFACTION ISSUES

Customer value and satisfaction research contributes to several aspects of business-to-business marketing practice. In particular, we discuss its application to three questions of importance to a supplier's managers:

(1) Who should be the supplier's target customers?
(2) How should a supplier go about learning about its customers' value and satisfaction perceptions and feelings? and
(3) How can a supplier manage customer value and satisfaction?

Determining the Supplier's Business-to-Business Customers

For business-to-business suppliers, this first question involves two different issues. One concerns which customer organizations to target, and the other concerns who in a customer organization are the key decision-makers. While each of these issues requires much more analysis than just customer value and satisfaction research, this research can play a role.

Quite likely, not all business-to-business customers provide the same potential sales opportunity for a supplier. Customer value research on functional versus relationship value suggests one way to look for these differences. Some customers may want only functional value. To them, supplier choice depends primarily on product availability and quality, delivery service quality, and competitive pricing. A supplier may want to group these customers into a segment to receive a marketing offer aimed at delivering functional value. Other business-to-business customers, perhaps the larger ones, may want a combination of functional and relationship value. These customers would be better placed in another segment to receive a different marketing offer. For them, specific relationship services may be essential. For example, a supplier may offer, to selected customers, results of market analyses of that customer's customers to help look for new product opportunities.

Similarly, changes in customer desired value might offer opportunities to segment customers. Some customer organizations are likely to be more dynamic in their desires than other organizations. The most dynamic customers may be changing what they desire from suppliers in dramatic leaps, in many parts of the organization, and quite frequently. Responding to this level of change demands extensive time and resources. These customers might be grouped into a segment different from those customer organizations that do not exhibit much change in the value they desire. For this latter segment, the important issue might be consistent and continuous delivery of the same level of value over time because making a change, even if the supplier feels the change might be an "improvement," disrupts customers' operations, creating negative and undesired consequences.

Within a customer organization, different personnel may be influential on supplier choice. The supplier competing for a share of this customer's business must identify who these personnel are. Once identified, the challenge for the supplier becomes understanding what each of these key personnel desire. Customer value research techniques can be used to determine what these value differences are. For instance, the customer value hierarchy in Fig. 3 may describe how a key operations manager sees value, but not be an adequate representation of value that a purchasing manager wants from the supplier. Currently, we know little about whether there are common customer value dimensions for different positions across organizations (e.g. purchasing, operations, quality assurance, new product development, etc.). And even when we understand more about the commonalties and differences among organizational positions, it is likely that individuals holding the same position type will sometimes desire different values. This becomes important when suppliers work with separate divisions, departments, and teams within customer organizations forcing them to interact with multiple purchasing managers, multiple operations managers, and multiple design engineers within the same company. Consequently, a supplier must find out for itself what key decision makers in its customer organizations want.

Learning About Business-to-Business Customer Value and Satisfaction

As the above discussion suggests, determining target customers becomes intertwined with measuring customer value and satisfaction. In essence, a supplier should target those customers for which it gains competitive advantage from superior value delivery. But, what are the issues that a business-to-business supplier faces when trying to find out what customers will perceive as superior value? We now discuss three such issues.

Customer Value Determination

To compete on superior customer value strategies, a business-to-business supplier must acquire and maintain competency in customer value and satisfaction measurement. Managers may be tempted to rely on "experience" with customers to determine what value they desire. But that temptation should be resisted. For one thing, a supplier's experience with a customer acquired only through periodic interactions may be quite inconsistent and incomplete. For example, one supplier had a meeting of its managers from marketing, sales, operations, product development, and logistics to discuss how to increase its share of business with one key customer. The meeting started well enough as each manager shared what he or she knew about that customer and its needs. By the end of the day, however, the meeting was in disarray, as it became clear that managers held conflicting opinions about the customer. Equally disturbing were questions about customer value that kept coming up that could not be answered by anyone.

Even when a supplier's managers agree on what a customer wants, that may not be adequate. Too often, there are important discrepancies between what a supplier's personnel think customers want and what the customers actually want (Idassi et al., 1994; Sharma & Lambert, 1994). Such discrepancies may involve differences in: (1) definitions of a desired value dimension; (2) perceived importance of a desired value dimension; and/or (3) whether or not a value dimension is desired by the customer. For example, a supplier may believe that on-time delivery to a priority customer means delivery within two weeks after the supplier receives the order. To the customer, however, on-time delivery may be defined as a tight delivery window ten days after the customer places the order. For example, one purchasing manager of a manufacturing customer explained to us that she expected products delivered within a twenty-four hour window. Both early and late arrivals were looked upon unfavorably because storage and loading dock space were at a premium, forcing her to create tight schedules for specific deliveries. Such instances strongly indicate that while managerial experience is certainly relevant and can play a role, a supplier should go directly to customers to find out what they want, as well as how satisfied they are.

Woodruff and Gardial (1996) discuss a systematic process for going directly to customers, called customer value determination (CVD). While it is beyond the scope of this chapter to review all that is involved in this process, we offer several observations on it. First, CVD systematically addresses four major customer value information needs. As shown in Fig. 6, the process responds to the first need by determining the specific value dimensions desired by customers. Most importantly, qualitative techniques must be used to obtain from customers what they perceive as value. Focus group interviews, in-depth personal interviews, observations and

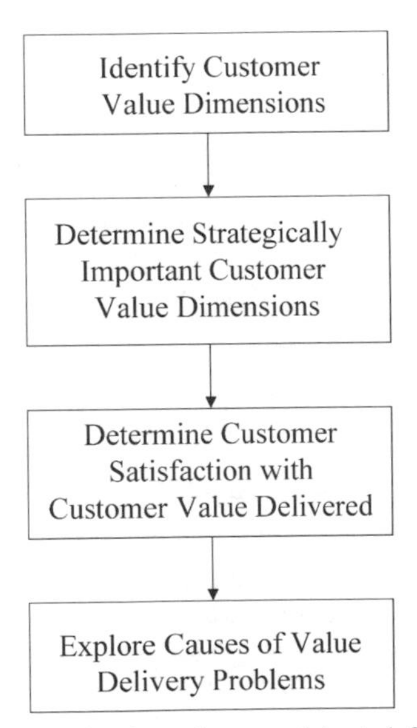

Fig. 6. Customer Value Determination. *Source:* Adapted from Woodruff and Gardial (1996), *Know Your Customer*.

related variations on qualitative research are available for this purpose (Anderson, Jain & Chintagunta, 1993; Griffin & Hauser, 1993).

To meet the second information need, value dimensions are evaluated for strategic importance as determined by criteria established by the supplier. For instance, a value dimension might be strategically important because a customer bases its supplier choice on it. Several research techniques have been used to determine these key value dimensions, including direct scaling of importance, conjoint measurement, and regression of buying criteria on various outcome measures such as share of purchases, loyalty, and likelihood of buying again. Other strategic importance criteria reflect additional considerations, such as the ability of the supplier to gain competitive advantage through special capability on a particular value dimension (Woodruff & Gardial, 1996).

The third information need concerns finding about customers' perceptions of received value. To meet this need, CVD shifts to customer satisfaction measurement. Typically, survey methods are used to go directly to customers to get their perceptions on how well the supplier delivers value to them (Dutka, 1994; Hayes, 1992; Woodruff & Gardial, 1996). The supplier receives a "report card" of ratings showing on which value dimensions that customers perceive that the supplier has performed well and on which ones that the supplier has not performed well. Consistent with the satisfaction theory framework discussed earlier, the value dimension performance rating should be measured relative to a standard, such as a competitor's performance. These relative measures are more likely to predict customer and market outcomes, such as overall satisfaction, repeat buying, loyalty, and share of purchases (Gale, 1994).

A report card only shows what customers think of the supplier, not why. So, the fourth information need addressed by a CVD process requires follow-up research to find out why customers believe that the supplier has problems delivering on the poorly-rated value dimensions. For instance, focus group interviews with a sample of customers from a segment may be used to explore reasons for poor ratings. Customer value strategy must focus on these reasons in order to improve value dimension performance and overall satisfaction ratings in the future.

As a second observation, the CVD process applies the customer value hierarchy concept to all information activities. Determining customer value means finding out about both attributes and consequences that customers desire. Similarly, customer satisfaction measurement should obtain customer ratings of strategically important attributes and consequences. Most importantly, all CVD activities consistently focus on the same customer value dimensions important to customers.

Thirdly, each CVD activity requires different research techniques to accomplish. CVD proficiency means that a supplier must develop competency in using results from both qualitative (e.g. personal interviews, observations) and quantitative research techniques (e.g. surveys). Further, each CVD activity builds on the previous one. For example, a satisfaction survey will be no better than the quality of the desired customer value dimensions data used to design the satisfaction questionnaire. Also, managers will have a difficult time using satisfaction "report card" data without access to the follow-up qualitative data exploring reasons for exceptionally high and low scores. The entire process reflects the supplier's customer information needs, not just any one activity.

Quantifying Customer Value

When determining customer value in business-to-business markets, a supplier may want to calculate the economic or dollar worth to the customer of its offer. These estimates may be very helpful in designing product service improvements, pricing the supplier's product/service, and in communicating to a key customer why that price is justified. While it is beyond the scope of this chapter to review the different ways this might be done, various methods have been suggested (e.g. Anderson, Jain & Chintagunta, 1993; Anderson & Narus, 1998). Essentially, a supplier must study the operations of the customer in which the supplier's product/services are being used in order to estimate cost savings or added value to the customer from that use over the life of the product or service. These dollar estimates sum to the economic worth of the supplier's offering.

Quantifying customer value can benefit from applying the customer value hierarchy model discussed above. To understand what a product/service is worth to a customer, the supplier must study the consequences of using the product in the customers' operations. Cost savings and value added come from these specific valued consequences of use (e.g. faster throughput, lower maintenance cost, reducing or eliminating quality inspections, lower inventory carrying cost, etc.). In addition, total economic worth will depend on the supplier's delivery of both functional value and relationship value. Of the two, functional value will typically be easier to estimate because it stems from observable customer operations. The psychological nature of relationship value makes it more difficult to quantify, though the supplier may present them to a customer and argue that they do have economic worth (Anderson & Narus, 1998).

Predicting Customer Value Change

Customer value determination as practiced today mostly focuses on how customers presently perceive value and satisfaction. Existing techniques are well developed for this purpose. A business-to-business supplier trying to become more responsive to their customers must start here. However, at some point, competency in the basic CVD process will lead to the question, "What is next?" One answer stems from the dynamic nature of customer value that we discussed earlier. A business-to-business supplier looking for a competitive edge may well find it by becoming competent at forecasting change in desired customer value. For that reason, Woodruff and Gardial (1996) included predicting customer value change as another activity in the full CVD process.

Virtually everyone agrees that what a customer values presently will change in the future. Similarly, most everyone agrees that a business should not wait for the change to become clearly apparent and then try to react by changing its core strategy. That may lead to playing catch up in a crisis situation. Interestingly, however, not everyone agrees on how to deal with customer value change. For example, a business might respond to change by designing flexibility into internal processes (Pine, 1993). The argument seems to be that the winners will be those businesses that respond the quickest after a market change becomes apparent. Others believe that a business can have foresight about the future (e.g. Flint, Woodruff & Gardial, 1997, 2002; Hamel & Prahalad, 1994; Woodruff & Gardial, 1996). In fact, various techniques, from use of futurists to trend analyses to scenario analyses, are available for this purpose. Much of this work, however, focuses broadly on macro-environments, markets, and industries. No research was found that studied or compared methods for predicting customer value change, specifically.

The emerging research discussed earlier on the nature and causes of customer value change suggest how such predictions might be made. This research demonstrates that customer value change does not happen randomly, but rather is caused by tension-inducing forces acting on a customer organization's personnel (Flint, Woodruff & Gardial, 1997, 2002). Consequently, it may be possible to predict customer value change by examining how the customer will likely respond to such forces and what role suppliers can, or will be expected to play (Woodruff & Gardial, 1996).

Figure 7 suggests a process that builds on this idea. It begins by using research to construct a customer value hierarchy describing how targeted customers currently perceive customer value. Then, market analyses are used to build one or more scenarios describing both forces that likely will shape the customer's business situation. These forces may be external, such as what is happening in the macro-environment (e.g. economic, social, and technological trends), within the customers' markets, or internal (e.g. new directives from top management, restructuring). These scenarios serve as a filter through which a supplier's managers can examine current customer value by asking a series of questions. Which customer value dimensions are likely to be filtered out because they will no longer be relevant for the customer if the scenario holds in the future? Which new value dimensions are likely to emerge for the customer to cope with the scenario? How will the strategic importance of remaining value dimensions

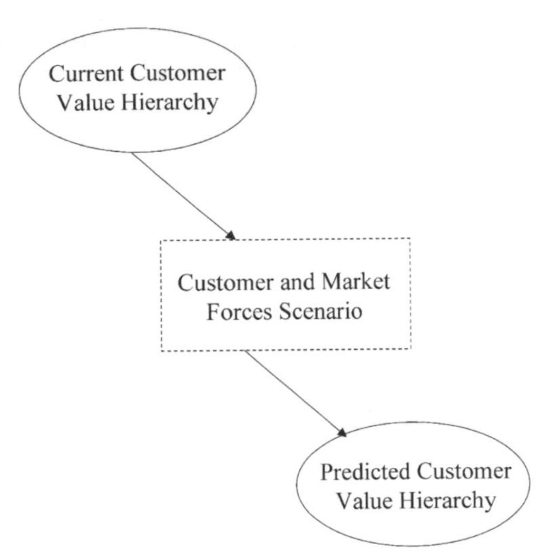

Fig. 7. Predicting Customer Value Change.

change in response to the scenario? What aspects of the scenario can suppliers influence?

This process has several advantages. First, it brings a supplier's managers directly into the predictions. That should help gain their commitment to acting on the predictions. Second, the process incorporates the customer value hierarchy concept of value. Predictions will not be restricted to attempts to forecast what supplier attributes that customers will want in the future. Most likely, much of the focus will likely be on consequences that customers want to have happen (or avoid) in their uses of a supplier's products and services. Third, the process uses data from customers for the predictions, but does not rely directly on those customers to make the predictions themselves. We cannot assume that customers are good at predicting their own future, particularly what they will want from a supplier. This is exactly why suppliers must understand how their specific customers tend to respond to and act within their environments in ways that result in changes in what they desire. By understanding how customers tend to behave, suppliers can make predictions that customers cannot make themselves.

Customer Information System

Business has made enormous commitment to information systems related to internal management needs, such as on product and process quality, costs, and product supply/logistics. The next major leap on this front will be in a customer information system designed to accumulate, integrate, and extract information from a supplier's many sources of data about customers (Parasuraman, 1997; Woodruff, 1997). While these systems will focus on all aspects of customers, customer value and satisfaction should play a central role in them. To realize this potential, customer information systems will have to deal with several issues.

A business-to-business supplier quite likely has many sources of data about its customers; some coming from research and others resulting from contacts with personnel inside a customer's organization (e.g. call reports). These data sources will yield data of different types, both qualitative (e.g. words in an interview or call report) and quantitative (e.g. ratings from a satisfaction survey). One challenge for information systems will be to integrate these different sources and data types into a common customer database for direct access by a supplier's managers.

In addition, a customer information system requires structure around which to organize different data on customer value and satisfaction. The customer value hierarchy can provide that structure. Most importantly, these systems must go beyond simply organizing data by product and service attributes. Instead, data that describe customer perceived consequences would be at least, if not more important. For example, call reports from salespersons should be analyzed for what they say about both attribute- and consequence-oriented problems that a customer mentions. That depends on having salespersons that are trained to listen for problems at all levels within a customer value hierarchy.

Third, a customer information system must be able to integrate data coming from all steps of a customer value determination process (recall Fig. 6). Data on customer satisfaction (third step) should be matched with strategically important customer value dimensions (first and second steps). Further, probable causes of satisfaction problems should be matched with the corresponding customer value dimension's customer performance ratings.

Finally, a customer information system must respond to different user needs. These users may vary across levels and functions within a supplier's organization. For instance, salespersons may need data kept separated by customer for use in planning sales calls. At the same time, a supplier's business planning team needs data aggregated by market segment to look for customer initiatives to incorporate into business strategy plans.

On a much larger scale, a supplier will eventually want a customer information system to mesh with other information systems. One particularly important opportunity lies in integrating cost data with customer value data, so that costs of delivering each strategically important customer value dimension may be estimated. That could be instrumental to a business strategy planning team trying to develop profitable customer initiatives to implement.

Managing Business-to-Business Customer Value

Judging by the business literature, the most pressing customer-oriented issues of the day revolve around managing customer value strategies in organizations (e.g. Gale, 1994; Naumann, 1995; Slywotzky & Morrison, 1997; Wayland & Cole, 1997). While many techniques exist to find out about customer value and satisfaction (Anderson, Jain & Chintagunta, 1993; Griffin & Hauser, 1993; Woodruff & Gardial, 1996), the resulting data only have payoff when they are utilized to create and successfully implement customer value strategies. Two types of issues arise in this regard: (1) need for a customer-oriented culture; and (2) implementing a customer value management process.

Everyone seems to agree that a supplier organization must be customer oriented to effectively compete on customer value strategies (e.g. Day, 1994; Narver & Slater, 1990; Slater, 1997). A customer orientation reflects a culture that places importance on the customer as stakeholder and encourages use of customer and market information in creating strategies (Slater, 1997). For that orientation to survive, top management must support it.

Having a customer-oriented culture in place is one thing, but knowing how to manage customer value effectively is quite another. The culture must facilitate carrying out the day-to-day responsibilities of a supplier's personnel central to competing on superior customer value. Figure 8 suggests the activities that lead to successful customer value strategies in practice (Woodruff, 1997). Each of these activities requires different skills and training.

Customer Learning

This activity broadly includes all the customer information system data and analyses needed to identify high value customers and to deeply understand what those customers value (Woodruff & Gardial, 1996). Much of the existing research on customer value and satisfaction has most potential to impact this activity. An impressive body of knowledge exists for business-to-business suppliers, as discussed in the previous section, which can help them, become proficient at customer learning. The immediate challenge for a supplier lies in developing and continually improving this proficiency.

With regard to customer value, the first priority must be to acquire CVD process capability. That means confidently answering such questions as: (1) What are

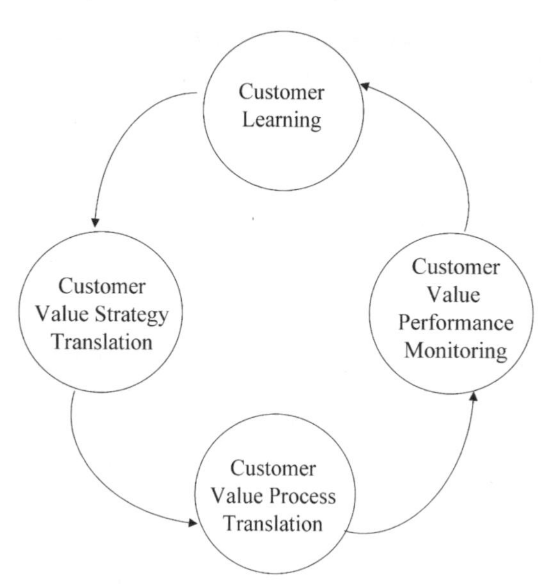

Fig. 8. Customer Value Management.

the specific functional and relationship value dimensions that targeted customers want? (2) Which of these value dimensions are most strategically important? (3) How satisfied are these customers with the specific value dimensions that they perceive that they receive from what we offer to them? and (4) for those value dimensions that customers rate low, what are the causes of these ratings? Most importantly, a supplier should answer these questions at least at the segment level, and perhaps, at the individual customer level. Customer value and satisfaction will differ by segment, and in some cases, even by individual customer.

As customer learning skills build, a business-to-business supplier will quickly find that customer information system concerns begin to take precedence. Managerial support for customer value and satisfaction data depends on the help it provides to them. While the systems issues are complex, users will expect to have the right data in a format that can be easily interpreted and distributed to them in a timely fashion.

As a supplier's customer information system becomes well accepted, more demands will be placed on it. For reasons discussed above, we think that one such demand quite likely will be to provide predictions of change in customer value among targeted customers. For instance, managers should want to examine trends coming from repeated customer value and customer satisfaction research.

The more sophisticated systems will go beyond these trends to base predictions on the factors that are driving change in customer value for individual segments and customers.

Customer Value Strategy Translation

The customer value and satisfaction literature has been preoccupied with measurement. Data are important, but the payoff comes only through translation into successful strategies. However, too often, suppliers are ill equipped to make those translations happen. Managers may not know how to extract information from the data or how to use that information as a guide to designing customer initiatives in business strategies. Part of the reason may be lack of training in translation skills. While customer learning is an analytical activity, strategy translation requires both analysis and creativity skills.

Customer value strategy translation requires thinking about the decisions that must be made. We look at the contributions of customer value and satisfaction research to the creative task of making four major strategy decisions. Translation begins with the decision on *which customers to target*. For business-to-business suppliers, these targets may be segments and/or individual customer organizations. The logic for targeting comes from a supplier's need to focus resources on high economic-worth customers. These customers are the organizations that, through their supplier selection, repeat purchases, and loyalty, are the source of highest return to the supplier. The targeting decision may be facilitated by actually calculating the economic worth of each potential customer.

Then, for each business-to-business customer target, the supplier should *set specific market objectives* to achieve. These objectives parallel each of the types of performances shown previously in Fig. 1. For instance, a supplier may want to set objectives describing desired levels of overall satisfaction ratings. Customer satisfaction data may be quite helpful here by establishing historical satisfaction levels and trends that suggest how realistic alternative proposed objectives are. Similarly, satisfaction survey data might suggest reasonable objectives to set for other kinds of performances, such as for customer loyalty and share of business. Here as well, suppliers must determine on which value dimensions they can realistically compete.

Next, customer value strategy should contain *value-positioning statements* for each customer target. These statements identify the essential customer value dimensions for which the supplier will depend for competitive advantage. For some targets, functional value may dominate, as may be the case with smaller customers who are price sensitive. Other targets, usually the highest economic-worth customers, may require positioning based more a combination of functional and relationship value. Customer learning provides the essential understanding of what value dimensions are important to the different targets. However, creativity plays an especially important role at this point. For example, managers may want to brainstorm alternative positioning statements to identify different ways to reach high economic worth customers.

Finally, customer value strategy lays out how the value positioning will be implemented through the supplier's *customer value offer*. In part, this offer consists of product quality, availability, delivery, information exchange, order process convenience, support services, customer service, customer contacts, and so forth. Further, the offer includes the communication needed to ensure that targeted customers know the kinds and level of value being delivered by the supplier. Certainly, the results of CVD processes can help. The supplier must consider offer alternatives that deliver the kinds of value most important to customers. However, this activity also is heavily dependent on implementing a creativity process, as managers consider the many possible ways that a supplier's capabilities may be used to meet customer needs.

Unfortunately, customer value and satisfaction research has concentrated almost entirely on the analytical aspects of strategy translation. We know a lot about the data and analyses regarding customers that should influence strategy decisions. We know less about how to incorporate these data and analyses into the creative activity needed to design customer value strategies. Certainly, there is a wealth of knowledge about creativity processes in general (e.g. Higgins, 1994), but very little on how to marry customer value and satisfaction learning with creativity for designing customer value strategies. At this point, we can recommend little more than for managers to receive training in creativity tools, such as brainstorming, and idea checklists.

Customer Value Process Translation

Customer value strategy serves as a blueprint for a supplier's internal processes. The critical processes are the ones required to deliver each of the customer value dimensions in the value positioning statement for targeted customers. For example, a chemical supplier targeted a large customer and decided to position based on selected relationship value dimensions. One dimension was "perceived by the customer as an expert in the customer's business." Managers thought about how to achieve this positioning by coordinating its business research and customer satisfaction research departments to focus on learning about this customer's markets and its targeted customers. This new process would enable the supplier to find opportunities for this target customer to increase business among its customers with new product opportunities. Success from this effort was intended to build satisfaction and loyalty from this customer, leading to an increased share of its purchases.

While a positioning statement captures the central value dimensions on which a supplier will achieve competitive advantage, other value dimensions may also be linked to achieving high satisfaction among the targeted customers. These may be components of functional value, relationship value, or both. For each one, a supplier must design and implement a corresponding process to deliver that value. In addition, communication processes are needed to ensure that targeted customers know about the value being delivered to them.

The customer value and satisfaction literature says little about how to identify, design, or implement customer-value-delivering processes. However, the quality literature does offer one tool for this purpose – quality function deployment (QFD). Essentially, QFD lays out a step-by-step process for translating selected customer value dimensions, typically customer-preferred attribute requirements, into product and operations process specifications (Griffin & Hauser, 1993; Hauser & Clausing, 1988).

Monitoring Customer Value Strategy Performance

Customer value management never stops. Once customer value delivery processes are in place and being implemented, management's responsibility shifts to assessment and improvement. Assessment means measuring how well the processes are performing to achieve objectives. Recalling Fig. 1, these objectives may be customer response performance, company market performance, and company financial performance. Customer response performance assessment begins with customer satisfaction measurement. Repeated satisfaction surveys enable managers to track customer-value-process performance as perceived by targeted customers. Most importantly, these surveys measure how well the customer believes that the supplier delivers each and every important value dimension, attributes and/or consequences, from the customer value hierarchy.

Customer satisfaction measurement should yield data that facilitate monitoring process performance success in achieving the other objectives as well. More specifically, a supplier, through analysis of past survey data, should ensure that satisfaction data predict customer loyalty, retention, share of customer purchases, profitability, and other kinds of performance objectives set during customer value strategy planning. For example, do individual customer satisfaction responses correlate with the supplier's share of that customer's business over time? These analyses are needed periodically to recheck on the extent to which satisfaction data do correlate with these other performance measures.

Customer Value Management Audit

How well does a supplier perform customer value management? That question can be answered by a customer value management audit. A thorough audit focuses

- **Customer Value Learning**
  - Does supplier do all customer value determination activities?
  - Are customer value and satisfaction measures based on sound theory?
  - Are analyses of customer data performed at the segment or individual customer level?
  - Does customer information system meet all users' needs?
  - Does supplier predict customer value change?

- **Customer Value Strategy Translation**
  - Do a supplier's business strategies include specific customer initiatives?
  - Are those initiatives consistent with customer value learning?
  - Does strategy translation properly combine customer value and satisfaction analyses with creativity techniques?

- **Customer Value Process Translation**
  - Is responsibility for each critical value dimension assigned to a process owner?
  - Are processes for delivering value cross-functional when necessary?
  - Are value communication processes focusing on the critical value dimensions being delivered?

- **Customer Performance Monitoring**
  - Does customer satisfaction measurement focus on critical customer value dimensions?
  - Do customer satisfaction measures predict market and financial performances?
  - Do performance measures trigger strategy and process translation improvements?

Fig. 9. Customer Value Management Audit.

on a supplier's performance of each of the activities shown in Fig. 8. While customer value and satisfaction literature provides no guidance on what the audit should cover, our discussion so far suggests a number of questions that the audit should address (see Fig. 9). These questions can be answered by examining key documents, from satisfaction questionnaires to strategic plans, and interviewing key personnel involved in customer value management (e.g. customer value delivery process owners).

A customer value management audit could benefit from standards that reflect state of the art or "best" practice for each of the four activities in Fig. 8. An auditor would like to compare a supplier's actual practice in answer to each question in Fig. 9 to a standard reflecting what the supplier "should be doing." Unfortunately, no research could be found that provides such standards. However, some help comes from frameworks and theories that represent newest thinking on customer value and satisfaction. For instance, the customer value hierarchy in Fig. 2 suggests the kinds of data that a business-to-business supplier should have about what targeted customers want or value from a supplier. The customer value determination process in Fig. 6 provides a normative framework for evaluating what a supplier's customer learning should cover. Such theories and frameworks will have to suffice until we know more about best practice in business-to-business customer value management applications.

IMPLICATIONS AND CONCLUSIONS

Customer value and satisfaction research is evolving from an almost total focus on other customer learning theories, concepts, and techniques to much more attention devoted to customer value management issues. That does not mean that opportunity to advance customer learning is gone. We have a lot to learn about how to find out about customer value and satisfaction, particularly for business customers. However, we have barely scratched the surface of knowledge about how to make customer value learning work in practice to improve an organization's performance. This imbalance in knowledge has implications for both business practice and future research.

Customer Value and Satisfaction in Business-to-Business Practice

Every business can improve on its application of customer value management. In particular, the state of the art of both customer value and customer satisfaction research are sufficiently advanced to offer practical guidance to business-to-business suppliers that want to learn more about their targeted customers. This research knowledge applies equally well to both the customer learning and customer value strategy performance monitoring activities. But, it does require that a supplier make a commitment to the time and resources needed to learn directly from its customers.

More challenging for business-to-business suppliers will be keeping up with new customer value and satisfaction measurement advancements. This topic attracts a lot of interest from universities, research firms, and consultants, and they keep pushing the state of the art forward for both measurement and information systems. To stay abreast of these advancements, a supplier may want to screen emerging literature on the topic, hire competent research firms and consultants, and/or participate in round-table discussions among "best practice" suppliers.

More often than not, suppliers that devote even moderate efforts to CVD data do get useful insights into their customers. If that does not lead to improved performance, the problem is much more likely to lie in poor application of the data during the customer value strategy and internal process translation activities. For this reason, perhaps the most attractive opportunity for many businesses to improve their capability to compete on customer value strategies lies in these two translation activities within customer value management (Fig. 8).

Translation of customer value and satisfaction data into strategy can get a boost from top management when they clearly communicate to business strategy planning teams that their strategies must include customer initiatives. That will go a long way toward ensuring that managing toward customers becomes a part of the organization's culture. With that expectation a reality, top management should challenge the organization to develop superior capability at customer value management. This capability can be a source of competitive advantage (Day, 1994).

The road toward building customer value management capability begins with an audit. Management must assess where the organization stands today. Just what is the business-to-business supplier doing well in managing toward customers, and equally importantly, where are the problems occurring? Objectively answering the questions listed in Fig. 9 will go a long way toward bringing out the supplier's capability strengths and weaknesses. Out of this assessment will come specific objectives for improving customer value management practice in the organization. For example, a business-to-business supplier might want to launch an effort to annually predict likely changes in value desired by selected customers. Another objective might be to improve application of creativity tools, such as brainstorming, to business strategy planning. Just as continuous improvement has become standard practice for quality management, and so it should be for customer value management.

Customer Value and Satisfaction Research Opportunities

Each activity in the customer value management process offers significant opportunity for new research to advance state of the art. While the opportunities are almost endless, we suggest some high priority research needs for each activity in Fig. 8.

Customer Learning Research Opportunities

Even though much of the customer value and satisfaction research focuses on this activity, we can get much better at it. As the field shifts from an attribute-based view of customer value to the customer value hierarchy theory, much more attention must be given to developing techniques for probing customer value at the consequences and desired end state levels. That will enable business-to-business suppliers to probe more into the motivation behind customer preferences for certain attributes.

Currently, business practice is overly dependent in general on interviewing customers to get these insights, and in particular on focus group interviewing. Interviewing relies on the verbal abilities of customers, and tends to present a more rational picture of customer value. For example, interviews may be better for drawing out perceptions of functional value than for relationship value. In the future, we need to better understand how to supplement verbal with non-verbal techniques for customer value learning, such as observational and participatory methods (Zaltman, 1996). Hopefully, such techniques will draw out the more emotional side of customer value, such as quite likely is a part of relationship value.

The need to predict change in customer value opens the door for a new direction in customer value research. As Flint, Woodruff and Gardial (1997, 2002) argue, the most important advancement on this topic will come from developing theory to explain how and why value change occurs. Business-to-business customers provide a particularly rich context for studying value change processes. Out of new value change theory can come new techniques for predicting that change in practice.

Another critical area for new research will be solving customer information systems issues. Perhaps the most difficult challenge will be designing such systems for quite different users within business-to-business supplier organizations. These users range from customer contact persons (e.g. sales persons) who need customer value and satisfaction data disaggregated by individual customer to strategic planning teams that must consider customer initiatives at the individual customer and segment level to top management who develop corporate wide strategy at the market level. A customer information system will have to accommodate data on individual customers as well as data coming from samples of customers representing targeted segments and markets. Research that designs and tests system characteristics to meet these needs can make an important contribution to advancing customer value management practice.

Customer Value Strategy and Process Translations

At these middle steps in the customer value management process, the greatest need is for tools to facilitate managers' efforts to create new strategy and processes for strategy implementation. As indicated above, much has been written about creativity tools and some organizations already train managers in their use. The challenge for the future is to learn how to use these tools to specifically help mangers create successful and innovative customer value and satisfaction initiatives in business strategy.

We can set some requirements for new research in this area. Translation tools must help managers use customer value and satisfaction data effectively without hindering their creativity. That is, strategy and process development should reflect new, innovative thinking but at the same time be grounded in what managers learn about their customers. Creativity processes, to meet this need, must incorporate tools that allow managers to bounce back and forth between creativity tools (e.g. brainstorming) and analytical tools (e.g. CVD analyses).

Customer Value Strategy Performance Monitoring
Finally, perhaps the greatest push toward implementing customer value management will come from demonstrating that customer value strategies lead to higher organization performance. Already, evidence is emerging supporting that linkage (Day, 1994; Narver & Slater, 1990). However, we know little about why that is so. For that reason, there is a tremendous need to understand how customer-value strategies work to increase shareholder value of an organization. New research must establish how this link from impact on customers to effect on shareholder value happens in actual practice. That will determine what kinds of performance measures are ultimately required to track how well customer value strategies are working.

CONCLUDING COMMENTS

Customer value and satisfaction research has enormous potential to help business-to-business suppliers become more customer-oriented. In part, this potential comes from the improved advancement already occurring in customer value and satisfaction theory, tools, and techniques that management can apply to better understand their customers and what they value. From a larger perspective, the future is bright for even greater opportunities to develop customer value management capabilities as research shifts more toward the application of customer value and satisfaction data to customer value strategy planning and implementation. In this chapter, we presented a framework for thinking about customer value management. We used this framework to discuss the opportunities to apply customer value and satisfaction to improvement organization performance in business-to-business markets. We also presented suggestions for new research to advance the practice of customer value management even further.

REFERENCES

Anderson, J. C., Jain, D. C., & Chintagunta, P. K. (1993). Customer value assessment in business markets: A state-of-practice study. *Journal of Business-to-Business Marketing*, *1*(1), 3–30.

Anderson, J. C., & Narus, J. A. (1998). Business marketing: Understand what customers value. *Harvard Business Review* (November–December), 53–65.

Birch, E. N. (1990). Focus on value. In: *Creating Customer Satisfaction* (pp. 3–4). New York: The Conference Board, Research Report No. 944.

Butz, H. E., & Goodstein, L. D. (1996). Measuring customer value: Gaining the strategic advantage. *Organizational Dynamics*, *24*(Winter), 63–77.

Cadotte, E. R., Woodruff, R. B., & Jenkins, R. L. (1987). Expectations and norms in models of consumer satisfaction. *Journal of Marketing Research*, *XXIV*(August), 305–314.

Dabholkar, P. A. (1995). The convergence of customer satisfaction and service quality evaluations with increasing customer patronage. *Journal of Consumer Satisfaction, Dissatisfaction, and Complaining Behavior*, *8*, 32–43.

Day, G. S. (1994). The capabilities of market driven organizations. *Journal of Marketing*, *58*(October), 37–52.

Dutka, A. (1994). *AMA handbook for customer satisfaction*. Lincolnwood, IL: NTC Business Books.

Dwyer, F. R., Schurr, P. H., & Oh, S. (1987). Developing buyer-seller relationships. *Journal of Marketing*, *51*(April), 11–27.

Flint, D. J., & Mentzer, J. T. (2000). Logisticians as marketers: Their role when customers' desired value changes. *Journal of Business Logistics*, *21*(2), 19–45.

Flint, D. J., Woodruff, R. B., & Gardial, S. F. (1997). Customer value change in industrial marketing relationships: A call for new strategies and research. *Industrial Marketing Management*, *26*(March), 163–175.

Flint, D. J., Woodruff, R. B., & Gardial, S. F. (2002). Exploring the phenomenon of customers' desired value change in a business-to-business context. *Journal of Marketing*, *66*(4), 102–117.

Gale, B. T. (1994). *Managing customer value*. New York: Free Press.

Garver, M. S., Gardial, S. F., & Woodruff, R. B. (1999). Customer value, loyalty, and revenge in buyer relationships with salespersons. Working Paper, The University of Tennessee Knoxville, College of Business Administration, Knoxville, TN.

Griffin, A., & Hauser, J. R. (1993). The voice of the customer. *Marketing Science*, *12*(Winter), 1–27.

Gutman, J. (1982). A means-end chain model based on consumer categorization processes. *Journal of Marketing*, *46*(Spring), 60–72.

Hamel, G., & Prahalad, C. K. (1994). *Competing for the future*. Boston: Harvard Business School Press.

Hauser, J. R., & Clausing, D. (1988). The house of quality. *Harvard Business Review*, *3*(May–June), 63–73.

Hayes, B. E. (1992). *Measuring customer satisfaction*. Milwaukee, WI: ASQC Quality Press.

Higgins, J. M. (1994). *101 creative problem solving techniques*. Winter Park, FL: New Management Publishing Company.

Homburg, C. (1998). On closeness to the customer in industrial markets. *Journal of Business-to-Business Marketing*, *4*(4), 35–72.

Idassi, J. O., Young, T. M., Winistorfer, P. M., Ostermier, D. M., & Woodruff, R. B. (1994). A customer-oriented marketing method for hardwood lumber companies. *Forest Products Journal*, *44*(July/August), 67–73.

Jones, T. O., & Sasser, W. E., Jr. (1995). Why satisfied customers defect. *Harvard Business Review*, *73*(November–December), 88–99.

Kaplan, R. S., & Norton, D. P. (1996). *The balanced scorecard*. Boston: Harvard Business School Press.

Mohr-Jackson, I. (1998). Managing a total quality orientation: Factors affecting customer satisfaction. *Industrial Marketing Management*, *27*(2), 109–126.

Narver, J. C., & Slater, S. F. (1990). The effect of a market orientation on business profitability. *Journal of Marketing*, *54*(October), 20–35.

Naumann, E. (1995). *Creating customer value*. Cincinnati, OH: Thompson Executive Press.

Oliver, R. (1997). *Satisfaction: A behavioral perspective on the consumer*. New York: McGraw-Hill Companies.

Parasuraman, A. (1997). Reflections on gaining competitive advantage through customer value. *Journal of the Academy of Marketing Science*, *25*(Spring), 154–161.

Patterson, P. G., Johnson, L. W., & Spreng, R. A. (1997). Modeling the determinants of customer satisfaction for business-to-business professional services. *Journal of the Academy of Marketing Science*, *25*(Winter), 4–17.

Pine, J. B., II (1993). *Mass customization: The new frontier in business competition*. Boston, MA: Harvard Business School Press.

Qualls, W. J., & Rosa, J. A. (1995). Assessing industrial buyers' perceptions of quality and their effects on satisfaction. *Industrial Marketing Management*, *24*(5), 359–368.

Sharma, A., & Lambert, D. M. (1994). How accurate are salespersons' perceptions of their customers. *Industrial Marketing Management*, *23*, 357–365.

Slater, S. (1997). Developing a customer value-based theory of the firm. *Journal of the Academy of Marketing Science*, *25*(Spring), 162–167.

Slywotzky, A. J., & Morrison, D. J. (1997). *The profit zone*. New York: Times Books.

Spreng, R. A., MacKenzie, S. B., & Olshavsky, R. W. (1996). A reexamination of the determinants of consumer satisfaction. *Journal of Marketing*, *60*(July), 15–32.

Srivastava, R. K., Shervani, T. A., & Fahey, L. (1998). Market-based assets and shareholder value: A framework for analysis. *Journal of Marketing*, *62*(January), 2–18.

Tanner, J. F., Jr. (1996). Buyer perceptions of the purchase process and its effects on customer satisfaction. *Industrial Marketing Management*, *25*(2), 125–134.

Wayland, R. E., & Cole, P. M. (1997). *Customer connections*. Boston: Harvard Business School Press.

Weitz, B., & Jap, S. D. (1995). Relationship marketing and distribution channels. *Journal of the Academy of Marketing Science*, *23*(4), 305–320.

Wilson, D. T. (1995). An integrated model of buyer-seller relationships. *Journal of the Academy of Marketing Science*, *23*(4), 335–345.

Woodruff, R. B. (1997). Customer value: The next source for competitive advantage. *Journal of the Academy of Marketing Science*, *25*(Spring), 139–153.

Woodruff, R. B., Clemons, D. S., Schumann, D. W., Gardial, S. F., & Burns, M. J. (1991). The standards issue in CS/D research: A historical perspective. *Journal of Consumer Satisfaction, Dissatisfaction and Complaining Behavior*, *4*, 103–109.

Woodruff, R. B., & Gardial, S. F. (1996). *Know your customer: New approaches to understanding customer value and satisfaction*. Cambridge, MA: Blackwell Publishers.

Zaltman, G. (1996). Metaphorically speaking: New technique uses multidisciplinary ideas to improve qualitative research. *Marketing Research*, *8*(Summer), 13–20.

Zeithaml, V. A. (1988). Consumer perceptions of price, quality, and value: A means-end model and synthesis of evidence. *Journal of Marketing*, *52*(July), 2–22.

Zeithaml, V. A., Parasuraman, A., & Berry, L. L. (1990). *Delivering quality service: Balancing customer perceptions and expectations*. New York: Free Press.

META-EVALUATION: ASSESSING ALTERNATIVE METHODS OF PERFORMANCE EVALUATION AND AUDITS OF PLANNED AND IMPLEMENTED MARKETING STRATEGIES

Arch G. Woodside and Marcia Y. Sakai

ABSTRACT

A meta-evaluation is an assessment of evaluation practices. Meta-evaluations include assessments of validity and usefulness of two or more studies that focus on the same issues. Every performance audit is grounded explicitly or implicitly in one or more theories of program evaluation. A deep understanding of alternative theories of program evaluation is helpful to gain clarity about sound auditing practices. We present a review of several theories of program evaluation.

This study includes a meta-evaluation of seven government audits on the efficiency and effectiveness of tourism departments and programs. The seven tourism-marketing performance audits are program evaluations for: Missouri, North Carolina, Tennessee, Minnesota, Australia, and two for Hawaii. The majority of these audits are negative performance assessments.

Evaluating Marketing Actions and Outcomes
Advances in Business Marketing and Purchasing, Volume 12, 549–663

ISSN: 1069-0964/doi:10.1016/S1069-0964(03)12009-1

Similarly, although these audits are more useful than none at all, the central conclusion of the meta-evaluation is that most of these audit reports are inadequate assessments. These audits are too limited in the issues examined; not sufficiently grounded in relevant evaluation theory and practice; and fail to include recommendations, that if implemented, would result in substantial increases in performance.

INTRODUCTION TO META-EVALUATION

In 1969 New York established the first state auditing and evaluation unit. A total of 61 such government departments exist in the U.S. – at least one in each of the 50 state legislatures in the United States (Brooks, 1997). Each of these auditing offices is assigned by their state legislature with the responsibility for conducting financial audits and performance audits of government departments and their specific programs. The mandate of these auditing offices is to provide answers to questions, including the following issues:

- is the audited department spending funds legally and properly in accordance to its legislative mandate, is the department's accounting and internal control systems adequate, are the department's financial statements accurate?
- is the department managing its operations efficiently?
- is the department achieving substantial impact in effectively accomplishing its goals?

Thus, the auditing work done for state legislative branches include two major categories of audits: (a) financial audits; and (b) performance audits. Some state audit manuals distinguish among program, operations, and management audits (e.g. *The Auditor*, 1994). For example, "a *program audit* focuses on how effectively a set of activities achieves objectives. A program audit can stand-alone or be combined with an operations audit. An *operations audit* focuses on the efficiency and economy with which an agency conducts its operations. In Hawaii the term *management audit* is used often to refer to an audit that combines aspects of program and operations audit. A management audit examines the effectiveness of a program or the efficiency of an agency in implementing the program or both" (*The Auditor*, 1994, pp. 1–2). In this study we use "management audit" and "performance audit" interchangeably.

The paper before you includes a meta-evaluation of available performance audits done by legislative audit offices in evaluating government tourism-marketing efforts. The term, "meta-evaluation" was created by Scriven (1969) to mean an evaluation of evaluations. We conducted a meta-evaluation on the performance audits of tourism-marketing programs completed for the state of Tennessee (1995), Missouri (1996), North Carolina (1989), Minnesota (1985), Hawaii (1987, 1993), and for Australia (1993). These performance audits address two central issues: (1) how well are the government tourism-marketing programs being managed; and (2) how effectively are the actions of the state tourism offices contributing to their goals.

Our meta-evaluation results in two main conclusions. First, the tourism-marketing performance audits spotlight serious problems in the performances of government tourism-marketing programs. Second, the audits themselves have major shortcomings; for example, they fail to include comprehensive reporting on topics relevant in performance auditing. With one exception (i.e. Hawaii, 1987), the audit reports are not grounded in relevant theory and empirical literature on best practices. Consequently, the recommendations in the reports are too limited in scope – while beneficial, implementation of the recommendations is unlikely to have a major impact on increasing the effectiveness of state government sponsored tourism-marketing programs.

META-EVALUATION OBJECTIVES

Meta-evaluations have three central objectives. First, meta-evaluations are syntheses of the findings and inferences of research on performance – both on the managing of programs and on the effectiveness of achieving goals of programs. Thus, meta-evaluations enlighten; they help increase our knowledge and insight about what works well and poorly in managing programs.

Second, meta-evaluations are reports on the validity and usefulness of evaluation methods. Meta-evaluations include guidance in the methods useful to apply for evaluating.

Third, meta-evaluations may provide strong inference on the impact, payback, and repercussions of enacting specific decisions. Consequently, the findings in meta-evaluations help to justify and increase the confidence of legislative members and program managers, in designs and implementation of specific decisions. This third objective meets the instrumental use criterion for evaluations described by Cook (1997), "Would those who pay for evaluation be satisfied if it creates enlightenment but did not feed more directly into specific decisions? I'm not sure they would." Related to a travel decision, Campbell (1969) provides a detailed meta-evaluation that fulfills this third objective in summarizing multiple studies on the impacts (e.g. reduction in the number of deaths) of the legislated requirement to wear safety helmets by motorcyclists.

We adopt these three objectives – enlightenment, method usefulness, and instrumental use – in preparing this report. Thus, presenting bad news is not our focus; our focus is on increasing useful "sensemaking" (Weick, 1995) and reducing "knowing what isn't so" (Gilovich, 1991) in evaluations. Sensemaking is creating, examining, and revising plausible explanations of events that have occurred; sensemaking is always retrospective: "People can know what they are doing only after they have done it" (Weick, 1995, p. 24). Auditing is one category of sensemaking.

Most retrospective processes used by humans cause problems: "People are extraordinarily good at ad hoc explanation" (Gilovich, 1991, p. 21). However, humans are biased strongly in favor of misinterpreting incomplete and unrepresentative data. "A fundamental difficulty with effective policy evaluation is that we rarely get to observe what would have happened if the policy had not been put into effect [implemented]. Policies are not implemented as controlled experiments but as concerted actions" (Gilovich, 1991, pp. 41–42). Without training, we rarely make and implement plans to find evidence disconfirming our beliefs. This observation applies to government tourism-marketing programs. Without collecting information about the attitudes and behavior toward the brand (e.g. Hawaii, North Carolina, or Australia) of targeted travelers *not* exposed to the marketing program, the success rate is evaluated – the awareness, attitudes, visits, and purchases among targeted travelers exposed to the marketing program – a research process inadequate to do the job of evaluating marketing performance.

In the second section we present a brief review of the theory-related literature on performance audits and evaluation research. Based on the conclusions drawn from the literature review, the third section offers propositions for planning and implementing a meta-evaluation of performance audits. The fourth section describes the method for collecting and analyzing performance audits of state tourism-marketing programs. Sections 5 through 11 present the findings from a meta-evaluation of the state performance audits. Section 12 provides conclusions and implications for research and practice on performance auditing of tourism-marketing programs, as well as effective strategies for managing such programs, are provided. Section 13 advocates five golden rules to apply for effective performance auditing of tourism-marketing programs.

PERFORMANCE AUDITS AND EVALUATION RESEARCH THEORY

The U.S. Office of the Comptroller General defines performance auditing as an objective and systematic examination of evidence for the purpose of providing an independent assessment of the performance of a government organization, program, activity, or function in order to provide information to improve public accountability and facilitate decision-making by parties with responsibilities to oversee or initiate corrective action. An audit is "inherently retrospective, concerned with detection of errors past – whereas many evaluative techniques can be applied retrospectively, concurrently, or prospectively" (Pollitt & Summa, 1997, p. 89).

While additional differences between audits and evaluations have been described (see Pollitt & Summa, 1997), the Association of Governmental Accountants emphasize, "Policy makers [want] reliable facts and sound, independent professional judgment, and they care little about . . . terminology. They use terms like performance auditing and program evaluation interchangeably. Their greatest concern is that they get answers to their most pressing questions about the performance of government programs and agencies" (AGA Task Force Report, 1993, p. 13, quoted from Brooks, 1997, p. 115). Thus, although recognizing that some differences exists, we use the terms "performance auditing" and "program evaluation" interchangeably.

Importance of Program Evaluation Theory

What are the key issues related to theories of performance auditing? In the case of government evaluations of state tourism-marketing programs, we believe the key issues include the following points: (1) recognition of the importance of theory in program evaluation; (2) learning the existing alternative theories of program evaluation; and (3) adopting a multiple perspective approach to program evaluation – use of multiple theories as foundations for tourism-marketing program evaluations.

Shadish, Cook and Leviton (1991, p. 20) emphasize that, "It is . . . a serious mistake to overlook the importance of theory in program evaluation." Theory is defined as a set of assumptions used for sensemaking (i.e. a "mental model" description of some topic, see Senge, 1990). "Theory connotes a body of knowledge that organizes, categorizes, describes, predicts, explains, and otherwise aids in understanding and controlling a topic" (Shadish et al., 1991, p. 30).

> Without its unique theories, program evaluation would be just a set of loosely conglomerated researchers with principal allegiances to diverse disciplines, seeking to apply social science methods to studying social programs. Program evaluation is more than this, more than applied methodology. Program evaluators are slowly developing a unique body of knowledge that differentiates evaluation from other specialties while corroborating its standing among them (Shadish et al., 1991, p. 31).

The fundamental reason theory is important for program evaluation in that the assumptions in the theory used are the rationale for the focus and method applied for the evaluation. All performance audits are based on implicit or explicit program evaluation theory. Program evaluation theory answers the questions of whether or not the sponsored program should achieve outcomes that would not have occurred without the program, and should the program be judged by whether or not certain policies are implemented.

For the seven performance audits of government tourism-marketing programs examined in this report, we find little explication of program evaluation theory selected for use, and we find a poor match to features found in "good theory for social program evaluation" (see Chapter 2 in Shadish et al., 1991). The majority of these reports do not explicitly address the issues of program evaluation theory beyond making a general statement early in the audits. Here is an example of one such statement:

> The objectives of the audit were to review the department's legislative mandate and the extent to which the department has carried out that mandate efficiently and effectively and to make recommendations that might result in more efficient and effective accomplishment of the department's legislative mandates .... The audit was conducted in accordance with generally accepted government auditing standards and includes: (1) review of applicable statues, regulations, policies, and procedures; (2) examination of document files, contracts, data, and reports, including information compiled by the U.S. Travel Data Center; (3) interviews with a contracted vendor, directors of regional tourist associations, and staff of the Departments of Tourist Development, Transportation, Health, and Environment and Conservation; and (4) site visits to the 12 welcome centers (State of Tennessee, 1995, pp. i–ii).

These statements fail to display knowledge and understanding of alternative theories of program evaluation; and the statements fail to indicate features of good program evaluation theory. Steps toward good program evaluation theory include detailed explicit statements regarding the objectives and methods applied in a performance audit, including the auditor's view on what constitutes a high quality evaluation of a tourism-marketing program.

For example, do the auditors agree with the views of Stake and his colleagues (1997) that "Programs designed to contribute to improvement in social well-being are to some degree meritorious just by existing." Or, do the auditors embrace Scriven's theory that a program's merit should be judged by the impact of its outcomes in comparison to other programs and written standards (evaluation checklists, see Scriven, 1995). Stake et al. (1997) make clear that these views can not be held simultaneously, "...we object to Michael Scriven's claim that the basic logic of evaluation is criterial [sic] and standards-based."

Because of such disagreements among both program evaluation theorists and practitioners, describing a detailed, explicit, statement of the theory of program evaluation being used by a performance audit team is a valuable step. Without such a detailed explicit statement, the different stakeholders affected by the audit are likely to assume different meanings for concept words such as, "*efficient*" and "*effective*."

Fundamental Issues in Development of Program Evaluation Theory

Related to program evaluation, fundamental issues of theory development and use include these questions:

(1) Social programming: what are the important problems this program addresses? Can the program be improved? Is this program worth improving? If not, what is worth doing (including the basic issue: should a state government be involved at all in tourism-marketing, see Bonham & Mak, 1996)?
(2) Knowledge use: how can I make sure my results get used quickly to help this program?
(3) Valuing: is this a good program? By which notion of "good"? What justifies this conclusion?
(4) Knowledge construction: how do I know the information is accurate? What counts as a confident answer? What causes that confidence?
(5) Evaluation practice: given my limited skills, time, and resources, and given the seemingly unlimited possibilities, how can I narrow my options to do a feasible evaluation? What is my role – educator, methodological expert, and/or judge – of program worth? What questions should I ask, and what methods should I use? (adapted from Shadish et al., 1991, p. 35).

Several well-formed alternative theories of evaluation have different – and often opposing – answers to these five issues. The details of these theories cannot be covered in depth here; however, we briefly describe the main assumptions held by a leading proponent of each theory.

To create a learning aid, we oversimplify these theories using two dimensions: valuing results vs. valuing activity, and logical (post)positivism vs. relativism. (See Fig. 1.) The specific placement of theorists in Fig. 1 is arbitrary to some extent, but does reflect the kernel assumptions of the cited theorists. The kernel theoretical propositions of each theorist are described below.

The two dimensions found in Fig. 1 address the issues: (1) what should be valued in program evaluation; and (2) what method should be used for doing evaluations? *Valuing results* is performance auditing built on the proposition that a (state tourism-marketing) program should achieve certain outcomes that would not have occurred from other programs – including having no program. *Valuing activities/processes* is performance auditing built on the proposition that a program should include activities or processes specified by one or more program stakeholder groups; these activities/processes may or may not result in desirable outcomes. *(Post)positivism* in performance auditing builds on the proposition that one reality does exist related to performance and that: (1) multiple accounts/perspectives of this reality are data that should be collected, including quasi- and non-experimental

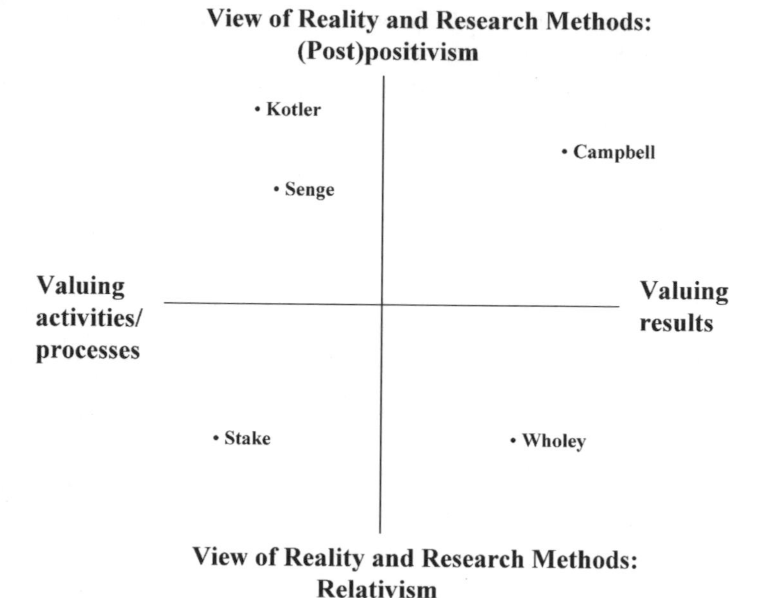

Fig. 1. Classifying Theories of Program Evaluation: Focus on What is Explained and How Reality Regarding Evaluation is Viewed.

data, for valid measurement of this one reality (the post-positivist view); or (2) the scientific method including test and control groups with high internal validity should be used to measure performance (positivist view). *Relativism* builds on the proposition that no one reality exists – therefore, data on multiple perspectives should be collected and used as indicators of performance.

Post Positivism and Focusing on Valuing Goal-Free Effects: Scriven

First, the body of work by Scriven (1967, 1974, 1986, 1995) focuses both on valuing effects resulting from the program implemented as well as judging the organization/processes performed. He defines evaluation as the science of valuing: "Bad is bad and good is good, and it is the job of evaluators to decide which is which" (Scriven, 1986). Scriven believes the evaluator's ultimate task is to try very hard to condense the mass of data collected into one work: good or bad. "Sometimes this really is impossible, but all too often the failure to do so is simply a cop-out disguised as, or rationalized as, objectivity" (Scriven, 1971, p. 53). Scriven coined the identical concepts "goal-free" and "needs-referenced" evaluations. "The evaluator's job," according to Scriven, "is to locate any and all program effects, intended or not, that might help solve social problems. Goals are a poor source of such effects. Goals are often vaguely worded to muster political support, and rarely reflect side effects that are difficult to predict" (quoted in Shadish et al., 1991, p. 80).

Scriven is a post positivist. He rejects positivistic ideas that reality is directly perceived without mediating theories and without perceptual distortion, that scientific constructs should be operationalized directly in observables, and that empirical facts are the sole arbiter of valid scientific knowledge (Shadish et al., 1991). Scriven argues that reality is independent of the observer, and it is possible to describe reality objectively. He does encourage the use of positivist research methods (e.g. experiments with test and control groups) to measure program impact; however, he stresses that practical constraints often limit the use of positivist research methods.

Note that in Fig. 1, Scriven is plotted right of center, toward valuing results. Scriven (1974, p. 16) himself provides the reason, "One way or another, it must be shown that the effects reported could not reasonably be attributed to something other than the treatment or product. No way of doing this compares well with the fully controlled experiment, and ingenuity can expand its use into most situations." Scriven also believes in valuing activities/processes: he has developed an eighteen-point "Key Evaluation Checklist" that includes measuring functions and processes done (see Scriven, 1980, pp. 113–116).

Scriven also advocates the use of multiple methods and observers to reach convergence for learning facts about objective reality. He acknowledges that no single scientist ever observes reality completely and undistorted; everyone observes through biased perceptual filters. Consequently, according to Scriven, using multiple perspectives and methods helps construct a more complete and accurate view of objective reality than can be accomplished by applying any one research method. He calls this view "perspectivism," and differentiates the view from relativism: "Perspectivism accommodates the need for multiple accounts of reality as perceptive from which we build up a true [objective] picture, not as a set of true pictures of different and inconsistent realities" (Scriven, 1983, p. 255). The original concept of "triangulation" (Denzin, 1984) is applied perspectivism: the use of multiple methods and observers to validate information about objective reality.

An important point to note here is that Denzin and other researchers (Denzin & Lincoln, 1994) have adopted a new definition for triangulation: the research process of using multiple observers and methods for assuring multiple viewpoints of reality (also see Stake et al., 1997). The proposition that an objective reality exists is *not* implied as part of this recent view of triangulation; this later view represents a shift toward viewing reality from a relativist perspective. To avoid confusion, we refer to triangulation as applied perspectivism as "triangulation for convergence" and triangulation for gaining access of multiple viewpoints as "triangulation for divergence."

Positivism: True and Quasi-experimentation for Valuing Results

Positivism focuses on achieving unbiased valuation of results of an implemented program. Positivism includes the assumptions: (1) that an objective reality exists; and (2) that we can measure accurately the effects of changing an independent variable (such as: expenditures levels of a marketing program or alternative advertising campaigns) on the levels of dependent variables (such as the number of inquiries, visits, length of stay, and total visitor party expenditures) can be accurately measured.

First fully developed by Fisher (1949), the purest form of positivism is the application of the "true experiment," that is, the use of treatment groups and control groups with random assignment of subjects among the groups to achieve "internal validity," a term coined by Campbell. "Random assignment promotes internal validity because it falsifies many competing interpretations – threats to internal validity – that could plausibly have caused an observed relationship *even if the treatment had never taken place*" (Shadish et al., 1991, p. 126, italics in original). Campbell and Stanley (1963) identify nine threats to internal validity in a true experiment. "Their explication is one of Campbell's major contributions to social science and to evaluation" (Shadish et al., 1991, p. 126).

Quasi-experiments (Campbell, 1969) are novel research designs that rule out threats to internal validity when true experiments are not used. Quasi-experiments often include measuring the occurrence of a target variable (such as requests for a catalog, visits, length of stay, expenditures) when the program implemented to cause the target variable is absent. Thus, many quasi-experimental designs fill the need to measure impact by measuring a target variable when a program is *present and absent*. Campbell's (1969) "Reforms as Experiments" is likely the most well known quasi-experiment.

Contrary to popular belief, true experiments to measure the impact of a program are used often for measuring the effect (i.e. impact) of marketing and advertising programs (see Banks, 1965; Caples, 1974; Woodside, 1996). Exhibit 1 presents an example of a true experiment applied to examine the effectiveness of a government tourism-marketing program. The data summarized in Exhibit 1 are fictitious because no such study exists in the literature on tourism advertising.

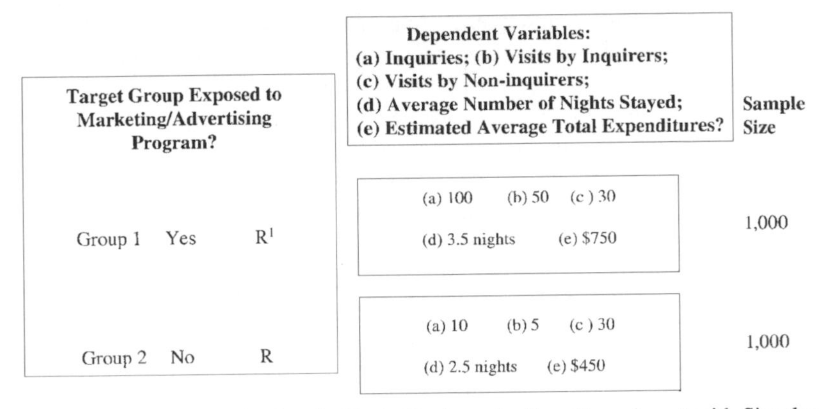

Exhibit 1. Positivism: Example of a Basic Design of a True Experiment with Simulated Data for a Government Tourism-Marketing Program. *Note:* R = random assignment of the 2,000 households to Group 1 or Group 2. A high inquiry rate occurs for Group 1: 100/1,000 = 10%, a lower inquiry rate occurs for Group 2: 10.1 = 1%. The "conversion rate" for Group 1 is 50/100 = 50%, the conversion rate for Group 2 is 5/10 = 50%. The average total number of nights for Group 1 is 40% higher than the average number of $m =$ nights for Group 2 visitors ((3.5 − 2.5)/2.5) = 0.40. The average total expenditures for visitors from Group 1 is $300 greater than the average total expenditures for visitors from Group 2 ($750 − $450 = $300).

Almost all government tourism-advertisements include an offer for free literature (e.g. a "visitor's information guide [VIG]"). The ads are created in part to gain inquiries, that is, requests, for this free literature. A VIG is an example of "linkage advertising." Such linkage-advertising offers provide a mechanism for advertisers to identify "qualified prospects," that is, people displaying an interest in buying the advertised product. Marketers sometimes use lists of such qualified prospects for additional unique marketing programs to convert qualified prospects into product buyers (see Rapp & Collins, 1994).

Note that the test displayed in Exhibit 1 is a comparison of five impacts of implementing the program vs. not implementing the program, for example, exposing target customers to advertisements to visit Texas vs. not exposing target customers to these advertisements. Because the two groups were created using random assignment with 1,000 persons per group, the two groups are identical statistically before the two treatment levels (i.e. program exposure vs. no program exposure) are administered. Adding more groups makes other comparisons possible, for example, between Program A vs. B vs. no program conditions. Such experiments are often referred to as "split-run" tests because a sample from a known population (e.g. a sample of a list of newspaper, magazine, or cable TV subscribers) is split randomly to create two or more equal groups; then each group is exposed to a different treatment (see Schoenbachler et al., 1995).

Given the results in Exhibit 1, was the program effective? That is, did the program increase the levels of the five dependent variables *that would not occur without exposure to the program*? To answer this question with respect to a government tourism-advertising campaign (the largest budget item for most state tourism-marketing programs), first note that the inquiry rate is much higher for Group 1 vs. Group 2 (10% vs. 1%). We conclude that the campaign was effective in increasing the inquiry rate. Second, note that the "conversion rate" of visitors from inquiries was the same for both groups: 50/100 = 5/10. Thus, the conversion rate did not increase for Group 1 vs. Group 2.

Third, note that the rate of total visitors did increase for Group 1 vs. Group 2: 80/1,000 > 35/1,000. The difference in visitor rates is 4.5% (8 – 3.5 = 4.5). The rate of visits among households exposed to the program was more than 100% above the visitor rate of households not exposed to the program. Fourth, note that the average number of nights per stay is much higher for Group 1 vs. Group 2: 3.5 > 2.5. Also, the average total expenditure is much higher for visitors from Group 1 vs. Group 2: $750 > $450.

Hundreds of such true experiments have been done in industries other than government tourism-marketing programs (see Caples, 1974), surprisingly none have been reported in the travel and tourism literature (Woodside, 1990). In relation to testing the impacts of advertising campaigns, the costs of such experiments are relatively small – most likely less than 3% of the total advertising budget.

Rather than using true experiments or well-designed quasi-experiments, the evaluation of the impacts of government tourism-advertising programs has used one-group conversion studies almost exclusively. These conversion studies use no non-exposed, randomly assigned group. Thus, such studies cannot answer the basic issue: did the advertising program increase the number of visitors and dollar expenditures vs. what would have occurred without the advertising campaign? Such one-group conversion studies are useful for comparing the conversion rates and average dollar expenditures of different groups exposed to different government tourism-advertising campaigns (e.g. see Woodside & Motes, 1981) and competing advertising media (e.g. see Woodside & Soni, 1990), but such studies do not address the more basic issue of did the advertising program cause results that would not otherwise have been experienced.

Another major problem with conversion studies sponsored to date by government tourism-marketing agencies is the failure to measure the impact of advertising on households exposed to the advertising who do not inquire – shown as variable *c* in Exhibit 1. Although the hypothetical example in Exhibit 1 indicates no increase in the visiting rate among non-inquirers exposed vs. not exposed to the advertising, a positive (and in some studies, a negative) impact is likely to occur.

Even though highly critical of both the activities and results of marketing programs, none of the seven performance-audit reports include mentioning the lack of high-quality research designs to measure advertising impacts. These performance audits do not include indications of knowledge of the core assumptions of positivism as a theory of program evaluation.

Shift Away from Positivism toward Valuing Activities and Relativism Methods: Weiss

Note in Fig. 1 that Weiss appears in two positions. "Weiss (1972) initially tried to use traditional experimental methods in evaluation, but the political and organizational problems she encountered in doing so sensitized her to the need to consider such problems more realistically" (Shadish et al., 1991, p. 183). In her later work on evaluation theory, she perceived evaluation as a political activity in a political context; she decided that evaluation is more useful if it suggests feasible action or challenges current policy.

> I would like to see evaluation research devote a much larger share of its energies to tracing the life course of a program: the structure set up for its implementation, the motivations and attitudes of its staff, the recruitment of participants, the delivery of services, and the ways in which services and schedules and expectations change over time, the response of participants and their views of the meaning of the program in their lives (Weiss, 1987, p. 45).

Note that, unlike Scriven and Campbell, Weiss is more of a descriptive theorist than a prescriptive theorist of valuing. She advocates evaluations that report how stakeholders, such as program administrators, value their activities to assess whether or not changes are warranted/possible in policy decisions. Weiss' primary goal for evaluations is the enlightenment of stakeholder groups – in thinking about issues, defining problems, and gaining new ideas and perspectives.

> Scriven and Campbell focus on studying [valuing] the effects of possible solutions to problems; Weiss makes a role for studying all activities associated with social problem solving. . . . Second, Weiss has a more realistic view of how social programs work and change. Scriven and Campbell wrongly assume that programs will be significantly responsive to rational information, that ineffective programs will be replaced and effective ones retained, and that programs can be treated as uniform wholes implemented similarly across sites and times. Weiss explicitly rejects such assumptions (Shadish et al., 1991, pp. 216–217).

Partnering with Program Management to Achieve Performance Goals: Wholey

Note that Wholey is located in the bottom right quadrant of Fig. 1. While in principal endorsing the view that an objective reality exists and results can be measured using positivistic tools, Wholey (1977, p. 51) observes, "Practically, however, we must settle for substantially less in virtually all federal programs. An assumption is considered 'testable' if there exists test comparisons that the manager/intended user would consider adequate indication that observed effects were attributable to program activities."

Wholey's theory of program evaluation includes working closely with program managers. He writes, "Evaluators and other analysts should place priority on management-oriented evaluation activities designed to facilitate achievement of demonstrable improvements in government management, performance, and results" (Wholey, 1983, p. 30). Wholey and his colleagues advocate the reporting by the evaluator to the program managers about progress and results at all steps in the evaluation. "Final reports and exit briefings ought not be the places where significant findings are revealed for the first time. Evaluators should continually be sharing their insights, findings, and conclusions with staff, managers, and policy makers" (Bellavita, Wholey & Abramson, 1986, p. 291).

Wholey prefers evaluators to limit their role as program critics and to be team players working with management to improve programs. "Although evaluators clearly should not be asked to be cheerleaders for government, there is a legitimate point that they can play a more positive role in both shaping public opinion about government and helping agency managers run their programs" (Bellavita et al., 1986, p. 286).

From our examining the seven audits described below, government auditors completing performance audits of their governments' tourism-marketing programs follow Wholey's assumptions regarding valuing results using relativistic methods. However, the work and reports by the auditors do not follow Wholey's views on partnering with program management. The state auditors in the U.S. did not follow Wholey's recommendation of providing on-going reports of findings during the audit processes – the tone of the reports are confrontational, an emotional state that Wholey avoids in his auditing work.

Evaluating Activities and Using Relativistic Methods: Stake

"Stake advocates that case study methodologies be used to improve local practice" (Shadish et al., 1991, p. 271). Stake's theory of program evaluation is closer to Weiss's enlightenment approach than to Wholey's partnering views. Unlike

Wholey, Stake does not emphasize concerns by legislators, although he describes such concerns in his evaluations.

Stake's empirical work concentrates on evaluation of education programs. His theory of program evaluation includes the beliefs that evaluations should increase practitioner understanding and allow the practitioners (i.e. program managers) to solve any problems uncovered by the evaluation. Major concerns for Stake include: (1) how can evaluations be made to facilitate their use; and (2) how can evaluation reports be made credible? He argues that credible reports are more likely to be used than reports believed to be false. He argues that case studies are the best way to produce credible reports (see Stake & Easley, 1978; Stake et al., 1997).

Consequently, Stake is a strong advocate of the use of case-study methods. These methods include the use of interviews, observation, examination of documents and records, unobtrusive measures, and investigative journalism, resulting in a case report that is complex, holistic, and involves many variables not easily unconfounded. These written cases are informal, narrative, with verbatim quotations, illustrations, allusions, and metaphors.

Stake distinguishes between a preordinate evaluation and a responsive evaluation. "Many evaluation plans are more 'preordinate,' emphasizing: (1) statement of goals, use of objective tests; (2) standards held by program personnel; and (3) research-type reports" (Stake, 1980, p. 76). He advocates the use of "responsive evaluation" in place of preordinate evaluation. A program evaluation is responsive evaluation: "(1) if it orients directly to program activities rather than to program intents; (2) if it responds to audience requirements for information; and (3) if the different value-perspectives of the people at hand are referred to in reporting the success and failure of the program" (Stake, 1980, p. 77).

Stake has moved away from attempting to find objective reality and the use of triangulation of methods to converge on a view of such reality. Earlier he wrote that, "One of the primary ways of increasing validity is by triangulation . . . trying to arrive at the same meaning by at least three independent approaches" (Stake & Easley, 1978, p. C:27). As we have seen earlier, he now views the use of triangulation as a means of learning the multiple perspectives of reality. He states that there are only the realities that we construct: valid knowledge is that which bests portrays the richness and individual variations of each person's created reality; and the case method is best suited to such portrayals.

Our meta-evaluation of the seven performance audits of tourism-marketing programs indicates that they do not match closely with the core propositions of Stake's theory of program evaluation: these audits do not include rich, detailed, case studies of the programs examined; they do not include a relativist perspective of reality. Although triangulation methods are the primary research methods used in these seven audits, the auditors use triangulation to achieve convergence rather than divergence.

Objective Reality and Valuing Activities: Kotler

Kotler's (see Kotler, 1997; Kotler, Gregor & Rodgers, 1977, 1989) work on marketing audits may be viewed accurately as a post positivist theory of program evaluation similar to Scriven's (1983, 1995) contributions: both Kotler and Scriven imply an objective reality exists and can be best learned using multiple research methods (i.e. triangulation to achieve convergence). However, Kotler focuses much more than Scriven on valuing activities rather than results. Thus, note the placement of Kotler in the upper left quadrant of Fig. 1.

Kotler (1997, p. 777) defines a marketing audit to mean "a comprehensive, systematic, independent, and periodic examination of a company's – or business units' [or, policy program such as government tourism-marketing programs] – marketing environment, objectives, strategies, and activities with a view to determining problem areas and opportunities and recommending a plan of action to improve the company's marketing performance." He reports, ". . . most experts agree that self-audits lack objectivity and independence . . . the best audits are likely to come from outside consultants who have the necessary objectivity, broad experience in a number of industries, some familiarity with the industry being audited, and the undivided time and attention to give to the audit" (Kotler, 1997, p. 777).

He continues, "The cardinal rule in marketing auditing is: Don't rely solely on the company's managers for data and opinion. Customers, dealers [i.e. firms carrying manufacturers' product lines], and other outside groups must also be interviewed. Many companies do not really know how their customers and dealers see them, nor do they fully understand customer needs and value judgments" (Kotler, 1997, p. 779). Similar to Scriven's "Key Evaluation Checklist," Kotler provides checklists, rating forms, and open-ended questionnaires useful for the marketing audit he describes (see Kotler, 1997, pp. 776–782). Kotler's "Components of a Marketing Audit" open-ended questionnaire focuses on evaluating the behavior of markets, customers [or clients], competitors, distribution and dealers, facilitators and marketing firms, and the public. Kotler's audit form includes questions relating to valuing how well the executives of the audited organization are scanning the organization's stakeholders (e.g. markets, industry, customers and competitors), planning decisions, implementing these decisions, and assessing profitability and costs.

In the following sections, we note specific areas of inactivity in most of the seven performance audit reports on how the government tourism-marketing programs are being managed. However, in our meta-evaluation, none of the seven performance-audit reports indicated use of checklists, rating forms, or open-ended questionnaires for valuing the scanning, planning, or implementing activities of the programs evaluated. The seven audits exhibit little evidence of systematic performance auditing for valuing the full range of management functions, as proposed by Kotler (1997).

Systems Thinking and Systems Dynamics Modeling as a Theory of Program Evaluation: Senge

While Senge (1990) and other systems researchers (e.g. Hall, Aitchison & Kocay, 1994) do not refer to program evaluation, their work clearly relates to valuing activities and outcomes of programs as systems. Similar to Wholey (1983) Senge does not suggest program evaluation to find fault. But Senge goes further than Wholey, "In mastering systems thinking, we give up the assumption that there must be an individual, or individual agent, responsible. The feedback perspective suggests *that everyone shares responsibility for problems generated by a system*" (italics in original, Senge, 1990, p. 78). Senge recommends developing a useful micro world (i.e. simplified representative view of the events and flows of relationships) of actual systems; he views everything in life as a system. A systems-dynamic modeling is an advanced form of creating micro worlds (see Hall, 1983). The view that customer behavior influences the behavior of program managers and program manager behavior influences the behavior of customers is the circular linkage included in a systems theory of program evaluation. Developing long-term relationship marketing programs with customers, including an interactive database marketing strategy with multiple contacts over several years, is a systems view of program valuing.

Note that Senge is located in a positivistic, valuing activity position in Fig. 1. However, systems researchers never advocate one reality as optimal; they advocate modeling the activity relationships among variables in a system and measuring the impacts of inputs and outcomes for each variable over time – using objectively-viewed indicators for each variable. Hall (1983, 1994) provides useful examples of such systems evaluation work.

A systems-thinking approach to tourism-marketing program evaluation includes identifying the participants, their activities, and the relationships among events occurring in implementing the program. Core assumptions in such systems-modeling work include the following propositions:

(1) all events are both causes and outcomes of other events;
(2) several feedback loops among two or more participants and events occur in the implementation of the program;

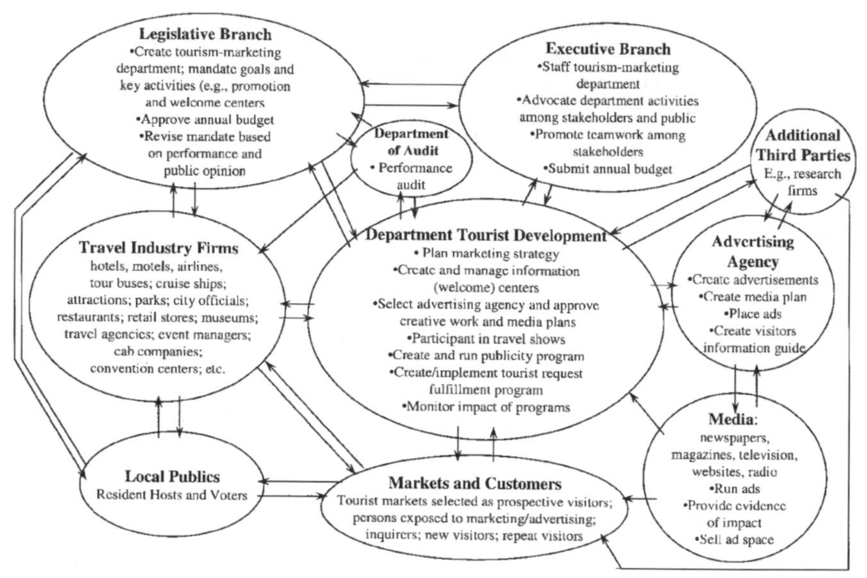

Fig. 2. Case Example of Participants Interested in Marketing Strategy Evaluation Participants and Core Events in Government Tourism-Marketing Programs.

(3) participants in managing and operating a tourism-marketing program have incomplete mental models of how the program functions – the human mind can not comprehend clearly the substantial complexities and nuances in a system;
(4) the longer, indirect, paths in systems often have a total impact opposite of the effect of shorter, more direct paths; and
(5) the principal objective of performance auditing of the program is to create useful systems models of the program and to test improvements in program outcomes via simulating the running of the program.

Figure 2 depicts the participants and main relationships (shown by arrows) for a systems theory of program performance auditing. Note that the main functions/activities performed by each of the eleven participant groups shown in Fig. 2 are identified. Note that the Department of Audits interacts chiefly with two principal parties: the Legislative Branch and the Department of Tourist Development. As Fig. 2 shows, audit researchers are likely to interact with other system principals, for example, travel industry firms.

The arrows in Fig. 2 indicate influence or scanning for information. Most relationships between principals in Fig. 2 indicate two-way influences. A few relationships in Fig. 2 are one direction to indicate information-scanning activities by a system participant. For example, under "Additional Third Parties: research firms," customer data are collected; the research firm analyzes these data and prepares reports for the clients of the research firm. As depicted in Fig. 2, these clients may include the Department of Tourist Development or the Advertising Agency, or both these organizations.

A program-evaluation case study by Hall and Menzies (1983) illustrates the detail in systems performance auditing. Hall and Menzies (1983) describe a performance audit using micro-systems modeling of a leisure-sports organization – a curling club in Winnipeg, Canada. Details are described briefly here. Figure 3 summarizes the events (i.e. variables) and relationships for their systems model. The plus and minus signs on the arrows indicate the valence of influence between events. For example, increasing the annual fee (box 10) increases members' dues (box 6) leading to increases in total profit (box 9). However, this short positive path (10–6–9) to increasing profit is misleading; increases in annual fees have strong, negative, and more indirect, links to profits. Please take a moment to identify all the indirect paths from annual fees (box 10) to total profit (box 9). (Hint: six of them occur starting with box 10 and going through box 7.)

Several simulations were run using the curling-club systems model for operating the leisure club for different decision scenarios. The simulation results included decreases in total profits following increases in annual fees, and increases in total profits with increases in bar and food revenue.

The propositions in a systems theory of program evaluation receive no specific attention among the seven performance audits reviewed in the meta-evaluation of government tourism-marketing programs. For example, none of the performance audits include evaluations concerning the lack of database marketing behavior for building long-term relationships with inquirers – inquirers responding to government offers of free travel information (e.g. visitor information guides). Because of the advances in the efficiency and effectiveness in database marketing programs and in system dynamics modeling (e.g. ALPHA/Sim, 1998), applications of systems thinking in performance auditing of government tourism-marketing programs are likely to occur – given that government performance auditors receive training in applications of database marketing and systems dynamics modeling.

Comprehensive Theories of Program Evaluation: Cronbach and Rossi

Two comprehensive theories of program evaluation are positioned in central locations in Fig. 1. These locations identify the contributions to program evaluation

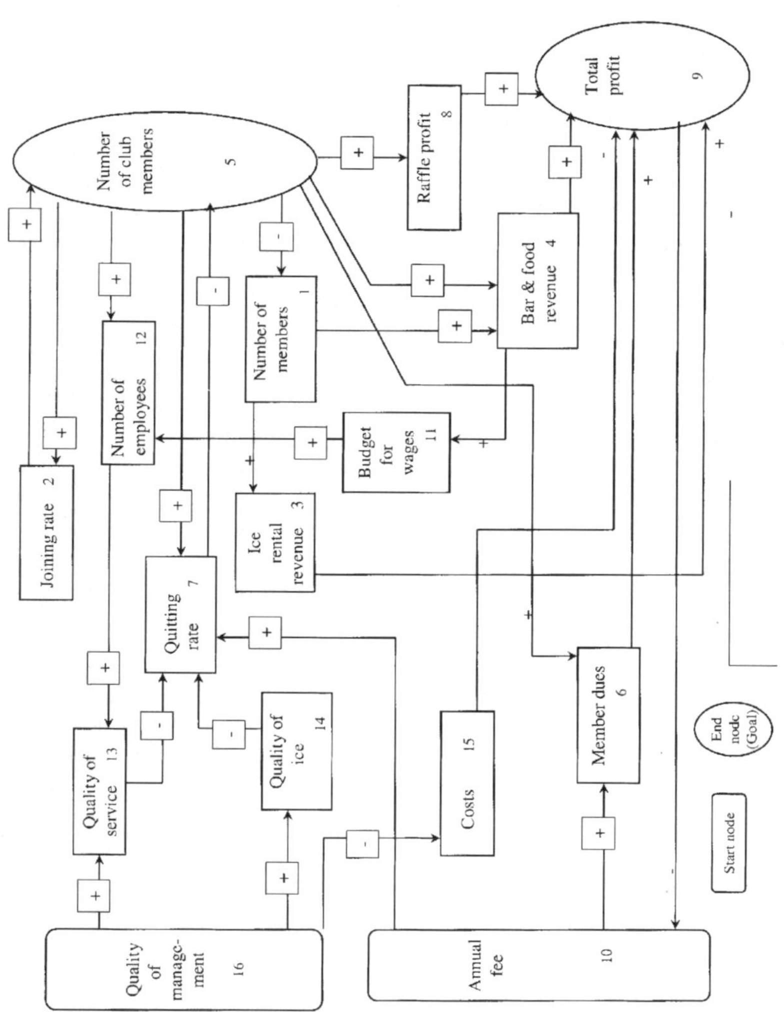

Fig. 3. Policy Map of an Ice Sports Club Operation (Adapted from Hall & Menzies, 1983).

theory made by Lee Cronbach and Peter Rossi. Both Cronbach's and Rossi's views have much in common: they both advocate the use of multiple methods including positivistic and relativistic research methods. They both believe that no single paradigm (i.e. set of assumptions representing a mental model) for knowledge construction has sufficient empirical or theoretical support to dominate the field. Both want evaluators to study program description, causation, explanation, and generalization, sometimes including all of them in the same study, albeit with different priorities depending on circumstances (Shadish et al., 1991, p. 318).

The proposition that dependable knowledge is contingent on multiple conditionals captures Cronbach's belief about objective reality. Rather than seeking a general estimate of objective reality, Cronbach seeks to reduce uncertainty and increase understanding of the complexities of relationships in given local contexts. Such complex, dependable, explanatory knowledge requires evaluators to place more weight on interpretive (relativistic), thick-descriptive methods rather than on methods traditionally espoused in social sciences – positivistic research methods, such as experimentation.

Cronbach and his colleagues (1980) make some highly perceptive observations about program administration and decision making in general in *Toward Reform of Program Evaluation*:

> ... rarely are decisions made rather than slipped into; rarely is there a single decision maker; rarely do data about optimal decisions take precedence over politics; rarely do all stakeholders give the project the same attention and meaning as evaluators; rarely is a report ready on time; rarely are decisions made on particular days; and rarely are evaluation results used to modify programs in ways that instrumentally link the modifications to the evaluation. Cronbach and his co-authors depict a world of politics and administration that undercuts the rational model and its image of clear command and optimal decision making (Shadish et al., 1991, p. 335).

This view of sensemaking matches well with the later work on sensemaking by Weick (1979, 1995). Weick views most decision making as following behavior rather than preceding behavior. He asks, "How can I know what I value until I see where I walk [what I've done]? People make sense of their actions, their walking, and their talking. If they are forced to walk the talk, this may heighten accountability, but it is also likely to heighten caution and inertia and reduce risk taking and innovation" (Weick, 1995, p. 183).

Consequently, Cronbach believes that evaluations must be crafted to the political system as it is rather than to an abstract model of how it should be. He suggests that it is unwise to focus on whether or not a project has "attained its goals" (Cronbach et al., 1980, p. 5). For Cronbach, evaluation is a pluralist enterprise. Evaluations should contribute to enlightened discussions of alternative paths for social action, clarifying important issues of concern to the policy-shaping community (Shadish et al., 1991, p. 338). Evaluators as educators and as knowledge resources reflect Cronbach's views of program evaluation.

Rossi's theory of program evaluation recognizes three kinds of studies by evaluators: (1) analysis related to the conceptualization and design of interventions (e.g. government social programs, such as tourism-marketing programs); (2) monitoring of program implementation; and (3) assessment of program utility (i.e. worth). Related to government tourism-marketing programs, the first kind of study would include asking the following question: what type of marketer-customer relationship should state government design and how should state government go about implementing such a design? Only one government audit, the Evaluation of the Australian Tourist Commission's Marketing Impact 1993, comes close to raising this issue.

Related to the program monitoring – the second issue – Rossi believes that, "Program monitoring information is often as important or even more so than information on program impact." He suggests that evaluators can help by monitoring "(1) whether or not the program is reaching the appropriate target population; and (2) whether or not the delivery of services is consistent with the program design specifications" (Rossi & Freeman, 1985, p. 139). Regarding program utility, Rossi and Freeman (1985, p. 40) state, "Unless programs have a demonstrable impact, it is hard to defend their implementation and continuation." Going much further than other evaluation theorists, Rossi and Freeman devote a chapter to cost-benefit analysis.

Note in Fig. 1 that Rossi is positioned above the midpoint on the (post)positivistic-relativistic dimension. This positioning is based in part on his view that "Assessing impact in ways that are scientifically plausible and that yield relatively precise estimates of net effects requires data that are quantifiable and systematically and uniformly collected" (Rossi & Freeman, 1985, p. 224). Rossi's theory of program impact measurement includes measuring effects in the absence of the program:

> Adequate construction of a control group is usually central to impact assessment. Such groups should consist of multiple units comparable to the treatment group [group exposed to the tourism-marketing program] in composition, experiences, and dispositions. The preferred option is use of randomized controls, something that is often more feasible with innovations than established programs. When randomization is not feasible, other options are (from most to least preferred): matched controls, statistically compared controls .... Rossi also advises increasing the precision of impact evaluations by adding multiple pre- and post-measures and using large sample sizes (Shadish et al., 1991, p. 403).

Rossi also believes that assessing how well program innovations are implemented is crucial. "The components of the delivery system must be explicated and criteria of performance developed and measured" (Rossi & Freeman, 1985, p. 77).

"This [measuring] includes assessment of the target problem and population; service implementation; qualifications and competencies of the staff; mechanisms for recruiting targets; means for optimizing access to the intervention, including location and physical characteristics of service delivery sites [e.g. visitor information guides; websites; interstate visitor welcome Centers]; and referral and follow-up" (Chen & Rossi, 1983; quoted from Shadish et al., 1991, p. 401).

Conclusions on Theories of Program Evaluation Applied to Performance Auditing of Government Tourism-Marketing Programs

The seven performance-audit reports do not include mention of alternative theories or methods for program evaluation. They all report using a triangulation of methods, including interviews, document analysis, and site visits. Although not stated formally, the goal for using multiple methods for ensuring for divergent viewpoint data is not implied in these reports. Rather, the two objectives for using a triangulation of research methods in most of these reports include: (1) uncovering major problems in the actions and outcomes of operating the government tourism-marketing program; and (2) achieving convergence for presenting a Scriven-type summary view of the one, objective, reality of program activities and performance outcomes (i.e. "good" or "bad").

For example, the "Audit Highlights," the first page in the Tennessee 1995 performance audit of the Department of Tourist Development, addresses one activity found to be a problem by the auditors. "The Cost-Effectiveness of Using State Personnel Instead of Outside Contractors Should Be Analyzed" (all capitals in the original, Tennessee 1995). Similarly, the 1994–1995 performance audit of Missouri's Department of Economic Development, Division of Tourism, reports, "Our review was made in accordance with applicable generally accepted government auditing standards and included such procedures as we considered necessary in the circumstances. In this regard, we reviewed the division's revenues, expenditures, contracts, and other pertinent procedures and documents, and interviewed division personnel."

The central theme of the Missouri audit report is on reporting improper activities and bad performance. The report includes written responses to these findings by the Missouri Division of Tourism. The North Carolina 1989 audit report of the Division of Travel and Tourism has the same theme: "In reviewing the activities of the division we found that the division had failed to answer 93,221 reader service inquiries during the first part of 1988" (North Carolina, 1989, p. 2). Note that this quote comes from the Executive Summary of the North Carolina State Auditor report. Although less adversarial, the 1985 Minnesota audit of tourism programs of the Department of Energy and Economic Development (DEED) focuses on identifying shortcomings in activities of the program – including DEED's lack of assessment of the return on investment of its tourism-marketing programs.

With the exception of the audit report for Australia's tourism-marketing program, the performance-audit reports do not include a comprehensive coverage of the activities and results of government tourism-marketing programs. Again, with the exception of the Australian audit, the audits do not attempt to identify best practices in assessing performance of tourism-marketing programs. The audits do not describe other possible theories of program evaluation as foundations for performance auditing. The audits include only brief mentions to the theoretical grounding in accounting for the audits.

We conclude that the substantial majority of performance audits of government tourism-marketing programs are summarized accurately as post positivistic and are oriented towards valuing both activities and results. Related to the theories of program evaluation described earlier, these audit reports most closely follow the assumptions of Scriven's theory of program evaluation. Figure 4 shows a summary of our meta-evaluations of the theories-in-use for the seven performance audits.

Notably, the only performance audit that was compiled jointly with an audit department and the department managing the tourism-marketing program is Australia's performance audit compiled by "The Department of Finance, Department of Arts, Sport, the Environment, Tourism and Territories & the Australian

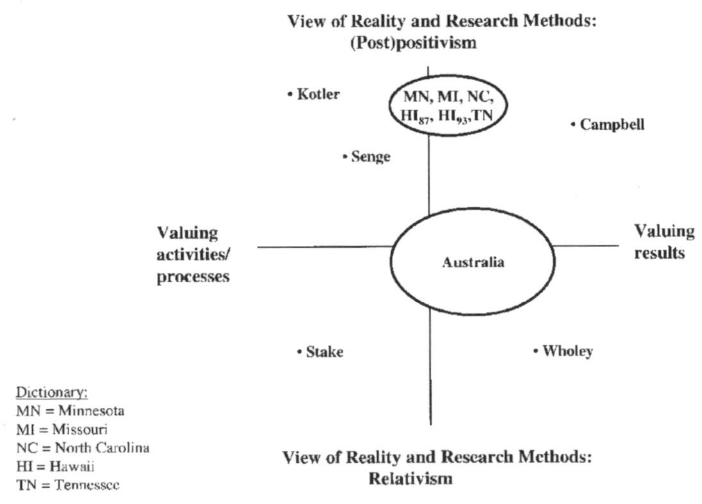

Fig. 4. Classifying Performance Audits of Government Tourism-Marketing Programs by Two Dimensions of Program Evaluation Theory.

Tourist Commission" (Australia, 1993, cover page). The executive summary of the Australian performance audit states, "This is the first comprehensive evaluation of the Australian Tourist Commission's (ATC) marketing and promotional activities" (Australia, 1993, p. 1).

The performance-audit partnering of departments and the attempt to achieve a comprehensive program evaluation are examples of the evidence we used to locate the Australian audit among the theoretical assumptions of Wholey, Weiss's later work, and Cronbach (see Fig. 4). However, several findings in the Australian audit report support the conclusion that the Australian auditors assume that an objective reality exists; the Australian report uses triangulation to achieve convergence, as well as divergence. The Australian audit report is the only one among the seven analyzed that includes references to tourism-marketing impact studies done by other governments. Thus, the theory of program evaluation covers Rossi's assumptions and recommendations for estimating objective reality.

PROPOSITIONS FOR PERFORMANCE AUDITING AND META-EVALUATIONS OF PERFORMANCE AUDITS OF GOVERNMENT TOURISM-MARKETING PROGRAMS

Our review of the marketing and strategic management literature (e.g. Buzzell & Gale, 1987; Clifford & Cavanaugh, 1985; Huff, 1990; Kay, 1995; Kotler, 1997) and the reviewed theories of program evaluation leads us to the following propositions. We argue that the performance audits of government tourism-marketing programs that include the following traits are most likely to result in achieving the goals of government and public stakeholders:

- being comprehensive by including both (post)positivistic and relativistic research methods and both objective and subjective views of realities;
- valuing both program activities and results;
- carried out using a "partnering" (Wholey, 1983) approach by a "working party" (Australia, 1993) of representatives of three or five organizations assigned full-time to complete the audit.

The accounting paradigm for performance auditing as practiced now in the United States state audit reports is performing poorly in assisting in achieving:

- the program goals mandated by the state legislatures in creating: (a) the tourism-marketing departments; and (b) the performance auditing program by offices of legislative auditors;

- knowledge of activities and processes performed in the government tourism-marketing department;
- use of scientific methods for measuring the economic impact of government tourism-marketing programs;
- the adoption and use of advanced database-marketing strategies in these tourism-marketing programs.

Accepting both objective and subjective realities is an example of applying Weick's (1979) recommendations for sensemaking. "In environments where multiple contingencies arise, responses that are appropriate at one point in time may be detrimental at another. Cause mappings (see Huff, 1990) valid at one point in time may be invalid at another. Flexibility to deal with environmental changes is maintained if opposed responses are preserved" (Weick, 1979, p. 220).

Advanced database-marketing strategies include several characteristics, including the following features:

- learning the names, addresses and telephone numbers of customers and entering them into a database;
- having on-line computer files for each individual customer that can be accessed when a customer telephones or contacts the tourism department by email;
- based on knowing an individual customer's leisure travel preferences and behavior, creating two-way communication strategies that include outbound contacts with customers beyond the one-time requests for brochures;
- and creating special member benefits for uniquely valuable visitors – such as frequent visitor programs.

Examples of such database marketing strategies exists in many travel-related industries, for example, car rental firms, credit card companies, and air lines.

Steps toward database marketing are indicated in a 1998 advertisement for Canada's Atlantic Coast. (See Exhibit 2.) This ad's design includes three specific features:

- four illustrations to create an image of the unique benefits offered by a trip to one or more of four Canadian Atlantic Coast provinces;
- an expensive free offer to gain inquiries for the "**FREE** 1998 Adventure Guide and Touring Planner. It's over 100 full-colour pages of all the ways we make people feel good";
- a tour package to gain direct sales for a touring visit including round-trip airfare, accommodation vouchers, compact car rental with insurance, and the "opportunity to extend your stay with extra-night rates."

Illustrations:
1. Beach, boulders and water at New Brunswick's Bay of Funday
2. Red-clay cliffs of Prince Edward Island
3. Waterfront buildings of Halifax, Nova Scotia
4. Map showing Canada's Maritime Provinces

Headline:
CANADA'S ATLANTIC COAST

Copy:
Where adventure runs as high as the tides. Walk in the wake of the world's highest tides or follow in the footsteps of the Vikings. Experience a seaside city filled with musical nights and *Titanic* sites.... **FREE** 1998 Adventure Guide and Touring Planner. It's over 100 full-colour pages of all the ways we make people feel good.

FOOTLOOSE & FANCY FREE
Here's one of the many thrilling adventure packages you'll discover on Canada's Atlantic Coast: [a trip package including air fare from Boston to Saint John, New Brunswick; accommodations for 3 nights; compact-car rental for 3 days; toll-free reservation to "pre-book"; opportunity to extend stay with extra night rates]

Direct Linkage Mechanisms:
1. Toll-free telephone number: 1-800-565-2627
2. E-mail address: canadacoast@ns.sympatico.ca
3. World-wide website: www.canadacoast.com

Exhibit 2. Summary Description of Advertisement of Canada's Atlantic Coast.

This ad to attract visitors to the Canadian Maritime Provinces includes three methods for inquiries by prospects: a toll-free telephone number; an email, and a website. Inquiries from this ad represent initial responses to the possibility of a highly interactive long-term relationship between the destination marketer and visitors. Several "outbound marketing" strategies can be used to reach inquirers, visitors, and repeat visitors for several years. A "frequency-marketing" program (e.g. frequent flyer memberships in the airline industry) could be created to stay in contact with frequent travelers and to promote trips highly profitable to the destination areas. Rapp and Collins (1988) provide an in-depth rationale for such strategies along with detailed descriptions of currently running database-marketing programs.

Create and implement database-marketing strategies because:

- a substantial share of tourist visits result in purchases to and around the destination area being promoted by governments;
- repeat-trip visitors represent an important target segment;
- advances in electronic technologies and software reduced the costs for operating such programs; and

- many firms in many industries have completed the shift successfully to such "mass-customized" programs.

In particular, asking three questions is appropriate here:

- Given that most U.S. resident households will spend more than $1,000 for each leisure trip to Canada, are such database marketing strategies likely to be effective, efficient, and competitive?
- Do a substantial share of leisure visitors to Florida, California, Hawaii, Minnesota, Pennsylvania, Texas, or other states, provinces, and countries spend more than $1,000 on their trips?
- Which state and provincial governments are using advertising and inquiry responses to build database marketing programs with the described program characteristics?

The answers most accurate to these questions appear to be: yes, yes, and none. Only the answer to the third question may be surprising.

Figure 5 summarizes five specific activities that occur, or should occur, in a government tourism-marketing program. These activities include administering, scanning, planning, implementing, and assessing the performance quality of activities and program-system outputs. Including details on each of the five activities in a performance audit is a useful step toward achieving comprehensive evaluations of government tourism-marketing programs. None of the seven audit reports reviewed in the following sections reflects such a comprehensive review (the details leading to this conclusion are provided for each section reviewing the seven audits).

Evidence of Scanning and Interpreting

Scanning includes both formal and informal collecting and interpreting data on the behaviors, attitudes, and intentions of relevant stakeholders: clients (e.g.

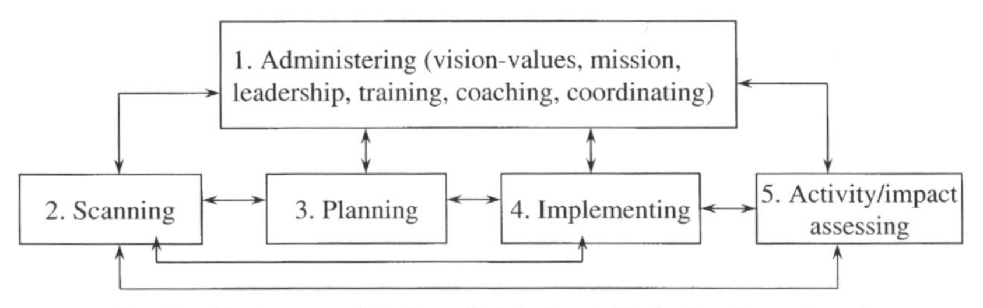

Fig. 5. Decisions and Actions Related to a Marketing Department.

households believed to be prospective visitors; known inquirers requesting the advertised free literature offer; households known to have visited the state recently), suppliers (including legislative and executive branches supplying funds and program objectives), competitors, distributors, internal marketing staff, and additional parties (e.g. advertising agencies and industry representatives). Evidence of lack of formal scanning of stakeholder groups, especially potential customer markets, is an indicator of poor performance, arrogance, and the lack of management expertise. Daft and Weick (1984) note that patterns of scanning, interpreting, and learning vary across organizations as a function of their willingness to act in order to learn and their willingness to accept that the environment is difficult to analyze.

Related to the issue of formal scanning of relevant stakeholders, the issue of interpretation includes whether or not formal overall evaluation of the tourism-marketing program and "products" (i.e. state tourism services, benefits delivered, and visitor-perceived trip value) has been completed. In short, have program managers completed a formal overall evaluation of the program's strengths, weaknesses, opportunities, and threats – a SWOT analysis? Is an annual SWOT analysis report prepared annually based on new scanning data?

Evidence of Planning

A formal, annual, marketing plan includes:

- an executive summary (usually prepared in 2–4 pages);
- a statement on the current marketing situation, including a SWOT and issue analysis, and the targeted segment of customers identified as potential visitors most likely to be influenced by the state's tourism-marketing programs (4–6 pages);
- objectives that have been converted into measurable goals – to facilitate implementation and impact measurement (1 page);
- a marketing strategy – including a "what if analysis" and indicating the major decisions in the plan, for example, target markets and advertising strategy (e.g. a 200 page, fact-filled visitor's information guide will be prepared and the offer for the free guide will be featured in all advertisements) (6–10 pages);
- planned actions – presents the details and schedules of what is to be done by whom, when, and how, and additional, related, special-events, marketing activities (4–6 pages);
- projected costs-benefits analysis – including the estimated expenditures by visitors in the state and state taxes collected from implementing the tourism-marketing program (2 pages);

- controls – indicating how the planned activities will be monitored to insure performance; including details of the methods used to measure the net impacts of the tourism-marketing program on inquiries, intentions, visits, revenues, and taxes (4 pages).

Does formal planning (i.e. a written statement explaining what you are going to do and why) really matter? If you believe the answer to be obviously yes, posing the question seems surprising. However, examining this issue is important – especially since none of the seven audit reports describes formal planning done by the tourism-marketing executives.

Armstrong (1982) conducted a meta-evaluation of all published research on evaluation of formal planning. In his evaluation, Armstrong asked, what is the value of formal planning? His analysis indicates:

> Formal planning was superior in 10 of the 15 comparisons drawn from 12 studies, while informal planning was superior in only two comparisons. Although this research did not provide sufficient information on the use of various aspects of the planning process, stakeholders [in the formal planning process] provided mild support for having participation. Formal planning tended to be more useful where large changes were involved, but, beyond that, little information was available to suggest when formal planning is most valuable (Armstrong, 1985, p. 582).

The Usefulness of Written Reflection

We suggest including a "retrospective commentary" (in 2–4 pages) in the annual written marketing plan. This retrospective commentary can provide answers to the following issues: how much does what happened last year match with our strategic plans and action plans?

- What was planned that was executed well and poorly? Why?
- What was planned that did not get done? Why?
- What noteworthy events got done that were unplanned? Why?
- What did not get done but should have been planned and done? Why?

By including a retrospective commentary, program managers force themselves to reflect – to examine last year's marketing plan and to ponder for a few moments about what happened and why (not necessarily, what things that went well and poorly, and why). Without a retrospective commentary, last year's annual plan may never get noticed, or used, after it is written. Weick (1979, 1995) argues persuasively for reflection. "Reflection is perhaps the best stance for both researchers and practitioners to adopt if the topic of sensemaking is to advance" (Weick, 1995, p. 192). Weick (1995, pp. 24–26) further observes that, ". . . people can know what they are doing only after they have done it . . . . The creation of meaning is an attentional process, but it is attention to that which has already occurred . . . Only when a response occurs can a plausible stimulus then be defined [to explain its occurrence]."

> The important point is that retrospective sensemaking is an activity in which many possible meanings may need to be synthesized, because many different projects are under way at the time reflection takes place (e.g. Boland, 1984). The problem is that there are too many meanings, not too few. The problem faced by the sensemaker is one of equivocality, not one of uncertainty. The problem is confusion, not ignorance. I emphasize this because those investigators who favor the metaphor of information processing (e.g. Huber, Ullman & Leifer, 1979) often view sensemaking, as they do most other problems, as a setting where people need more information. That is not what people need when they are overwhelmed by equivocality. Instead, they need values, priorities, and clarity about preferences to help them be clear about which project matters. Clarity on values clarifies what is important in elapsed experience, which finally gives some sense of what that elapsed experience means (Weick, 1995, pp. 27–28).

This meta-sensemaking view fits with the theory and empirical work in strategic management by Mintzberg (1978). Mintzberg defines strategy as observed patterns in past decisional behavior. Retrospection and research similar to Mintzberg's studies on realized strategies lead naturally into mapping strategy (see Huff, 1990) and program evaluation using strategy mapping tools (e.g. Stubbart & Ramaprasad, 1988). Consequently, retrospection may be viewed usefully as a valuable major tool for both sensemaking and program evaluation theory and research.

Figure 6 is intended to be a useful sensemaking tool for retrospective analysis – in comparing planned strategies with realized strategies. Note that the bottom right-hand cell reflects the use of future perfect thinking applied to the immediate past: what did not happen that was unplanned but might have (should have) occurred?

In one study on using future perfect thinking for sensemaking, Boland (1984), gathered a group of film-lending executives in 1980, provided them with accounting reports prepared for 1982–1985, and asked them to imagine it was July 21, 1985, and then to discuss what the film service had become and why. Boland reported that a major outcome of the study was that in trying to understand what had been done in an imaginary future, participants discovered that they had an inadequate understanding of an actual past. The study uncovered disagreements about the nature and meaning of past events that people did not realize had impeded their current decision making (from Weick, 1995, p. 29).

Compared with not using such a sensemaking tool, answering the issues raised in all four cells in Fig. 6 promotes reflective thinking and likely will result in more useful, deeper, sensemaking. "All the mysteries become clearer [and some are solved] when we pay more attention to how people plumb the past and to what outcomes they have *in hand* when they do so" (italics added, Weick, 1995, p. 185).

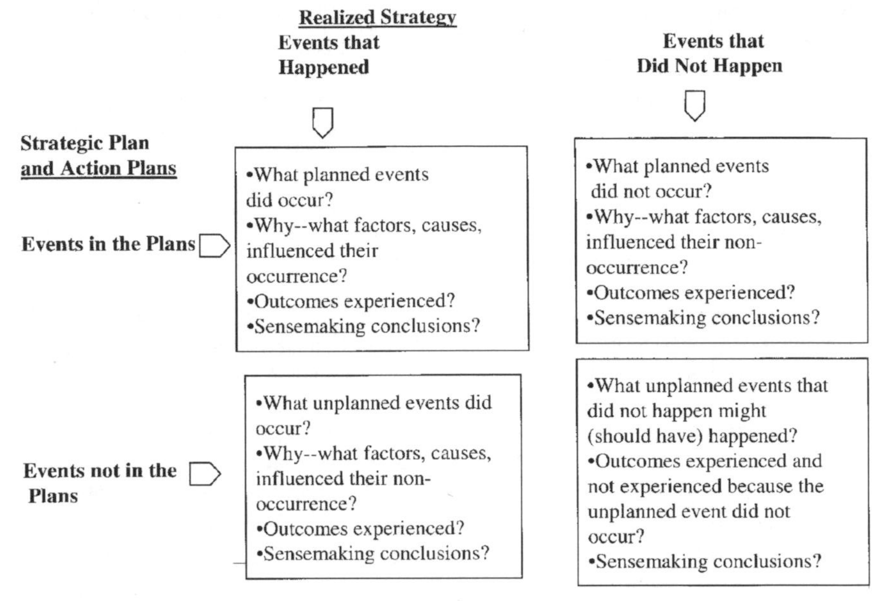

Fig. 6. Sensemaking Tool for Preparing Retrospective Commentary in the Annual Marketing Plan.

Evidence of Implementing

If Mintzberg (1988) is correct in proposing that realized strategies rarely are the same as planned strategies, then both the program administrators and performance auditors need to attend to what actually gets done. Implementing includes accomplishing and not accomplishing the completion of events scheduled in the marketing plan. Implementation questions include:

- What was done?
- By whom?
- Where?
- When?
- How was it done (i.e. the detailed steps and with what observed levels of skills)?
- What occurred next (i.e. the event chain following the event)?

Marketing implementation has been defined to mean the process that turns marketing plans into action assignments and ensures that such assignments are executed in a manner that accomplishes the plan's stated objectives (see Bonoma, 1985; Goetsch, 1993; Kotler, 1997). This definition is narrow and prescriptively oriented to effective marketing implementation. Our definition of marketing implementation focuses on what gets done – accomplishing and not accomplishing the completion of events scheduled and not scheduled in the marketing plan.

For purposes of sensemaking and performance auditing in program evaluation, three concerns about implementation are central: (1) learning the details done, and not done, in the streams of behavior that were actually implemented; (2) comparing these realized actions to the actions planned in the marketing plan; and (3) learning the immediate and downstream outcomes to the observed stream of implemented decisions and behaviors. The intended shift here is from prescription toward description.

For both program management and program evaluation, specific issues to be examined for describing implementation include the following points – implementation mapping, critical incidents, coaching and coordination, and nonevent analysis:

- Have service-performance-process maps (e.g. see Shostack, 1987), or causal policy maps (e.g. see Hall, Aitchison & Kocay, 1994), been created that accurately describe implemented behavior, including the downstream events occurring as a result of implementations? Such mapping tools help us to learn the intricacies and nuances that have occurred in recent implementations.
- What critical incidents (e.g. see Binter, Booms & Mohr, 1994; Price, Arnould & Tierney, 1995) occurred during the implementations of the program? What were the immediate and downstream influences (outputs) of these critical incidents? What combination of events (inputs) resulted in the critical incident?
- How many times did on-the-job coaching and coordination actions occur during program implementation? How many times did senior program executives and middle managers participate in on-line implementation of day-to-day activities in different departments of the program: weekly, monthly, quarterly, once per year?
- What unplanned events did not occur that should have been planned and implemented? Which of these unplanned non-events would have likely resulted in substantial favorable vs. unfavorable outputs for the program?

In summary, we emphasize three points. First, use ethnographic research tools to describe what is happening. Several worthwhile implementation research tools are available for creating "thick descriptions" (Geertz, 1973) of realized strategies. Several industries use such tools to achieve deep understanding – improved sensemaking – of implemented strategies (see Arnould & Wallendorf, 1994). These tools are helpful for both program administrators and performance auditors.

Second, in reviewing the seven performance audits of the tourism-marketing programs, we find little evidence is found in using such tools. The reports do not include detailed descriptions of the activities actually performed, or the policy maps of the decisions implemented.

Third, the argument that members of the legislative branch will not read such thick descriptions of implementation misses the more important point: the reason for implementation research is to improve sensemaking and decision making among program administrators and performance auditors. Meta-evaluation of implementation research helps to overcome Cook's (1997, p. 48) impression: "My impression, perhaps mistaken, is that we evaluators are less deliberate and systematic than others in accumulating substantive findings and in drawing inferences about the programs and classes of programs that seem to work best."

The U.S. findings of empirical studies on the reading and use of performance audit reports is consistent for both state and federal legislative members. The evidence is remarkably clear: these reports will not be read by legislators, with few exceptions (for a review of these studies, see Patton, 1997, especially Chapter 1). Even when read by legislators, the reports yield little to no impact. If the reports are going to be read and used to cause improvements in program performance, several stakeholders need to "buy in" to the findings, conclusions, and recommendations in the report. Possibly the only effective buy-in, performance auditing tool that works is Wholey's (1983) partnering approach. The Australian (Australia, 1993) performance audit is an example of this partnering approach applied to tourism-marketing program evaluation.

Impact Assessment of the Performance Quality, Conformance Quality, and Quality-in-Use of Program Activities and Program System Outputs

We use the term, "performance quality" to refer to a program's effectiveness and efficiency in delivering benefits to stakeholders, including clients (e.g. tourists and prospective tourists), public and tourist-industry related firms (in the form of improved services, wise use of tax-funded program expenditures, and increases in revenues and taxes paid by tourists), legislative and executive brands, and program managers and operators. Performance quality is distinct from conformance quality; conformance quality refers to product manufacturing and service delivery that meets mandated standards. Also, performance quality is distinct from "quality-in-use," that is, customer beliefs about the quality of the services they experienced. See Garvin (1987) for a discussion on the multiple dimensions of quality.

All three dimensions of quality are useful for evaluating the impacts of a tourism-marketing program. Though not described in the seven audit reports reviewed in this meta-evaluation, quality-in-use data (i.e. customer beliefs about quality experienced) have been collected in government-sponsored evaluation research of tourism-marketing programs (e.g. see Dybka, 1987 for a Canadian study on U.S. residents' quality-in-use beliefs about their visits to Canada). Conformance quality data are included in some of the seven audit reports; one example has been described briefly (in the North Carolina 1989 audit report: the non-delivery of the advertised visitor guide to 93,221 inquirers requesting the guide).

For achieving both triangulation for convergence and triangulation for divergence, using multiple indicators is best for measuring each of the three dimensions of quality. Indicators for measuring the effectiveness and efficiency of activities and systems in a tourism-marketing program are summarized in Exhibit 3.

Indicator (Variable Measured)		**Effectiveness Indicators**	**Efficiency Indicators**
1. Conformance measures: activities done in compliance to standards set (e.g., mandated) the first time in every instance activities attempted?		X	
2. Cost-benefit analyses, e.g., comparisons of cost per inquiry (CPI) ratios of different media vehicles used in the advertising campaign?		X	
3. Return on investment or net profit estimates, where net profit equals total tourist-related taxes minus government costs for the tourism-marketing program?		X	
4. Revenues for tourist-related industry firms generated due to the tourism-marketing program?	X		
5. Revenue per inquiry (RPI) estimated from tourism-advertising conversion studies?	X		
6. Government revenues (taxes) generated from tourism-related industries due to tourism expenditures?	X		
7. Tourist visits to the county, state, or country due to the tourism-marketing program?	X		
8. Quality-in-use measure of tourists' beliefs of quality of services experienced during their visits and their intentions to return and/or recommend visiting?	X		
9. Ratio-of-ratio comparison, e.g., CPI/RPI?	X		
10. Tourist awareness, attitude, and intentions to visit?	X		

Exhibit 3. Impact Assessment Indicators of Tourism-Marketing Programs.

Exhibit 3 describes four indicators of efficiency and six indicators of effectiveness. Effectiveness refers to the performance quality of the activity, the quality-in-use of the program, and the value of revenues generated from the program (sales and taxes resulting from the government tourism-marketing program). Efficiency refers to conformance quality, profits, and ratio comparisons of results due to the program.

The first indicator in Exhibit 3 refers to conformance measures, as described earlier. The second measure is cost per inquiry (CPI). CPI is equal to the cost assigned to running a given advertisement in a given media vehicle in a given time period divided by the total inquiries generated by this ad placement. Also, CPI estimates are reported sometimes in comparing tourism ad placements in competing media vehicles (Woodside & Soni, 1990) and competing categories of media (e.g. see Woodside & Ronkainen, 1982). Estimates of the net profit to government resulting from tourism-marketing programs are available in the tourism program evaluation literature (e.g. Minnesota, 1985; Woodside & Motes, 1981; Chap. 4 in Woodside, 1996).

Tourists' visits generated by the government tourism-marketing program (the seventh indicator in Exhibit 3) are estimated by research sponsored by the program managers for most programs. However, the performance audit report for Minnesota (1985, p. xii) notes that such research was *not* being linked to the return on investment (the third indicator) for Minnesota's tourism advertising campaign.

Faulkner (1997) offers an additional indicator of tourism-marketing performance that is not included in Exhibit 3. Faulkner advocates computing two difference-score indexes:

(1) "market bias index," that is, the share of a particular market compared to the share of a destination's visitors overall; market bias equals zero if the two measures are equal;
(2) an "index of change" in the visitors received from each market relative to the change in that market overall; this change index equals zero if the growth rate of visits from an origin is equal to the overall growth rate of trips made by that origin.

Faulkner (1997) displays the computed indexes for each origin in two-dimensional exhibits.

However, we do not recommend Faulkner's approach for measuring tourism-marketing performance. Using indexes based on difference scores causes severe measurement problems; consequently, most psychometric experts recommend that difference scores not be used (see Teas, 1993). For a given origin, such indexes do not provide estimates of market share or rate of growth. Neither do such indexes provide a measure of impact (i.e. effect size) of the tourism-marketing program. Thus, for Australia as a destination, a conclusion that Japan as a market bias index equal to 5.3 and an index of change equal to 0.6 provides little information and misleading implications. The impacts of presence vs. absence of specific marketing actions and expenditures on market share and change are not included in such index analyses.

Moreover, market share is *not* a particularly useful dependent or independent variable in performance measurement. "You're better off if you create and develop [profitable market] niches" (Clifford & Cavanaugh, 1985, p. 17). When focusing on market share, judge the impact of changes in market share relative to a destination's two key competitors – with and without the presence of specific market actions in a given tourism-marketing program. For example, the following steps can be used with market share as the dependent variable.

(1) Identify ten specific markets by a combination of origins and lifestyles.
(2) Estimate market shares for a destination for each of the ten markets for the three years prior to administering marketing actions to specifically attract visits from five of the ten markets.
(3) Randomly assign five of the ten markets to receive unique marketing programs to attract visitors from these markets to the destination; the other five markets represent "control" markets – they receive no unique marketing programs to stimulate visits.
(4) Continue to estimate market shares for each of the ten markets for the year that the marketing actions are implemented in the five randomly assigned markets.
(5) Continue measuring market share for fifth and sixth year even if the marketing actions are implemented only in year four.
(6) Using multiple regression analysis, estimate the effect size of the impact of the marketing actions on market share.

Figure 7 summarizes the results of a hypothetical example of following these five rules. Note in Fig. 7 that market share for destination X averages about 5% over years 1–3 for both the five treatment group markets and the five control group markets. The average market share for X is close to 11% during years 4–6 for the treatment group markets exposed to unique marketing actions in these markets. However, the average market share for X is still about 5% during years 4–6 for the five markets in the control group. An analysis of the findings in Fig. 7 for competitor Y confirms the view that the marketing actions performed by X during years 4–6 influenced market shares: Y's market share declined systematically in years 4–6 compared to previous years in the treatment group markets but not the control group markets.

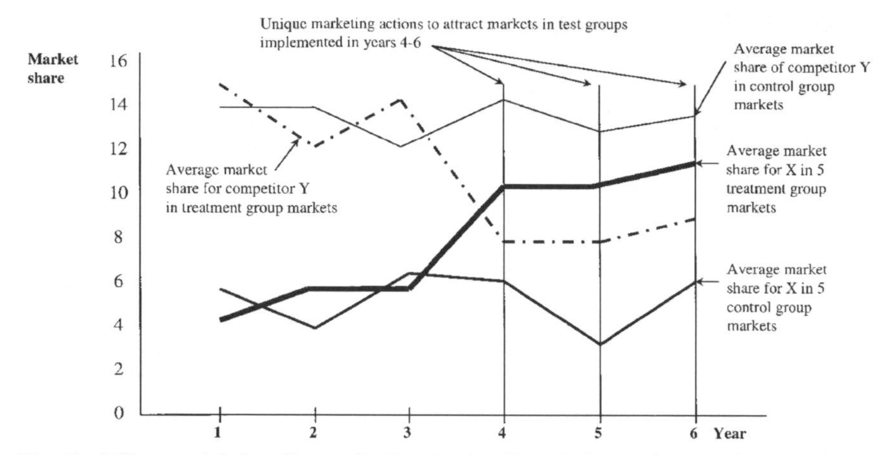

Fig. 7. Effect on Market Shares for Destination X and Competitor Y of the Marketing Actions by Destination X. *Note:* In this hypothetical example, systematic increases in market share for X are associated with the introduction of the marketing actions in years 4–6 in the treatment markets vs. years 1–3; in the markets in the control group no systematic change in market share for X occurred over the six years. Systematic market share decreases for competitor Y occurred in the treatment group but not the control group. Such findings represent strong evidence that the marketing actions implemented by X were effective in increasing market share for X and decreasing market share for Y.

Scientifically Measuring the Impact of Advertising and Marketing on Visits

The Minnesota (1985) performance audit of the state's tourism-marketing program included more knowledge of positivistic research for valuing results compared to the other six performance audits. For example, the Minnesota audit identified severe shortcomings of return-on-investment findings in the budget requests made by the tourism-marketing program director; these shortcomings are described in the results section. The ROI findings, communicated by the program director, were based on the results of a conversion research study done: Following the tourism season in Minnesota, a questionnaire was mailed to inquirers asking if they had visited Minnesota. Two shortcomings noted in the audit report are worth emphasizing here:

> The study assumed that all persons who said they had vacationed in Minnesota had come as a result of the office of tourism's promotion campaign. As a result, the study overestimated the return on investment because many would have undoubtedly vacationed in Minnesota regardless of the tourism promotion campaign. Furthermore, even for those who did not decide until after receiving tourism brochures, it is difficult to measure how much influence the brochures and advertising had on their vacation planning. . . . the study did not attempt to measure how many people vacationed in Minnesota as a result of the state's advertising campaign even though they

> did not request information from the tourism office. In addition, the study did not attempt to measure future benefits due to repeat visits or word-of-mouth advertising (Minnesota, 1985, pp. 91–92).

Unfortunately, the Minnesota performance-audit report does *not* indicate knowledge of true and quasi experiments for separating the impact of advertising from other factors on visits and expenditures (the positivistic theory valuing results and the empirical studies of Banks, 1965; Campbell, 1965; Caples, 1974; Ramond, 1966). In fact, the Minnesota performance audit reinforces the inaccurate and harmful myth that "It is not possible to separate the effects of tourism promotion from many other factors which affect tourism or travel spending, such as changing vacation preferences and changing business vacation patterns" (p. 90). True experiments are designs used to make such separations scientifically, to separate the impact of a new drug from other influence factors and medical research uses such designs widely (e.g. Salk and Sabin polio vaccine drug tests in the 1950s, see Woodside, 1990). In addition, such true experiments are used widely for measuring the unique impact of advertising on purchases; Caples' (1974) work is the seminal work on this topic.

Unfortunately, the use of true experiments has never been reported for measuring advertising effectiveness or efficiency by government tourism-marketing programs. The most likely reasons for not using scientific designs to measure impact include the following points.

(1) Lack of training and knowledge of the program evaluation literature in general among both program managers and performance auditors, especially lack of knowledge of the literature on measuring advertising effectiveness.
(2) The argument that the research costs are prohibitive – a false argument because true experiments can be built into the planned advertising campaign; moreover, the cost of a running a true experiment are no greater than the research costs of conversion research studies.
(3) Comfort with conversion research methods and "aided recall" advertising effectiveness studies.
(4) Fear and high-perceived risk of using another type of research study – especially if the results might indicate the advertising had little impact on causing visits that would not have occurred without the advertising.

By using a true experiment, program evaluators can scientifically estimate the separate impacts of exposures both to scheduled-media placed ads and to the visitor information guide. Such measurement includes estimating several influences: (1) being exposed vs. not being exposed to the advertising; (2) being exposed vs. not being exposed to a visitor's information guide (VIG); (3) the interaction effect of both being exposed and receiving a VIG vs. the additive effects of ad-exposure only or VIG-exposure only. The required true-experiment design to measure these impacts includes intentionally *not* exposing some identified prospects to the ads and the VIG, while exposing others, identified prospects to the ad only or the VIG only or both the ads and VIG. This research design is an expanded version of Exhibit 1. The expanded version is shown as Exhibit 4.

Because program and advertising managers may be unwilling intentionally *not* to fulfill requests from prospective visitors for free literature, two literature packets should be used: a minimal packet (a skinny brochure that is as close as allowable to a *no information-treatment condition*) and the regular packet (a thick, 50–200 pages, visitor's information guide full of details on special events, accommodations, restaurants, shopping, historical places, local touring routes, etc.). Using the research design displayed in Exhibit 4 answers three central questions.

- How many visits (and revenues, taxes, profits, etc.) resulted from the prospect-inquirers receiving the regular thick VIG vs. receiving the skinny brochure?
- How many visits resulted from the advertising program among persons exposed to the advertisements that did *not* request or receive the free VIG?
- How many visits (and revenues, taxes, profits, etc.) occurred because of the advertising program that would not have occurred without the program?

Taxes paid resulting directly from visits of tourists are government tourism-related revenues. What are the total direct tax-revenues resulting from the tourism advertising program? The tourism advertising impact on tax revenues can be estimated from Exhibit 4. Total tax revenues for each group:

Group 1:	(4 visits × $12) = $48	
	(50 visits × $9) = $450	total tax revenues: <u>$498</u>
Group 2:	(20 visits × $24) = $480	
	(100 visits × $18) = $1,800	total tax revenues: <u>$2,280</u>
Group 3:	(30 visits × $42) = $1,260	
	(200 visits × $18) = $3,600	total tax revenues: <u>$4,860</u>

Consequently, total tax revenues due to the regular advertising campaign vs. no advertising campaign equals $4,362 (4,860 − 498 = 4,362). The advertising campaign net tax revenues are $4.36 per 1,000 target customers exposed vs. not exposed to the advertising campaign (4362/1000 = 4.362). The combination of the ad plus the minimal brochure offer also increases tax revenues compared to the tax revenues resulting from no advertising campaign (2,280 − 498 = 1,782). The ad plus minimal brochure campaign results in a net tax revenue gain of $1.78

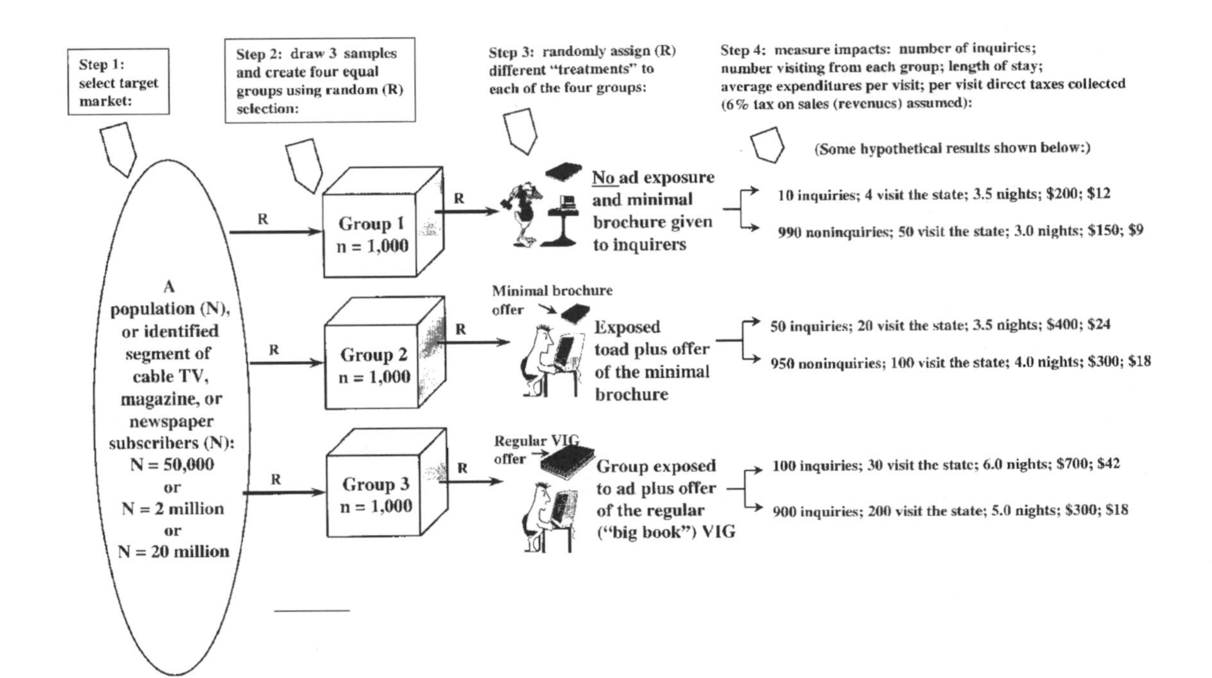

Exhibit 4. True Experiment Design to Measure the Effectiveness of Schedules-Media Advertising and Brochure Literature Sent to Prospects. *Note:* VIG = visitor's information guide, e.g. 100 pages. Note that the "conversion rate" of inquirers into visitors for each of the three groups is greater than 29%: 4/10 = 40% for Group 1; 20/50 = 40% for Group 3; 30/100 – 30% for Group 3. The conversion rate is lower for non-inquirers for each group: 50/90 – 5.3% for Group 1; 10.5% for Group 2; 22% for Group 3. Did advertising cause inquiries that would not have been made without the advertising? Yes: the inquiry rate was 1.0% for Group 1; 5.3% for Group 2; and 10.0% for Group 3. Use of this design requires high response rates (i.e. >50%) to a questionnaire sent to all members of the 3 groups; the questionnaire should not bias respondents by identifying the state sponsoring the study. Information regarding visiting several different states should be collected in the study to increase interest in answering the survey; to hide identity of the state sponsoring study.

per 1,000 target customers exposed compared to not having a tourism advertising campaign.

Which of the two advertising campaigns should be used? The answer depends in part on the cost of using the regular VIG vs. the lower cost of using the minimal brochure. We recommend performing a return-on-investment (ROI) analysis of each campaign. What is the net tax profit to the state for each of the three groups? Make some reasonable assumptions concerning costs. For example, assume that the total cost for creating, handling, and postage for the VIG is $4.00 per mailing, the per unit total costs for the minimal brochure is $1.25, and the cost of advertising in the schedule-medium is $400 per 1,000 subscribers for each ad. Addressing the ROI question and related issues (e.g. the increase in number of local areas toured by visitors receiving a VIG vs. visitors not receiving a VIG) is possible when powerful research designs are used to measure the impacts of tourism-marketing programs. (See Woodside, 1996 for a detailed real-life example of such effectiveness and for efficiency estimates for the tourism-marketing campaign for Prince Edward Island.)

Note that economic multiplier effects would need to be utilized in estimating the total impacts on revenues and taxes of the advertising campaign. Because the purchases by tourists result in purchases by tourism-industry service providers, and thus, a ripple effect of purchases occurs, we suggest a multiplier of 2.0–2.5 for estimating the total impact on revenues.

An additional point is worth noting from the hypothetical results shown in Exhibit 4. The impacts of a thick (100+ pages), detailed, VIG include substantial increases in the:

- average length-of-stays of visitors;
- average total expenditures of visitors;
- average number of local regions visited;
- intentions to return;
- and satisfaction with the services provided and the quality of the visit (for empirical evidence supporting these conclusions, see Chapter 4 in Woodside, 1996).

Evidence on Administering

As shown in Fig. 5, evidence on administering refers to vision-values, mission, training, coaching, and coordinating. Evaluations of success and failure in business organizations conclude that these administering issues are important (e.g. see Clifford & Cavanaugh, 1985; Scott, 1998). In a large-scale U.S. study of successful midsize companies, the senior executives in the firms studied:

> . . . set forth vividly the company's guiding principles – defining the ways value is to be created for customers, the rights and responsibilities of employees, and, most important, an overall affirmation of 'what we stand for. Philosophy and values are hard to measure – but their value can't be overstated. A statement of beliefs alone will not, of course, make a successful enterprise. Credos are an articulation of culture – not a substitute for it (Clifford & Cavanaugh, 1985, p. 13).

Clifford and Cavanaugh do, however, provide some concepts for measuring vision and values:

- earned respect – a sense that the enterprise is special in what it stands for;
- evangelical zeal – honest enthusiasm for creating and implementing plans;
- habit of dealing people in – communicating just about everything with everybody;
- view of profit and wealth-creation as inevitable by-products of doing other things well;
- leadership – executives show extraordinary commitment to the business, often to the point of obsession.

Related to these concepts, and the issues of training, coaching, and coordinating, are specific actions by program executives for achieving high levels of occurrence and to learn how administering impacts the four variables shown in the middle of Fig. 1. The senior program executives might ask:

- Do we have a written statement of vision and values unique for our state's tourism-marketing program?
- Do our employees, visitors, and legislators perceive us to have evangelical zeal?
- Do we communicate just about everything with everybody?
- What formal training programs have employees and executives completed this past year? What impacts result because of their participation in these training programs?
- Are we coordinating enough? Are we meeting weekly for coordination and for sensemaking? How often do executives communicate personally with prospective visitors and actual visitors?
- Have managers completed formal training on measuring program performance?
- Are managers and employees sometimes uncertain about what they really know about markets, prospective visitors, visitor behavior and attitudes? Or, do they think and act as if they have the answers and external data on markets and customers are unnecessary?

What activity results in improving coordination and sensemaking? Answer: participating – sharing and revising thoughts – in meetings. Sound sensemaking practice requires that people need to meet more often.

> That implication arises from the reluctance with which people acknowledge that they face problems of ambiguity and equivocality, rather than problems of uncertainty . . . . What will help them is a setting where they can argue, using rich-data pulled from a variety of media, to construct fresh frameworks of action-outcome linkages that include their multiple interpretations. The variety of data needed to pull this difficult task are most available in variants of the face-to-face meeting (Weick, 1995, pp. 185–187).

Thus, increase the quality of performance audits by measuring how often team meetings are held in the various departments of the tourism-marketing program, and the decisions and actions that follow from such meetings.

Work of the Performance Audit Team

For a comprehensive performance-audit report, include evidence that the work of the program-evaluation team included data collection on and interpretation of all five issues summarized in Fig. 5. Figure 8 illustrates this recommendation.

Note that Fig. 8 includes a sixth topic area: "Scanning other States and Industries by the Audit Team to Learn Best Practices." We are suggesting here that a comprehensive performance audit include such scans by the audit team. Only two of the seven audits below (i.e. Australia, 1993; Minnesota, 1985) included such scanning.

A comprehensive meta-evaluation would include evaluations on several comprehensive performance-audit reports – including describing the presence or absence of each of the six areas described. A comprehensive meta-evaluation would include commentaries on:

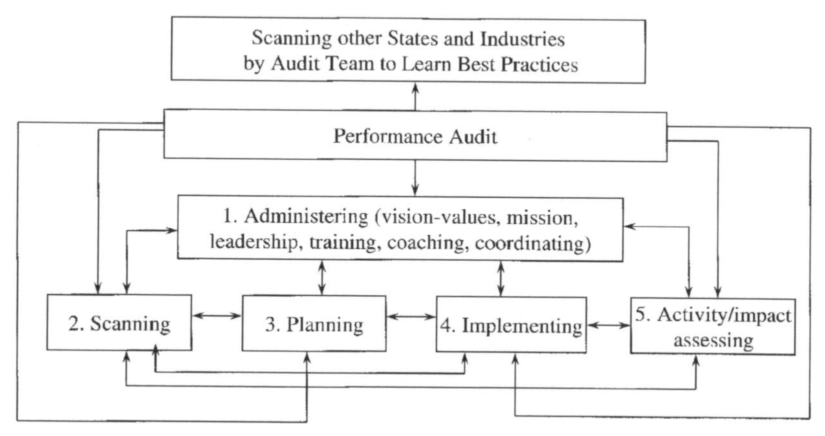

Fig. 8. Performance Audits of Marketing Department.

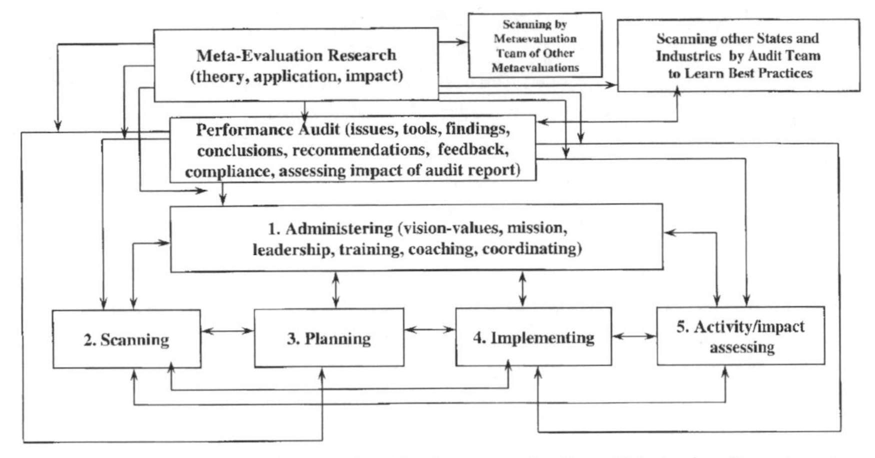

Fig. 9. Meta-evaluation Research of Performance Audits of Marketing Departments.

- the theory-in-use as displayed in each performance audit report;
- the scanning done by the audit team of other states and markets;
- the comprehensiveness of the performance audit reports;
- the subsequent impact of each performance audit report.

Figure 9 summarizes such comprehensive work by meta-evaluation researchers. Note that the box at the top in the middle of Fig. 9 refers to scanning by the meta-evaluation research team of other meta-evaluations (e.g. the meta-evaluation by Armstrong [1982]).

Reviews of additional meta-evaluation research studies on issues in program evaluation are likely to result in additional insights and nuances for theory and practice of program evaluation. Damson's review (reported in Patton, 1997) of the legislative use of 21 program evaluations is another example of scanning to uncover other meta-evaluations.

METHOD FOR ASSESSING PERFORMANCE AUDITS OF GOVERNMENT TOURISM-MARKETING DEPARTMENTS AND PROGRAMS

This section describes the development of two instruments for content analyzing the performance audit reports. This section also describes the procedural details of acquiring and analyzing the audit reports.

Content Analysis Measures

Based on the literature review of theories of program evaluation and propositions developed for evaluating performance audits of government tourism-marketing programs, we developed two separate content measurement forms for analyzing such performance audit reports. Content analysis is research techniques for the objective, systematic, and quantitative description of the manifest content of communication (see Berelson, 1952; Kassarjian, 1977 for descriptions of content-analysis studies).

Exhibit 5 shows the first content-measurement form, covering five categories of issues. The issues in Exhibit 5 are questions that are answered from reading and content analyzing each performance audit report.

Most of the specific issues included in Exhibit 5 have been described earlier. However, one concept in the "other meta-evaluations" section deserves elaboration: "meta-analysis." Meta-analyses are tools and theory used to accumulate knowledge across studies. "Scientists have known for centuries that a single study will not resolve a major issue. Indeed, a small sample will not even resolve a minor issue.

Theory in Use:
1. Program evaluation theory used/indicated by the performance audits?
2. Knowledge displayed in the audit report of availability of alternative theories?

Classification of Audit:
1. Comprehensive or limited in scope?
2. Focuses on evaluating which of the department activities shown in Fig. 5?
3. Within the focus of the performance audit, provides thick descriptions of activities done and strategies realized?
4. Evaluates research method used by department to measure effectiveness and efficiency of program? Audit conclusion sound?
5. Evaluates the findings of program impact provided by the department?
6. Reports whether or not this audit is the first, second, or third performance audit done of this department?
7. If previous audits done of this department, reviews findings and use of previous audit reports?
8. Reports scanning done by audit team of tourism-marketing program audits done by other states?

Assessment of Value and Impact of Audit:
1. Audit covers core issues related to effectiveness/efficiency impacts of activities and results of the program?
2. Audit includes explicitly, concrete, recommendations for change/improvement?
3. Response by program executives favorable or unfavorable toward recommendations in audit report?
4. Recommendations in report, if adopted, likely to result in substantial impacts on the quality of activities performed by the department?
5. Central issues raised but inadequately addressed in performance audit report?
6. Core issues not raised in the audit report?

Systems-Thinking and Database Marketing:
1. Systems thinking display in audit report, e.g., policy mapping?
2. Report covers implementation of database marketing?

Other Metaevaluations:
1. Was a search conducted to find other metaevaluations?
2. Any found? Focus issues in these metaevaluations?
3. Methods used for metaevaluation?
4. Findings? Was a metaanalysis performed on the reported findings? Specific meta-analysis results?
5. Value/relevancy of these studies for current metaevaluation?

Exhibit 5. Categories of Issues in a Meta-evaluation of Performance Audits of Government Tourism-Marketing Departments and Programs.

Thus, the foundation of science is the cumulation [sic] of knowledge from the results of many studies" (Hunter & Schmidt, 1990, p. 13).

Meta-analyses often include numerical estimates of the "effect size," the size of the impact, of an independent variable on a dependent variable. For example, one effect size would be the estimated share of inquirers responding to the advertised offer of a free VIG who converted into visitors who would not otherwise have visited. Given the large number of conversion studies in the scientific tourism literature, surprisingly no meta-analysis of reported conversion shares has been reported. Even though these studies do not attempt to measure the unique impact of advertising vs. no advertising inquiry-to-visitor conversions, they do report shares of conversion that can be analyzed using meta-analysis tools (e.g. see Chapter 7 in Hunter & Schmidt, 1990).

The second content analysis form includes multiple items for each activity/decision topic described earlier in the section on propositions for performance auditing tourism-marketing programs. These topics include: scanning; planning; implementing; program self-assessment of activities and impacts; and administering. Exhibit 6 includes the specific items included for the content analysis of each of these five topics.

Related to items in Exhibit 6, a core issue related to a meta-evaluation of performance audits is whether or not data are collected. One reason for not collecting such information is that the performance audit was planned to be limited in scope – what Wholey (1983) refers to as a "rapid feedback evaluation."

However, our suggestion is that all performance audits of tourism-marketing programs can and should include some data collection, evaluation, and reporting for all five major activities summarized in Exhibit 6. Even with a performance audit deadline of less than two weeks, such data can be collected, analyzed, and reported – while the performance auditors still focus mainly on very specific issues for a rapid feedback evaluation. In particular, we offer the suggestion because performance audits of tourism-marketing programs are usually completed less than once per decade, and dramatic changes and improvements in tourism-marketing programs are unlikely to occur without deep sensemaking learned from formal, broad-coverage, performance audits.

Procedure

We conducted searches of U.S. state government websites to learn of the completion of performance audits of tourism-marketing programs. We found six reports. We contacted state audit offices and requested copies of each of the reports. In addition, in August 1993, Peter Valerio, Manager, Strategic Analysis Unit of the

Activity/Decision	Did Auditor Collect?	Absence Noted?	Level Observed L1	Level Observed L2	Level Observed L3	Accuracy?
Scanning:						
• Evidence of formal scanning reports?						
• Formal SWOT assessment done?						
• Market-by-market analysis: S/D?						
• Trends identified formally?						
• In-depth research & knowledge of customer segments/targets?						
Planning:						
• Up-to-date written marketing plan for tourism-marketing program?						
• Marketing plan includes written milestones of schedule actions?						
• Evidence of "what if" reflection/analyses?						
• Formal testing of alternatives? Before and/or after execution?						
• Formal projections of cost/benefit analysis of planned actions?						
Implementing:						
• Evidence of on-the-job coaching and coordination and/or lack?						
• Have service-performance-process maps been created/ used?						
• Written monitoring guidelines established/followed?						
• Vital competencies done in-house?						
• Formal system for noting/responding to critical incidents?						
Activity/impact assessing:						
• Realized/planned strategy match?						
• Multiple dependent impact measures?						
• Cost/benefit analysis?						
• True experiment of marketing impact?						
Administering:						
• Written mission statement prepared?						
• Annual quantitative objectives established?						
• Evidence of training, coordination, coaching by senior executives?						

Exhibit 6. Key Issues in a Meta-evaluation of Performance Audits of State Tourism Development Departments. *Note:* SWOT = strengths, weaknesses, opportunity, threats; S/D = supply and demand analysis; Levels: L1 = found to be poorly done; L2 = found to be done.

Australian Tourist Commission (ATC) sent a copy of *Evaluation of the Australian Tourist Commission's Marketing Impact* – a 73 pages report with an additional eight appendices.

The seven performance audit reports were evaluated using two content-analysis instruments. Two researchers, trained in theories and research methods of program evaluation, prepared written answers in response to the written questions in the two instruments. Two researchers worked separately in reading the reports and writing their answers. They then compared their answers and resolved conflicts through discussions and referring to the reports until they reached consensus. Complete agreement among the written answers was ranged 77–92% across the ten topics in Exhibits 5 and 6.

Note that the items included in "Other Meta-evaluations" in Exhibit 5 represent a self-assessment among the meta-evaluation researchers; not surprisingly, agreement was 92% for the written answers to these items. Part of the concern with self-assessments is the limited range of observed findings. This concern can be addressed by requesting additional researchers, knowledgeable about program evaluation theory but not involved directly in the current study, to serve as external judges, Each of these external judges would read some (or all) of the performance audits, review the completed content analysis instruments, and provide comments on the accuracy of the completed forms. This external judging procedure was not used in the study reported here, and not including such an external review step represents a limitation of the study.

A META-EVALUATION OF A LIMITED FOCUS AUDIT

The 1996 Missouri Performance Audit Report of the Division of Tourism

We begin with an evaluation of an audit report limited in focus. Margaret Kelly, CPA, State Auditor, signed the letter of transmittal for this report. The contents of the letter included the following statements. "We have conducted a special review of the Department of Economic Development, Division of Tourism... As part of our review, we assessed the divisions' management controls to the extent we determined necessary to evaluate the specific matters described above and not to provide assurance on those controls."

The report itself includes the following sections:

- history and organization of the Division of Tourism (3 pages). This section includes a statement of purpose: "The purpose of the division is to promote the state's travel industry by encouraging visits by out-of-state vacationers and by encouraging Missourians to vacation in their home state";

- a summary of findings (1 page);
- a "presidential library funding" section (3 pages); summary: $1 million grant made improperly by the division; "... state lost potential interest revenue by disbursing the grant monies before expenditures were incurred...." The not-for-profit corporation receiving the $1 million from the Division invested more than $800,000 of these funds into fixed income investments;
- a "sales tax funding" section (3 pages); summary: funding to the Tourism Supplemental Revenue Fund exceeded the statutory limit by $4 million;
- an "expenditures and contracts" section (5 pages); summary: several different payments irregularities found; bid documentation insufficient;
- a "sponsorships" section (2 pages); summary: "Written contracts were not obtained to specify what the division would receive in exchange for paying over $33,000 for promotional items and sponsorship fees to promote tourism at various events."

Each section included a "WE RECOMMEND" subsection. For example, for the first section, "WE RECOMMEND the division seek reimbursement of the unexpended portion of the [$1 million] grant."

Each of these sections included specific findings and recommendations by the auditor. Also, "AUDITEE'S RESPONSE" subsections were included for each of the sections. The first of these responses begins by implying that the Division of Tourism did not commit an error in paying the $1 million: "The Division of Tourism clearly received $1 million for the purpose of funding the Truman Presidential Library in Independence, Missouri and made the payment to the library in compliance with the state of Missouri procedures." However, the next sentence in the division's response indirectly addresses the auditor's concerns about making grants, "In the future, the Missouri Division of Tourism will ensure that the contracts are written for all significant expenditures and that disbursements as a result of these contracts are made only as expenditures are incurred." Nothing is stated in the division's response to the auditor's recommendation for the "division to seek reimbursement of the unexpended portion of the grant."

Claims and counterclaims are made by the auditor and auditee in the second section, "Sales Tax Funding" with both quoting portions of state regulations. For example, the "AUDITOR'S COMMENT" following the "AUDITEE'S RESPONSE" is the following statement:

> While Section 620.467, RSMo does contain the language the division quotes, Section 620.467.2 RSMo 1994, states "[H]owever, such transfer in any fiscal year other than fiscal year 1996 shall not exceed three million dollars." Therefore, not only is there no authority to transfer anything in excess of $3 million to the Tourism Supplemental Revenue Fund, in any fiscal year other than 1996, that is expressly prohibited by statute (Missouri, 1996, pp. 12–13).

The next section of the 1996 Missouri performance audit is entitled: "FOLLOW-UP ON STATE AUDITOR'S PRIOR RECOMMENDATIONS." This section reviews several control problems noted in the 1994 audit report and the actions taken, if any, by the Division of Tourism between 1994 and 1996.

For example, the 1994 report includes the following findings. (1) Overpayment "by at least $50,876 to the advertising agency and the division made various payments to the advertising agency before the payments [made by the advertising agency to media firms] were actually done." (2) "The Department of Corrections (DOC) provided telemarketing services to the division. The division did not request copies of the actual telephone bills paid by the DOC to determine the validity of the amount reimbursed to the DOC by the division."

The auditee's responded to these criticisms by stating that the "division received a $20,823 credit from the advertising agency." The remaining amount of overpayments found by the auditor "was attributable to personnel classification errors" (p. 22). Also, the division discontinued the telemarketing contract with the DOC.

The final section of the audit report include three appendices: a schedule of appropriations; a comparative schedule of expenditures for five years (1992 through 1996); and a "comparative schedule of disbursements, transfers and changes in cash and cash equivalents" for 1995 and 1996. Appendix 2 shows advertising expenditures as 76% of the 1995 expenditures: $(\$5,950,181)/(\$7,854,393) = 75.76\%$. For 1996, if the $1 million library grant is excluded as non-recurring special funding, the advertising to total-expenditures ratio is 70%.

Meta-evaluation of the 1996 Missouri Performance Audit Report

The 1996 performance audit of the Division of Tourism by the Offices of the State Auditor of Missouri is concisely written with specific recommendations included. The report is exceptional in describing the history, organizational structure, several annual budgets, and expenditures. Moreover, the report is unique in providing a follow-up report on the findings, recommendations, and actions taken and not taken in a related previous performance audit report. Carrying out two performance audits of a state agency in less than five years has several potential benefits that outweigh related limitations. In particular, such nearly back-to-back audits provide necessary information for developing a theory of utilization of program evaluations (see Patton, 1997).

However, the Missouri performance audit report is limited to measuring efficiency for some relatively minor issues in the Division of Tourism. The big issues are not addressed in the 1994 and 1996 audit reports.

For example, the audit report does not address the effectiveness of the Missouri's tourism advertising program. The 1996 state tourism-advertising budget was $4.7 million, slightly more than 70% of total budget minus the one-time only $1 million library grant. Despite the funds spent on advertising, the report never asks what number of visits to Missouri resulted that would not have been made without these advertising expenditures?

	Metaevaluation	**Rationale/additional comments**
Theory in Use:		
1. Program evaluation theory used/indicated by the performance audits?	• Accounting theory of program evaluation	•A highly confrontational audit report and response by the tourism division.
2. Knowledge displayed in the audit report of availability of alternative theories?	• Other theories of program evaluation not mentioned	•Referring to Scriven, audit report indicates "bad" control by the division
Classification of Audit:		
1. Comprehensive or limited in scope?	• Limited in scope	• A ". . . special review"
2. Focuses on evaluating which of the department activities shown in Fig. 5?	• Focused on one topic	". . . we assessed the division's management controls . . . we assessed control risk"
3. Within the focus of the performance audit, provides thick descriptions of activities done and strategies realized?	• Yes	•E.g., grant made for $1 million by division to corporation before expenses incurred; division's response: future behavior will be different.
4. Evaluates research method used by department to measure effectiveness and efficiency of program? Audit conclusion sound?	• Documented analyses; interviewed division personnel	•Most conclusions sound as indicated by agreement and actions of division subsequent to the report
5. Evaluates the findings of program impact provided by the department?	• No. Topic not included in this special audit	
6. Reports whether or not this audit is the first, second, or third performance audit done of this department?	• Yes. This is the second audit of the division. First in 1994	• Serious problems noted as resolved; some found to still exist. E.g., division overpaid advertising agency $51,000 in 93/94; following 1994 audit and finding the Division received $21,000 credit; remainder "attributable to personnel classification errors" according to Division
7. If previous audits done of this department, reviews findings and use of previous audit reports?	• Yes	
8. Reports scanning done by audit team of tourism-marketing program audits done by other states?	• No	•A limited purpose audit
Assessment of Value and Impact of Audit:		
1. Audit covers core issues related to effectiveness/efficiency impacts of activities and results of the program?	• Partially	•Audit focused on cost controls, efficiency but not effectiveness
2. Audit includes explicitly, concrete, recommendations for change/improvement?	• Yes	•$3 million limit in transfer of sales tax revenues violation/recommendation
3. Response by program executives favorable or unfavorable toward recommendations in audit report?	• Highly unfavorable	Auditor author: "We will request Attorney General's opinion regarding the proper interpretation...."
4. Recommendations in report, if adopted, likely to result in substantial impacts on the quality of activities performed by the department?	• No.	•Issues address program efficiencies only, not effectiveness.
5. Central issues raised but inadequately addressed in performance audit report?	• No.	
6. Core issues not raised in the audit report?	• several core issues not raised	•Not raised: scanning; planning; administrating issues; control issues related to effectiveness
Systems-Thinking and Database Marketing:		
1. Systems thinking displayed in audit report, e.g., policy mapping?	• No.	No evidence of knowledge of systems thinking, policy mapping, database marketing tools.
2. Report covers implementation of database marketing?	• No.	

Exhibit 7. Results: Answers for the Issues in a Meta-evaluation of the 1996 Performance Audit of Missouri's Tourism-Marketing Program and the Division of Economic Development, Division if Tourism.

Activity/Decision	Did Auditor Collect?	Absence Noted?	Level Observed L1 L2 L3	Accuracy?
Scanning				
• SWOT assessment?	No	No		NA
• Market analysis: S/D?	No	No		NA
• Trends?	No	No		NA
• Research: customer segmenting?	No	No		NA
Planning				
• Written marketing plan?	No	No		NA
• Evidence of "what if" analyses?	No	No		NA
• Formal testing of alternatives?	No	No		NA
Implementing				
• Milestones established?	Yes	Yes	L1	Accurate
• Written monitoring guidelines established/followed?	Yes	Yes	L1	Accurate
• Vital competencies done in-house?	Yes	No	L1	Accurate
Activity/impact assessing				
• Realized/planned strategy match?	NA	No		NA
• Multiple dependent impact measures?	Yes	Yes	L1	Accurate
• Cost/benefit analysis?	Partial: costs	Yes	L1	Accurate
• True experiment of marketing impact?	No	No		NA
Administering				
• Quality of written mission statement?	Yes	No absent	Not evaluated	NA
• Annual quantitative objectives established?	No	No		NA
• Evidence of training, coordination, coaching?	Yes	Yes	L1	Accurate

Exhibit 8. Key Issues in a Meta-evaluation of the 1996 Missouri Performance Audit. *Note:* SWOT = strengths, weaknesses, opportunity, threats; S/D = supply and demand analysis; Levels: L1 = found to be poorly done; L2 = found to be done adequately; L3 = found to be done in superior manner and/or superior performance outcome noted. Accuracy = meta-analysis evaluation of validity of auditor's conclusion on activity/decision.

The issues that are addressed in the 1996 audit report represent less than $200,000 in expenditures, including the estimated loss of $50,000 in interest on the $800,000 invested of the prepayment of the $1 million library funds. Nearly all of the audit report's focus relates to potentially unnecessary expenditures and to loss of funds that total less than 3% of the 1996 budget.

The potential public relations disaster of using prisoners and the state Department of Corrections (DOC) for the division's tourism telemarketing program is not raised in the follow-up section of the 1996 audit report. Would the inquirer knowing that she or he was going to be giving her or his name and address to a prison inmate affect response rates to a tourism-marketing program? Would knowledge of the use of DOC personnel and a legal consultant for such telemarketing work be likely to increase or decrease response rates? How much of an impact would such knowledge have? No information is included in the follow-up section that these issues were addressed.

Exhibits 7 and 8 are summaries of our meta-evaluation of the 1996 Missouri performance audit report on the Division of Tourism. Based on the discussion presented above, our summary evaluation of the Missouri audit report includes the following categories and descriptions:

- coverage of history and budget: excellent
- clarity of writing: excellent
- theory applied: accounting (cost efficiency)
- coverage of issues: limited to minor issues
- style: confrontational
- likely size of impact: little to none
- overall performance quality of report: low.

A META-EVALUATION OF A PERFORMANCE AUDIT FOCUSED ON NON-PERFORMANCE OF CORE ACTIVITIES

The 1989 Performance Audit of the North Carolina Department of Commerce, Division of Travel and Tourism

The 1989 North Carolina performance audit is 21 pages including the letter of transmittal written by Edward Renfrow, State Auditor. The following discussion is limited to the portion of the report concerning the Division of Travel and Tourism but not the Film Office.

The letter by Mr. Renfrow implies a comprehensive audit, "The purpose of this audit was to examine the management and operation of the Division of Travel and Tourism and the Film Office and to provide you [the letter is addressed to Governor, Secretary of Commerce, and members of the General Assembly] with our findings and recommendations designed to continue the improvement of those areas."

The 1989 performance audit report for the North Carolina Division of Travel and Tourism includes a three page executive summary, a half-page of audit objectives, a half-page on audit methods and scope, eight pages of findings, and a three-page letter (dated November 16, 1988) of responses to the audit by Claude E. Pope, Secretary, North Carolina Department of Commerce.

The Executive Summary

The following are the key sentences referring to the Division of Travel and Tourism in the executive summary:

> The division was established to promote the growth and development of the state's travel and tourism industry. The division has an annual budget in excess of $6 million. Over $2 million fund the advertising program that promotes North Carolina as a travel destination. ... One of the division's major functions is mailing promotional literature [biggest expense: a visitor's information guide (VIG)] to people requesting travel information. The division's conversion study has shown that 51% of those who request information come to North Carolina within the next 12 months and their parties spend an average of $442 while in the State.

Unlike the 1996 Missouri audit report, the 1989 North Carolina audit report gives no detailed information on funding the division or on a schedule of expenses. The conclusion implied by the 51% conversion ratio is that an advertising-and-free-literature combination was the prime influence for the visits and for the subsequent average expenditure of $442. The audit report does not examine details of the conversion study. The audit report does not include questions on the scientific value of the research method used in the conversion study. Telling questions not raised in the North Carolina audit report include:

- what share of inquirers-who-visited North Carolina would have come if they had not been exposed to the advertising and had not inquired?
- what share of inquirers-who-visited NC would have come if they had not received the free literature they requested, or, if they had received a one-sheet, four-sided brochure, rather than a 100+ page VIG?
- what share of non-inquirers who were exposed to the advertising was influenced to visit North Carolina because of the advertising?
- how much additional trip expenditures occurred in North Carolina among visitors exposed to the advertising who received the free literature vs. (a) North Carolina

visitors not known to be exposed to the advertising and who did not request the ad-offer of free literature; and (b) North Carolina visitors who were exposed and requested, but did not receive, the free literature?

Although the one-group conversion research design (i.e. no control group) used in North Carolina conversion studies cannot answer these questions, true-experiment designs can, as described by Campbell and Stanley (1963) in program evaluation, Banks (1965) in marketing, Caples (1974) in advertising testing, and Woodside (1990) in tourism-advertising program evaluation (for details, see Exhibit 3).

Because of the main finding on the first page of the executive summary: "... that the division had failed to answer 93,221 reader service inquiries during the first part of 1988," the final question just raised is of particular interest. (Reader service inquiries are requests for ad-offered free literature that are received from the reader service postcards, also known as "bingo cards," in magazines.)

> Readers may request information from advertisers by circling the advertiser's number on the postcard. These requests are then forwarded to the division. It has been assumed that reader service inquiries are less valuable than the other forms of inquiries (toll-free telephone calls and coupons), because less effort is involved. Nonetheless, even if they are assumed to be half as valuable, the conversion study suggests that the potential travel revenue associated with these unanswered inquiries exceed $10.5 million (North Carolina, 1989, p. 2).

The Department of Commerce responded to this critique by arguing that, "Based on the division's survey of 70 people, the delay in responding had very little, if any, impact on their vacation plans" (North Carolina, 1989, p. 3). In reply the state auditor argued that:

> In our opinion, the critical finding is the division's failure to take action to remedy the problem until it accumulated 93,221 unanswered inquiries. The division's statement that the delay in responding had little short-term effect on vacations is inconsistent with the results of the conversion study. The conversion study was conducted by the advertising agency retained by the department. *If that study is inaccurate, that raises the question of whether the state benefits from generating and responding to these inquiries* (italics added, North Carolina, 1989, p. 3).

The immediate concern is put clearly by the auditor: how can the division both claim North Carolina tourist visits due to inquiries generated by the advertising and claim no impact due to inquiries generated by the advertising? Unfortunately, the even bigger issue is not addressed: no indication is found in the 1989 North Carolina audit report that the auditor is aware of the more powerful, true-experiment designs for estimating advertising-program impact in comparison to one-group conversion study designs.

The auditor's theory that some share of reader service inquiries is associated with visits is supported empirically. In a conversion study for Louisiana, 26%

of reader-service card inquirers compared to 36% of direct-response coupon inquirers, and 43% of toll-free call inquirers, converted into visitors (Woodside & Soni, 1988, p. 33).

Two additional points are worth emphasizing here. First, the main impact of the free literature sent to inquirers (i.e. the VIG) is likely to be on affecting the behavioral characteristics of the trip (e.g. on increasing the average expenditures made on the trip, the number of NC regions visited, building word-of-mouth advertising, the overall quality visitor beliefs about the visit, the satisfaction with the total visit experience); the impact of the VIG on generating visits is likely to be positive but much smaller than its impact on the behavioral characteristics of the trip. The combined net profit payoff from creating and providing such detailed information to ad-generated inquirers is likely to be substantial. Unfortunately, while strong evidence is available on the power of VIGs to influence visit characteristics, the theory is not tested via a true-experiment design.

Second, most conversion research studies include sending a sample of inquirers a cover letter and questionnaire that tell the reader that he or she is known to have requested literature. This procedure most likely causes decreases in response rates among inquirers who did not convert into visitors compared to the response rate among inquirers who did convert. Strong empirical evidence supports this proposition (e.g. Woodside & Ronkainen, 1984). The resulting conversion rate, when the sponsor of the study is the state from which the surveyed household member requested information will be very high, often 50% or higher. In comparison, conversion survey studies that do not identify a particular state as sponsoring the study result in lower conversion estimates, for example, 31% overall for the 1987 Louisiana conversion study (Woodside & Soni, 1988).

If the questionnaire and cover letter to refer to travel to several states (e.g. 5–8 southern states), the inquirer receiving the questionnaire does not need to be informed that he or she is known to have requested literature from any one state. Also, additional valuable information on customer intentions and behavior toward competing tourist destinations can be collected in the questionnaire.

The remainder of the executive summary highlights other problems found by the auditor and presents the response by the department. These highlights included the following excerpts.

> In the first half of 1988 the division exhausted its supplies of the calendar of events, the accommodations directory, and the state map. As a result 175,000 inquiries received only the attractions study and a form letter stating that the other materials would be sent later. The division failed to provide complete and timely responses to inquiries which potentially could result in some $39 million of travel revenue . . . . A major factor leading to the exhaustion of inventories was the division's failure to reorder publications on a timely basis" (North Carolina, 1989, p. 3).

The following statement is the complete response made by the department, "The division has established a printing schedule" (North Carolina, 1989, p. 3).

The final concern by the auditor was the lack of comprehensive written agreements with industry co-sponsors of business trips to cities outside North Carolina to promote visits to North Carolina. The department's response to this concern was that, "The division is executing written agreements with the co-sponsors of the travel missions" (p. 4).

Audit Objectives

The performance audit carried out does not reflect the audit objectives set forth in the report. The half-page statement of audit objectives presents the following statement first, "Under the North Carolina General Statues, the State Auditor is responsible for reviewing the economy, efficiency, and effectiveness of state government programs" (p. 5). The largest part of the half-page statement describes the method used to conduct the audit. The audit reports states, ". . . we reviewed and analyzed: the divisions' management of its publishing program, including the timeliness of its reorders; the effectiveness and efficiency of the division's promotional advertising program; and the division's Welcome Center operation." The report provides no description of findings or evaluations of the NC welcome center operation.

Audit Methods and Scope

The next section in the report is titled, "AUDIT METHODS AND SCOPE." The first sentence references the theory frame for the study:

> Our performance audits are conducted in accordance with the Standards for Audit of Government Organizations, Programs, Activities, and Functions issued by the Comptroller General of the United States and endorsed by the American Institute of Certified Public Accountants, except that no reviews of general or application controls of the data processing system were completed during this audit (p. 6).

The statement further describes the method used for the audit: review of policies and procedures and interviews with management, staff, and other parties. The last two sentences in this section point out that "This audit does not provide a review of any financial statements of the department. Such a review is incorporated in our comprehensive annual financial report of the State of North Carolina" (p. 6).

The report describes no other theories of program evaluation. The description of the method used for the audit lacks the necessary detail for learning what actions the audit office took in collecting data for the audit. The usefulness of the audit report is restricted severely because the report does not include a review of financial statements of either the department or Division of Travel and Tourism.

The method description indicates that a shallow review was performed to measure the effectiveness and efficiency of the division.

The Main Body of the Report

Following the audit methods and scope section, eight pages of description of the activities of the Division of Travel and Tourism provide more details of the problems described above and more responses by the department. These eight pages are entitled, "Chapter 1 – Division of Travel and Tourism." Page eight mentions that, "The programs described above are in line with the efforts of other states. The vast majority of states use advertising to promote their state to travelers and assist travel writers." However, no details support this comparison to other states. The chapter includes no bench marking or specific comparisons to best practices; elsewhere in the report no such comparisons occur.

The main body of the report includes a footnote on page nine to the following statement made in the division's *Resource Manual* (1988): "The life-blood of the travel industry is response to advertising, and a phenomenal 51% of those who directly answer North Carolina's ads (i.e. they inquire at this office) travel to the state within 12 months. They (average party of 2.7 persons) spend more than $442 while here." The footnote:

> These statistics come from the 1984 conversion study done for the division by the advertising agency. The agency surveyed a sample of people who had asked the division for travel information. The survey also provided a demographic profile of the people who are traveling in North Carolina (North Carolina, 1989, p. 9).

The audit report does not raise the issue of lack of independent participation in evaluating the work of the advertising agency. Might the advertising agency be biased in designing an advertising effectiveness study to examine its own performance? Earlier in the audit report the following statement appears: "The division oversees the overall direction of the advertising program. Actual design and implementation of the program is the responsibility of the advertising agency of McKinney and Silver of Raleigh, North Carolina."

The main body of the audit report identifies lower productivity in the inquiry fulfillment section as the primary cause for not responding in a timely manner to 93,221 inquirers. "A primary cause of the problem has been a fall-off in the productivity of the inquiry section during the first five months of 1988" (North Carolina, 1989, p. 10).

The Response by the Secretary

A letter response from Claude E. Pope, Secretary, Department of Commerce, is appended to the audit report. In his letter Mr. Pope mentions that "approximately

70,000" ad-generated inquiries were from reader service cards, and he reiterates that 70 persons were contacted from these cards "to find out what impact the delay in providing the requested material had on their vacation plans for the summer season just ended." He reports "... very little, if any, effect ..." Mr. Pope's letter includes no details on the follow-up study, such as the following issues:

- How many attempts to interview persons were necessary to result in 70 interviews?
- What problems occur with generalizing results for a population greater than 70,000 based on responses from 70 persons?
- What biases might occur when persons are asked if the information requested and not sent/received affected their summer travel plans?
- What factors account for the difference of more than 23,000 inquiries in the 70,000 reader service inquirers not receiving the requested literature and the 93,221 mentioned in the audit report?
- Why does Mr. Pope make no response on the incomplete response mailing sent to 175,000 additional inquirers?

Mr. Pope makes no reference in his letter to the estimated loss of revenues due to the lack of fulfilling the literature requests.

Mr. Pope writes, "The productivity of the individuals working at the Inquiry Section during the first five months of 1988 was not down; there were two fewer people working at Inquiry during this period than in the previous years" (p. 11). The report does not discuss the possibility of hiring temporary workers to work in the inquiry section. The auditor felt compelled to reply to Mr. Pope's response: "In our opinion, this response of the department requires a reply. Our concern with the response is its tendency to divert attention from the most critical issue: which is the division's failure to take action to remedy the problem until the 93,221 inquiries accumulated" (p. 12).

Meta-evaluation of the 1989 North Carolina Performance-Audit Report

The 1989 North Carolina performance audit identifies substantial numbers of inquiries unanswered and partially answered by the division in 1988. The report includes the suggestion that the estimated loss of these problems exceeds \$49 million in travel revenues (\$10.5 million + \$39 million). Although how much travel revenue was lost is debatable, most likely some amount was lost due to poor coordination of literature fulfillment.

The audit report does not provide estimates of the total value of the state's tourism-marketing program. Is North Carolina's tourism-marketing program profitable (i.e. the taxes net of expenses of the program are a substantial return on the funds invested)? The audit report does not include any specific benchmarking analysis or best practice analysis of government tourism-marketing programs. The report indicates no awareness of theories of program evaluation other than accounting audits, for example, Campbell's (1969) theory applied to the need for scientific tests of advertising impact and how to conduct such tests are not covered in the report.

The North Carolina audit report presents nothing on the scanning activities or planning activities of the Division of Travel and Tourism. Consequently, a discussion of discrepancies between planning and reaching milestones is limited to the failure to respond to inquiry requests from the advertising program. The audit report describes performing a review and analysis of the division's welcome center operation. Neither review nor analysis of this operation is included in the report.

Exhibits 9 and 10 provide a summary of several points in this meta-evaluation of the NC audit report. Based on the discussion presented above, our summary evaluation of the North Carolina audit report includes the following categories and descriptions:

- coverage of history and budget: poor
- clarity of writing: excellent
- theory applied: accounting only
- coverage of issues: limited to a few issues
- style: confrontational
- likely size of impact: little to none
- overall performance quality of report: low

The style of the 1989 North Carolina audit report is confrontational. The report focuses on estimated losses of travel revenues caused by poor coordination in responding to ad-generated inquiries. The response by Mr. Pope, Secretary of the Department of Commerce, conflicts on substantive issues of the size of the problem; he does not respond to the estimated lost revenue impacts. He does attempt to divert attention from the auditor's findings and minimize problems.

The audit report does not describe activities the division is doing particularly well, if any. The report does not describe recent improvements, if any, in the operations of the division. The report includes no suggestions for improving the division's scanning or planning activities. The audit report does not discuss the lack of a modern database-marketing program in the division.

The audit report does not discuss the potential public relations and ethical problems of "using women prison inmates to help package responses" (p. 3). Would news stories on this fulfillment program affect the number of inquiries and, therefore, visits to the state?

	Meta-evaluation	Rationale/ additional comments
Theory in Use:		
1. Program evaluation theory used/indicated by the performance audits?	• Accounting theory of program evaluation	• A confrontational audit report and response by the tourism division.
2. Knowledge displayed in the audit report of availability of alternative theories?	• Other theories of program evaluation not mentioned	• Referring to Scriven, audit report indicates "bad" control by the division
Classification of Audit:		
1. Comprehensive or limited in scope?	• Limited in scope	Focus is on poor administrating and loss of potential visits due to the failure to fulfill literature requests
2. Focuses on evaluating which of the department activities shown in Fig. 5?	• Focused on a ltd number of topics	
3. Within the focus of the performance audit, provides thick descriptions of activities done and strategies realized?	• No	
4. Evaluates research method used by department to measure effectiveness and efficiency of program? Audit conclusion sound?	• Partially	• Audit identifies conflicting assumptions made by the division on the value of converting inquiries into visitors
5. Evaluates the findings of program impact provided by the department?	• Partially	• Only for inquiries not answered
6. Reports whether or not this audit is the first, second, or third performance audit done of this department?	• No	
7. If previous audits done of this department, reviews findings and use of previous audit reports?	• NA	
8 Reports scanning done by audit team of tourism-marketing program audits done by other states?	• No	Claim is made that such scanning was done but unsupported in the report
Assessment of Value and Impact of Audit:		
1. Audit covers core issues related to effectiveness/efficiency impacts of activities and results of the program?	• Partially, focused on effectiveness in program causing revenues	• No efficiency analyses; central issue not raised how many more visits occurred that would not have occurred without the advertising
2. Audit includes explicitly, concrete, recommendations for change/improvement?	• Yes	
3 Response by program executives favorable or unfavorable toward recommendations in audit report?	• Unfavorable	• response only partially matches with problems raised
4. Recommendations in report, if adopted, likely to result in substantial impacts on the quality of activities performed by the department?	• Yes	Timely fulfillment program likely to influence number and quality of visits
5. Central issues raised but inadequately addressed in performance audit report?	• Yes	• Limited value of conversion study; welcome center operation?
6. Core issues not raised in the audit report?	• Yes	• What are the net tax revenues and costs/benefits of tourism-marketing program?
Systems-Thinking and Database Marketing:		
1. Systems thinking display in audit report, e g, policy mapping?	• Very little	
2. Report covers implementation of database marketing?	• No	

Exhibit 9. Results: Answers for the Issues in a Meta-evaluation of the 1989 Performance Audit of North Carolina's Tourism-Marketing Program.

Activity/Decision	Did Auditor Collect?	Absence Noted?	L1 L2 L3	Accuracy?
Scanning				
• SWOT assessment?	No	No		NA
• Market analysis: S/D?	No	No		NA
• Trends?	No	No		NA
• Research: customer segmenting?	No	No		NA
Planning				
• Written marketing plan?	No	No		NA
• Evidence of "what if' analyses?	No	No		NA
• Formal testing of alternatives?	No	No		NA
Implementing				
• Milestones established?	Partially	Yes	L1	Accurate
• Written monitoring guidelines established/followed?	Partially	Yes	L1	Accurate
• Vital competencies done in-house?	Yes, indirectly	No	L1	Accurate
Activity/impact assessing				
• Realized/planned strategy match?	Yes	No	L1	NA
• Multiple dependent impact measures?	Yes	Yes	L1	Accurate
• Cost/benefit analysis?	Partial: costs	Yes	L1	Accurate
• True experiment of marketing impact?	No	No		NA
Administering				
• Quality of written mission statement?	Yes	Not evaluated		NA
• Annual quantitative objectives established?	No	No		NA
• Evidence of training, coordination, coaching?	Yes	Yes	L1	Accurate

Exhibit 10. Key Issues in a Meta-evaluation of the 1989 North Carolina Performance Audit of the Division of Travel and Tourism. *Note:* SWOT = strengths, weaknesses, opportunity, threats; S/D = supply and demand analysis; Levels: L1 = found to be poorly done; L2 = found to be done adequately; L3 = found to be done in superior manner and/or superior performance outcome noted. Accuracy = meta-analysis evaluation of validity of auditor's conclusion on activity/decision.

A META-EVALUATION OF AN AUDIT FOCUSING ON RESEARCH METHOD

The 1985 Performance Audit of the Minnesota Tourism Development Program

The preface in the 1985 Minnesota performance-audit report includes the signatures of James R. Nobles, Legislative Auditor, and Roger A. Brooks, Deputy Legislative Auditor. The preface states, "The Department of Energy and Economic Development [DEED] is the principal state agency responsible for state economic development and tourism promotion programs and is funded primarily through the state's General Fund" (Minnesota, 1985, p. iii).

The Minnesota audit report includes a 12-page executive summary followed by a 236-page report. The second half of the report focuses on the "Iron Range Resources and Rehabilitation Board [IRRRB]." We limit the following analysis to the first half of the report that is focused on DEED. However, several activities of the IRRRB relate directly to tourism-marketing and are described negatively in the audit report; for example, "The IRRRB should halt plans to construct the science museum, the historic village and railroad, and the golf course, and other amenities for the proposed hotel development" (p. xvi).

The following are the main tourism-marketing related findings in the audit report sections focused on DEED:

> The department's expanded tourism promotion efforts appear to have been generally successful in stimulating the state's economy, although not as successful as program advocates suggested. Our best estimate is that the additional state expenditures on promotion and advertising are generating on average, at least $5 in additional tourism spending per tax dollar spent.
>
> The effectiveness of the 1985 television campaign needs to be closely monitored to ensure that future promotional expenditures are targeted to those markets and media that bring the highest return. We also recommend that the Minnesota Office of Tourism develop a methodology for estimating the return on investment for major advertising campaigns, particularly those conducted outside Minnesota. The office has in place some of the studies and surveys needed to estimate the return that the state is receiving for its tax dollars. However, the office needs to combine survey results with a methodology similar to but more extensive than the one we used in this report. We believe it is important to know which of the various expanded tourism promotional efforts have been worth their costs. (p. xii)

Note that this statement in the Minnesota audit report is a more sophisticated view (i.e. it uses a theory of tourism-marketing performance auditing) than found in the Missouri (1996) and North Carolina (1989) reports. Focusing on estimating the return on investment of tax dollars of tourism-marketing expenditures sets the Minnesota report apart from the Missouri and North Carolina report. However, note that the quotation above does not estimate net tax dollars generated by the promotion and advertising expenditures but rather estimates $5 in additional tourism spending per tax dollar spent.

Chapter 1 in the report starts with a background section. This section describes the state's approval of a $6.1 million increase in the Office of Tourism's biennial budget for use in attracting additional tourists to Minnesota. This increase was one outcome of "a severe economic recession and increased competition among states for new business development . . ." (Minnesota, 1985, p. 1).

The second section of Chapter 1 in the report implies that the audit is comprehensive, not limited in scope. "Instead, this report focuses on whether economic development programs have been designed and administered so that the state obtains the maximum benefits possible." The report does not describe stakeholders other than the state.

The audit report includes neither a method section nor does the report refer to a particular grounding in theory of program evaluation. However, the numerous tables on revenues and costs, and the ROI discussions, imply an accounting and financial foundation for the study.

The budget data in Table 2.1 of the audit report indicate substantial increases in tourism-marketing expenditures:

Year	Budget in $ Millions
1983	$1.2
1984	$3.9
1985	$4.7

The following theoretical assumptions and audit issues conclude Chapter 2 of the report. "It is believed that with the increased awareness about Minnesota as a vacation area, more Minnesotans will vacation in the state and more tourists from other states and countries will be attracted to Minnesota, thus stimulating economic activity for tourism-related businesses" (Minnesota, 1985, p. 15). "In reviewing these efforts [DEED's tourism-marketing program], we asked:

- How successful are promotional activities in stimulating economic activity?
- How much benefit is the state receiving from its investment in promotion?
- Are the programs targeted to audiences that are likely to be influenced to consider Minnesota in their vacation or business plans?" (p. 15)

Note that the issue of target audiences is a sophisticated question not raised in the Missouri and North Carolina audit reports. Given that performance may vary substantially among different tourism-marketing activities, such a question indicates thoughtful sensemaking by the Minnesota auditors.

Chapter 6, "Tourism Promotion," in the Minnesota (1985) report is the main section on the state's tourism-marketing program. Chapter 6 includes "Rationale for State Tourism Promotion;" this section describes two theories: one supporting tourism promotion and one opposed. The bottom-line reason for advocates of tourism-marketing is that "The net increase in spending within Minnesota [caused by tourism-marketing] will stimulate the state's economy."

A negative reason for tourism-marketing expenditures is that opponents "question whether benefits gained at the expense of other states will last. Increasing Minnesota's tourism promotion budget may cause other states competing for the same tourists to increase their budgets. The net effect might not be beneficial to any of the states because the impact of their promotional campaigns would offset one another" (Minnesota, 1985, p. 87). The theoretical causal chain in this argument assumes implicitly that tourism-marketing does not affect the total tourism-revenue pie, that all states have benefits equally attractive for tourists, and that all states have equal abilities in spending tourism-marketing funds. All three of these implicit assumptions are likely to be inaccurate.

The report poses an important concern, "... we believe it is useful to focus the debate on whether increased tourism promotion has produced significant benefits for Minnesota and whether further increases [in tourism-marketing expenditures] are in the best interests of the state" (Minnesota, 1985, p. 88). What is valuable in particular about the Minnesota audit report is that it describes two sets of assumptions (i.e. two theories) on the impacts of tourism-marketing programs. Posing such a debate indicates that the auditors do not assume one view is always accurate – a valuable stance for achieving improved sensemaking.

The following comments are a brief summary of the two auditors' central observations and findings. First, the auditors examined the tourism office 1982 conclusion that the state tourism promotion returned $43 in benefits for each dollar invested during fiscal year (FY) 1981. After developing a "modified inquiry-conversion method," and after calculating "The average return to tourism advertising during FY 1984 ... by dividing tourism expenditures of $32.4 to $36.4 million by costs of $2,270,000, the auditors conclude that the estimated average return is between $14.30 and $16.00."

The audit report describes "several major limitations" in the previous study that resulted in the 43:1 return on advertising expenditures. These limitations are summarized here; they include the following points (Minnesota, 1985, p. 91):

- the conversion study did not measure the marginal return from the increased state spending on tourism promotion;
- the study failed to consider the effect of non-response bias and, thus, substantially overestimated the percentage of persons who vacationed in Minnesota;
- the results are based on responses to a single mailing;
- the study assumed that all persons who said they had vacationed in Minnesota had come as a result of the office of tourism promotion campaign;
- the study did not consider other factors that would have resulted in a higher estimated rate of return; for example, the study did not attempt to measure how many people vacationed in Minnesota as a result of the state's

advertising campaign *even though they did not request information from the tourism office*.

The audit team applied conversion survey data for 1984 to estimate tourism expenditures. "After each advertising season, the tourism office surveys a sample of persons who telephone or write to the office and request tourism brochures" (Minnesota, 1985, p. 111). The audit team did its own data analysis of the surveys. "Tourism expenditures were calculated by multiplying the following factors together:

- the number of inquires;
- the percentage of those surveyed who vacationed in Minnesota;
- the percentage of those surveyed who said that the brochure helped them decide to vacation in Minnesota;
- the average amount those surveyed said they spent during their vacation in Minnesota" (Minnesota, 1985).

Although this method has improvements over almost all tourism-marketing conversion studies; nevertheless, it includes telling weaknesses compared with the use of a true experiment design with advertising treatment and control groups. Among others, these weaknesses include the following points:

- assuming that respondents are able to know whether or not the brochure helped them decide (a leading question);
- the biases in asking only about travel to Minnesota and reminding the survey respondents that they are known to have requested the brochure;
- the inability to measure the impact of the advertising on persons exposed to it who did not request the brochure;
- and, not adjusting for the over-estimate in conversion caused by the non-response bias, due in part to the use of a single mailing of the questionnaire.

The audit includes a section on "planning and research activities" of the office of tourism. The report includes giving credit to the office for engaging in scanning and planning activities (pp. 102–103). However, the audit report does not state whether or not the office of tourism prepares a written marketing plan. The report does point out improvements made by the office in using conversion research to measure advertising effectiveness: "Beginning in 1983, the tourism office reduced non-response bias by using two [survey] mailings instead of one mailing. Since the [tourism research] literature indicates that non-response bias for surveys with two mailings can still be significant, the tourism office needs to do more analysis of non-response bias to improve accuracy of the survey results" (Minnesota, 1985, p. 104).

	Meta-evaluation	Rationale/additional comments
Theory in Use:		
1. Program evaluation theory used/indicated by the performance audits? report	• Accounting theory of program evaluation	• A confrontational audit and response by the tourism division
2 Knowledge displayed in the audit report of availability of alternative theories?	•No specific foundation report for program evaluation cited	•Referring to Scriven, audit indicates "bad" control by the division
Classification of Audit:		
1. Comprehensive or limited in scope?	• Wide but not comprehensive in scope	Focus is on improving the quality in estimating
2. Focuses on evaluating which of the economic department activities shown in Fig. 5? program	• Focused on measuring results, less on scanning and planning	impact of advertising
3. Within the focus of the performance audit, provides thick descriptions of activities has done and strategies realized? document	•No	Unclear if Office of Tourism a marketing planning
4. Evaluates research method used by department to measure effectiveness and advertising efficiency of program? Audit conclusion sound?	•Yes, works hard to improve method for measuring impact	• Two theories of described with a suggested resolution of conflict
5. Evaluates the findings of program impact provided by the department? estimated	•Yes	• Substantially reduces impact
6 Reports whether or not this audit is the first, second, or third performance audit done of this department?	• No	
7 If previous audits done of this department, audits reviews findings and use of previous audit reports?	• NA	• No mention of previous on Office of Tourism
8 Reports scanning done by audit team of tourism- made by marketing program audits done by other states? extensive	•No	• No claims of scanning audit team but they did scanning as shown in their references to the advertising effectiveness literature
Assessment of Value and Impact of Audit:		
1. Audit covers core issues related to net effectiveness/efficiency impacts of activities and results of the program?	• Partially	• No efficiency analyses on taxes caused by advertising; central issue not raised how many more
2. Audit includes explicit, concrete, not recommendations for change/improvement?	• Yes	visits occurred that would have occurred without the advertising
3. Response by program executives favorable or unfavorable toward recommendations in audit report?	• None included	
4. Recommendations in report, if adopted, likely to result in substantial impacts on the quality of activities performed by the department?	• More than a little but not substantial	• How to scan and plan not covered; scientific testing advertising not covered
5. Central issues raised but inadequately addressed in performance audit report? revenues	• Yes, how to measure impact scientifically	• What are the net tax and costs/benefits of tourism-marketing program? Database marketing, systems thinking?
6. Core issues not raised in the audit report? • Yes		
Systems-Thinking and Database Marketing:		
1. Systems thinking display in audit report, e g. policy mapping?	• Very little	
2. Report covers implementation of database marketing?	• No	

Exhibit 11. Results: Answers for the Issues in a Meta-evaluation of the 1989 Performance Audit of Minnesota's Tourism-Marketing Program.

Meta-evaluation of the 1985 Minnesota Performance Audit Report

The 1985 Minnesota tourism-marketing performance audit focuses on valuing results from a post positivism Scriven (1980) perspective. The auditors spent the majority of their efforts on improving the estimation of the return on investment of the state's tourism-marketing program. However, they do devote a little reporting space to describe briefly the scanning, planning, and implementing activities of the state's Office of Tourism. The audit report is less confrontational than the Missouri (1996) and North Carolina (1989) reports. Surprisingly and disappointingly, the Minnesota audit does not contain a response from DEED.

Activity/Decision	Did Auditor Collect?	Absence Noted?	L1 L2 L3	Accuracy?
Scanning				
• SWOT assessment?	No	No		NA
• Market analysis: S/D?	No	No		NA
• Trends?	No	No		NA
• Research: customer segmenting?	No	Yes	L1	NA
Planning				
• Written marketing plan?	?	No	L2	NA
• Evidence of "what if' analyses?	No	No		NA
• Formal testing of alternatives?	No	No		NA
Implementing				
• Milestones established?	?	No		
• Written monitoring guidelines established/followed?	No	No		
• Vital competencies done in-house?	Yes		L2	Accurate
Activity/impact assessing				
• Realized/planned strategy match?	No	No		NA
• Multiple dependent impact measures?	Yes		L1	Accurate
• Cost/benefit analysis?	Partially	Problems L1		Accurate
• True experiment of marketing impact?	No	No		NA
Administering				
• Quality of written mission statement?	No	NA		NA
• Annual quantitative objectives established?	No	No		NA
• Evidence of training, coordination, coaching?	Somewhat	Yes	L2	Accurate

Exhibit 12. Key Issues in a Meta-evaluation of the 1989 Minnesota Performance Audit of the Division of Travel and Tourism. *Note:* SWOT = strengths, weaknesses, opportunity, threats; S/D = supply and demand analysis; Levels: L1 = found to be poorly done; L2 = found to be done adequately; L3 = found to be done in superior manner and/or superior performance outcome noted. Accuracy = meta-analysis evaluation of validity of auditor's conclusion on activity/decision.

As described earlier, the Minnesota performance audit of the Office of Tourism's activities and impacts raises more sophisticated issues than found in the Missouri and North Carolina audit reports. However, the auditors do not indicate knowledge of the literature on true experiment designs in marketing and advertising. Consequently, their recommendations for estimating effectiveness and efficiency impacts are limited in value; the recommendations do not illustrate the paradigm shift to the use of true experiments.

Exhibits 11 and 12 provide summaries of our meta-evaluation of the 1985 Minnesota audit report of the state's tourism-marketing program. Based on the discussion presented above, our summary evaluation of the Minnesota audit report includes the following categories and descriptions:

- coverage of history and budget: useable
- clarity of writing: excellent
- theory applied: accounting only
- coverage of issues: limited: one issue in depth
- style: non-confrontational
- likely size of impact: modest
- overall performance quality of report: good, steps in right direction.

A META-EVALAUTION OF A NON-AUDIT AUDIT

The 1995 Tennessee Performance Audit Report of the Department of Tourist Development

This 13-page report includes a transmittal letter (dated July 28, 1995) to several members of the Senate and House of Representatives, State of Tennessee, from W. R. Snodgrass, Comptroller of the Treasury. In his letter, Mr. Snodgrass states, "This report is intended to aid the Joint Government Operations Committee in its review to determine whether the department should be continued, abolished, or restructured" (Tennessee, 1995, p. i).

Purpose and Authority for the Audit

Several sentences early in the report imply that the 1995 Tennessee performance audit was intended to be a comprehensive study. "The objectives of the audit were to review the department's legislative mandate and the extent to which the department has carried out that mandate efficiently and effectively and to make recommendations that might result in more efficient and effective accomplishment of the department's legislative mandates" (p. ii). The report gives four objectives for the audit. These objectives include:

> to evaluate the efficiency and effectiveness of the department's administration of its programs; to develop recommendations, as needed, for department or legislative actions that might result in more efficient and/or more effective operation of the department.

"Under Section 4-29-217 of that statue, the Department of Tourist Development is scheduled to terminate June 30, 1996. . . . This audit is intended to aid the Joint Government Operations Committee in determining whether the Department of Tourist Development should be abolished, continued, or restructured" (Tennessee, 1995, p. 1).

The 1995 Tennessee audit includes the following sections:

- Purpose and Authority for the Audit (1/4 page);
- Objectives of the Audit (1/4 page);
- Scope and Methodology of the Audit (1/4 page);
- Organization and Statutory Duties (1(2/3) pages);
- Economic Impact of Tourism (1/3 page);
- Observations and Comments (1(2/3) pages);
- Findings and Recommendations (4 pages), includes response comments.

In the scope and method section, the report states that, "The current activities and procedures of the department were reviewed, focusing on procedures and conditions in effect during field work, May through July 1994. The audit was conducted in accordance with generally accepted auditing standards and included . . ." The method reported included: (1) review of statues, regulations, policies, and procedures; (2) document analysis of departmental files; (3) interviews with a "contracted vendor," directors of regional tourist associations, and staff members; (4) site visits to the 12 welcome centers (p. 1).

Comments on the Method Described in the Report

The report includes no details of the method used. No findings on the policies and procedures are described. No details on how many documents were reviewed; the contents of documents reviewed are not discussed in the report. Who was interviewed and how many persons were interviewed are not reported. The details on the data collected during the welcome center visits are absent.

A Review of the "Organization and Statutory Duties" Section of the Report

The report states, "For the year ended June 30, 1994, the department had expenditures of $9.3 million and revenues and appropriations of $9.5 million . . . . During fiscal year 1994, the department had 152 staff positions, 108 of which were at the welcome centers" (p. 2). Neither this section nor any other section of the report includes financial statements. The number of staff positions appears excessive but this issue is not raised; the welcome center operation includes five regional mangers: one each for the northeast, middle, west, southeast, and middle east.

The Main Information in the Report

The heart of the report appears on 1/3 of a page; the section is "Economic Impact of Tourism." This section makes the following points without supporting evidence presented anywhere in the report:

> The department collects and analyzes data on the economic impact of tourism on the state. "According to the department, the 38.9 million tourists traveling to and through Tennessee in 1993 spent $6.77 billion that generated $517 million in state and local taxes.... In 1992, approximately 460,000 people visited Tennessee from Canada and nearly 2,000,000 visited from overseas countries" (Tennessee, 1995, p. 5).

Main Findings in the Report

The report describes two findings. First, "The cost-effectiveness of using state personnel instead of outside contractors should be analyzed. Since January 1994, the department has contracted with Wessan Interactive Network of Omaha, Nebraska, to answer the state's 800 tourist information number, mail requested information, and provide a summary of caller characteristics" (p. ii). "The Department of Tourist Development conducted no formal study and gathered no empirical data to support the need to go outside state government for these services" (p. 7).

Second, "the Department needs to ensure that staff comply with water quality statues and rules" (p. 11). This finding relates to hygiene management of the 12 welcome centers.

We offer the following observations about the first finding. An organization cannot achieve the objective of getting close to its customers by contracting out contacts with customers. Effective marketing practices requires direct interactions with customers. Nuances about customer beliefs, preferences, and behaviors are learned by talking with them. This principle about getting and staying close to customers is described in the strategic management literature (e.g. see Peters, 1987; Peters & Waterman, 1982). Both Tennessee auditors and the state's tourism executives appear to be unaware of the principle. Contracting out answering customers' telephone calls is likely to be more efficient initially, but it is also less effective than building an in-house direct-customer contract program.

To use an analogy often found in the *Economist*, the second finding is a "small beer" compared to major issues that are not addressed in the report. For example, the findings in the report do not include a discussion of the need for database marketing. The findings do not discuss effectiveness issues related to developing department relationships with inquirers. Given that the high value in establishing and building relationships with prospects and customers is a central principle of effective marketing strategy (e.g. see Kotler, 1997), then several more powerful reasons can be made for answering inquiry telephone calls, sending literature, and communicating with customers from Nashville, not Omaha.

Potential "cost-effectiveness" (really, cost-efficiency) is one reason for bringing the literature-fulfillment program home to Tennessee, especially given the initial likely higher efficiency expertise of Wessan Interactive Network. However, for reasons related to effectiveness (e.g. revenues, visit quality, customer satisfaction with service encounters), Tennessee should perform the operation by full-time Department staff members.

The "Management Comment" to the first main finding was that, "The department believes there is adequate information available to justify its decision to establish the Wessan contract. All state procedures were followed in the award of the contract" (p. 9). Surprisingly, the audit report includes no attempt to actually evaluate the "cost-effectiveness of contracting these services" but simply states such an evaluation should occur.

Meta-evaluation of the 1995 Tennessee Performance Audit Report

Exhibits 13 and 14 summarize the details of our meta-evaluation of the 1995 Tennessee performance audit. The 1995 Tennessee performance audit is unique in its lack of measurement and evaluation of the activities and impacts of the state's tourism-marketing programs. Almost no information is provided on the scanning, planning, implementing, and impact assessment, and administering activities of the Department of Tourist Development. The audit report expresses a possible problem with using an outside contractor for the state's literature-fulfillment program because the department did not evaluate the possibility this work could have been done in Tennessee with lower costs. The report does not describe why the auditor did not conduct such a cost-analysis evaluation.

However, we do not mean to suggest that this cost-efficiency issue is a valuable one to be resolved. As described above, other, more powerful, reasons are available for answering customer inquiries in Tennessee, from Nashville, and in the offices of the department. Also, we estimate that the potential cost savings of running the existing fulfillment program represent less than one percent of the department's annual budget. Also, the cost savings of Tennessee vs. Wessan fulfillment programs are likely to be negative, because of the substantial experience-curve impact (i.e. cost-then-price reductions based on accumulated fulfillment experience) in a professional fulfillment organization, such as Wessan Interactive Network.

	Meta-evaluation	**Rationale/additional comments**
Theory in Use:		
1. Program evaluation theory		
used/indicated by the performance audits?	• Accounting theory of	• A milsly confrontational
audit		
2 Knowledge displayed in the audit	program evaluation	report and response by the
report of availability of alternative theories?		• No specific foundation
		tourism division.
	report for program evaluation	
	cited	
Classification of Audit:		
1. Comprehensive or limited in scope?	• Comprehensive in scope	Focus is on improving the
2. Focuses on evaluating which of the		quality in estimating
economic		
department activities shown in Fig. 5?	• Focused on possible cost,	impact of advertising
program		
3. Within the focus of the performance audit,	savings in fulfillment	
provides thick descriptions of activities	• No	• Little to no information on
the		
done and strategies realized?		managing and operating
4. Evaluates research method used by		activities in the report.
department to measure effectiveness and	•No	• No coverage of this topic
efficiency of program? Audit conclusion sound?		
5. Evaluates the findings of program impact		
provided by the department?	•No	• No coverage
6 Reports whether or not this audit is the first,		
second, or third performance audit done of	• No	
this department?		
7 If previous audits done of this department,	• NA	• No mention of previous
audits		
reviews findings and use of previous audit reports?		on Office of Tourism
8 Reports scanning done by audit team of tourism-	• No	• No claims of scanning
made by		
marketing program audits done by other states?		audit team; no scanning
		apparent in report
Assessment of Value and Impact of Audit:		
1. Audit covers core issues related to	• No	• No efficiency analyses on
net		
effectiveness/efficiency impacts of activities and		taxes caused by advertising;
results of the program?		central issue not raised how
		many more
2. Audit includes explicit, concrete,	• Only minor topic recommendations	visits occurred that would
not		
recommendations for change/improvement?		have occurred without the
3. Response by program executives favorable or	• Unfavorable	advertising
unfavorable toward recommendations in audit report?		
4. Recommendations in report, if adopted, likely to	• No	• How to scan and plan not
result in substantial impacts on the		covered; scientific testing
quality of activities performed by the department?		advertising not covered
5. Central issues raised but inadequately addressed in	• No	
performance audit report?		• What are the net tax
revenues		
6. Core issues not raised in the audit report? • Yes		and costs/benefits of tourism-
		marketing program?
		Database
Systems-Thinking and Database Marketing:		marketing, systems thinking?
1. Systems thinking display in audit report,		
e g. policy mapping? • None		
2. Report covers implementation of database		
marketing?	• No	

Exhibit 13. Results: Answers for the Issues in a Meta-evaluation of the 1989 Performance Audit of Tennessee's Tourism-Marketing Program.

Activity/Decision	Did Auditor Collect?	Absence Noted? L1 L2 L3	Accuracy?
Scanning			
• SWOT assessment?	No	No	NA
• Market analysis: S/D?	No	No	NA
• Trends?	No	No	NA
• Research: customer segmenting?	No	No	NA
Planning			
• Written marketing plan?	No	No	NA
• Evidence of "what if' analyses?	No	No	NA
• Formal testing of alternatives?	No	No	NA
Implementing			
• Milestones established?	No	No	
• Written monitoring guidelines established/followed?	No	No	
• Vital competencies done in-house?	Yes	Yes	Accurate
Activity/impact assessing			
• Realized/planned strategy match?	No	No	NA
• Multiple dependent impact measures?	No	No	Accurate
• Cost/benefit analysis?	No	Yes (Possible problem noted)	Accurate
• True experiment of marketing impact?	No	No	NA
Administering			
• Quality of written mission statement?	No	NA	NA
• Annual quantitative objectives established?	No	No	NA
• Evidence of training, coordination, coaching?	No	No	NA

Exhibit 14. Key Issues in a Meta-evaluation of the 1989 Tennessee Performance Audit of the Division of Travel and Tourism. *Note:* SWOT = strengths, weaknesses, opportunity, threats; S/D = supply and demand analysis; Levels: L1 = found to be poorly done; L2 = found to be done adequately; L3 = found to be done in superior manner and/or superior performance outcome noted. Accuracy = meta-analysis evaluation of validity of auditor's conclusion on activity/decision.

The Tennessee performance audit may be prototypical of a non-audit audit. The performance of such a minimal audit may be particularly disappointing given the circumstances stated early in the report. Stated again: "This report is intended to aid the Joint Government Operations Committee in its review to determine whether the department should be continued, abolished, or restructured" (p. i). The report offers no evidence that the activities of the department lead to tourism activities, expenditures, or net tax profits for the state that would not have occurred if the department had not existed. Consequently, the report does not provide evidence in support of continuing the department; the report does not address this issue directly in its findings – another indicator of the report's uniqueness as a non-audit audit.

Based on the discussion presented above, our summary evaluation of the 1995 Tennessee audit report includes the following categories and descriptions:

- coverage of history and budget: poor
- clarity of writing: good
- theory applied: accounting (cost efficiency)
- coverage of issues: limited to minor issues
- style: mildly confrontational
- likely size of impact: little to none
- overall performance quality of report: very low.

A META-EVALAUTION OF A NEGATIVE COMPREHENSIVE AUDIT

The 1987 Hawaii Management Audit Report of the Hawaii Visitors Bureau and the State's Tourism Program

This 1987 Hawaii performance audit report is comprehensive in covering the topics of scanning, planning, implementing, measuring results, and administering by the Hawaii Department of Planning and Economic Development (DPED) and the Hawaii Visitors Bureau (HVB). Following Scriven's (1980) theory of evaluation, the 1987 Hawaii audit is a prototype of a very good audit indicating very bad performance overall by both the DPED and the HVB. The conclusion that the quality of the 1987 Hawaii audit report is substantially better than the 1995 Tennessee audit understates the differences in quality of the two reports.

Primarily because of historical reasons, the HVB performs most of the decisions and actions for Hawaii's tourism-marketing program, a private non-profit organization. Funding to run the HVB comes mostly from the state of Hawaii, "Today, legislative appropriations account for nearly 80% of HVB's operating budget" (Hawaii, 1987, p. 33). Legislative statues mandate that tourism-marketing funds be appropriated to the state agency charged with tourism development, DPED, "that agency would then enter into a contract with HVB for the promotion and development of tourism" (p. 33).

During the 1970s and 1980s the Hawaiian Legislature passed a number of statues to get the DPED to take charge of directing the state's tourism-marketing program and requiring that the HVB follow the directives of the DPED. Substantial evidence is found throughout the 1987 audit report that both the DPED and HVB have ignored these statues. The report concludes that much effort and time are spent attempting to communicate and coordinate actions between DPED and HVB administrators and staff, the resulting quality of managing Hawaii's tourism-marketing program is very low.

> The bureau [the HVB] enjoys the operating freedom of a private organization with virtually guaranteed, substantial state funding but with no need to produce profits, unhampered by the reviews and controls of a regular state agency, and accountable to no one for organizational effectiveness. As pointed out in Chapter 4, the Department of Planning and Economic Development (DPED) has not been aggressive in enforcing its contract with HVB. Theoretically, the State could contract with any other organization or organizations for all or part of its tourism marketing services. At one time, the Legislature directed DPED to contract directly for tourism advertising. But over time, HVB has convinced state decision makers that it is the best entity to market tourism for the State. Except for the review of its annual budget request at the Legislature, HVB is not held to account (Hawaii, 1987, p. 97).

The Contents of the 1987 Hawaii Performance Audit Report

This audit report contains 221 pages. The audit is organized into ten chapters followed by comments (letter plus 4 pages) by Roger A. Ulveling, Director of the DPED, and by comments (17 pages in a letter) by Walter A. Dods, Jr. (HVB Chairman of the Board) and Stanley W. Hong (HVB President).

Chapter 1: Introduction (2 pages). Chapter 1 states three objectives for the audit: examine and evaluate the role of the DPED in managing the State's tourism program; assess the effectiveness of the programs and operations of the HVB; and to make recommendations.

Strong evidence is found in the audit report to support the auditor's view that, "The audit took a broad approach to the industry and its place in Hawaii's economy. We reviewed the trends in worldwide tourism because Hawaii competes in a global economy." This claim is supported with a thorough literature review and references in later chapters in the report.

Chapter 2: The Tourist Industry Worldwide (12 pages). This chapter includes benchmarks and best practices data in government tourism-marketing programs. For example, the chapter describes the New Zealand's database-marketing program. "The New Zealand tourist office keeps a computerized list of the 50,000 to 100,000 inquiries it receives each year as a result of its mail-in advertising coupons. The list is then used for direct mail promotion and is also provided to travel operators and agents" (Hawaii, 1987, p. 10, quoted from Nick Verrasto in *ASTA Travel News*, June 15, 1986, p. 25). While not noted in the audit report, Hawaii does not have a similar database tourism-marketing program – one indication of the lack of systems-thinking that is widespread in government tourism-marketing programs in all U.S. states.

Chapter 2 includes a report on trends in travel products and markets. For example, "To stimulate demand, many areas are creating new travel products and purposes. These cater to special interests or hobbies, e.g. special hiking tours or adventure trips" (Hawaii, 1987, p. 14). Chapter 2 includes a total of 34 footnotes to data and literature on tourism-markets.

A description of the Canadian-government study of U.S. travelers supports the proposition that "Market researchers recognize that travelers are not a homogeneous group" (Hawaii, 1987, p. 9). While seemingly obvious, this proposition is important to support empirically. The proposition leads to several related, less obvious propositions, such as:

- for a given destination, different visitor segments seek different destination experiences;
- some prospective visitor segments are not worth spending government tourism-marketing funds to attract because these segments do not want to visit or if they did, they will not spend enough money to justify attracting.

Chapter 2 earns only a "good," not a "very good," rating because the chapter does not describe nor identify specific competitors with Hawaii. Consequently, the unique features of Hawaii products-markets are not compared to the competitor's unique products-markets. Still, recognizing the need for such coverage as the contents of Chapter 2 goes well beyond that found in any state's performance audit of tourism-marketing programs.

Chapter 3: Tourism in Hawaii (19 pages). Chapter 3 reports a history of tourism behavior in Hawaii, including benefits and costs. "In 1986, there was a 14.8% increase in the number of visitors over 1985 with total expenditures estimated at $5.8 billion. Given the multiplier effect of tourist expenditures, the visitor industry ultimately contributes to over half of the gross state product and supports more than one in three jobs in Hawaii" (p. 19).

The chapter describes some costs. On the negative side, the auditors perceive that substantial profits from tourism leave the state and benefit only the multinational corporations that control more and more of Hawaii's tourist industry. "One economist estimates that almost half of the $4 billion that was spent in Hawaii in 1984 left the State almost immediately" (p. 21). Two points are clear from a 1986 survey of Hawaiian residents: (1) they agreed that the positive economic benefits of the [tourism] industry outweighed the negative social impacts; and (2) an overwhelming majority felt that the economic benefits were not more important than protecting the environment.

Chapter 3 reports detailed profiles of Hawaii's westbound visitors (i.e. from California and other states) and eastbound visitors (i.e. mainly from Japan). Here is one comparison: Japanese tourist parties are estimated to spend $250 per day in Hawaii compared to U.S. tourist parties who are estimated to spend $98 per day (Hawaii, 1987, p. 23).

Chapter 3 describes government funding of tourism-marketing and related programs. The point is made that for the first 40 years of its existence (since 1903) the HVB received $1 in government support for every $2 it received in [membership] subscriptions from the business community. However, the ratio changed after World War II. "Today, legislative appropriations account for nearly 80% of HVB's operating budget" (p. 33).

During the 1970s the state developed a state plan for tourism. "The Hawaii State Plan, codified as Section 226, Hawaii Revised Statues, was enacted in 1978. It establishes the overall theme, goals, objectives, and policies for the State and a state planning process" (p. 34).

Wow! States rarely have a plan for tourism let alone a tourism plan enacted into law. "The State Tourism Functional Plan was approved by the Legislature in 1984 ... State government is responsible for administering the objectives and policies in the functional plan" (p. 35). A "tourism branch" in the DPED is currently responsible for coordinating activities in implementing the state's tourism plan. The head of the tourism division reports to one of the two deputy directors at DPED.

> The branch has three full-time positions: a tourism-program officer, a tourism specialist, and a secretary. The branch sees itself as having two main responsibilities. The first is the administration of the State's contract with HVB including preparing and assisting in negotiating the annual contract. It sees its responsibility in this area as primarily fiduciary. Branch personnel report that they attend many of the HVB committee meetings to keep on top of HVB activities (p. 35).

What is implied here becomes clear after reading the rest of the report: state government in Hawaii has organizationally buried the governance and leadership of its tourism plan and the state's tourism-marketing responsibilities. Also, the allocation of two government professional positions for tourism managing responsibilities is a further indicator that, to a great extent, the state has abdicated its participation in statewide tourism-marketing programs.

Consequently, in the following chapter, the auditor recommends that "An office of tourism be established in the Department of Planning and Economic Development that is headed by a deputy director ... The office should also strengthen the State's role in budgeting, including the restoration of the tourism program as a separate and identifiable program in the executive budget" (Hawaii, 1987, p. 69). Given the substantial number of statutes passed by the Hawaiian government requiring the state itself to manage its tourism programs, including the state's tourism-marketing programs, and given the serious shortcomings in the marketing efforts of the HVB reported by several sources, the Director of the DPED might be expected to support the auditor's recommendations for an Office of Tourism and a Deputy Director of Tourism. However, Mr. Ulveling and the DPED disagreed: "It [the response by Mr. Ulveling and the DPED] does not agree with our recommendation that an office of tourism be established. It also does not see the need to maintain the State Functional Plan for Tourism" (Hawaii, p. 197).

Close reading of DPED's response to the management audit strongly supports the conclusion that the state has abdicated its leadership and management responsibilities in tourism. For example, the DPED response includes the following statement in response to the recommendation that an Office of Tourism and Deputy Director position be created in the DPED:

> The report comments unfavorably on the current DPED economic development organization. The implication is that creation of a branch within a line division rather than the former staff office is detrimental to the tourism mission of the department. While it may appear that the change in reporting level decreases visibility and departmental support, in fact the opposite is true. With the addition of a division head and staff, the tourism branch now has more people available to assist and support it than ever. The division head and professional staff backs up the branch manager in preparing various documents, attending meetings, and administering contracts (Hawaii, 1987, p. 203).

This response does not support the view that DPED is responsible for leading and managing the state's tourism-marketing programs. The response does not suggest that tourism behavior and industries are central to the economic well being of the state. Having the state's senior, full-time, head of tourism operations be branch manager who reports to a division head, who reports to a deputy director, who reports to the director of DPED, more than suggests that the DPED has decided to be minimally involved in tourism. Note that the branch manager's role is described to be one of "preparing various documents, attending meetings, and administering contracts." The DPED's response indicates little responsibility for senior management activities regarding tourism-marketing. Also, note that in 1987 the head of this three-person branch (including a secretary) administers the contract with the 89-member staff of the HVB.

Chapter 4: The state's tourism program (32 pages). Poor performance of the DPED in all areas or responsibility is the main finding in Chapter 4.

> We find the following: (1) Despite repeated studies and agreements over the years about the State's goals, objectives, and policies for tourism, there has been little consideration or concerted efforts towards attaining these objectives. (2) Although the Department of Planning and Economic Development (DPED) is the State's lead agency for tourism, the State has yet to implement a tourism program because the department has played a reactive and passive role. There is an absence of leadership and focus to its efforts. (3) The department has yet to clarify its responsibilities for the tourism program *vis-a-vis* those of the Hawaii Visitors Bureau

> (HVB). The lack of clarity has lead to problems in administering and monitoring the State's contract with the bureau. (4) Because of the department's lack of initiative and its failure to identify and clarify its responsibilities for the State tourism program, some important government responsibilities for tourism are overlooked" (Hawaii, 1987, pp. 37–38).

Chapter 4 clearly indicates agreement "that the State has a broader role to play than HVB. The State is responsible for the entire tourism program. This [responsibility] would include coordinating infrastructure needs, monitoring the industry and its impact on the community [and the environment], and ensuring the [high] quality of the visitor industry generally" (Hawaii, 1987, p. 39).

Several tourism studies conducted, and then ignored by the State of Hawaii, are described briefly in Chapter 4. One study in particular is eye catching, a 1970 study commissioned by DPED with Mathematica of Princeton, New Jersey, to assess the costs and benefits of Hawaii's tourist industry. Mathematica is a firm noted for its expertise in policy mapping and systems dynamics modeling. Its work for Hawaii is one indicator supporting the proposition in Chapter 4 that "The State has conducted numerous studies of considerable value. For the most part, however, the studies have been ignored. They have not been used productively to build on the work that has gone on before" (Hawaii, 1987, p. 39).

Chapter 4 reviews several reports by the DPED and details of the "State Tourism Functional Plan" passed by the Legislature supports the view that the public sector is the lead organization to administer State objectives and policies of the State Tourism Functional Plan. However:

> Despite this long history of repeated demands for a larger government role and the existence today of an agency that says it is responsible for being a lead organization for tourism, the same complaints are still being heard . . . . The State appears to have gone full circle. The same complaints that were being made in 1957 are still being heard today. The need for coordination remains. Infrastructure continues to be a problem. The impact of the visitor industry on the State is of continuing concern. And government is still being prodded to assume its responsibilities for tourism (pp. 47–49).

Specific lack of managing activities (see Fig. 5) by the DPED are reviewed in Chapter 4. For example, "DPED has not budgeted funds for tourism promotion or for a tourism program. Instead, DPED merely forwards HVB's budget requests without analysis [to the Legislature]. The department testifies before the Legislature on HVB's budget request, but there is no analysis of whether the amounts requested for promotion are insufficient, adequate, or too much" (p. 51).

The following false view (i.e. theory of advertising program evaluation) expressed by the 1986 Director of the DPED, Kent M. Keith, to the Legislature reflects the primary cause for the lack of high quality studies on advertising effectiveness of tourism-marking programs. Note that in its response to Mr. Keith's propositions, the audit report identifies the solution, even though the audit report does not describe the availability of scientific tests to measure the impact of advertising and other marketing actions.

> In 1985, the director [of the DPED, Kent M. Keith] commented in his testimony that while it was impossible to estimate the share of the additional $300 million increase in visitor expenditures for 1983 for which HVB was responsible, even a relatively small share would have handsomely repaid the State's investment. The director believed that the investment was repaid many times over.
>
> This kind of testimony is of little assistance to legislative decision makers. It reflects DPED's general lack of concern [about] or information on the impact of funds appropriated to HVB and the effectiveness with which they are expended. The department still has not developed a system for oversight of HVB nor has it identified any more useful measures of effectiveness. Consequently, it has no means for assessing existing promotional efforts or evaluating potential new markets (Hawaii, 1987, pp. 51–52).

The rest of Chapter 4 provides evidence supporting the conclusion that DPED is, at best, incompetent. The topics covered include the following items. "Failure to take actions relating to physical [tourism infrastructure] development" (p. 53), "Failure to take actions relating to employment and career development" (p. 54), "Failure to coordinate and monitor" (p. 55), "Failure to maintain plan" (p. 57), and "Current organization inadequate" (p. 58). Non-compliance of HVB activities of written requirements in its contract with the DPED and the lack of enforcement of these requirements by the DPED are described in detail on page 67 in the audit report.

Consequently, Chapter 4's last page offers substantial overwhelming evidence supporting the auditor's recommendations. Here are the first two recommendations: "We recommend the following: (1) An office of tourism be established in the Department of Planning and Economic Development that is headed by a deputy director; and (2) The office of tourism make the State Functional Plan for Tourism the starting point for its tourism program" (Hawaii, 1987, p. 69).

However, these recommendations are unrealistically optimistic that a paradigm shift toward competency in managing Hawaii's tourism program can be achieved within the existing DPED and HVB organizations. Because the incompetence is so pervasive and historical in both the DPED and HVB, such a paradigm shift may be possible to achieve only by creating a new organization to manage the state's tourism program unique and separate from both the DPED and HVB. More recent work (i.e. the Hawaii, 1993 audit report) supports the view that the DPED and HVB have been unable to create an effective and efficient tourism-marketing program for the state.

Chapter 5: The [HVB] board of directors (21 pages). This Chapter starts by describing how unique the relationship is between the State of Hawaii and the

HVB. "The Hawaii Visitors Bureau (HVB) is both similar to – and different from – government agencies responsible for tourism promotion throughout the world. It is only one of two private, non-profit organizations promoting tourism in the United States; all other states have government-operated tourism agencies" (Hawaii, 1987, p. 71). However, Colorado voters abolished their state's tourism board in 1993 (Bonham & Mak, 1996); such an action is one indicator that bold action can be achieved in making changes in a state's tourism program.

The first page of the Chapter 5 summarizes the main findings:

We find the board's effectiveness is undermined in the following respects:

- The Board of Directors does not play a fully meaningful role in the governance of the bureau. The bureau has not given the board the support and attention it needs to function effectively . . .;
- The board has been unable to carry out adequately its responsibilities for policy planning, budget review, and evaluation;
- The selection processes for board directorships and leadership positions do not encourage or permit equal participation from bureau members (p. 71).

The chapter presents detailed evidence in support of these findings. The evidence is not reviewed here, but it is compelling and profound.

Chapter 6: The management of the Hawaii visitors bureau (23 pages). The main finding in Chapter 6 is stated on the first page of the chapter: "We find that: (1) There is a lack of direction and attention to several basic responsibilities in organizational management in the bureau." Here are a few specifics to support this finding:

> We find the HVB administration does not focus sufficient attention on the organization and management of the bureau. It is not being managed as a coherent whole. The basic ingredients of sound management are lacking. The bureau has no long-range plan to give it direction, it has no formal organization chart that delineates the functions of its many units to support attainment of its goals, and its budgeting practices are hurried and slipshod. These basic deficiencies have significant ramifications for the programs it operates (Hawaii, 1987, pp. 97–98).

Chapter 6 provides detailed evidence to support these observations. The evidence includes unrealistic estimates of the private income from HVB memberships – the actual amounts are always substantially less than the budgeted amounts for 1983–1987. "The bureau compounds the problem by disregarding the actual collections for the year in budgeting for the next year – and doing this year after year" (Hawaii, 1987, p. 110).

The main recommendation of the auditor concerning management improvements of the HVB is for HVB to develop an organization plan based on a strategic plan that sets the overall direction of the program (Hawaii, 1987, p. 115). This recommendation is highly unlikely to become a reality given the overall finding of the auditor that the HVB is "negligent in its management responsibilities." The primary response to the main HVB related findings of the audit report support this view of reality:

> While some of the findings and recommendation of the preliminary [audit] report have merit, many of the recommendations would be more appropriately addressed to a government bureaucracy rather than a private, non-profit organization. The report demonstrates an inordinate concern for developing "formal" policies and other "red tape," and thus ignores the dynamic environment within which the HVB operates and the need for immediate, opportune response to market and media stimuli (Signed response by Walter A. Dods, Jr., Chairman of the Board, and Stanely W. Hong, President [of HVB], p. 206).

Possibly the only effective means of moving toward an effective state tourism program for Hawaii includes the complete termination of the state's relationship with the HVB. Such a decision and action would be met with stiff resistance given the size of the HVB and the focus of its full-time staff on maintaining the status quo. Thus, Hawaii may want to consider funding the HVB with annual reductions of 25% per year from the current level of state support for four years – or reach some other solution resulting in the dissolution of its relationship with the HVB. Even though the state now could contract with a private, non-profit organization other than the HVB to operate its tourism-marketing program, the State of Hawaii would likely develop an effective tourism-marketing program by housing such a program in a separate Department of Tourism Programs.

The Hawaii Legislature should end its attempt to legislate competent behavior of the DPED and the HVB by passing additional statutes. The course of action likely to be effective includes continuing the current DPED actions of *not* giving the HVB direction and to continue *not* monitoring the activities and results of the HVB. The time has come for the State of Hawaii to create its own organization to manage the state's tourism program.

Chapter 7: Marketing program of the Hawaii Visitors Bureau (26 pages). Chapter 7 provides substantial evidence supporting the findings offered in the first page of the chapter. "We find the following: (1) there is only limited marketing coordination among the bureau's various departments, committees, and section offices; (2) The bureau lacks formal standards or procedures for selecting an advertising agency – its most important marketing tool and its largest cost item; (3) The bureau's marketing program has not developed or implemented any formal monitoring and evaluation procedures or mechanisms to determine the effectiveness of its marketing activities" (p. 118).

As far as measuring the impact of the advertising and marketing efforts of the HVB, the audit report makes clear that the HVB does not follow its own *Strategic Marketing Plan* guidelines for creating monitoring systems.

> However, despite the acknowledged importance of a monitoring/evaluation system, the HVB marketing program does not contain an integral evaluation system. Instead HVB relies on such vague ideas as the health of the industry and such haphazard methods as what people tell them (p. 138).

Even the bureau's advertising program, the largest and the most important component of its marketing effort, is not subject to any kind of formal and ongoing monitoring or evaluation (Hawaii, 1987, p. 139). The audit report goes on to suggest several methods for evaluating advertising effectiveness – including the use of experimental designs (the Campbell & Stanley, 1963 true experiment approach, e.g. the studies reviewed by Caples, 1974 used experimental designs), as well as recall and recognition tests (these are invalid but popular methods for measuring advertising effectiveness; they are discussed in more detail below).

Given the low level of performance documented in Chapter 7, the audit report recommendations to coordinate marketing activities and to develop formal and written standards and procedures offered at the end of the chapter, while commendable, are unlikely to be accepted or adopted by the HVB. What the auditor asks for is neither "red tape" nor excessive, but rather, sound activities to achieve an effective tourism-marketing strategy.

Chapter 8: Market research program of the Hawaii Visitors Bureau (20 pages). The chapter includes a benchmarking discussion of tourism research programs in the world. The main finding reported in Chapter 8: "While the Hawaii Visitors Bureau's market research program provides the State, the bureau, and the visitor industry with valuable data, it has a number of problems which contributes to its inefficiency in certain areas, reduce its effectiveness, and limit its ability to conduct more sophisticated market research needed to improve Hawaii's competitive edge in the world tourism-market" (Hawaii, 1987, p. 145). The audit report points out that the HVB research program is too heavily involved in statistical compilation instead of more relevant and sophisticated market research (p. 154).

The serious problems in the HVB market research extend into 1997 with the HVB commissioned study on advertising accountability study (Longwoods, 1997). This study is based on aided-recall measurement tools to measure advertising effectiveness. The results of the 1997 study indicate spectacular high performance: $75.5 million in taxes generated by visitor spending resulting from a $7.87 million advertising investment. But two points are clear from reading the scientific literature on measuring aided-recall tests: (1) they are invalid predictors of purchase; and (2) valid predictors of purchase are available (e.g. unaided top-of-mind-awareness measures and true experiments). The research design used in the HVB commissioned 1997 study is non-scientific and incorporates measurement procedures known to be invalid for estimating purchase (see

Axelrod, 1968; Caples, 1974; Haley & Case, 1979). Consequently, the findings in the HVB commissioned 1997 study are likely to be gross overstatements of Hawaii's advertising impact on visits to the state.

Chapter 9: Visitor satisfaction and Community Relations Program (16 pages). The first finding reported in Chapter 9 is that "the visitor satisfaction and community relations program is less effective than it could be because the program lacks definition and is not integrated into the bureau's overall marketing efforts." The auditor recommends that the HVB develop a precise framework for the operation of its visitor satisfaction and community relations program.

Chapter 10: Personnel management 10 (pages). The main finding reported in Chapter 10 is that "The bureau has not carried out its personnel functions in a responsible and effective manner, because it does not have an ongoing personnel program" (Hawaii, 1987, p. 185). Chapter 10 presents several details to support this finding.

Responses by the DPED and HVB

The primary response by the DPED has been discussed: the DPED disagreed with the principal recommendations in the audit report. The main response by the HVB board chairman and the president has been described: the HVB is a private non-profit organization that should not be bound by bureaucratic procedures and red tape. Other relevant points are that 80% of HVB funds came from the state; more importantly, the audit report supplies ample evidence in support of its findings, and the report's recommendations match well with the components described earlier for effective and efficient government, tourism-marketing, programs.

Meta-evaluation of the 1987 Hawaii Management Audit Report

This report includes a useful (even though incomplete) review of relevant management and research literature pertaining to government tourism-marketing programs. The 1987 Hawaii Management Audit Report offers ample and specific evidence in support of its findings. Its recommendations are systematic in covering scanning, planning, implementing, measuring results, and administrating Hawaii's tourism program. The recommendations are sound and their adoption would very likely increase the effectiveness and efficiency of the state's tourism program. However, the recommendations are unlikely to be adopted, given the strong negative responses to them by the DPED and HVB. Exhibits 15 and 16

	Meta-evaluation	Rationale/additional comments
Theory in Use:		
1. Program evaluation theory used/indicated by the performance audits? audit	• Scriven-type postpositivistic	• A highly confrontational report and response by the
2 Knowledge displayed in the audit report of availability of alternative theories?		• None described DPED and HVB.
Classification of Audit:		
1. Comprehensive or limited in scope?	• Comprehensive in scope	Focus is on improving the quality in estimating
2. Focuses on evaluating which of the economic department activities shown in Fig. 5? program	• Focused on possible cost, savings in fulfillment	impact of advertising
3. Within the focus of the performance audit, provides thick descriptions of activities the done and strategies realized?	• No	• Little to no information on managing and operating activities in the report.
4. Evaluates research method used by department to measure effectiveness and efficiency of program? Audit conclusion sound?	• No	• No coverage of this topic
5. Evaluates the findings of program impact provided by the department?	• No	• No coverage
6 Reports whether or not this audit is the first, second, or third performance audit done of this department?	• No	
7 If previous audits done of this department, audits reviews findings and use of previous audit reports?	• NA	• No mention of previous on Office of Tourism
8 Reports scanning done by audit team of tourism-made by marketing program audits done by other states?	• No	• No claims of scanning audit team; no scanning apparent in report
Assessment of Value and Impact of Audit:		
1. Audit covers core issues related to net effectiveness/efficiency impacts of activities and results of the program?	• Yes, with one issue not raised fully →	• No efficiency analyses on taxes caused by advertising; central issue not raised how many more visits occurred that would not have occurred without the advertising
2. Audit includes explicit, concrete, recommendations for change/improvement?	• Yes, very comprehensive	
3. Response by program executives favorable or unfavorable toward recommendations in audit report?	• Unfavorable	
4. Recommendations in report, if adopted, likely to result in substantial impacts on the quality of activities performed by the department?	• Yes	• How to scan and plan not covered; scientific testing advertising not covered
5. Central issues raised but inadequately addressed in performance audit report? revenues	• No →	• What are the net tax and costs/benefits of tourism-marketing program?
6. Core issues not raised in the audit report?	• Yes	Database marketing, systems thinking?
Systems-Thinking and Database Marketing:		
1. Systems thinking display in audit report, e g. policy mapping?	• Yes, but inadequately	
2. Report covers implementation of database marketing?	• Yes, but inadequately	

Exhibit 15. Results: Answers for the Issues in a Meta-evaluation of the 1989 Performance Audit of Hawaii's Tourism-Marketing Program.

Activity/Decision	Did Auditor Collect?	Absence Noted?	L1 L2 L3	Accuracy?
Scanning				
• SWOT assessment?	No	No		NA
• Market analysis: S/D?	Yes	Yes	L1	Accurate
• Trends?	Yes	No	L2	NA
• Research: customer segmenting?	Yes	No	L2	NA
Planning				
• Written marketing plan?	Yes	No	L1	NA
• Evidence of "what if' analyses?	No	No		NA
• Formal testing of alternatives?	Yes	No	L1	NA
Implementing				
• Milestones established?	Yes	Yes	L1	Accurate
• Written monitoring guidelines established/followed?	Yes	Yes	L1	Accurate
• Vital competencies done in-house?	Yes	Yes	L1	Accurate
Activity/impact assessing				
• Realized/planned strategy match?	Yes	Yes	L1	Accurate
• Multiple dependent impact measures?	No	No		NA
• Cost/benefit analysis?	No	Yes	L1	Accurate
• True experiment of marketing impact?	Yes	Yes		Accurate
Administering				
• Quality of written mission statement?	Yes	NA	L3	Accurate
• Annual quantitative objectives established?	No	Yes	L1	Accurate
• Evidence of training, coordination, coaching?	Yes	Yes	L1	Accurate

Exhibit 16. Key Issues in a Meta-evaluation of the 1989 Hawaii Performance Audit of the Division of Travel and Tourism. *Note:* SWOT = strengths, weaknesses, opportunity, threats; S/D = supply and demand analysis; Levels: L1 = found to be poorly done; L2 = found to be done adequately; L3 = found to be done in superior manner and/or superior performance outcome noted. Accuracy = meta-analysis evaluation of validity of auditor's conclusion on activity/decision.

summarize our meta-evaluation of the Hawaii 1987-management audit report are provided in Exhibits 15 and 16.

The course of action most likely to move the State of Hawaii's tourism-marketing program away from its pervasive ineffectiveness and inefficiency is more a dramatic step: creating a separate Department of Tourism and ending the state's relationship with the HVB. Given the unusually high importance of tourism for Hawaii compared to other states, and given the evidence in the 1987 audit report and more recent evidence, warrants a new state department of tourism. Some states have made dramatic changes in administering their tourism programs (e.g. Colorado in 1993, see Bonham & Mak, 1996 and our earlier discussion); consequently, enacting and implementing such a paradigm shift by Hawaii is feasible.

Based on the discussion presented above, our summary evaluation of the 1987 Hawaii audit report includes the following categories and descriptions:

- coverage of history and budget: excellent
- clarity of writing: excellent
- theory applied: a Scriven-type, post positivistic audit
- coverage of issues: comprehensive
- style: highly confrontational
- likely size of impact: little to none
- overall performance quality of report: high.

A META-EVALUATION OF A REPEAT PEFORMANCE AUDIT

The 1993 Management and Financial Audit of the Hawaii Visitors Bureau

The 1993 performance audit of the State of Hawaii's tourism-marketing program was performed under the direction of Marion M. Higa, the State Auditor. This 32-page report includes two chapters and responses by executives of the HVB and the Department of Business, Economic Development, and Tourism (DBEPT).

Although less comprehensive, the findings and recommendations in the 1993 report are remarkably similar to the findings and recommendations in the 1987 report. However, the 1993 report notes that potential improvements in managing the tourism-marketing program were made after the 1987 audit report:

> In 1990, the Legislature created the Office of Tourism in DBEDT to coordinate and plan tourism development. State funds for tourism marketing activities are channeled through this office. The office contracts with HVB and other tourism promotion programs. Currently, the Office of Tourism has separate contracts with the HVB and with each of the HVB sections on the islands of Hawaii, Kauai, and Maui. The office is responsible for monitoring HVB and performing annual reviews to ensure the effective use of state funds (Hawaii, 1993, p. 2).

Chapter 1: Introduction (3 pages). "As Hawaii's designated tourism-marketing organization, HVB in FY 1992–1993 received over 90% of its $20 million budget from the State. In 1993, the Legislature appropriated nearly $60 million for fiscal biennium 1993–1995 to fund tourism promotion projects (Hawaii, 1993, p. 1). Note the state's funding share of HVB has incurred from 80% in 1986–1987 to over 90% in 1992–1993. Bonham and Mak (1996) describe the "free rider" behavior by tourism industry members that is associated with substantial HVB expenditures and management efforts to solicit membership dues each year. Also:

> Lobbying, or rent seeking, is also costly. HVB has three full-time political lobbyists, excluding its corporate officers who also actively lobby at the legislature. Moreover, HVB's increasingly aggressive lobbying at the legislature for money has tarnished its public image and reputation. Indeed, one critique of HVB recently observed that 'the HVB remains focused mostly not on building tourism but on building a budget' (Bonham & Mak, 1996, p. 6; Rees, 1995, p. 5).

"To determine whether HVB has used public funds effectively, the 1993 Legislature, in House Concurrent Resolution No. 284, requested the State Auditor to conduct a management and financial audit of HVB" (Hawaii, 1993, p. 1). The 1993 audit report does provide some degree of independent confirmation of findings in the 1987 report because a different person serving as state auditor presents the findings in the 1993 audit report. However, considering that the 1993 report finds the same high level of negligence as in the 1987 report, the continued use of the HVB throughout the 1990s by the State of Hawaii is surprising. But this apparent incompetence may be explained in part by the strong tendency of most legislators not to read audit reports and by the strong, effective, lobbying efforts of the HVB.

Chapter 2: The Hawaii Visitors Bureau and its relationship with the state (12 pages). The following statement is the complete "Summary of Findings" in Chapter 2 in the 1993 audit report:

> Both the Hawaii Visitors Bureau (HVB) and the Department of Business, Economic Development and Tourism (DBEDT) have fallen short in fulfilling their respective responsibilities for the State's tourism program.

(1) The HVB board of directors has been weak and exercised little oversight over HVB.
(2) Unclear functions and underutilization of its own market research information weaken HVB's marketing efforts. HVB has yet to resolve the status and roles of its regional offices on the mainland and its sections on the neighbor islands.
(3) HVB reports do not comply with requirements in its contract with DBEDT and do not show whether public funds are properly and effectively utilized.
(4) In the absence of strong HVB board and management leadership, DBEDT has begun to direct HVB to undertake certain programs. This conflicts with DBEDT's responsibilities for monitoring the HVB contract (Hawaii, 1993, p. 5).

What should become clear from close reading of the 1987 and 1993 audit reports on Hawaii's tourism-marketing programs is that the state has an unnecessary and costly layer of management in these programs. Much of the effort spent in coordinating and monitoring activities between the DBEDT and the HVB could be spent more profitably elsewhere – on effectively and efficiently managing government tourism-marketing programs. Even without the continuing poor tourism-marketing performance of the HVB, the State of Hawaii would benefit by ending its unique arrangement of contracting out for vital competencies that need to be mastered within the Office of Tourism.

The recommendations in Chapter 2 include reducing the number of directors on the HVB board, providing written guidelines for committee members, and developing a (written) strategic plan. The recommendations include the DBEDT "submitting annual reports to the Legislature that contain the information requested by the Legislature on tourism promotion programs and their effectiveness" (Hawaii, 1993, p. 17).

The HVB and DBEDT Responses and the Auditor's Response to the Responses
The HVB president commented that many of the audit report's recommendations reflect inordinate concern for developing "formal" practices. The auditors' response, "We note that a certain minimum level of written procedures and instructions is required for any organization to operate effectively.

The director of the DBEDT did not respond directly to recommendations. "Permit me to offer the following general comments, rather than a point-by-point rebuttal," stated Mufi Hanneman, Director of the DBEDT, by letter on December 27, 1993 (Hawaii, 1993, p. 28). The audit report criticizes the DBEDT for having the HVB undertake tourism initiatives developed by the DBEDT. The director's response defends the department's authority to undertake tourism initiatives regardless of whether they are planned or not. The auditor's responds, "We believe, however, that the department should undertake these initiatives in its own name and not under that of HVB" (Hawaii, 1993, p. 21).

Meta-evaluation of the 1993 Hawaii Management and Financial Audit Report on the HVB and Its Relationship with the State

The 1993 report supports the main findings strongly using detailed evidence. However, the range of topics covered in the report is disappointing. Whereas the 1987 Hawaiian audit report is comprehensive in coverage, the 1993 audit report is limited in coverage. Unlike the 1987 audit report, the 1993 report does not include the attempt to identify best practices in government tourism-marketing programs. The 1993 report does not include detailed information on the scanning and planning activities of the HVB and the DBEDT. The 1993 audit report does not

	Meta-evaluation	**Rationale/additional comments**
Theory in Use:		
1. Program evaluation theory used/indicated by the performance audits? audit	• Accounting theory of	• A mildly confrontational
2 Knowledge displayed in the audit report of availability of alternative theories?	program evaluation	report and response by the tourism division.
Classification of Audit:		
1. Comprehensive or limited in scope?	• Limited in scope	
2. Focuses on evaluating which of the department activities shown in Fig. 5?	Focused on lack of coordination and DBEDT directing HVB	
3. Within the focus of the performance audit, provides thick descriptions of activities done and strategies realized?	•No	
4. Evaluates research method used by department to measure effectiveness and efficiency of program? Audit conclusion sound?	•No	•No coverage of this topic
5. Evaluates the findings of program impact provided by the department?	•No	•No coverage
6 Reports whether or not this audit is the first, second, or third performance audit done of this department?	• Yes	
7 If previous audits done of this department, reviews findings and use of previous audit reports?	• Yes	
8 Reports scanning done by audit team of tourism- made by marketing program audits done by other states?	•No, none apparently done	• No claims of scanning audit team; no such scanning apparent in report
Assessment of Value and Impact of Audit:		
1. Audit covers core issues related to net effectiveness/efficiency impacts of activities and results of the program?	• No	• No efficiency analyses on taxes caused by advertising; central issue not raised how many more
2. Audit includes explicit, concrete, not recommendations for change/improvement? very limited	• Yes, but	visits occurred that would have occurred without the advertising
3. Response by program executives favorable or unfavorable toward recommendations in audit report?	• Unfavorable	
4. Recommendations in report, if adopted, likely to result in substantial impacts on the quality of activities performed by the department?	• No	• How to scan and plan not covered; scientific testing advertising not covered
5. Central issues raised but inadequately addressed in performance audit report? revenues	• Yes	• What are the net tax
6. Core issues not raised in the audit report? • Yes		→ and costs/benefits of tourism-marketing program? Database marketing, systems thinking?
Systems-Thinking and Database Marketing:		
1. Systems thinking display in audit report, e g. policy mapping? • No		
2. Report covers implementation of database marketing?	• No	

Exhibit 17. Results: Answers for the Issues in a Meta-evaluation of the 1993 Performance Audit of Hawaii's Tourism-Marketing Program.

include a financial audit report, even though "financial audit" appears in the title of the report. The 1993 audit report fails to examine the effectiveness and efficiency of the HVB's tourism-marketing programs. Other than reporting that the "HVB reports do not comply with requirements in its contract with DBEDT and do not show whether public funds are properly and effectively utilized (Hawaii, 1993, p. 5)," the 1993 audit report fails to report HVB's own attempts, if any, to measure the effectiveness and efficiency of its tourism-marketing programs.

In brief, the 1993 audit report is substantially lower in quality compared with the 1987 audit report. Exhibits 17 and 18 summarize our meta-evaluation of the 1993 audit report.

Activity/Decision	Did Auditor Collect?	Absence Noted?	L1 L2 L3	Accuracy?
Scanning				
• SWOT assessment?	No	No		NA
• Market analysis: S/D?	No	No		NA
• Trends?	No	No		NA
• Research: customer segmenting?	No	No		NA
Planning				
• Written marketing plan?	Yes	Yes	L1	Accurate
• Evidence of "what if' analyses?	No	No		NA
• Formal testing of alternatives?	No	No		NA
Implementing				
• Milestones established?	Yes	Yes	L1	Accurate
• Written monitoring guidelines established/followed?	Yes	Yes	L1	Accurate
• Vital competencies done in-house?	No	No		NA
Activity/impact assessing				
• Realized/planned strategy match?	Yes	Yes	L1	Accurate
• Multiple dependent impact measures?	No	No		NA
• Cost/benefit analysis?	No	No		NA
• True experiment of marketing impact?	No	No		NA
Administering				
• Quality of written mission statement?	Yes	NA	L3	NA
• Annual quantitative objectives established?	No	No		NA
• Evidence of training, coordination, coaching?	Yes	Yes	L1	Accurate

Exhibit 18. Key Issues in a Meta-evaluation of the 1993 Hawaii Performance Audit of the Division of Travel and Tourism. *Note:* SWOT = strengths, weaknesses, opportunity, threats; S/D = supply and demand analysis; Levels: L1 = found to be poorly done; L2 = found to be done adequately; L3 = found to be done in superior manner and/or superior performance outcome noted. Accuracy = meta-analysis evaluation of validity of auditor's conclusion on activity/decision.

The following categories and descriptions are a further summary to the 1993 Hawaii management audit report of the HVB:

• coverage of history and budget:	incomplete
• clarity of writing:	good
• theory applied:	a Scriven-type, post positivistic audit
• coverage of issues:	limited
• style:	confrontational
• likely size of impact:	little to none
• overall performance quality of report:	medium.

A META-EVALUATION OF A PERFORMANCE AUDIT OF A NATIONAL TOURISM-MARKETING ORGANIZATION

The 1993 Australian Audit Report of the Australian Tourist Commission's Marketing Impact

The 1993 Australian audit report, *Evaluation of the Australian Tourist Commission's Marketing Impact*, is 73 pages followed by more than 100 pages of appendices. The Australian audit report is the only one of the seven reviewed that was "compiled" by three organizations, rather than one government audit organization. The three compilers are The Department of Finance (DOF); Department of Arts, Sport, the Environment, Tourism and Territories (DASETT); and the Australian Tourist Commission (ATC).

The Australian audit report has an executive summary and eight chapters. The report is the only example of a partnering-type performance audit among the seven audits examined in this meta-evaluation (see Wholey, 1983). The main conclusion of the audit is a positive one: "The report concludes that ATC has been effective in meeting its marketing objectives as set out in its Act" (Australia, 1993, p. 1).

The other main findings are positive as well:

- ATC provides important value added and, therefore, contributes to tourism receipts which benefit tourism industries generally and the community as a whole;
- The impact of ATC marketing, though difficult to quantify, is positive;
- Continued Government funding of international tourism promotion is justified, in conjunction with – where possible – increased private sector contribution to joint activities (Australia, 1993, p. 1).

The audit was intended to be the "first comprehensive evaluation of the Australian Tourist Commission's (ATC) marketing and promotional activities." Unfortunately, the report does not evaluate many of the activities of the ATC directly. For example, the audit report does not report whether or not the ATC prepares a written marketing plan. Consequently, the report does not address the quality of planning done by the ATC before the audit.

The audit report fails to describe how the ATC is organized; the size of the ATC; and its budget and disbursements. How did the ATC evaluate its impact before the auditing team did its study? The report does not address these issues. Detailed descriptions and rationales of the specifics of Australia's marketing and advertising campaigns are not provided in the 1993 Australian audit report. The report provides a conglomeration of examples of numbers of consumers reached by Australian's tourism advertising for 1989, 1990, and other years (no annual breakouts are provided). The reports describes conversion studies without comparing the relative performances of different media vehicles, or different advertising-segment campaigns – the most appropriate use for conversion studies (see Woodside & Motes, 1981).

What is the tax profit from the Australian tourism-marketing program net of expenditures for the program? The report does not provide estimates in response to this issue. Although the 1993 Australian report is attractive in appearance, it fails to provide a cost-benefit analysis of the Australian tourism-marketing program.

The auditors writing the report have a naïve understanding of "control experiment" (p. 52), "One of the problems throughout this evaluation has been the difficulty in conducting a controlled experiment by manipulating the variables in the decision-making process. All these variables, such as disposable income, airfares and exchange rates are beyond the control of researchers" (p. 52). True-experiment designs for examining the effects of advertising include manipulating one or more advertising variables – using one or more treatment groups and a control group. See the discussion of true experiments presented earlier in this meta-evaluation for a detailed description of true experiments in relation to measuring effectiveness of tourism-marketing and advertising programs.

The audit reports discusses the use of aided recall for measuring the impact of Australian's tourism advertising. For example, the question is raised, "have you seen advertising for visiting Australia?" Among respondents in a study in Japan, 19% reporting Australia in first-mention responses to the question on intent to visit Australia in the next three years versus 10% of respondents in the "not seen" first-mention category (p. 53). Given the inherent biases in aided-recall questions, given the lack of validity of the method in measuring advertising effectiveness in the literature, and given the long-time availability and wide applications of scientific methods to measure advertising effectiveness (see Banks, 1965;

	Meta-evaluation	Rationale/additional comments
Theory in Use:		
1. Program evaluation theory used/indicated by the performance audits? departments	• No specific grounding	• Three government
2 Knowledge displayed in the audit report of availability of alternative theories?	• None	participated in preparing the audit
Classification of Audit:		
1. Comprehensive or limited in scope? to be	• Limited in scope	• Claim made early in report
2. Focuses on evaluating which of the department activities shown in Fig. 5?		comprehensive
3. Within the focus of the performance audit, provides thick descriptions of activities done and strategies realized?	• No	
4. Evaluates research method used by department to measure effectiveness and valid	• No	• Accepts methods used as
efficiency of program? Audit conclusion sound?		• No coverage of this topic
5. Evaluates the findings of program impact provided by the department?	• No	
6 Reports whether or not this audit is the first, second, or third performance audit done of this department?	• Yes, the first audit	
7 If previous audits done of this department, reviews findings and use of previous audit reports?	• NA	
8 Reports scanning done by audit team of tourism-marketing program audits done by other states?	•No	
Assessment of Value and Impact of Audit:		
1. Audit covers core issues related to net effectiveness/efficiency impacts of activities and results of the program?	• To a limited degree	• No efficiency analyses on taxes caused by advertising; central issue not raised how many more
2. Audit includes explicit, concrete, not recommendations for change/improvement?	• Yes, but very limited	visits occurred that would have occurred without the advertising
3. Response by program executives favorable or unfavorable toward recommendations in audit report?	• Favorable; ATC co-author of report	
4. Recommendations in report, if adopted, likely to result in substantial impacts on the quality of activities performed by the department?	• Few specific recommendations in report	
5. Central issues raised but inadequately addressed in performance audit report?	• Yes	→ • How to scan and plan not
6. Core issues not raised in the audit report?	• Yes, several	covered; scientific testing advertising not covered
Systems-Thinking and Database Marketing:		
1. Systems thinking display in audit report, e.g. policy mapping?	• No	• What are the net revenues and costs/benefits of tourism-marketing program?
2. Report covers implementation of database marketing? Database	• No	marketing; systems thinking?

Exhibit 19. Results: Answers for the Issues in a Meta-evaluation of the 1993 Performance Audit on the Australian Tourist Commission.

Caples, 1974; Ramond, 1966), the use of aided recall measures by the ATC is surprising. As noted above, this use indicates a lack of knowledge of and training in scientific measurement of program impact (see Banks, 1965; Campbell, 1969; Woodside, 1990).

Meta-evaluation of the 1993 Australian Audit Report of the Australian Tourist Commission's Marketing Impact

The report presents no history on the creation, budgets, and performance history of the ATC. The report does not describe specific annual marketing plans. The report

Activity/Decision	Did Auditor Collect?	Absence Noted?	L1 L2 L3	Accuracy?
Scanning				
• SWOT assessment?	No	No		NA
• Market analysis: S/D?	Yes	Present	L3	Accurate
• Trends?	Yes	Present	L3	Accurate
• Research: customer segmenting?	Yes	Present	L3	Accurate
Planning				
• Written marketing plan?	No	No		NA
• Evidence of "what if" analyses?	No	No		NA
• Formal testing of alternatives?	No	No		NA
Implementing				
• Milestones established?	No	No		NA
• Written monitoring guidelines established/followed?	No	No		NA
• Vital competencies done in-house?	No	No		NA
Activity/impact assessing				
• Realized/planned strategy match?	No	No		NA
• Multiple dependent impact measures?	Yes	Present	L3	Inaccurate
• Cost/benefit analysis?	No	No		NA
• True experiment of marketing impact?	Yes	Yes	L3	Inaccurate
Administering				
• Quality of written mission statement?	No	No		NA
• Annual quantitative objectives established?	No	No		NA
• Evidence of training, coordination, coaching?	Yes	Present	L2	Accurate

Exhibit 20. Key Issues in a Meta-evaluation of the 1993 Australian Performance Audit Report. *Note:* SWOT = strengths, weaknesses, opportunity, threats; S/D = supply and demand analysis; Levels: L1 = found to be poorly done; L2 = found to be done adequately; L3 = found to be done in superior manner and/or superior performance outcome noted. Accuracy = meta-analysis evaluation of validity of auditor's conclusion on activity/decision.

does not describe how scanning activities are used to create marketing plans. The report does not describe the present organization and responsibilities of persons in management positions. How did the results of prior advertising-impact studies influence current decisions of the ATC? What are the cost-benefit estimates for specific advertising campaigns in specific countries? Does the ATC have overseas offices? If so, how are the activities of these overseas offices evaluated for their effectiveness and efficiency? The 1993 Australian audit report does not address these issues.

Exhibits 19 and 20 provide summaries of our meta-evaluation on the Australian 1993 audit report. Based on the discussion presented above, our summary evaluation of the 1993 Australian audit report includes the following categories and descriptions:

- coverage of history and budget: poor
- clarity of writing: poor
- theory applied: no specific grounding
- coverage of issues: limited to minor issues
- style: partnering
- likely size of impact: moderate
- overall performance quality of report: low.

CONCLUSIONS AND IMPLICATIONS FOR META-EVALUATION RESEARCH ON PERFORMANCE AUDITS OF TOURISM-MARKETING PROGRAMS

For the most part the government performance audits reviewed indicate that comprehensive audits were not performed of either the activities or impacts of their governments' tourism-marketing program. With the exception of the 1987 Hawaii audit report, three vital questions in tourism-marketing stand out as *not* being addressed adequately by either the government agency responsible for the tourism-marketing program or by the auditors:

- What are the profits in net taxes resulting from the government's tourism-marketing program?
- Are the activities required (see Fig. 5) for achieving an effective and efficient tourism-marketing program being performed well by the government agency responsible for administering the program?
- What are the environmental and community impacts of the tourism-marketing program and the entire state or country tourism program?

The audit reports reviewed for this meta-evaluation report do *not* indicate training in or knowledge of the field of program evaluation. References to the leading theories of program evaluation do not form the foundation of any of the audit reports reviewed. Although the Australian (1993) audit report claims that the audit team completed reviews of state-of-the-art evaluation techniques, the report does not include evidence of even rudimentary awareness of the journals and literature of the field of program evaluation. A statement favored for theoretically grounding the performance audits reviewed is this one: "Our work was conducted . . . in accordance with generally accepted government auditing standards" (Hawaii, 1993, p. 3). If accurate, then the generally accepted government auditing standards are out-of-date, limited in scope, and lack requirements for cost-benefit analyses.

As found in this meta-evaluation, the depth of performance audits is a continuum from very limited to very comprehensive. While limited performance audits may be necessary for some specific situations, depth of performance audits is not independent from audit quality. Everything else being equal, the performance audit covering and *reporting detailed findings* of all managing topics (i.e. scanning, planning, implementing, measuring results, and administrating) will be more useful and higher in quality than the audit limited in scope. The worry is that limited-focused audits tend to result in the non-audit-audit as illustrated by the 1995 Tennessee audit. Thus, although possibly more efficient, the limited-focused audit is more often less effective than the comprehensive audit.

Little evidence was found in this meta-evaluation that audit findings and recommendations have immediate impact on changing behavior. At best, they appear to provide some enlightenment to alternative views of reality. However, some changes other than enlightenment do result from audits; the creation of the Office of Tourism in Hawaii in 1990 is likely due, in part, to the attention paid by the Hawaiian Legislature to the results of the 1987 audit. Including a detailed description in the 1993 Hawaii audit of what happened in 1987 through 1990 by government agencies using the 1987 performance audit report would have improved the quality of the 1993 report. Ultimately, increasing enlightenment and perseverance among a few persons (e.g. Legislature members and tourism-marketing program directors) sometimes results in improved sensemaking and effective and efficient performance.

However, both audit teams and legislators might consider creating organizational mechanisms to evaluate recent performance audits (a self-meta-evaluation) and to revise and revising and implementing a few recommendations in the audit. These steps are missing now from performance audit practice. We suggest creating an audit-report-utilization "working team" of three parties; the team would include three to seven persons chosen from the legislature, the tourism-marketing program, and the audit office. We have no time for such an additional audit-utilization step – is the likely cry by auditors, program directors, and legislators. However, the alternative is continuing to focus on doing more audits that no one reads as opposed to shifting to doing fewer audits that get used.

Future research on the meta-evaluations of performance audits of government tourism-marketing programs will benefit from direct interviews with audit team members doing the fieldwork for the program evaluations. In addition, meta-evaluation research studies are needed to estimate the relative impacts of alternative advertising campaigns.

Although single-group conversion-study designs do not address the issue of how many visitors exposed to the advertising and who also visited would have come without being exposed, such no-control group conversion studies are valuable for examining the relative impacts of competing advertising messages and advertising media. Meta-analyses (see Hunter & Schmidt, 1990) are needed for estimating conversion rates across studies; and for testing different moderator variables, for example, the effects of advertising themes, different direct-response campaigns, different customer segments, and different years of program operation on conversion rates. Detailed reporting of conversion rates for different media buys, markets, ads, and customer segments would provide valuable data for such meta-analysis studies (e.g. see Woodside & Motes, 1981).

Is the performance audit agency improving in its auditing activities and the impacts of its reports? Is the auditing agency up-to-date in grounding its research and reporting in the literature on program evaluation? We need additional meta-evaluation studies to answers these issues.

Because state agencies and all strategic business units are responsible for self-assessments of activities and impact, performance auditors need to accept greater responsibility for training the agency being audited on the methods of program evaluation – including the relevant literature of program evaluation. The auditors need to display the expertise necessary for sound performance audits.

BUILDING A NEW PARADIGM: FIVE GOLDEN RULES FOR PERFORMANCE AUDITING OF TOURISM-MARKETING PROGRAMS

This section describes five golden rules for performance auditing of tourism-marketing programs. These golden rules are intended to be a composite practical model of performance auditing of government tourism-marketing programs.

Golden Rule 1: Embrace On-going Formal Training in Program Evaluation

On-going formal training in program evaluation theory and method is necessary for both auditors and executives of tourism-marketing programs. A major finding of the present meta-evaluation is the absence of references and understanding of program evaluation knowledge and literature by the program evaluators and tourism-marketing executives.

Some of the audits we have reviewed make limited reference that standard accounting audit procedures were applied. None of the audits provide detailed descriptions of the coverage in accounting audits. More importantly, the reports offer no theoretical ground of what constitutes sound marketing practice and performance outcomes. The performance audits reviewed do *not* explicitly state and build on a framework (i.e. a mental model or set of assumptions) of what constitutes sound tourism-marketing practice or impact.

The seven audits include no references to the scientific literature of program evaluation. Consequently, useful theoretical and practical foundations of program evaluation knowledge (see Shadish, Cook & Leviton, 1991) are absent from these reports.

Consequently, most of the reports do not address the epistemology for doing sound performance audits and implementing sound marketing strategies. The 1985 Minnesota audit and the 1993 Australian report come closest to explicating an epistemology of evaluating tourism-marketing programs. Both the Minnesota and the Australian reports raise a fundamental issue in epistemology: how to build into our marketing programs the comparisons of outcomes of planned marketing action vs. planned non-action. However, neither these two reports nor the others go very far in identifying the necessary improvements in tourism-marketing strategies and program evaluation methods – the reports do not advocate split-run testing (i.e. true experiments in marketing, see Banks, 1965; Woodside, 1990).

Embracing formal training in program evaluation should include reading seminal works in program evaluation (e.g. Shadish et al., 1991) and scientific evaluating methods of marketing actions (e.g. Banks, 1965; Caples, 1974). Reading meta-evaluations of tourism-marketing evaluations will be useful as well. Acquiring deep knowledge about the literature on program evaluation leads to building sound explicit models of program evaluation beyond the naïve approaches in use in unwritten implicit models.

Golden Rule 2: Audit both Program Activities (Implemented Strategies) and Impacts on Planned Objectives

For several reasons measuring outcomes only is not good enough. Unfortunately, outcome measurement of tourism-marketing programs is done without planned control-group comparisons. The relatively larger sizes of environmental factors may make systematic program influences difficult to uncover. Sampling and non-sampling errors may occur in measuring outcomes that degrade the information and increase the level of "noise" in the outcome data. Finally, using multiple indicators of performance of actions taken provides convincing evidence of impact.

Program performance measurement needs to be inclusive of both process and outcome. The core issue to raise is whether or not the government's tourism-marketing program generated visits that would not have occurred without the program? The performance audit should raise another core issue: what actions were implemented in administering, scanning environments, planning, implementing, and self-assessing of performance that caused the impact?

Creating an explicit, written, plan of activities helps to improve thinking and action. Writing an annual marketing plan forces executives to make explicit their implicit models of how tourism-marketing strategy is done effectively. Writing a marketing plan requires reflection and deepens sensemaking (see Weick, 1995). Written plans include important implementation details often overlooked in verbal plans. Detailed mapping of the actions and feedback loops lead to dramatic improvements in systems thinking. Such mapping provides knowledge that can be gained in preparing detailed written plans.

However, only the audits for Hawaii's tourism-marketing programs raised issues related to the preparation and use of a marketing plan. None of the seven audits raised the fundamental issues of:

- What constitutes a high-quality marketing plan?
- What is the dominant mental model displayed in the marketing strategy actually implemented in the current tourism-marketing program?
- Should the present dominant model be replaced with a new paradigm? If yes, what are the details in this new mental model?

Detailed descriptions of what was done in the planning and implementing of the tourism-marketing program should be an integral part of the performance audit. Whether or not a written annual marketing plan was prepared should be included in such detailed descriptions. If a plan was prepared, the audit should address the quality of the plan. The detailed description should include details on what actions were implemented and the degrees of fit of the implemented strategy with the planned strategy. Also, the audit should ask the program participants in implementing the strategy to evaluate their actions and actions not taken that should have been taken.

> While participants are asked to generate a lot of data in program evaluations, rarely are they directly asked to evaluate the program, to judge its adequacy, to advise on its continuance, discontinuance, dissemination or modification. Rather than evaluating programs, participants are usually asked about themselves and their own adequacy. We are thus wasting a lot of well-founded opinions (Campbell, 1978 in Overman, 1988, p. 374).

The seven audits reviewed do not include the participants' views on the quality of the implemented strategies as integral parts of the audit reports. With the exception of the Australian audit report, senior program executives provided comments only on the accuracy of the audit and only after reading the written audit report. This procedure is not the same as seeking the views on the quality of the strategy implemented by participants as integral to the auditing process. Also, the detailed descriptions need to include the views on program performance of section managers and service workers. Applying the old saw, "God is in the details," requires learning the details about implementation unlikely known senior program executives – as well as details about how implementation could have been done better.

Qualitative assessment of activities serves to confirm and extend evaluation of quantitative results. Designing both evaluation methods into performance audits is likely to confirm the major findings on their shared dimensions. Adopting this golden rule should include embracing the following view by Campbell. If *disconfirmation* of findings on shared dimensions is found, "we should consider the possibility that the quantitative procedures are in error. If I will concede this, why would I be reluctant to see the qualitative procedures used without the quantitative? It is because I believe that the quantitative, when based on firm and examined qualitative knowing, can validly go beyond the qualitative, and can produce subtleties that the qualitative would have missed" (Campbell, 1978 in Overman, 1988, p. 375).

Golden Rule 3: Embrace a New Multi-theory Based Paradigm for Performance Audits of Tourism-Marketing Programs that Includes Stakeholder Participation

Among the seven audits reviewed, the dominant paradigm displayed in conducting the performance audits of tourism-marketing programs is confrontational. The lone exception is the Australian 1993 audit – a performance audit planned and implemented by executives of three agencies including executives of the agency being audited.

The case for adopting a new paradigm that includes stakeholder participation in designing the audit includes several dimensions. First, Stake (1980) provides a well-developed theory with rationales for adopting what he refers to as a "responsive evaluation" (as we described in Section 2). Second, Campbell expressed one of the core rationales of Stake: stakeholders are likely to provide unique and valuable information on implementation problems and opportunities – as well as creative insights on improving the planning of next year's tourism-marketing programs. Third, earlier sections provide strong evidence in this meta-evaluation that supports the following conclusions: (1) tourism-marketing program executives fight against making the changes and adopting improvements recommended by auditors; and (2) the executives win these fights. The result is the implementation of only minor changes. Witness the battle over a number of years between the Hawaii audit office vs. the HVB and the leaders of the Hawaii's tourism-marketing program. Even with the heavy weight of the Hawaii legislative branch supporting the conclusions and recommendations made by the auditor, no major improvements in the activities, performance outcomes, or measurement methods has occurred in the Hawaii tourism-marketing program.

Embracing Golden Rule 3 implies partnership, not audit leadership, by the executives of the tourism-marketing program. The objective of such an audit becomes less on passing judgment on the lack of planning and poor performance of individuals and more on documenting the strategy paradigm implemented in the current tourism-marketing program – and comparing the strategy implemented with alternative, potentially more effective and efficient, strategies. See Golden Rule 4 below.

Golden Rule 4: Transform Government Tourism-Marketing Strategy from Transactional to Relationship Marketing

Figure 10 summarizes the dominant strategy paradigm in use in government tourism-marketing programs. Over 50% of the total marketing funds budgeted is spent in steps one and three in Fig. 10. In step one, the placement of advertising in scheduled media, mostly in newspapers, magazines, television, usually represents half or more of the total tourism-marketing budgets.

Note that Fig. 10 includes one contact only with inquirers responding to the offer of free literature. After fulfillment of the requests, the names, addresses, and related information (e.g. media vehicle) are discarded. This strategy includes no efforts or budget for development of an on-going relationship with the inquiring prospective visitors. The strategy includes making no additional contacts after mailing the free "linkage advertising" (Rapp & Collins, 1988) literature. Almost all government tourism-marketing programs do not practice database marketing, even though some departments provide inquiry names to industry trade associations.

Transforming tourism-marketing programs from the currently dominant transactional paradigm to a relationship-marketing paradigm is justified on several grounds. First, inquirers are persons who have identified themselves as having a high interest in buying the product (i.e. visiting the destination). Even if most do

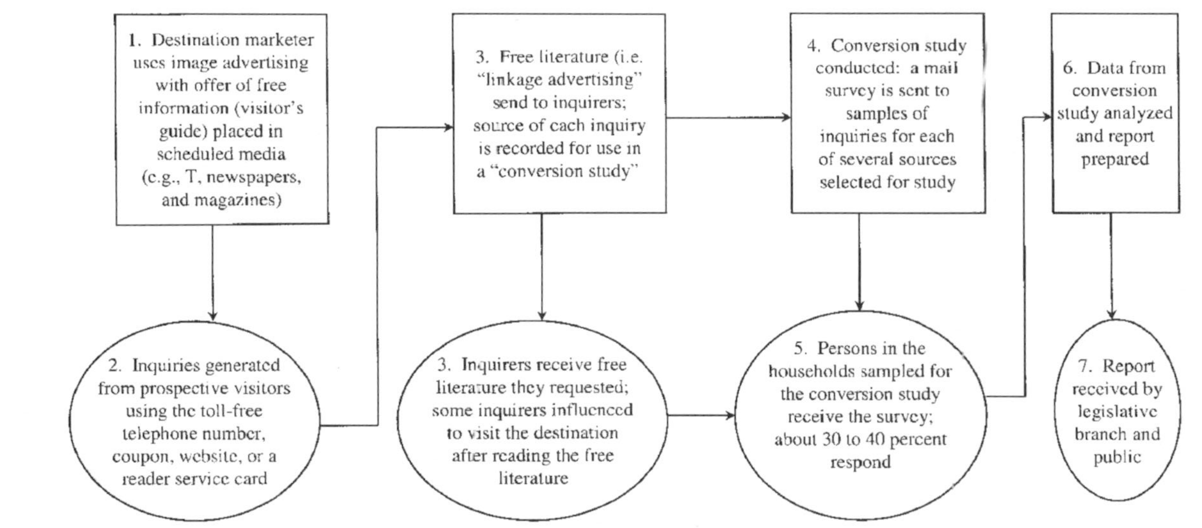

Fig. 10. The Dominant Two-Step Marketing Strategy Now Being Used in Government Tourism-Marketing Programs. *Note:* The strategy is "two-steps" because the marketer makes two attempts to attract visitors: (1) image advertising followed by; (2) the linkage advertising of the free visitor's guide. The typical conversion study following the implementation of such strategies are *not* valid for measuring the share of destination visitors exposed to the image and/or linkage advertising who would have not otherwise have visited the destination.

not buy during the current buying season, the strong potential exists that some may be converted into product buyers (i.e. visitors) if contacted more than once over the next two buying seasons. Tourism-marketing executives can test this proposition scientifically – with test and control groups – and then, estimating the profitability of multiple-contacts customers is possible.

Second, the current dominant marketing strategy includes substantial image and linkage advertising to identify inquirers by names, addresses, and inquiry sources, as well as to fulfill their requests for information. Estimating the long-term net benefits from these expenditures is impossible if no database is created and used that contains detailed information by individual customer files.

Third, because of the advances in computer technology and reductions in software costs, the costs of data handling and storage has declined dramatically over the past three decades. In 1993, Bulkeley made this point dramatically, "High-end pc's with two-gigabit hard drives – 20 times faster than the 100 megabyte drives most home users buy now – can hold several million customer names on hardware that costs about $10,000 now. The software to manage such data starts as low as $15,000" (B6).

Fourth, the strategy paradigm for database or "maximarketing" is well developed now (see Rapp & Collins, 1988; Woodside, 1996). Most large private enterprises have transformed their marketing strategies to database marketing, for example, Sears, Nordstrom, General Motors, Papa John's Pizza, Pfizer, Schwab, and Delta Air Lines. Consequently, transforming a government tourism-marketing program to database marketing does not have to be a shot-in-the-dark: many firms in many industries completed the transformation years ago.

For some of the media vehicle sources, the *average* trip expenditure made by inquirers who do visit the advertised destination is greater than $500 for the one trip to the destination – for some media vehicles the average expenditures per converted inquiry party is greater than $1,000. A 1987 inquiry study for Louisiana notes that the average expenditures in Louisiana among inquirers from an ad placed in *Travel and Leisure* who converted into visitors was $1,208 in 1987 dollars (see Woodside, 1996, p. 133). Consequently, developing a relationship-marketing program aimed at identifying and nurturing special long-term relationships with such valuable customers makes more sense than continuing the use of one-time transactional marketing.

Figure 11 includes some details of the start-up of one possible tourism relationship-marketing program. Note that in steps 6 and 8 in Fig. 11 includes making multiple contacts with first inquirers. The destination marketer provides these inquirers with the option of ending all future contacts. A core proposition in such database marketing is that "mining" inquiries by learning the individual travel interests of inquirers and designing specialized marketing appeals to

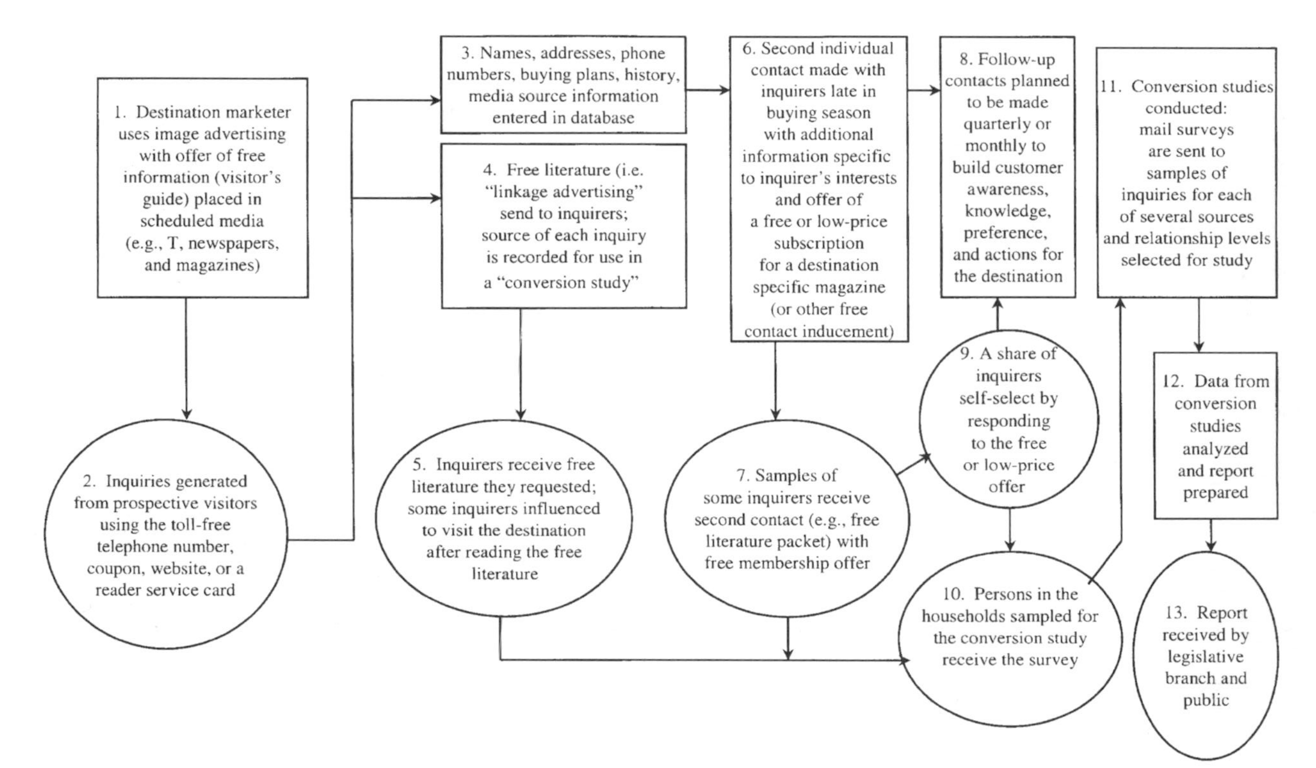

Fig. 11. Steps in Building Toward a Relationship-Marketing Program. *Note:* The strategy includes multiple contacts with inquirers by creating a database marketing program. Prospects are segmented into by interest groups and purchase response categories for more individualized future mailings. Prospects who agree to participate in the long-term relationship but do not respond after 3–5 contacts with a purchase (i.e. a visit to the destination) may be dropped from the program or may be asked to pay a premium for their continued participation.

special interest groups (SIGs) is highly profitable – more profitable than making one response only to inquirers. The literature on database supports this core proposition (e.g. Rapp & Collins, 1988).

Given a set budget constraint, the shift from transactional marketing to database relationship marketing does result in shifting some share of a scheduled media budget to nurturing one-on-one relationships. Even though advertising agencies have fought such shifts in expenditures (see Cappo, 1992; Woodside, 1996), most firms in most industries made the transformation without overwhelming reductions in scheduled media budgets.

Also, the tourism-marketing program managers can expect a decline in first-time inquiries with a transformation to relationship marketing. The focus on the value of individual customers for the destination becomes much greater with relationship marketing compared to transactional marketing.

Transforming the marketing program to relationship marketing requires providing evidence of its greater profitability compared to the traditional two-step only marketing program. Conversion studies can be transformed into treatment group vs. control group experiments to examine the value of relationship marketing programs. Given that individual names, addresses, and relationships with the destination are known, random samples can be created to receive special marketing offers or no offers at all. Such research designs are the basis for scientific testing of the impact of marketing programs on sales and profits (see Banks, 1965).

Golden Rule 5: Do Split-Run Tests of Specific Marketing and Advertising Actions

Given that the primary interest regarding tourism advertising and marketing of government legislative members is whether or not the advertising caused visits that would not otherwise have occurred, the following conclusion may be surprising. *No scientific tests of effectiveness have been done on the impact of government tourism advertising on influencing visits.* Scientific testing of a treatment's influence on a dependent variable requires direct comparisons of a treatment group with a comparable control group.

The lack of scientific testing of advertising effectiveness is even more surprising for the following additional reasons. First, the details on how to perform such tests methods are widely known (see Caples, 1974; Faulkner, 1997; Woodside, 1996). Second, most government tourism-marketing departments spend substantial funds each year for measuring the impact of advertising on influencing visits. Third, each year legislators raise the primary issue regarding government tourism advertising: did the advertising generate visits that would not have occurred otherwise?

We can speculate on at least four reasons for our lack of scientific data on tourism advertising's influence on visits. First, tourism-marketing executives, tourism-advertising researchers, elected representatives, and performance auditors may lack the knowledge of and training in how to estimate advertising-to-sales causation scientifically – even though such knowledge and training is readily available (see Banks, 1965; Caples, 1974; Woodside, 1990).

Second, non-scientific testing, such as frequently used conversion studies and aided recall surveys, appears to provide useful information for answering the core issue (e.g. Longwoods International, 1997; Woodside & Soni, 1990). However, the available conversion studies and aided recall surveys do not meet the necessary requirement of scientific testing: such studies include no planned comparisons of persons exposed to the advertising vs. comparable persons not exposed to the advertising. Thus, the typical conversion study report usually ends with the misleading implication that advertising resulted in visits by 40 or 50% of the inquirers generated from the tourism-advertising program who visited the destination after receiving the free literature they requested.

Asking respondents to indicate whether or not the scheduled media advertising and free literature influenced their decision to visit is not a scientific measurement of cause-and-effect relationships. Self-reports of causal attribution are often inaccurate and distorted by psychological processes (see Plous, 1993).

Third, in a number of scientific tests, detecting the impact of advertising on sales is not possible but the impact is very small. Or worse: the finding is reached that advertising decreases sales – such a result has been found to occur in at least one scientific test (see Ackoff & Emshoff, 1975). Consequently, why take the chance of doing a scientific study when the results may be unfavorable?

Thus, fear of unwanted results may be a fourth reason why scientific testing of advertising's impact on visits has not occurred. Scientific test results may result in a poor performance evaluation of the advertising program, and by association, poor performance by the advertising agency and the tourism-marketing executives. By comparison, the widely used single-group conversion study and aided-recall surveys result in feel-good findings. Such studies nearly always result in favorable findings that advertising influences visits and huge expenditures by tourists, for example, 50% inquiry-to-visits conversion and $200 expenditures by tourists for every $1 of scheduled media advertising.

To accomplish the paradigm shift in thinking necessary to start using scientific tests of tourism advertising effectiveness, we suggest taking the following steps. First, one big scientific test should not be run. Designing and implementing small scale tests decreases the risk that the results of any one study provide inaccurate findings regarding the effectiveness of advertising generating visits. Applying the counsel on quantitative program evaluation offered by Campbell (1978 in Overman, 1988, p. 372) is wise:

> Quantitative experimental studies involve so many judgmental decisions as to mode of implementation, choice and wording of measures, assembly of data in analysis, etc., they too should be done in replicate. Our big evaluations should be split into two or more parts and independently implemented. When the results agree, the decision implications are clear. When they disagree, we are properly warned about the limited generality of the findings.

Second, a research design using advertising free standing inserts (FSIs) in magazines and newspapers offer a low cost application for scientific testing. For example, we offer the following suggested steps: select 2,000 subscribers in each of three magazines and three newspapers systematically. For the study, create a full-page advertisement as a FSI. Assign half of the subscribers (1,000) in each of the six media groups randomly to receive the FSI in their regular issue of the media vehicle – the other half (1,000) receives no FSI in the same regular issue of the media vehicle. Thus, for each media vehicle, two scientifically comparable groups are created, with the treatment group exposed to the tourism advertising and the control group not exposed to the tourism advertising. Because of the random assignment of households to the two groups, we can eliminate sources of influence other than the advertising as generating found differences in visits. Exhibit 1 shows this scientific research design for one of the media vehicles in the hypothetical test.

Third, we should expect to find small not large positive effect sizes of advertising on generating visits. Conversion studies support the following proposition: most persons responding with inquiries for free literature offered in advertising report having partial or complete plans to visit the destination. However, tourism advertising is more likely to generate other, very important, positive influences related to visitors receiving the free literature vs. not receiving the free literature.

The likely most substantial impact of tourism advertising is *not* on generating visits but in generating higher-quality visits than would otherwise happen. The findings of a quasi-experimental test support this proposition. Expenditures at the destination (Prince Edward Island, a Canadian Province), the number of activities, the perceived quality of the visit, the overall satisfaction with the trip, and the intention to return all increased substantially among visitors receiving vs. visitors not receiving the free advertising literature, a 130-page visitors' information guide (see Woodside, MacDonald & Trappey, 1997). An unexpected additional finding in the PEI study was that the increases in visit quality influenced by the free literature were greater among repeat vs. first-time visitors. Possibly the free literature aided in getting repeat visitors to try new, unknown, or forgotten activities in the destination area in addition to their planned destination activities. Consequently, the repeat visitors using the current free visitor's guide had the highest dollar expenditures and greatest satisfaction with their visits compared to thc other three groups (first time visitors using and not using the guide and repeat visitors not using the guide).

In addition, tourism advertising often has a small, positive, systematic impact on visits. Several rationales and findings from the advertising-to-sales literature support this proposition. Increases in advertising (from none to some advertising) often result in increasing unaided retrieval of the product advertised, and increases in unaided retrieval are associated with increases in sales (literature is reviewed in Woodside, 1996). Most linkage advertising (e.g. free visitor's catalog) dramatically reduces the effort of competing and increases knowledge of how to complete many details in planning a visit to the destination. In fact, several studies in consumer psychology support the finding that the availability of such problem-solving tools influences choice decisions (see Bettman, Luce & Payne, 1998).

CONCLUSIONS

We use the term "golden rules" to generate discussion and cause action. A core conclusion from the meta-evaluation presented is that few explicit attempts have been made to explicitly improve the methods of evaluating of government tourism-marketing programs. The Australian 1993 audit deserves recognition as an attempt to create and apply a new paradigm for a tourism-marketing program evaluation. In part, our hope is that this study will serve as a foundation for additional attempts to transform the discipline.

Certainly, additional "golden rules" need to be developed. For example, scientific measurement of the long-run impacts of advertising on generating visits and attracting high-quality visits need to become standard practice in tourism-marketing program evaluations. Such scientific measurements of long-run impacts require creating additional treatment groups to enable measuring once only for each group – to avoid the introduction of bias from answering surveys.

Another golden rule would be to perform only comprehensive performance audits of government tourism-marketing programs. "Limited-focus" audits do not represent substantial savings in effort. The trade-off of accuracy and detail for greater speed results in unfounded and misleading confidence that a comprehensive audit is unnecessary. Combining, replicating, and extending the findings from qualitative and quantitative measurement is practical only in comprehensive audits. Useful comprehensive audits can be completed quickly (within two to three months), a time period not much greater than a limited focus audit.

REFERENCES

Ackoff, R. L., & Emshoff, J. R. (1975). Advertising research at Anheuser-Busch, Inc. (1963–1968). *Sloan Management Review*, *16*(4), 1–15.

ALPHA/Sim (1998). Burlington, MA: ALPHATECH, Inc., alpha.sim@alphatech.com

Armstrong, J. S. (1982). The value of formal planning for strategic decisions: Review of empirical research. *Strategic Management Journal*, *3*, 197–211.

Armstrong, J. S. (1985). *Long-range forecasting: From crystal ball to computer*. New York: Wiley.

Arnould, E. J., & Wallendorf, M. (1994). Market-oriented ethnography: Interpretation building and marketing strategy formulation. *Journal of Marketing Research*, *31*(November), 484–503.

Australia (1993). *Evaluation of the Australian tourist commission's marketing impact*. Sydney: Australian Tourist Commission (March).

The Auditor (1994). *Manual of guides*. Honolulu, State of Hawaii, June.

Axelrod, J. N. (1968). Attitude measures that predict purchase. *Journal of Advertising Research*, *8*(1), 3–17.

Banks, S. (1965). *Experimentation in marketing*. New York: McGraw-Hill.

Bellavita, C., Wholey, J. S., & Abramson, M. A. (1986). Performance-oriented evaluation: Prospects for the future. In: J. S. Wholey, M. A. Abramson & C. Bellavita (Eds), *Performance and Credibility: Developing Excellence in Public and Non-profit Organizations*. Lexington, MA: Lexington Press.

Berelson, B. (1952). *Content analysis in communications research*. Glencoe, IL: Free Press.

Bettman, J. R., Luce, M. F., & Payne, J. W. (1998). Constructive consumer choice processes. *Journal of Consumer Research*, *25*(December), 187–217.

Binter, M. J., Booms, B. H., & Mohr, L. A. (1994). Critical service encounters: The employee's view. *Journal of Marketing*, *58*(October), 95–106.

Boland, R. J., Jr. (1984). Sense-making of accounting data as a technique of organizational diagnosis. *Management Science*, *30*, 868–882.

Bonham, C., & Mak, J. (1996). Private vs. public financing of state destination promotion. *Journal of Travel Research*, *35*(2), 3–10.

Bonoma, T. V. (1985). *The marketing edge: Making strategies work*. New York: Free Press.

Brooks, R. A. (1997). Evaluation and auditing in state legislature. In: E. Chelimsky & W. R. Shadish (Eds), *Evaluation for the 21st Century: A Handstudy* (pp. 109–120). Thousand Oaks, CA: Sage.

Bulkeley, W. M. (1993). Marketers mine their corporate databases. *Wall Street Journal* (June 14), B6.

Buzzell, R. D., & Gale, B. T. (1987). *The PIMS principles: Linking strategy to performance*. New York: Free Press.

Campbell, D. T. (1969). Reforms as experiments. *American Psychologist*, *24*, 409–429.

Campbell, D. T. (1978). Qualitative knowing in action research. In: M. Brenner, P. Marsh & M. Brenner (Eds), *The Social Contexts of Method* (pp. 184–209). London: Croom Helm. Reprinted in: E. S. Overman (Ed.), *Methodology and Epistemology for Social Science* (pp. 360–376). Chicago: University of Chicago.

Campbell, D. T., & Stanley, J. C. (1963). *Experimental and quasi-experimental designs for research*. Chicago: Rand McNally.

Caples, J. (1974). *Tested advertising methods* (4th ed.). Englewood Cliffs, NJ: Prentice-Hall.

Cappo, J. (1992). [Editorial], *Advertising Age* (November 16), 11.

Chen, H., & Rossi, P. H. (1983). Evaluating with sense: The theory-driven approach. *Evaluation Review*, *7*, 283–302.

Clifford, D. K., Jr., & Cavanaugh, R. E. (1985). *The winning performance*. Toronto: Bantam.

Cook, T. D. (1997). Lessons learned in evaluation over the past 25 years. In: E. Chelimsky & W. R. Shadish (Eds), *Evaluation for the 21st Century*. Thousand Oaks, CA: Sage.

Cronbach, L. J., Ambron, S. R., Dornbusch, S. M., Hess, R. D., Hornki, R. C., Phillips, D. C., Walker, D. F., & Weiner, S. S. (1980). *Toward reform of program evaluation*. San Francisco: Jossey-Bass.

Daft, R. L., & Weick, K. E. (1984). Toward a model of organizations as interpretation systems. *Academy of Management Review, 9*, 284–295.

Denzin, N. K. (1984). *The research act*. Englewood Cliffs, NJ: Prentice-Hall.

Denzin, N. K., & Lincoln, Y. S. (1994). *Handstudy of qualitative research*. Newbury Park, CA: Sage.

Dybka, J. M. (1987). A look at the American traveler: The U.S. pleasure travel market study. *Journal of Travel Research, 25*(Winter), 2–4.

Fisher, R. A. (1949). *The design of experiments* (5th ed.). New York: Hafner.

Garvin, D. A. (1987). Competing on the eight dimensions of quality. *Harvard Business Review, 65*, 101–109.

Geertz, C. (1973). *The interpretation of cultures*. New York: Basic Studys.

Gilovich, T. (1991). *How we know what isn't so*. New York: Free Press.

Goetsch, H. W. (1993). *Developing, implementing and marketing and effective marketing plan*. Chicago: American Marketing Association; Lincolnwood, IL: NTC Business Studys.

Haley, R. I., & Case, P. B. (1979). Testing thirteen attitude scales for agreement and brand discrimination. *Journal of Marketing, 43*(4), 20–32.

Hall, R. I. (1983). The natural logic of management policy making: Its implications for the survival of an organization. *Management Science, 30*(8), 905–927.

Hall, R. I., Aitchison, P. W., & Kocay, W. L. (1994). Causal policy maps of managers: Formal methods for elicitation and analysis. *System Dynamics Review, 10*(4), 337–360.

Hall, R. I., & Menzies, W. B. (1983). A corporate system model of a sports club: Using simulation as an aid to policy making in a crisis. *Management Science, 29*(1), 52–64.

Hawaii (1987). *Management audit of the Hawaii visitors bureau and the state's tourism program*, Report No. 87-14. Honolulu: The Legislative Auditor of the State of Hawaii (February).

Hawaii (1993). *Management and financial audit of the Hawaii visitors bureau*, Report No. 93-25. Honolulu: The Auditor, State of Hawaii (December).

Huber, G. P., Ullman, J., & Leiffer, R. (1979). Optimum organization design: An analytic-adoptive approach. *Academy of Management Review, 4*, 567–578.

Huff, A. S. (Ed.) (1990). *Mapping strategic thought*. Chichester, UK: Wiley.

Hunter, J. E., & Schmidt, F. L. (1990). *Methods of meta-analysis: Correcting error and bias in research findings*. Newbury Park: Sage.

Kassarjian, H. H. (1977). Content analysis in consumer research. *Journal of Consumer Research, 4*(June), 8–18.

Kay, J. (1995). *Why firms succeed*. New York: Oxford Press.

Kotler, P. (1997). *Marketing management* (9th ed.). Upper Saddle River, NJ: Prentice Hall.

Kotler, P., Gregor, W. T., & Rodgers, W. H. (1977). The marketing audit comes of age. *Sloan Management Review, 18*(Winter), 25–43.

Kotler, P., Gregor, W. T., & Rodgers, W. H. (1989). Retrospective commentary. *Sloan Management Review, 30*(Winter), 59–62.

Longwoods International (1997). Hawai'i [Advertising] Accountability Research, Toronto, Ontario: Longwoods International.

Minnesota (1985). *Economic development*. Saint Paul, MN: Program Evaluation Division, Office of the Legislative Auditor, State of Minnesota.

Mintzberg, H. (1978). Patterns in strategy formation. *Management Science*, *24*, 934–948.

Missouri (1996). *Special review of the department of economic development, Division of tourism*, Report No. 96-86, Jefferson City, Missouri: Offices of the State Auditor of Missouri (November 14, 28 pp).

North Carolina (1989). *Performance audit: Department of commerce, Division of travel and tourism and North Carolina Film Office*. Raleigh, NC: Office of the State Auditor (January, 21 pp).

Patton, M. Q. (1997). *Utilization-focused evaluation* (3rd ed.). Thousand Oaks, CA: Sage.

Peters, T. J. (1987). *Thriving on chaos*. New York: Alfred Knopf.

Peters, T. J., & Waterman, R. H., Jr. (1982). *In search of excellence: Lessons from America's best-run companies*. New York: Harper & Row.

Plous, S. (1993). *The psychology of judgment and decision making*. New York: McGraw-Hill.

Pollitt, C., & Summa, H. (1997). Performance auditing. In: *Evaluation for the 21st Century*. Thousand Oaks, CA: Sage.

Price, L. L., Arnould, E. J., & Tierney, P. (1995). Going to extremes: Managing service encounters and assessing provider performance. *Journal of Marketing*, *59*(April), 83–97.

Ramond, C. (1966). *Measurement of advertising effectiveness*. Menlo Park, CA: Stanford Research Institute.

Rapp, S., & Collins, T. (1988). *Maxi-marketing*. New York: McGraw-Hill.

Rapp, S., & Collins, T. (1994). *Beyond maxi-marketing*. New York: McGraw-Hill.

Rees, R. M. (1995). The HVB: Private gain vs. public good. *Pacific Business News* (February 13), 5.

Schoenbachler, D. D., di Benedetto, C. A., Gordon, G. L., & Kaminski, P. F. (1995). Destination advertising: Assessing effectiveness with split-run technique. *Journal of Travel and Tourism Marketing*, *4*(2), 1–21.

Scott, M. C. (1998). *Value drivers*. Chichester, UK: Wiley.

Scriven, M. S. (1967). *The methodology of evaluation, AERA Monograph series on curriculum evaluation* (Vol. 1). Chicago: Rand, McNally.

Scriven, M. S. (1974). Evaluation perspectives and procedures. In: J. W. Popham (Ed.), *Evaluation in Education: Current Application* (pp. 3–93). Berkeley, CA: McCutchan.

Scriven, M. S. (1980). *The logic of evaluation*. Inverness, CA: Edgepress.

Scriven, M. S. (1983). Evaluation ideologies. In: G. F. Madaus, M. Scriven & D. L. Stufflebeam (Eds), *Evaluation Models: Viewpoints on Educational and Human Services* (pp. 229–260). Boston: Kluwer-Nijhoff.

Scriven, M. S. (1995). The logic of evaluation and evaluation practice. In: D. M. Fournier (Ed.), *Reasoning in Evaluation: Inferential Links and Leaps* (pp. 49–70). San Francisco: Jossey-Bass.

Senge, P. (1990). *The fifth discipline*. New York: Doubleday.

Shadish, W. R., Jr., Cook, T. D., & Leviton, L. C. (1991). *Foundations of program evaluation*. Newbury Park: Sage.

Shostack, G. L. (1987). Service positioning through structural change. *Journal of Marketing*, *51*(January), 34–53.

Stake, R. E. (1980). *Program evaluation, particularly responsive evaluation, in rethinking educational research*. In: W. B. Dockrell & D. Hamilton (Eds). London: Hodder and Stoughton.

Stake, R. E., & Easley, J. A. (1978). *Case studies in science education*. Champaign, IL: University of Illinois, Center for Instructional Research and Curriculum Evaluation.

Stake, R., Migotsky, C., Davis, R., Cisneros, E. J., DePaul, G., Dunbar, C., Jr., Farmer, R., Feltovich, J., Johnson, E., Williams, B., Zurita, M., & Chaves, I. (1997). The evolving syntheses of program value. *Evaluation Practice*, *18*(2), 89–103.

Stubbart, C., & Ramaprasad, A. (1988). Probing two chief executives' schematic knowledge on the U.S. steel industry by using cognitive maps. *Advances in Strategic Management*, *5*, 139–164.

Teas, R. K. (1993). Expectations, performance evaluation, and consumers' perceptions of quality. *Journal of Marketing*, *57*(4), 18–34.

Tennessee (1995). *Performance audit: Department of tourist development*. Nashville, TN: State of Tennessee, Comptroller of the Treasury, Department of Audit, Division of State Audit.

Weick, K. E. (1979). *The social psychology of organizing*. New York: McGraw-Hill.

Weick, K. E. (1995). *Sensemaking in organizations*. Thousand Oaks, CA: Sage.

Weiss, C. H. (1972). *Evaluation research: Methods for assessing program effectiveness*. Englewood Cliffs, NJ: Prentice-Hall.

Weiss, C. H. (1987). Evaluation social programs: What have we learned? *Society*, *25*(1), 40–45.

Wholey, J. S. (1977). Evaluability assessment. In: L. Rutman (Ed.), *Evaluation Research Methods: A Basic Guide*. Beverly Hills, CA: Sage.

Wholey, J. S. (1983). *Evaluation and effective public management*. Boston: Little, Brown.

Woodside, A. G. (1990). Measuring advertising effectiveness in destination marketing strategies. *Journal of Travel Research*, *29*(2), 3–8.

Woodside, A. G. (1996). *Measuring the effectiveness of image and linkage advertising*. Westport, CT: Quorum Studys.

Woodside, A. G., MacDonald, R., & Trappey, R. J., III (1997). Measuring linkage-advertising effects on customer behavior and net revenue. *Canadian Journal of Administrative Sciences*, *14*(2), 214–228.

Woodside, A. G., & Motes, W. H. (1981). Sensitivities of market segments to separate advertising strategies. *Journal of Marketing*, *45*(Winter), 63–73.

Woodside, A. G., & Ronkainen, I. A. (1982). Travel advertising: Newspapers vs. magazines. *Journal of Advertising Research*, *22*(June), 39–43.

Woodside, A. G., & Ronkainen, I. A. (1984). How serious is non-response bias in advertising conversion research? *Journal of Travel Research*, *23*(Spring), 28–33.

Woodside, A. G., & Soni, P. K. (1988). Assessing the quality of advertising inquiries by mode of response. *Journal of Advertising Research*, *28*(4), 31–37.

Woodside, A. G., & Soni, P. K. (1990). Performance analysis of advertising in competing media vehicles. *Journal of Advertising Research*, *30*(1), 53–66.